Green IT: Technologies and Applications

Jae H. Kim and Myung J. Lee (Eds.)

Green IT: Technologies and Applications

Springer

Editors

Dr. Jae H. Kim
Boeing Research & Technology
The Boeing Company
P.O. Box 3707
MC 7L-49
Seattle, WA 98124-2207
USA
E-mail: jae.h.kim@boeing.com

Prof. Myung J. Lee
City University of New York
Department of Electrical Engineering
Convent Avenue at 140th Street
New York, NY 10031
USA
E-mail: lee@ccny.cuny.edu

ISBN 978-3-642-43055-8
ISBN 978-3-642-22179-8 (eBook)

DOI 10.1007/978-3-642-22179-8

Typeset & Cover Design: Scientific Publishing Services Pvt. Ltd., Chennai, India.

Printed on acid-free paper

9 8 7 6 5 4 3 2 1

springer.com

Preface

As the world's economy and population ever expands, the question of sustainable growth comes to the foreground of matters regarding energy. Even with advanced technologies, the exploration of non-renewable sources struggles with its cost and engendered environmental impacts, such as carbon emission and ecological danger. A variety of environmentally and socially responsible and economically viable solutions to energy has recently emerged, addressing energy conservation and renewable sources. The energy conservation or "greening effort" has focused on the efficient use of electricity at home and industry and fuel-efficient transportation, attaining certain fruition. It is only recently that the energy expenditure of the IT sector has received attention on both the corporate and federal levels, due to the explosive growth of Internet and mobile communications. Currently, the number of PCs worldwide will surpass 2 billion units by 2014 (Gartner) and more than half of the world population will own mobile phones. The transition of user traffic toward data-intensive video contributes tremendously to the rise of IT traffic, which requires the use of powerful IT systems for high-capacity user devices, wired and wireless networks, and data centres. These high-capacity IT systems in turn call for high energy expenditure; therefore, the issue of green IT is brought forth. The IT sector is unique in that IT is part of the problem but at the same time a key to the solution. In many sectors, quantum leaps in energy saving are often attributed to innovative applications of IT technology.

With heightened research and development efforts in green IT and numerous reports of technical, policy, and standard issues and solutions through conferences, workshops, and journals, it is timely to compile these findings into a book for interested readers to find a comprehensive view on Green IT technologies and applications. Although several books are already published with similar titles, the majority of them are essentially non-technical and only deal with the green computing aspect. To the best of our knowledge, this book distinguishes itself from others in two important aspects. First, it brings together in a single volume both green communications and green computing under the theme of Green IT. Second, it focuses on the technical issues of green IT in a survey style to enhance readability. The chapters of this book are written by researchers and practitioners who are experts in the field. This book will be an excellent source of information for graduate/undergraduate students, researchers, engineers, and engineering managers in IT (EE, CSC, CompEng, Information Science) as well as interdisciplinary areas such as sustainability, environment, and energy.

Organization of the Book

The book is organized into three parts: Green Communications, Green Computing, and Smart Grid and Applications.

Part I presents Green Communications in ten chapters. The first two chapters introduce an overall mobile communication architecture and energy metrics for wireless communications respectively. The beginning chapter describes how mobile communication architectures are evolving in regard to energy efficiency. It also illustrates how a comprehensive view of component technologies enables a strategy for future energy efficient wireless communications. Chapter 2 provides an overview of the metrics for energy efficiency in wireless networks. Then, it addresses approaches to optimize energy efficiency and explains a tradeoff existing between energy expenditure and achieved performance; a cognitive raido network is used as an example.

From the next chapter on, the discussions deal with component technologies, starting from the PHY layer to the application layer aspect. The state-of-the-art energy aware link adaptation protocols for MIMO-based cognitive systems are presented in Chapter 3. In Chapter 4, the management algorithms in energy-saving MAC protocol are classified into two categories: asymmetric single-hop MAC and symmetric multi-hop MAC. The chapter then discusses characteristics, advantages, and disadvantages of protocols of each category with representative MAC protocols. Chapter 5 presents energy efficient routing in Delay-Tolerant Networks (DTN) with radios of short range transmission. Performance studies based on Markovchains favor algorithms exploiting the information of residual battery energy. Chapter 6 uses the concept of cooperative relay in reducing the energy consumption in a cellular network, based on a microeconomic approach of incentive offering to the relay nodes. Then, the problem is formulated as a multi-objective optimization problem with two objective functions of energy and cost. The similar cooperative relay concept is further explored in Chapter 7 in the context of wireless sensor networks with energy harvesting nodes. An optimal policy is theoretically established for the decision process for sender and relay node using the Partially Observable Markov Decision Process (POMDP) model. Chapter 8 studies two types of energy efficient parallel packet forwarding architectures: shared data structure and duplicated data structure. The focus of the design is to minimize the overall power consumption while satisfying the throughput demand.

The next three chapters discuss energy issues in systems and applications. Chapter 9 examines experimentally the energy expenditures of different components in a mobile phone such as screen, CPU, and wireless network interface card. Then, it introduces energy saving techniques for streaming video, dynamic decoding, screen control, and some hybrids of these techniques for a system level energy minimization. Chapter 10 reports the design of an energy efficient localization algorithm using two built-in functions in mobile phones: the compass and accelerometer. It also shows the implementation and compares its performance with other approaches. Part I concludes in Chapter 11 which introduces various energy efficient game theoretic frameworks based on the status of available information that can serve to solve diverse wireless network problems. It also describes an application scenario for a radio equipped with multiple wireless interfaces.

Part II presents Green Computing in six chapters, each dealing with various energy issues in data centers, computing storage, and optimization techniques. Chapter 12 reviews the state-of-the-art data center architectures and addresses the issue of their power consumption. It also presents the trade-off between the power consumption and the performance of the data center. Chapter 13 also discusses energy management in data centers, focusing on scaling-down approaches considered to enhance energy efficiency. A survey on existing energy management techniques has also been presented, followed by GreenHDFS (Hadoop Distributed File System) with its simulation studies. For the optimization of power consumption of data centers, Chapter 14 promotes a distributed system approach for high-performance computing and deals with the multi-objective optimization problem for energy consumption and latency given operational conditions such as voltage and frequency scaling. Chapter 15 introduces storage technologies in regard to energy and performance. It surveys component technologies and their power-management features, followed by system-level technologies and associated dynamic power management approaches and challenges. Chapter 16 discusses the Environmentally Opportunistic Computing (EOC) concept and its implementation.

Part III presents Smart Grid and Applications in five chapters. Chapter 17 describes an overview of the smart grid including its motivation, goals, and benefits. It further discusses the smart grid's conceptual architecture, key enabling technologies, standard efforts, and global collaboration. Chapter 18 discusses research challenges and open problems in building various information systems' elements as well as applications enabled by the smart grid. A potential approach to tackle some of those issues is also presented. Chapter 19 introduces an architectural model for the integration of semantic information for smart grid applications, which draws from diverse sources ranging from smart meters to social network events. Chapter 20 presents a new method of modeling urban pollutants arising from transportation networks. Introducing Markov chain to model transportation networks, it extends the transition matrix to model pollutant states. Chapter 21 presents a WSN architecture based on long range radios Long Term Evolution (LTE) waveform. Several new green communication applications that become possible by the proposed architecture are described, along with field experiment results. In concluding the book, Chapter 22 overviews the standardization activities to reduce the impact of telecommunication on climate change. It also presents IT standards for the smart grid as a typical example of using IT to improve the environmental friendliness of other industry sectors.

Acknowledgements

The Editors would like to acknowledge the generous sponsorship of this book project from the Korean-American Scientists and Engineers Association (KSEA). The book is the first volume of the KSEA TechBook Series.

The Editors would like to thank the chapter authors for their technical contributions and timely support of making each manuscript to the publication schedule. The Editors especially thank Preston Marshall, Emanuele Crisostomi, László Gyarmati, Weirong Jiang, Hamidou Tembine, Tuan Anh Trinh, Hrishikesh Venkataraman who also assisted the peer review process.

Special thanks to the members of Korean Computer Scientists and Engineers Association (KOCSEA) who assisted the chapter review process: Jihie Kim (USC ISI), Taekjin Kwon (Telcordia), Yoohwan Kim (UN Las Vegas), Yoonsuck Choe (Texas A&M), Eunjin Jung (Univ. of San Francisco), Minkyong Kim (IBM), Yongdae Kim (Univ of Minnesota), Bong Jun Ko (IBM), Eunjee Song (Baylor Univ.), Taehyung Wang (CSU Northridge), and Jongwook Woo (CSU Los Angeles).

Lastly, the Editors are grateful to Thomas Ditzinger, Jessica Wengrzik, and Prabu Ganesan at Springer Verlag for their help and patience throughout the preparation of this book.

Dr. Jae H. Kim
Boeing Research and Technology

Dr. Myung J. Lee
CUNY

About the Authors

Editors

Dr. Jae Hoon Kim is a Senior Technical Fellow / Executive of The Boeing Company. He is working with Boeing Research & Technology in the area of Communications and Network Systems Technology, focusing on wireless mobile ad-hoc network (MANET), Sensor network, and Cognitive radio-based NetOps. Dr. Kim is also an Affiliate Professor and Graduate Faculty of Electrical and Computer Engineering Department, University of Washington, Seattle, WA. He has served as an IEEE Associate Editor for a monthly Technical Journal, Communications Letters for the last decade. Dr. Kim has been a Principal Investigator and Program Manager for a number of U.S. Department of Defense (DoD) contract programs from DARPA, Army CERDEC, AFRL, and ONR. Prior to Boeing, Dr. Kim has been a Task Manager and Member of Technical Staff as Senior Research Scientist at the California Institute of Technology, NASA Jet Propulsion Laboratory.

Dr. Kim received his Ph.D. degree from Electrical and Computer Engineering of the Univeristy of Florida, Gainesville. He is an author/co-author of 80+ publications, holds 4 U.S. patents granted with 1 patent pending, has received 10 NASA Technical Innovation Awards, and 25+ Boeing Technology Awards for outstanding technical performance recognition. Dr. Kim received an Asian-American Engineers of the Year 2009 Award from the Chinese Institute of Engineers in USA. Dr. Kim is currently serving as the 39th President of the Korean-American Scientists and Engineers Association (KSEA). Dr. Kim can be reached at jae.h.kim@boeing.com.

Dr. Myung Jong Lee is a Professor and Director, Advanced Wireless Networks Laboratory in the department of Electrical Engineering of City University of New York. Dr. Lee received B.S. and M.S. degrees from Seoul National University, Korea, and a Ph.D degree in electrical engineering from Columbia University, New York. He was visiting scientists to Telcordia and Samsung Advanced Institute of Technology. His research interests include wireless sensor networks, ad hoc networks, CR networks, wireless multimedia networking, and Internet. He published extensively in these areas (140 plus) and holds 30 U.S. and International Patents (pending incl.). His researches have been funded by government agencies and leading industries, including NSF, ARL, Telcordia, Panasonic, Samsung, ETRI, etc.

Dr. Lee actively participates in international standard organizations (the chair of IEEE 802.15.5). His research group contributed NS-2 module for IEEE 802.15.4, widely used for wireless sensor network researches. He is an associate editor for IEEE communications magazine, and co-awarded the best paper from IEEE CCNC 2005. Dr. Lee also received CUNY Excellence Performance Award. Dr. Lee is serving as the Vice President for Technology Affairs of the Korean-American Scientists and Engineers Association (KSEA). Dr. Lee can be reached at lee@ccny.cuny.edu.

About the Authors

Chapter Contributors

Preston F. Marshall (**Chapter 1**) is the Deputy Director of the Information Sciences Institute (ISI) and a Research Professor at Ming Hsieh Dept. of Electrical Engineering, University of Southern California's Viterbi School of Engineering. Formerly, he was a Program Manager with DARPA Strategic Technology Office (STO) for many of the Wireless, Cognitive Radio and networking programs. He is the author of the recently released "Quantitative Analysis of Cognitive Radio and Network Performance" by ARTECH House. Dr. Marshall can be reached at pmarshall@isi.edu.

Wei Wang, Zhaoyang Zhang, and Aiping Huang **(Chapter 2)** are with the Department of Information Science and Electronic Engineering, Key Laboratory of Integrate Information Network Technology, Zhejiang University, P.R. China. Dr. Wang is the editor of a book "Cognitive Radio Systems" (Intech, 2009). Dr. Wang can be reached at wangw@zju.edu.cn.

Eren Eraslan, Babak Daneshrad**,** Chung-Yu Lou and Chao-Yi Wang **(Chapter 3)** are with UCLA and Silvus Technologies. Dr. Daneshrad is a Professor of Dept. of Electrical Engineering at UCLA and a Co-Founder of Silvus Technologies. Eren Eraslan, Chung-Yu Lou and Chao-Yi Wang are currently working with Wireless Integrated System Research (WISR) Laboratory at UCLA. Dr. Daneshrad can be reached at ereneraslan@ucla.edu.

Tae Rim Park and Myung Jong Lee **(Chapter 4)** are with the Department of Electrical Engineering of City University of New York (CUNY). Dr. Lee is a Professor and Director of the Advanced Wireless Networks Lab at CUNY, and a Chair of IEEE 802.15.5 TG. Dr. Park a senior research engineer at Signals and Systems Lab in Samsung Advanced Institute of Technology and is an active contributor to IEEE 802.15.4e, IEEE 802.15 and ZigBee standard. Dr. Park can be reached at taerim.park@samsung.com.

Zygmunt J. Haas and Seung Keun Yoon **(Chapter 5)** are with the School of Electrical and Computer Engineering at Cornell University. Dr. Haas is a Professor and Director of the Wireless Network Laboratory of the School of Electrical and Computer Engineering at Cornell University. Dr. Seung Keun Yoon has recently joined as a research engineer Samsung Advanced Institute of Technology. Dr. Jae H. Kim is with Boeing Research & Technology. Dr. Haas can be reached at zhaas@cornell.edu.

Nicholas Bonello **(Chapter 6)** is with the University of Sheffield, U.K. He co-authored a book entitled "Classic and Generalized Low-Density Parity-Check Codes" (Wiley 2011). Dr. Bonello can be reached at NBonello1@sheffield.ac.uk.

Huijiang Li, Neeraj Jaggi and Biplab Sikdar **(Chapter 7)** are with the Department of Electrical Engineering and Computer Science, Wichita State University. KS. Dr. Sikdar is an Associate Professor and Huijiang Li is a Ph.D. candidate of the Dept. of Electrical, Computer and Systems Engineering of Rensselaer Polytechnic Institute, Troy, NY. Huijiang Li can be reached at lih4@rpi.edu.

Weirong Jiang and Viktor K. Prasanna **(Chapter 8)** are with the University of Southern California. Dr. Prasanna is a professor of Ming Hsieh Department of Electrical Engineering and the director of Center for Energy Information. Dr. Jiang is a Member of Technical Staff in Juniper Networks Inc. Dr. Jiang can be reached at weirongj@acm.org.

Hrishikesh Venkataraman **(Chapter 9)** is a senior researcher with the performance engineering laboratory at the Irish national research center – The RINCE Institute, in Dublin City University (DCU), Ireland. Dr. Venkataraman can be reached at hrishikesh@eeng.dcu.ie.

Injong Rhee **(Chapter 10)** is a Professor of Computer Science at North Carolina State University. His area of research interest is computer networks and network protocols on multicast problems, and multimedia networks such as video or voice transmission over the Internet, more recently focusing on fixing TCP problems for high speed long distance networks and concurrent MIMO MAC protocol design. Dr. Rhee can be reached at rhee@ncsu.edu.

Manzoor Ahmed Khan and Hamidou Tembine **(Chapter 11)** are with Ecole Superieure d'Electricite (Supelec, Gif-sur-Yvette, France). Dr. Tembine is an Assistant Professor. Dr. Tembine can be reached at tembine@ieee.org.

László Gyarmati and Tuan Anh Trinh **(Chapter 12)** are with the Network Economics Group of the Department of Telecommunications and Media Informatics at Budapest University of Technology and Economics (BME), Hungary. Dr. Trinh founded and is a coordinator of the Network Economics Group at BME. Dr. Trinh can be reached at trinh@tmit.bme.hu.

Klara Nahrstedt and Rini T. Kaushik **(Chapter 13)** are with the Computer Science Department at University of Illinois at Urbana-Champaign. Their research interests lie in enhancing energy-efficiency in storage and compute clouds, and in operating systems, file systems and storage systems. Dr. Nahrstedt can be reached at klara@cs.uiuc.edu.

Alexandru-Adrian Tantar **(Chapter 14)** is a PostDoctoral Researcher at the Computer Science and Communications (CSC) Research Unit, University of Luxembourg (AFR Grant). He is currently involved in the GreenIT (FNR Core 2010-2012) project. Dr. Samee U. Khan is an Assistant Professor of Electrical and Computer Engineering at the North Dakota State University, Fargo, ND. He is the founding director of bi-institutional and multi-departmental NDSU-CIIT Green Computing and Communications Laboratory (GCC Lab). Dr. Tantar can be reached at Alexandru.Tantar@uni.lu.

Hyung Gyu Lee and Jongman Kim **(Chapter 15)** are with Georgia Institute of Technology. Dr. Kim is a professor. Dr. Lee worked at Samsung Electronics as a senior engineer (2007 to 2010), and is currently a Postdoctoral Research Fellow at Georgia Tech. Dr. Lee can be reached at hyuggyu@gatech.edu.

In-Saeng Suh and Paul R. Brenner (**Chapter 16**) are with University of Notre Dame. Dr. Suh is a Research Associate Professor of Dept. of Physics, University of Notre Dame, and High Performance Computing Engineer at the Center for Research Computing, University of NotreDame. Dr. Suh can be reached at

George W. Arnold (**Chapter 17**) is a Deputy Director and National Coordinator for Smart Grid Interoperability at the National Institute of Standards and Technology (NIST). He is responsible for leading the development of standards underpinning the nation's Smart Grid. Dr. Arnold served as Chairman of the Board of the American National Standards Institute (ANSI) in 2003-2005, President of the IEEE Standards Association in 2007-2008, and is currently Vice President-Policy for the International Organization for Standardization (ISO). Prior to Joining NIST, he also served as a Vice-President at Lucent Technologies Bell Laboratories. Dr. Arnold can be reached at george.arnold@nist.gov.

Hans-Arno Jacobsen and Vinod Muthusamy (**Chapter 18**) is a professor of Computer Engineering and Computer Science at the University of Toronto. He holds the Bell University Laboratories Endowed Chair in Software and the Chair of the Computer Engineering Group. He directs and leads the research activities of the Middleware Systems Research Group (http://msrg.org). Dr. Jacobsen can be reached at jacobsen@eecg.toronto.edu.

Yogesh Simmhan, Qunzhi Zhou and Viktor Prasanna (**Chapter 19**) are with Computer Science Department, Viterbi School of Engineering, University of Southern California (USC). Dr. Simmhan is a Senior Research Associate at the Center for Energy Informatics at USC, and project manager for the software architecture for the Los Angeles Smart Grid project. Dr. Simmhan can be reached at simmhan@usc.edu.

Emanuele Crisostomi (**Chapter 20**) is currently a PostDoctoral Fellow at University of Pisa and has been a visiting researcher at Univesity of Cambridge, University of Leicester and Hamilton Institute, and National University of Ireland. S. Kirkland, A. Schlote, and R. Shorten are with the Hamilton Institute, National University of Ireland – Maynooth, Co. Kildare, Ireland. Dr. Crisostomi can be reached at emanuele.crisostomi@gmail.com.

Bong-Kyun "Bo" Ryu and Hua Zhu (**Chapter 21**) is currently with Argon ST, a wholly owned subsidiary of The Boeing Company, and is responsible for managing internal research projects in the areas of mobile ad hoc networking, sensor networks, satellite networking, information assurance and network security. Previously, he was the head of Network Analysis and Systems department at HRL Laboratories. He is author/co-author of 30+ publications, and holds 5 U.S. patents issued. Dr. Ryu can be reached at Bo.Ryu@argonst.com.

Ghang-Seop Ahn, Jikdong Kim, and Myung Jong Lee (**Chapter 22**) are with the Department of Electrical Engineering of City University of New York (CUNY). Dr. Ahn is a Research Assistant Professor at CUNY, and a Technical Editor of IEEE 802.15.4e. Dr. Kim is a senior research associate. Dr. Lee is a Professor and Director of the Advanced Wireless Networks Lab at CUNY, and a Chair of IEEE 802.15.5 TG. Dr. Ahn can be reached at gahn@ccny.cuny.edu.

Contents

Part 1: Green Communications Technology

Part 1
Green Communications Technology

Chapter 1
Evolving Communications Architectures: More Local Is More Green

Preston F. Marshall

Abstract. Current mobile communications architectures have increasing energy costs on both a unit, and often a per bit, basis. As requirements for bandwidth to mobile devices increases, it is important that the linear growth in energy consumption be avoided, or mobile communications will become an increasing source of energy consumption, and visual obstruction. This chapter discusses the potential evolution of mobile architectures that can be both more responsive to usage growth and provide lower energy consumption through more localized access to mobile points of presence. This transition will dependent on a number of technology developments that can be foreseen in the reasonable future. These technologies include cognitive radio and networking, interference tolerance , and advanced wireless system architectures.

1.1 Introduction

Much of the attention on energy savings has focused on large-scale, visible consumers of energy, such as heating, air conditioning, and transportation. While these are certainly important energy consumers, there are a variety of other energy consuming devices and systems that must be considered significant contributors to overall energy consumption, and therefore, important in any strategy for energy reduction.

Information technology products are increasing their demand for energy. New technology in this area is solving many of the fundamental design constraints that are otherwise imposed by thermal management constraints, thereby enabling these devices to actually increase their demand for energy as a consequence of newer

Dr. Preston F. Marshall
Information Sciences Institute & Ming Hsieh Department of Electrical Engineering
Viterbi School of Engineering University of Southern California
Los Angeles & Marina del Rey, CA USA
e-mail: pmarshall@isi.edu

technology[1]. New applications are increasing the demand for processing and communications resources exponentially, while technology is reducing the energy per computer operation or transmitted bit, at best linearly, (if at all in the case of communications)[2]. Thus, while other technology sectors are examining technology to reduce absolute energy consumption, the IT area is often looking to improve thermal design, in order to increase the energy that can be utilized in processing and communications systems, in order to address this growth in demand.

One area where this is particularly true is in wireless communications. The evolution of the mobile, or cellular phone, from a telephony device to a "smart phone" is estimated to also result in a 30–50 times increase in user bandwidth usage. Yet, the energy cost to deliver a bit is not something that can be continually reduced by technology. Seminal work by Shannon [26] established that there was a minimal amount of energy required to reliably communicate a bit of information in the presence of noise. The Shannon capacity relationship is given by:

$$C = B \log_2\left(1 + \frac{S}{N}\right) \tag{1.1}$$

Where:

C	Channel capacity	in bits/second
B	Bandwidth	in hertz
S	Total signal energy	
N	Total noise energy	same units as S

No technology can significantly reduce the amount of energy required to transmit a bit omni-directionally[3]. The lowest noise power is fixed by the temperature of the environment, and modern equipment can closely approximate these temperatures. The bandwidth available is finite, and the spectrum usable for mobile communications is highly constrained by physics and technology. Modern signal processing is capable of processing waveforms whose performance is very close to this bounding limit.

Therefore, it is likely that fundamental improvements in energy consumption will not be achieved through new technology in the networks's underlying physical links. Significant energy savings (orders of magnitude) can only come from changes in the architecture by which mobile services are delivered. Investigating how these orders of magnitude savings can be realized is the thrust of this chapter.

[1] In other words, instead of learning to use less energy, engineers are learning how to handle the heat generated by increased energy consumption.

[2] Information theory establishes that there is a minimum ratio of signal energy to noise energy that is required to reliably communicate a bit over a given bandwidth. SInce some modern equipment is environmental noise-dominated, there is an irreducible amount of energy required to transmit a bit over a Radio Frequency (RF) link; a value that can not be reduced through technology.

[3] Cellular systems are making extensive use of antenna directionality, primarily to increase capacity, not to decrease energy consumption.

1.2 Mobile Communications Service Architectures

Current mobile communications architectures have increasing energy costs on both a unit, and often a per bit, basis. As the bandwidth requirements of mobile devices increase, it is important that linear growth in energy consumption be avoided. Otherwise, mobile communications will become an increasing source of energy demand and visual obstruction.

This chapter discusses the potential evolution of mobile architectures that can be both more responsive to usage growth and provide lower energy consumption through more localized access to mobile points of presence. This transition will dependent on a number of technology developments that can be foreseen in the reasonable future. These technologies include cognitive radio and networking, interference tolerance, and advanced wireless system architectures. Fortunately, these technology opportunities will not only benefit the environment, but will have capacity, economic, and cost-reduction benefits to both the user and wireless operator communities.

Mobile access has become one of the most pervasive new technologies that emerged in the late 20th Century. It is worth considering how the current mobile/cellular architecture evolved. Early cellular systems were focused solely on voice communications. The interface between them and the Plain Old Telephone System (POTS) was a unique protocol and standard set that were in use only within telephone systems, and therefore the entire support structure of the early cellular was built around the internal methods of the telephone system control.

Early cellular systems were sparsely deployed, and ensuring complete area coverage was a major concern. Therefore, cellular towers were tall, utilized high power transmitters, and generally were very noticeable and unattractive. The growth of cellular services has generally followed this model, with infilling between towers to provide more bandwidth. Additionally, due to the significant cost differences between wireless and wired services, the use of mobile wireless devices was generally limited to when a fixed infrastructure service was not available.[4]

This approach was adequate when the growth of voice cellular bandwidth was primarily through additional users and usage. Even ten times more users meant a requirement for ten times the bandwidth, which was accommodated through additional sites, new frequencies, and advanced technologies that managed the spectrum more effectively. Doubling of spectral effectiveness, doubling of spectrum, and more sites accommodated linear voice growth, but is an inadequate strategy to address exponential data growth.

The problem this cellular architecture faces is that continued growth in bandwidth demand is not linear with, or driven by increases in the user population. Instead, it is driven by networking bandwidth, such as social networks, web services, location-based services, video conferencing and delivery, and the transition of traditional broadcast media to cellular delivery.

[4] Whereas now mobile services are used in residences or workplaces to avoid the inconvenience of switching between the fixed and mobile service, and to provide a seamless communications experience.

Increasingly, mobile users have expectations that mobile devices will be capable of not just replacing voice telephones, but also can provide many of the services of laptop and desktop computers. These functions include high levels of social networking, video upload and download, high definition television (HDTV) and graphics-rich news and entertainment. The industry has been satisfying the demand, generally through extension of the existing architectures, but it is clear that significant changes in architecture will be needed to address the exponential growth that Internet services will necessitate.

Since current equipment is highly efficient in terms of the bits/Hertz achieved for the power delivered to the receiver, it is unlikely that a significant increase in capability will be achieved through increased modulation or waveform efficiency. In the US, the Obama administration is seeking to provide cellular and other broadband services through an additional 500 MHz of spectrum [10], but even this spectrum is not a significant increase compared to the required growth in capacity.

Cooper [9] makes an effective case that most of the growth in radio services have been achieved not through spectrum efficiency, but through spectrum reuse Spectrum reuse allows the same spectrum to be reused many times over by separating users geographically so that adjoining users receive a signal level sufficiently low as not to cause harmful interference. In a simple model, if the spacing between systems (or base stations in the cellular architecture) using the same spectrum can be reduced by half (and thus its interference area reduced by a factor of four) potentially four times as much bandwidth can be provided to devices within that same service area and spectrum.

More localized base station access is also a natural evolution of other forces. Considering the characteristics of early cellular architectures, and current trends, there are a number of reasons to believe new wireless architectures will have to evolve, or be developed:

Range. Early cellular voice and data services were expensive, and were typically used only when the alternative fixed infrastructure was not available. As the cost of wireless service dropped, these mobile services have become a primary access mode for voice communications. Therefore it is possible to meet much of the mobile bandwidth demand with equipment located in very close proximity to the user's location. Just as much cellular utilization is while users are at their primary locations (residence and workplace), wireless access will increasingly be in these same locations, regardless of the potential availability of fixed infrastructure.

Bandwidth Needs. Users can now purchase unlimited wireless plans that can support multiple devices, essentially creating an alternative to wired connections, and placing the entire communications burden on wireless systems.

Location. Large base stations had to be located far from the typical user, forcing high power and highly elevated deployments. However, once it is possible to collocate cellular infrastructure with much of the user population, these undesirable traits can be mostly eliminated, and the devices can have similar footprints to existing communications equipment, such as cable modems, and Wi-Fi hubs.

Backhaul. First and second generation cellular systems (such as Advanced Mobile Phone System (AMPS) and GSM) were based on telephone switching, signaling and interoperability standards. Therefore, these base stations required access to the unique communications infrastructure generally associated with Telephone Company (TELCO) systems. With the advent of fourth generation cellular systems (Long Term Evolution (LTE), LTE–Advanced,, and Worldwide Interoperability for Microwave Access (WiMax)) the transition to Internet-based architecture is completed, and therefore cellular infrastructure can be provided through any suitable Internet access method.

Digital Processing Cost. Cellular signaling and control protocols are complex. In the early generations of base station equipment, processing was a significant cost driver, comparable to the cost of the analog elements, such as antennas, amplifiers, and high performance receivers. The cost of these elements has been rapidly reduced to the point where the digital control can be provided by inexpensive chips (as in femtocell devices). There is thus a financial incentive to build extremely low-cost cellular access by forgoing the necessity for higher transmit power and performance base stations, when users are in close proximity to the cellular base station.

Usage Patterns. The use of cellular telephones as a primary means of residential and workplace communications has been a consequence of the major reductions in cost. Residential "cordless phones" were a transition point, providing mobility with a small roaming region. The trend is clearly for voice users to abandon the use of both wired and local wireless services in favor of consolidation/unification to a single wireless service. The same is possible for broadband access, depending on cost, rather than technology. Many residences and workplaces use unlicensed Wi-Fi to provide mobility to a wired Internet service, but this could also be a transitional modality, just as "cordless phones" were a transition from wired "landline" telephone service, if the cost of wireless broadband could be sufficiently competitive to the wired access products.

1.3 Evolving Mobile/Cellular Architectures

The convergence of wireless technologies has been accelerating over the last decade. We can see some of these effects in examining the transition of cellular and wireless broadband access architectures. These evidence a transition of access from a completely centralized model, to one that provides local augmentation of services via non-tower solutions. For most cellular users, the concept of a cellular tower and cellular Radio Base Station (RBS) is synonymous; but it is challenging this very assumption that will be the basis for much of rest of this chapter. One major driver of this transition is the shift of service focus from voice to broadband data services.

The fundamental change from voice to data is evidenced in the evolution of the communications architectures. Fig. 1.1 illustrates wireless access architectures of the 80's and early 90's.

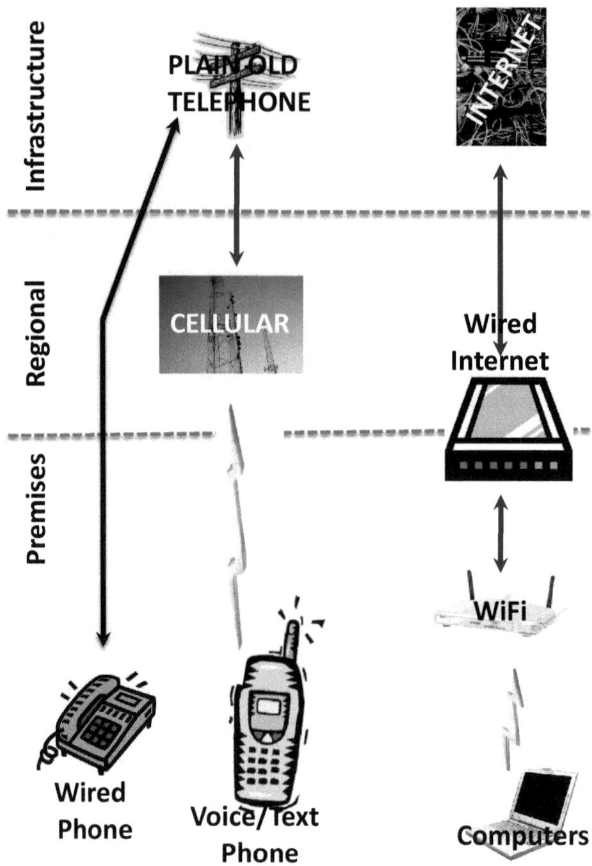

Fig. 1.1 "Stovepiped" Communications Architectures.

In these legacy architectures, the wireless and Internet paths were standalone and isolated. Interconnection, if any, occurred at deep levels in the TELCO infrastructure. It is very reasonable to treat these systems as independent, since their only interaction was through common application-level services. The telephony functions (wired and wireless) are more tightly coupled to each other than the two wireless services (telephony and data) were to each other, despite the common delivery mechanism. In fact, much of the data delivery is wireless, but using customer premise equipment, such as Wi-Fi.

In contrast, the architectures that are emerging have wireless architectures that are highly coupled, as shown in Fig. 1.2. Telephony is no longer the driver of these architectures, and voice network access is similar to that emerging for terrestrial access through Voice Over Internet Protocol (VOIP) services.

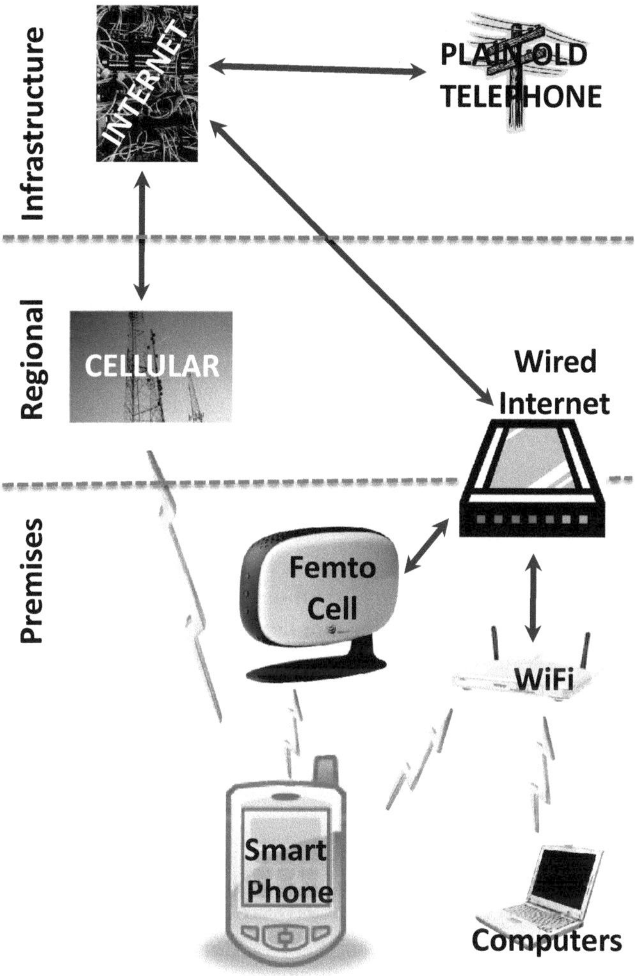

Fig. 1.2 Emerging Communications Architectures.

Both cellular voice and data use the Internet as the core distribution mechanism. While the only interaction between the cellular and premises Internet was in the network core in the stovepipe architecture, in the emerging architecture, the premises Internet supports mobile devices in parallel with the cellular services. Smart phones shift seamlessly from the cellular provider infrastructure to that provided in the premise as an extension of the fixed infrastructure At least one provider automatically shifts its cellular demand to local Wi-Fi to reduce the burden on its wireless infrastructure [3].

Even more integrated, the introduction of femtocells makes this access path invisible to network users, as the wireless service now is supported through both its own access to the Internet, and that provided by the distributed infrastructure supporting the femtocell devices. The success of this locally augmented architecture is evident in the rapid explosion of user-deployed femtocells , which were reported to exceed the number of carrier cells in late 2010 [11][5].

The limiting case of this architecture is to have the cellular base stations eliminated completely, and operate through the femtocells solely. Of course such a solution, as depicted in Fig. 1.3 is not practical in all conditions, such as coverage in obstructed urban areas, or remote areas, but we can imagine that most of the traffic passing over cellular networks is compatible with that vision, and that the role of these high footprint cellular infrastructure base stations can be limited to the spare regions where density is not a factor in any case. Statistics show that both cellular data and voice traffic is not dominated by remote regions, but primarily generated in residential and work-place locations.

There is a significant difference between this potential future architecture and the current use of femtocells. Current femtocell deployment is provided to enable premise owners to enhance their own personal cellular device performance. The services provided by these devices are not generally available to other wireless device users, and importantly, are not serving to offload general traffic from the local cellular wireless infrastructure. The carriers are not using them with any assumption of coverage, or as an alternative to building and operating conventional cellular infrastructure.

These examples show the necessity to approach the emerging complex, dense, and highly adaptable wireless systems with much more generality than the current architectures. The next section discusses the energy consumption implications of such strategies.

1.4 Energy Implications of Localized Access

The nature of wired and wireless communications is such that wireless communications require considerably more energy per bit than an equivalent distance wired connection. This is due to the control of transmission energy within a fixed media that electrical or photonics communications provides. Therefore backhaul communications can be made to use much less energy than the wireless modalities. This chapter will therefore focus on the energy consumption of the wireless elements of mobile access.

The energy (Path Loss) loss over a distance from transmitter to receiver is given by [20]:

$$L_{Path} = P_{exp} 10 \log_{10} \frac{4\pi r}{\lambda} \tag{1.2}$$

[5] It was reported that in the US there were more then 350,000 femtocells, compared to 256,000 carrier cells.

Where:

L_{Path}	Path Loss	in dB
P_{exp}	Propagation Exponent	Propagation at $1/r^{P_{exp}}$; typically 2 for free space, and up to 4 for diffracted paths
r	Range	Expressed in the same units as λ
λ	Wavelength	λ = speed of light/frequency

Reduction in range (r) leads directly to exponential reductions in required transmit energy, which is a major component in the energy budget of high-power, high duty-cycle wireless devices, such as cellular base stations.

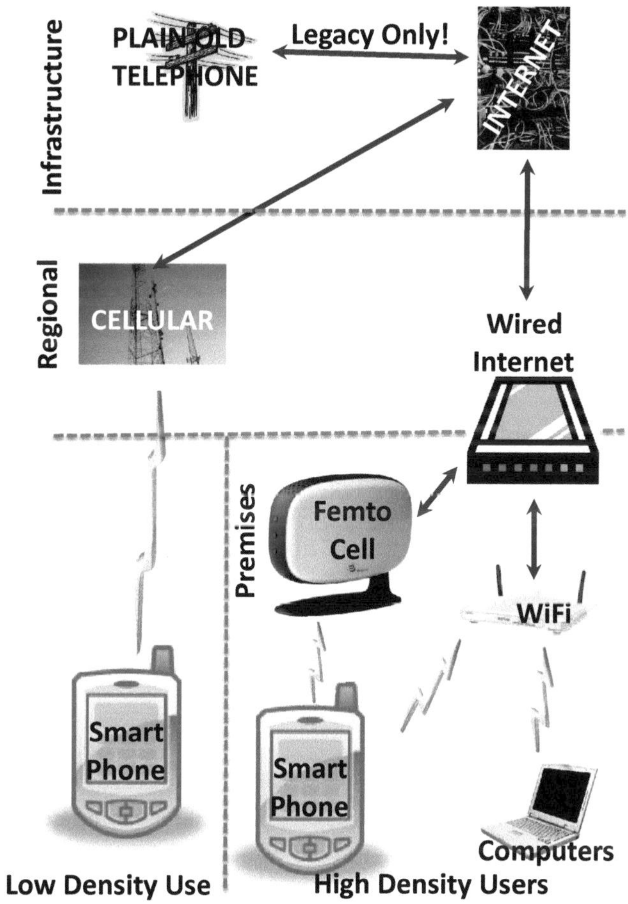

Fig. 1.3 Potential Future Communications Architectures.

For a typical cell phone system at 2.0 GHz and with free space (unobstructed) propagation, the path loss over a 1.5 km distance from a tower to a mobile device is approximately 102 dB[6]. If we can reduce the transmission distance of a wireless device to 200 meters, we would anticipate a path loss of 84.5 dB in the same propagation conditions.[7]

The reduction in range will enable (with an equivalent waveform and modulation) a total transmit energy reduction of a factor of over 55 times (17.5 dB)[8]! Of course, the lower height may have other effects, but it is clear that reduction in range is the most powerful tool to reduce the energy required to operate cellular systems. Fig. 1.4 illustrates the energy savings over a range of distances compared to a 1,500 meter distance from the cellular base station.

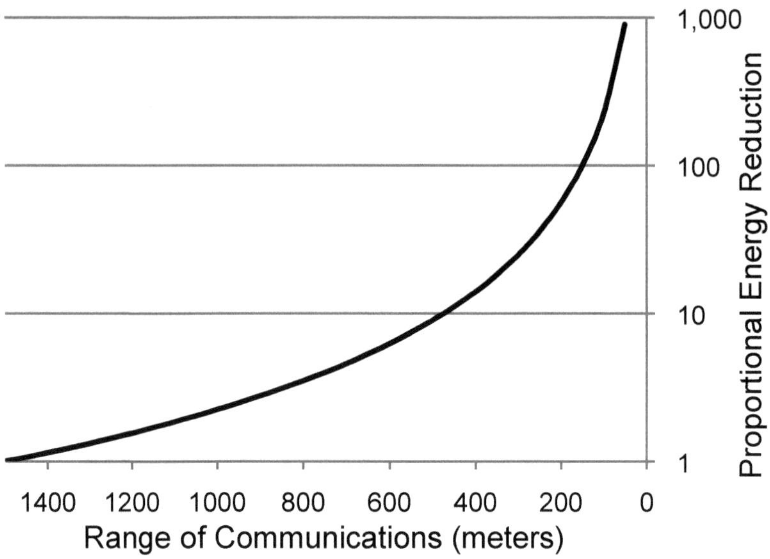

Fig. 1.4 Potential Transmit Energy Savings by Reduction in Path Loss through Distance from Base Station.

[6] Approximately $\frac{1}{16,000,000,000}$ of the transmitted energy reaches the receiver.

[7] This also makes the point of why we are not focused on the infrastructure backhaul energy consumption. As shown, a typical wireless link over 1.5 km has a loss of $10^{10.2}$. The same distance of photonic fiber (Singlemode 9/125m fiber @ 1310nm) has a loss of less than 25%. RF applications inherently have higher energy requirements due to loss of energy during propagation.

[8] It would allow for a 98% reduction in the energy required to transmit the same amount of information.

There is another benefit of this energy reduction. Estimates of likely cellular bandwidth increases vary widely from 25 to over 100 (see for example, the US Federal Communications Commission National Broadband Plan [10]). This growth in bandwidth is very likely to require a similar increase in transmission energy at both the handset and tower. A 3G cellular tower today typically transmits up to 6 sectors. 12 to 24 radios may be used per sector for Global System for Mobile Communications (GSM), although far fewer are typical for broadband.

This is a significant energy consumer, as the efficiency of the transmitters is in the 30–50% range. A single tower often hosts multiple carriers. With this much energy consumption, there is a consequent problem of heating, due to the processing and amplifier inefficiency. Although the power utilized by a small, sparsely loaded site can run as low as 20–40 watts, a dense, a multi-carrier site may run from 35–70 kW when fully loaded.Taken across 250,000 towers, the total power could be as high as 500–1,000 megawatts to power base station transmission of cellular traffic alone, even before build-out of 4G technology or additional "in-building". By comparison, a typical US residence has an average power consumption of 1.2 KW[9]. Cellular system energy is a significant contributor to per capita energy usage. An estimate provided by Bell Labs estimates that the global base station greenhouse gas emissions are on the order of 18,000,000 metric tons [1].

1.5 Technology Needs

The massive decrease in energy usage, as well as the concomitant bandwidth increases will not be achievable merely through the introduction of additional devices within the current technology framework. It will take additional technology development to achieve these objectives. Fortunately, industry has a vested interest in developing and deploying these technologies, since they are also key to the exponential increase in bandwidth needed to meet the likely customer demand.

1.5.1 Interference Management

The goal of spectrum management has generally been to maximize the degree to which spectrum dependent device operation is independent, or orthogonal, to other users. The limited frequency selection and tuning range of cell phones requires that femtocell devices operate on the same frequencies as the towers, and on the same general frequency plan. Therefore the femtocells form what is essentially an underlay network. There has been some initial interest in the application of Dynamic Spectrum Access (DSA) to cellular systems [7, 27]. DSA has typically been considered regimes in which it avoids any interference to other users, but in this example, we do not necessarily require that DSA operation by interference-free, but can be focused on the self-protection of the DSA device from interference.

[9] Source: US Energy Administration Residential Average Monthly Bill by Census Division, and State, 2009.

Considerable attention has been given to managing the interference between the femtocell and the tower systems [12]. This focus is an appropriate one when the relationship is close to one-to-one between cellular base stations and the femtocell [11] population. With this density, Femtocell-to-femtocell interference is not a major consideration, since the probability of interference between two femtocells is obviously low, due to their low power. However, as the density increases, this assumption becomes unreasonable. As the spacing of the femtocells approaches a distance of approximately twice their operating range, the likelihood of some level of mutual effects between femtocells increases from a possibility to a likelihood.

The problem of solving the interference progresses from a "one-to-many" (a single base station to a set of independent femtocells) to a "many-to-many" problem (many femtocells to many femtocells, as well as the base station to femtocells). As the complexity of finding non-interfering frequency and power combinations increases, the likelihood of finding a perfect, interference-free solution becomes more unlikely, if not impossible.

1.5.2 Interference Tolerance

A recent advance in RF communications concepts has been the introduction of Dynamic Spectrum Access (DSA) [14, 16, 22]. This technology attempts to sense unoccupied spectrum, and uses that spectrum with an assurance that it will not cause interference to other users, and will relinquish the spectrum to a more privileged user.

While this process is more flexible and adaptive than manual planning, it retains the objective of making each spectrum user independent of any other spectrum user. At some density point, it is unlikely that all of these devices can be de-conflicted, with finite spectrum and at the required density . The author has advocated a transition from interference-free to interference tolerant operation as a necessary step to achieve high spectrum density and network density and scalability [19, 21].

The US FCC Spectrum Policy Task Force proposed interference temperature as an approach [28] to interference management of underlay networks. It introduced the concept that regulation should not necessarily be focused on the complete elimination of interference, and this approach was shown to be tractable [4], however, there was no regulatory or technical momentum to introduce this form of spectrum sharing or management as a policy initiative.

Recently, the idea of interference tolerance, rather than avoidance did become a necessary consideration in the design of commercial mobile communications [2, 5, 12, 23, 24], particularly in the context of emerging LTE/LTE-A architectures. However, although some of the interference tolerance approaches have been embedded within the cellular 4G system design, they are not an explicit principle, so a fundamental shift in spectrum policy is not as evident as it would otherwise be. Nor is this interference tolerance margin accessible by other noncooperative users of the same spectrum.

The many-on-many spectrum deconfliction problem forces reconsideration of the absolute necessity to avoid any interference. There are numerous examples that demonstrate interference free operation is not an essential condition of wired or wireless wireless communications. The widely used Ethernet Local Area Network (LAN) operates on the principle of allowing, and randomizing interference. The popular Wi-Fi Wireless LAN technology often operates quite effectively in the presence of high degrees of collisions and interference. Transition from the concept of interference-free operation to a regime that can address mutual interference is an essential element of any strategy to support the increase in device density. Interestingly, that very problem, increased density, is both the problem to be addressed, and also the condition to be leveraged to achieve the energy savings developed previously.

There is no reason to believe that these interference-tolerant regimes must have a negative effect on the user experience. The author has shown [21] that even when the density is increased to relatively high levels of interference probability, the aggregate capacity that is created can be significantly more than is achieved when interference must be effectively precluded, as shown in Fig. 1.5.[10] Note that while current spectrum management targets interference probabilities are in the region below 10^{-4} (equivalent to less than 1 hour of interference effect per year), in this example, a communications system that is interference tolerant actually operates at its maximal aggregate capacity at orders of magnitude more interference probability than would be contemplated, or permissible, in current spectrum management and policy thinking.

To achieve this range of benefits, physical layer interference must be considered to be more than a challenge for the electrical engineers that design the physical communications layer. A fundamental shift in communications philosophy will require research and development at all layers of the communications stack. Fig. 1.6 illustrates some of the potential technology needs to achieve this integrated approach to capacity and interference within modern communications systems.

A necessary advance is that wireless systems and applications must become more robust to short-term disruption, and capable of segmenting traffic that does not require immediate delivery. One concept being investigated by the Internet research community is the concept of Delay Tolerant Networking (DTN) [8]. DTN does not mean that video must be interrupted. Instead, it argues that wireless systems must be aware of the delivery needs of each type of information that is presented, and should be capable of managing interruptions in the delivery more intelligently than wired Internet Protocol systems[11].

Email, E-Book delivery, and software updates are all examples of applications that can operate reliably without persistent end-to-end connections. Even a ten second disruption can cause the current Internet technology, Transmission Control Protocol (TCP), to abandon a transfer and start all over again. Even short disruptions are confused with congestion, and cause the transfer rate to be reduced, even after the full link operation is restored.

---

[10] These results are very situation and scenario dependent, so they should be considered as examples only. The case shown is for a Time Division Duplex (TDD), Push to Talk (PTT) mobile radio network.

[11] Which, by design, drops any not immediately deliverable packets.

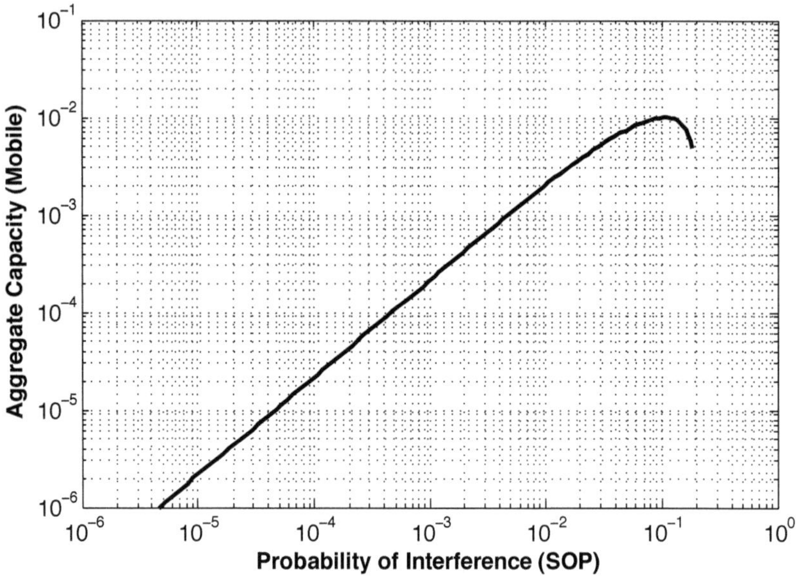

Fig. 1.5 Interference Probability vs. Aggregate Capacity for Land Mobile Scenario.

To address the interference-tolerance needs of future networks, it is essential that the wireless networks (rather than their applications) be able to manage wireless resources with the awareness of time-critical and delay tolerant traffic. Interruptions due to interference or collisions thus can be addressed through diverse routes, without impact on time-critical applications.[12] The author has proposed mechanisms by which DTN concepts can be embedded within wireless systems [15, 18].

The transition necessary to support a high density of devices is a fundamental technology need for the next generation of wireless systems. The typical design approach has been to concentrate on performance in the more benign conditions, and ensure performance was met in these benign conditions. In any other condition, "best efforts" would be provided to achieve some level of degraded performance, but these contended regimes were not the focus for the design objectives.

This must be reversed. The benign condition will become increasingly rare, and not the environment for which performance is stressed. Instead, the nominal case must become the highly interference-constrained one, and the benign condition considered a rare, but irrelevant condition. This is a fundamental shift in wireless design practice; away from focusing the design on environments that are noise limited, to those that are interference limited.

[12] For example, if a software upload transfer had to be interrupted, it would be handled within the wireless network itself. In contrast, under the Internet's TCP/IP Protocol, the transfer would fail after a number of seconds when the endpoint hosts detected the interruption, and would have to restarted when the connection was restarted.

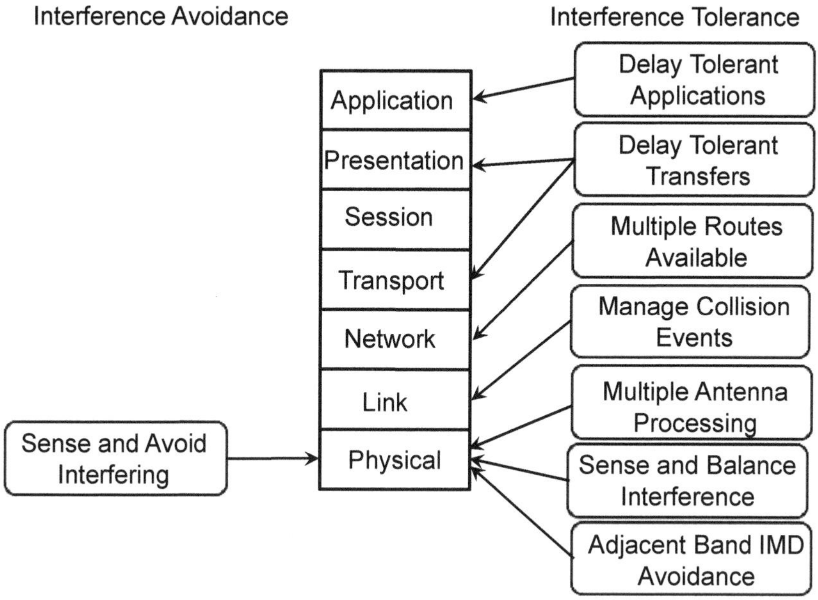

Fig. 1.6 Interference Tolerant Communications Requires New Technology throughout the Network Stack.

1.5.3 Adjacent Band and Co-location Effects

Density has an obvious effect on the likely interference within a channel. It also has significant effects on the operation of the devices, even if the channel itself is not shared with other users. This effect is due to the non-linear response of the receiver front-end, particularly the Low Noise Amplifier (LNA) and the mixer stages of the receiver. Signals entering the receiver front-end must be low enough in power so that they do not introduce intermodulation. Intermodulation distortion in these stages distorts the input signal so much that it may introduce mixing products (referred to as spurs) onto the same frequencies as the signal that is to be received.

Although specific situations vary, the receivers in most low or moderate duty-cycle, short range wireless systems are often the dominant energy consumer. Communications engineers have focused on efficient communications waveforms, but the receiver is often the major consumer of electrical energy. Much of this is required to create sufficient receiver linearity to ensure that strong signals do not overload the receiver, and cause false signal artifacts (referred to as intermodulation spurs) throughout the range of frequencies that would be mixing[13] in the RF and Intermediate Frequency stages of the receiver.

[13] Signal mixing creates artifacts at the sum and difference frequencies. If there are input signals f_1 and f_2, they mix to form $f_1 \pm f_2$ and $f_2 \pm f_1$. Often the more important products are the third order mixing, which are signals such as $2f_1 \pm f_2$ and as $2f_2 \pm f_1$; which, if f_1 and f_2 are close in frequency, some of the artifacts will be close to the original signals.

An example of intermodulation overload is shown in Fig. 1.7. This example shows the frequency domain of an input signal with a signal amplitude equal to the Third Order Input Intercept Point (IIP3) of the LNA, and the output spectrum resulting from the intermodulation. This point is typically in the range of –8 dBm for low cost consumer equipment.

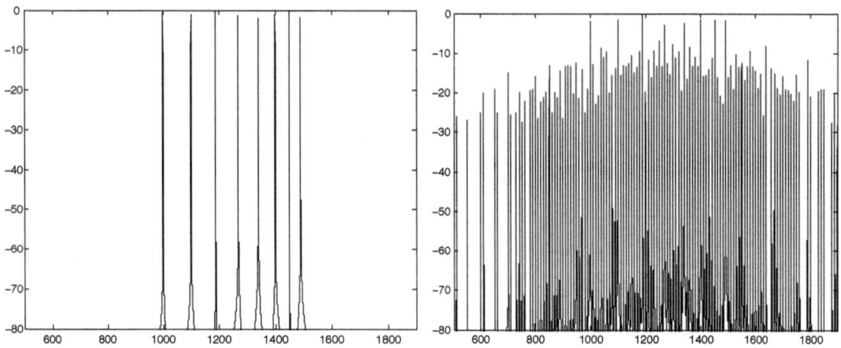

(a) Original Input Signal (b) Distorted Output Signal

Fig. 1.7 Effect of Third-Order Intermodulation Effects when Input Power (P_{in}) is equal to Amplifier IIP3 (P_{IIP3}).

This problem is not an academic one. Some equipment in the US public safety infrastructure was on frequencies adjacent to high power cellular base stations or towers by cellular provider NEXTEL. The effect of these allocations was that public safety receivers in the presence of these base stations were overloaded, and could not operate due to the intermodulation in their receivers. The US had to perform an expensive spectrum relocation of both cellular and public safety systems to resolve this adjacent channel overload-induced interference, even though the systems were operating on different frequencies [13].

The engineering solution to this problem is to increase the linearity of the receiver front-end. But increasing the linearity has an essentially linear effect on energy consumption. Enabling the device to tolerate other devices positioned twice as close will result in at four times more energy input to the linearity constrained stages, which will then require at least four times more linearity in the receiver RF stages.[14] Since linearity is proportional to energy consumption, this requires a proportional increase in receiver energy consumption. Conversely, reduction in receiver front-end linearity can provide the opportunity for equivalent reductions in the energy consumption of the RF components.

[14] Typically measured as the intercept point where distortion energy output equals signal output. Note that if this additional linearity is not provided, the third order effect is such at that a 6dB increase (4 times) in input energy times 3 (third-order) equals an 18 dB (or 4^3) increase in the generated distortion noise, or a factor of 64 times more noise.

This problem will increase as devices become more densely packed. The current Frequency Division Duplex (FDD) system used by most cellular providers have multiple handsets transmit on one frequency (uplink), which is received by the cellular base station tower. All of the handsets receive the single signal from the tower (downlink)[15]. Handsets do not have to "listen" to other handsets, and thus do not have to deal with a wide range of received signal energy.

The up and down links are sufficiently separated in frequency so that low cost fixed frequency RF filters can isolate the signals, so that no handset is impacted by even closely spaced adjacent handsets. As we increase the density of the networks, this simple assumption will no longer be valid. Femtocells will be in close proximity, and will be subject to overload by adjacent femtocells, and will also have to receive multiple handset signals of widely different amplitudes.

The author showed that a radio with a sufficient choice of frequencies could have high confidence in its ability to locate frequencies without adjacent signals that were capable of causing receiver front-end overload [17]. This operation is dependent on flexible Radio Frequency (RF) filtering. Unfortunately, the technology to perform this is not available at the price point needed for large-scale adoption in consumer products. A number of technologies show promise, such as Microelectromechanical systems (MEMS) [6, 25], but these have not yet been matured to the point where they are broadly and economically viable for low-cost RF filtering.

1.5.4 Security Technology Needs

An additional enabling technology is the security required to devolve centralized service and management functions down to devices that are not under the physical control of infrastructure providers. The same architectures described here introduce opportunities for a number of networking and service exploits, such as:

Emulation Attack. Cellular access devices can emulate authorized infrastructure networks and thereby capture confidential authorization data as devices attempt to connect.

Man-in-the-Middle Attack. Properly authorized devices can be modified to capture confidential data as it transits through the network.

Billing Fraud. Billing information is distributed across local devices, so the local operation is potentially vulnerable to manipulation to delete, modify, clone, or create improper billing data.

Personal Tracking. Location data associated with the invariant characteristics of wireless signaling, such as the permanent International Mobile Subscriber Identity (IMSI) would enable personal tracking to very fine resolutions, due to the local nature of each point of presence.

[15] Different standards enable the tower and handset to separate the received signal(s). Time Division Multiple Access (TDMA) systems separate the signals by assigning them to individual time slots. Code Division Multiple Access (CDMA) systems separate the signals by assigning them independent "spreading codes" which can be isolated by applying each code to the received signals.

Specific solutions to these security issues are beyond the scope of this volume, but these issues are not unique to wireless, and therefore should evolve and be deployed based on similar needs throughout the Internet.

1.6 Summary

A future consideration is that emerging applications of wireless network may have significantly more local transport requirements. Applications such as Machine-to-Machine connectivity, smart grid, and other local control applications will accelerate these benefits, as in most cases they can provide device-to-device paths that do not need to transit the cellular infrastructure, while still ensuring the reliability that is associates with these systems.

The progression to short range wireless access will require a significant change in the technology, equipment, and the technical and engineering assumptions of today's wireless practice. We can identify two major opportunities to reduce the energy consumption of wireless systems:

1. The first is to locate the cellular infrastructure along existing Internet access paths, in close proximity to the community which they serve. This will have an exponential effect on the transmission energy required from the cellular base-station utility sources, as well as from the mobile devices batteries.
2. The second opportunity is to fundamentally advance current receiver design so that receivers can adapt their operation, in order to avoid strong signals, rather than the current design approach, which is to provide high levels of amplifier linearity in order to achieve immunity to overload effects. Again, this has potentially a linear effect on overall energy consumption in the handset, and thus both operating time and reduction of energy demand from the charging infrastructure.

References

1. Alcatel-Lucent: Alcatel-Lucent maps the future of mobile technology. Press Release 002331 (2011), `http://www.alcatel-lucent.com/wps/portal/newsreleases`
2. Astely, D., Dahlman, E., Furuskar, A., Jading, Y., Lindstrom, M., Parkvall, S.: LTE: The evolution of mobile broadband. IEEE Communications Magazine 47(4), 44–51 (2009), doi:10.1109/MCOM.2009.4907406
3. AT&T Inc.: AT&T Wi-Fi network usage soars to more than 53 million connections in the first quarter. Press Release (2010)
4. Bater, J., Tan, H., Brown, K., Doyle, L.: Maximizing access to a spectrum commons using interference temperature constraints. In: 2nd International Conference on Cognitive Radio Oriented Wireless Networks and Communications (2007)
5. Boudreau, G., Panicker, J., Guo, N., Chang, R., Wang, N., Vrzic, S.: Interference coordination and cancellation for 4G networks. IEEE Communications Magazine 47(4), 74–81 (2009), doi:10.1109/MCOM.2009.4907410
6. Brank, J., Yao, J., Eberly, M., Malczewski, A., Varian, K., Goldsmith, C.: RF MEMS-Based Tunable Filters. John Wiley & Sons, New Yark (2001)

7. Buddhikot, M., Ryan, K.: Spectrum management in coordinated dynamic spectrum access based cellular networks. In: First IEEE International Symposium on New Frontiers in Dynamic Spectrum Access Networks, pp. 299–307 (2005), doi:10.1109/DYSPAN.2005.1542646
8. Cerf, V., Burleigh, S., Hooke, A., Torgerson, L., Durst, R., Scott, K., Fall, K., Weiss, H.: Delay-Tolerant Network Architecture, IETF RFC 4838. Internet Engineering Task Force (2007)
9. Cooper, M.: The myth of spectrum scarcity: Why shuffling existing spectrum among users will not solve America's wireless broadband challenge. A Martin Cooper Position Paper (2010)
10. Federal Communications Commission: Connecting America: The National Broadband Plan (2010)
11. Informa Telecoms & Media: The shape of mobile networks starts to change as femtocells outnumber macrocells in US. London, UK (2010)
12. Lopez-Perez, D., Valcarce, A., de la Roche, G., Zhang, J.: OFDMA femtocells: a roadmap on interference avoidance. IEEE Communications Magazine 47(9), 41–48 (2009), doi:10.1109/MCOM.2009.5277454
13. Luna, L.: NEXTEL interference debate rages on. Mobile Radio Technology (2003)
14. Marshall, P.F.: DARPA Progress in Spectrally Adaptive Radio Development. In: Software Defined Radio Forum Technical Conference (2006)
15. Marshall, P.F.: Progress towards affordable, dense and content focused tactical edge networks. In: IEEE Military Communications Conference (MILCOM), pp. 1–7 (2008)
16. Marshall, P.F.: Recent progress in moving cognitive radio and services to deployment. In: Proceedings of the 2008 International Symposium on a World of Wireless, Mobile and Multimedia Networks, pp. 1–8. IEEE Computer Society, San Diego (2008), doi: http://dx.doi.org/10.1109/WOWMOM.2008.4594812
17. Marshall, P.F.: Cognitive Radio as a Mechanism to Manage Front-end Linearity and Dynamic Range. IEEE Communications Magazine 47(3), 81–87 (2009)
18. Marshall, P.F.: Extending the Reach of Cognitive Radio. Proceedings of the IEEE 97(4), 612–625 (2009)
19. Marshall, P.F.: Dynamic Spectrum Access as a Mechanism for Transition to Interference Tolerant Systems. In: IEEE 4th International Symposium on New Frontiers in Dynamic Spectrum Access Networks, Singapore (2010)
20. Marshall, P.F.: Quantitative Analysis of Cognitive Radio and Network Performance. ARTech, Norwood (2010)
21. Marshall, P.F.: Research towards a cognitive network Eco-System. In: The 2010 Military Communications Conference, Networking Protocols and Performance Track (MILCOM 2010-NPP), San Jose, California, USA (2010)
22. McHenry, M., Livsics, E., Thao, N., Majumdar, N.: XG dynamic spectrum access field test results. IEEE Communications Magazine 45(6), 51–57 (2007)
23. Osseiran, A., Hardouin, E., Gouraud, A., Boldi, M., Cosovic, I., Gosse, K., Luo, J., Redana, S., Mohr, W., Monserrat, J., Svensson, T., Tolli, A., Mihovska, A., Werner, M.: The road to IMT-advanced communication systems: state-of-the-art and innovation areas addressed by the WINNER + project. IEEE Communications Magazine 47(6), 38–47 (2009), doi:10.1109/MCOM.2009.5116799
24. Pedersen, K., Kolding, T., Frederiksen, F., Kovacs, I., Laselva, D., Mogensen, P.: An overview of downlink radio resource management for UTRAN long-term evolution. IEEE Communications Magazine 47(7), 86–93 (2009), doi:10.1109/MCOM.2009.5183477

25. Rebeiz, G.M.: RF MEMS: Theory, Design, and Technology. Wiley Interscience, New York (2003)
26. Shannon, C.E.: A mathematical theory of communication. Bell System Technical Journal, 79–423, 623–656 (1948)
27. Subramanian, A., Al-Ayyoub, M., Gupta, H., Das, S., Buddhikot, M.: Near-optimal dynamic spectrum allocation in cellular networks. In: 3rd IEEE Symposium on New Frontiers in Dynamic Spectrum Access Networks, pp. 1–11 (2008), doi:10.1109/DYSPAN.2008.41
28. United States Federal Communications Commission: Spectrum Policy Task Force Report. ET Docket No. 02-135 (2002)

Chapter 2
Towards Green Wireless Communications: Metrics, Optimization and Tradeoff

Wei Wang, Zhaoyang Zhang, and Aiping Huang

Abstract. This chapter provides an overview on the current research progress in the green wireless communication field. Three key issues are discussed on green wireless communications. First, the metrics and the corresponding evaluation methods are discussed as the basis of energy efficiency optimization for wireless networks. Second, several methods are adopted to optimize the energy efficiency. For reducing the energy consumption, the methods can be categorized into two types, decreasing the equipment usage and adjusting the resource allocation. Third, the energy consumption increases and the energy efficiency decreases with the increasing transmit data rate. It is necessary to balance the tradeoff between energy and performance in practical wireless networks. Finally, the technical challenges are discussed towards green wireless communications.

2.1 Introduction

With the quick development of wireless communication technologies, the network scale increases amazingly. In recent years, more and more wireless applications are changing the human life. However, the enormous energy consumption in wireless networks brings incredible problems to both the energy sources and environment.

For the current energy consumption situation of wireless networks, the green wireless communication technologies [1][2] attract large interest from the researchers from both academia and industry. Decreasing the energy consumption in wireless networks can not only cut down the cost of communications, but also benefit the long-term development of wireless communications.

In this chapter, we discuss three key issues on green wireless networks, including metrics, optimization and tradeoff. The metrics are discussed firstly as the basis of energy efficiency optimization for wireless networks. Then, the methods

---

Wei Wang · Zhaoyang Zhang · Aiping Huang
Department of Information Science and Electronic Engineering, Key Laboratory of Integrate Information Network Technology, Zhejiang University, P.R. China

for energy efficiency optimization are presented by two types, decreasing the equipment usage and adjusting the resource allocation. Saving energy requires the cost of degraded performance, so the tradeoff between energy and performance is discussed and an example to balance the tradeoff is provided. Finally, the technical challenges are discussed on these three topics.

The remainder of this chapter is organized as follows. The metrics of energy efficiency are discussed and the evaluation methods are suggested in Section 2.2. Section 2.3 provides the strategies for optimizing the energy efficiency, including decreasing the equipment usage and adjusting the resource allocation. In Section 2.4, the tradeoff between energy and performance is discussed from both the information theoretic perspective and a specified example in cognitive radio networks. Following this, the technical challenges towards green wireless communications are discussed in Section 2.5. The conclusions are addressed in Section 2.6.

2.2 Metrics and Evaluation of Energy Efficiency

The energy efficiency metrics [3] are necessary to evaluate how much the wireless networks are green, which provides the objective for energy efficiency optimization. In this way, the techniques in wireless networks could be evaluated from an energy efficient perspective. There have been several definitions of energy metrics proposed by some international organizations and companies.

The International Performance Measurement and Verification Protocol (IPMVP) [4] is proposed by IPMVP committee of Efficiency Valuation Organization (EVO) in 2007. In IPMVP, the metrics and evaluation methods are proposed to measure the energy conservation, as well as the low-cost implementation of measurement. Four options for measurement are provided, including partially measured retrofit isolation, retrofit isolation, whole facility and calibrated simulation. These measurement methods can be considered as an important reference for the measurement and evaluation of energy efficiency in wireless networks.

To reduce the power consumption and energy costs, Verizon establishes a series of Telecommunications Equipment Energy Efficiency Ratings (TEEER) [5], which defines the formulas of energy efficiency and provides their measurement methods at various utilization levels. These metrics are applicable to broadband, video, data-center, network and customer-premises equipments. They are provided to the manufacturers for energy efficiency improvement.

The average power saving (APS) can be calculated as

$$APS = \left(\left(TEEER - Baseline \right) / Baseline \right) * SystemPower \tag{2.1}$$

In the above formula, the baseline parameter presents the average known performance of legacy equipments, and the TEEER parameter is the ratio of energy consumption and the corresponding system output. Define P_{Total} as the total consumed power, the TEEER can be calculated based on different cases respectively as follows.

For transport, the transport throughput T is considered as the system output.

$$TEEER = -\log\left(P_{Total} / T\right) \quad (2.2)$$

For switch and router, the forwarding capacity C is considered as the system output.

$$TEEER = -\log\left(P_{Total} / C\right) \quad (2.3)$$

For access, let L to denote the number of access lines,

$$TEEER = \left(L / P_{Total}\right) + 1 \quad (2.4)$$

For power amplifiers,

$$TEEER = \left(P_{TotalOut} / P_{TotalIn}\right) * 10 \quad (2.5)$$

where $P_{TotalIn}$ and $P_{TotalOut}$ are the input power and the output power of the amplifiers, respectively.

The energy and performance assessment [6] is proposed by Energy Consumption Rating (ECR) Initiative. ECR provides a common expression as the ratio of the energy consumption E and the effective system throughput T for different scenarios in communication networks.

$$ECR = \frac{E}{T} \quad (2.6)$$

Similarly, the Telecommunication energy efficiency ratio (TEER) [7] metric is proposed by Alliance for Telecommunications Industry Solutions (ATIS). This TEER metric is calculated by the ratio of the useful work to consumed energy.

A major problem in most of existing energy efficiency metrics is that they are designed for parts of wireless networks rather than the entire networks. In addition, they are not primarily designed considering various factors in different environments (e.g. suffered interference, traffic load). Due to the above issues, we propose the following considerations on the metrics and evaluation of energy efficiency for wireless networks.

- The metrics for green communications should consider the carbon emission eventually, rather than just the energy efficiency, especially for the cases evaluating the green property of entire networks. Actually, the carbon emission is not simply linear to the energy consumption, because consuming the energy from various sources causes totally different carbon emission. Therefore, towards the green wireless networks, the metrics should include the carbon emission as an important factor, which is directly related to how much the networks are green.
- The conventional per-bit energy consumption in information theory does not always represent the energy efficiency in practical wireless networks. The energy efficiency has close relationship to the specified situation.

(e.g. interference from other system, traffic load, etc.). For example, some protocols and algorithms work well and achieve mentionable energy efficiency when the traffic load is light, and the energy efficiency performance degrades quickly with the increasing load. Therefore, it is necessary to evaluate the energy efficiency based on the traffic load or the classic scenarios for providing more practical energy efficiency performance evaluation.

- To monitor the energy efficiency all over the networks, the modules for predicting, monitoring, evaluating the energy efficiency and finally decision making should be deployed in the equipments including core network, base station, relay node and user equipments. The corresponding signaling is also necessary for exchanging the energy information and control information between equipments in the networks. In the layered protocol architecture, the new modules should be coordinate to the existed protocols well, which would become a major property of green wireless networks. By collecting the information about energy, the green equipments can predict the energy consumption caused by different strategies and make the decision for improving the energy efficiency of entire networks.

2.3 Optimization of Energy Efficiency

In this section, we present the existing approaches that save the energy consumption and optimize the energy efficiency by decreasing the equipment usage and adjusting the resource allocation. The energy efficiency optimization schemes are illustrated in Fig. 2.1 and the following subsections describe each category respectively.

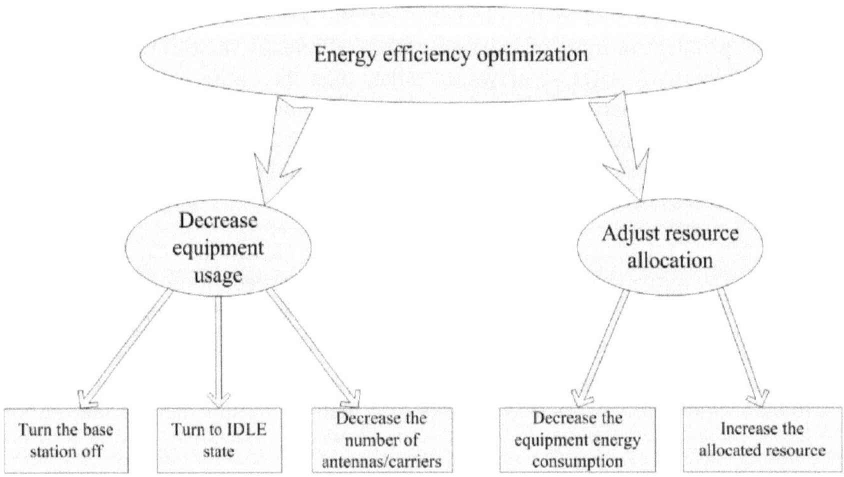

Fig. 2.1 Categorization of energy efficiency optimization

2.3.1 Decrease the Equipment Usage

The communication equipments can be turned off or reconfigured for saving energy when the traffic load is not heavy in the networks. The most usual methods for decreasing the equipment usage include turning the base station off, changing the state of the base station to IDLE and decreasing the number of antennas and frequency carriers [8][9].

1. Turn the base station off

   If the coverage areas of multiple base stations are overlapped and the traffic load of the overlapping area is relatively light, it is possible to turn off a part of the base stations for saving energy consumption.

   Using this kind of methods, the energy conservation can be maximal because we can save all the energy that was supposed to be consumed by the base stations if they are not turned off. However, the base station is allowed to be turned off only when its coverage area can be covered by other working base stations. For the partly overlapping case, which base station is turned off should be determined according the traffic distribution. In order to avoid the coverage hole after turning off some base stations, the base stations nearby should be reconfigured to compensate the coverage hole. In addition, the neighbor relation information needs to be updated to decrease the handover delay.

2. Change the base station state to IDLE

   When there is no active user in the coverage area of the base station, the state of the base station turns to IDLE, in which only the least necessary functions are reserved. Alternately, the discontinuous reception (DRX) can be adopted instead of the state transition. Using DRX, the energy is also saved by decreasing the transmission of unnecessary reference signal (RS). It is noted that the users have to measure the control channel more frequently for overcoming the possible synchronization problems caused by insufficient RS. It is noted that this method is only suitable for the case that there is no active user in the cell.

3. Decrease the number of antennas and carriers

   In almost all previous wireless communication systems, the number of active antennas is always fixed. The user equipment can obtain the number of antennas from the control information. The cyclic redundancy code (CRC) of physical broadcast channel (PBCH) is generated according to the antenna configuration of the corresponding base station, so the user would not receive the information from the base station successfully if the antenna configuration changes without notifying the user.

   The system information (SI) update notification is one of the feasible methods for the notification of the number change of antennas. The new number of antennas can be obtained by the user at the next update period. For both saving energy consumption and satisfying the traffic requirement, the number of antennas should be adjusted frequently. When the number of antennas changes, the user should obtain the information and reconfigure immediately. However, it is difficult to change the number of antennas and update the user's configuration synchronously, which causes that the user can not decode the information from the base station. In order to synchronize the antenna information at the based station and the user equipment, a default configuration is adopted for PBCH if the information can not be decoded successfully when the number of antennas changes. In this way, the related parameters can be reconfigured based on the updated number of antennas without the synchronization problem.

The similar method can be adopted for the case that the frequency carrier configuration is adjusted. By either the SI update notification or the detection of configuration change, the transmission on some carriers can be stopped for energy saving.

2.3.2 Adjusting the Resource Allocation

The energy efficiency can be optimized by adjusting the resource allocation. One kind of adjustment is decreasing the equipment energy consumption, and the other kind is increasing the allocated resource.

1. Decrease the energy consumption of equipments

Having some nodes stay in sleep mode is an efficient way to decrease the energy consumption of nodes. In the sleep mode, the nodes just remain the minimum necessary functions so that the energy can be saved. In [10], the nodes in the networks are divided into multiple disjoint sets, which sleep in turn. Only the nodes in one of the sets work, while the other nodes sleep for energy saving. A distributed algorithm is proposed in [10] for easy implementation. IEEE 802.15.4 [11] is the first communication standard which adopts the node sleeping scheme. Furthermore, the slot-based sleeping strategies TRAMA [12] and SERENA [13] schedule the nodes into sleeping according to the traffic.

By energy efficient routing and scheduling, the energy can be saved from a network perspective. In [14], the route with the lowest energy consumption is selected, and in [15], the route with the largest residual energy is selected. Considering both energy consumption and residual energy, the tradeoff between the above two factors is balanced in [16]. In [17], a packet scheduling algorithm is proposed considering the battery properties.

Another way to decrease the energy consumption of equipments is decreasing the transmitted information, including data aggregation and unnecessary information avoidance. Data aggregation is investigated mainly in wireless sensor networks. LEACH [18] and MLDA [19] are the classic data aggregation approaches. LEACH aggregates data based on clustering, in which the data from the whole cluster are aggregated at the cluster head node. MLDA establishes a tree for determining the data aggregation strategy. For avoiding the unnecessary transmission, the connected dominating set is adopted in [20].

2. Increase the allocated resource

Besides decreasing the energy consumption of equipments, decreasing the transmit data rate can increase the energy efficiency too. With the constraint of delay requirement in most practical scenarios, the energy efficiency can be improved by increasing the allocated radio resource, which also decreases the data rate in each unit resource. By data rate splitting, the nodes choose appropriate channels from the radio resource pool and transmit with lower data rate per channel.

The node cooperation can also improve the energy efficiency because the cooperation between nodes can increase the utilization of given resources, e.g. more power, more space freedom, etc. There are usually two types of node cooperation. One is cooperative relaying [21]. In most cases, the relay nodes are nearer to the destination than the source node, which provides additional channels with better channel quality essentially. The other is virtual multiple input multiple output (VMIMO) [22]. Because of the equipment size and hardware complexity limitation, most of current user equipments have only one antenna. For improving the transmission and energy efficiency, the user with single antenna transmits the information from not only itself but also the nearby users. In this way, its own antenna and the antennas from other nodes compose multiple antennas for VMIMO transmission.

For minimizing the energy consumption, one or more appropriate nodes are selected for node cooperation, and the transmit data rates between nodes are adjusted for load balancing. The energy consumed by nodes is composed by two parts. One part is the energy used for data transmission, which is related to the data rate. The other part is consumed by the equipment circuit and control signaling exchange. Even if the node does not transmit any data, the latter part of energy can not be saved. In that case, it is not always a good choice to utilize all the available nodes for cooperation.

2.4 Tradeoff between Energy and Performance

In this section, we discuss the tradeoff between energy and performance from the information theoretic perspective firstly. The energy can be saved with the degraded performance. Then, the energy-optimal sensing-transmission tradeoff is presented as an example to show the relationship between energy and performance.

2.4.1 Information Theoretic Perspective

The Shannon formula for band-pass channel in information theory is written as

$$C = W \log\left(1 + \frac{E_b R}{N_0 W}\right) \tag{2.7}$$

where W is the channel bandwidth, R is the data rate of transmission, E_b is the energy consumption per bit and N_0 is the power spectral density of thermal noise. Define $\eta = R/W$, the following formula can be obtained.

$$\frac{E_b}{N_0} \geq \frac{2^\eta - 1}{\eta} \tag{2.8}$$

The relationship of energy and performance based on (8) is shown in Fig. 2.2. It can be obtained that for a fixed mount of data, the energy consumption decreases with the decreasing data rate.

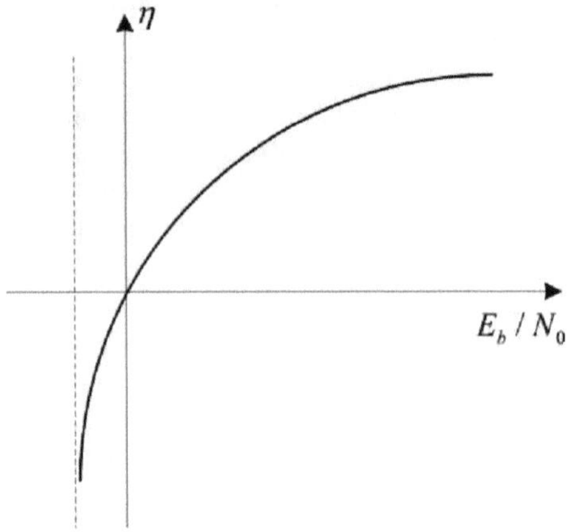

Fig. 2.2 The relationship between energy and performance

There exist the packet delay constraints in some cases, especially for the real-time services. In those cases, the data rate could not be too slow. It is obvious that there is a tradeoff between energy consumption and transmission performance. One of the methods to balance the tradeoff is adopting as slow data rate as possible on the condition that the delay requirement of the traffic is satisfied.

2.4.2 Energy-Optimal Sensing-Transmission Tradeoff in Cognitive Radio Networks

In this subsection, we take the energy consumption for both spectrum sensing and data transmission in cognitive radio networks as an example for the tradeoff between energy and performance. The performance of cooperative spectrum sensing depends on the cooperative strategies, which affects the energy consumed during cooperative spectrum sensing. The merit of cooperative spectrum sensing mainly lies in the sensing diversity gain provided by multiple secondary users (SUs). With the increasing number of cooperative SUs, the sensing performance is improved, which means that the environment information can be provided more accurately to discover more spectrum opportunities. In this case, the energy consumption of transmission can be saved when more spectrum bands are available.

On the other hand, the cooperative spectrum sensing with a large number of SUs induces a lot of extra communication overhead that causes more energy consumption.

There have been a few publications on the sensing-transmission tradeoff but not from the energy perspective. In [23], a tradeoff between the sensing duration and the transmission duration is investigated for maximizing the throughput. The tradeoff is designed for each single user and the length of both the sensing duration and the transmission duration are not the same for different users.

A few schemes of energy-efficient spectrum sensing are newly proposed. Spectrum sensing in cognitive sensor networks is studied in [24] to minimize the energy consumed subject to the constraints on both detection probability and false alarm probability. In [25], a cluster-and-forward based spectrum sensing scheme is proposed to save energy. The above works focus on the energy consumption only for sensing and reporting local sensing results. Actually, the performance of spectrum sensing has significant effect on the transmission. It is necessary to investigate the energy consumption of sensing, reporting and transmission jointly.

In [26], an energy-optimal cooperative strategy for cooperative spectrum sensing is proposed in the network scenario shown in Fig. 2.3. The energy consumption includes three parts. For sensing, if the *i*-th SU has been chosen to participate in cooperative spectrum sensing, it implements the spectrum sensing independently first, which consumes the energy E_s. The SU reports its local sensing results to the fusion center, which consumes energy E_{ri} for user i. It is reasonable to assume that the signal detection energy E_s for each SU is identical, however, E_{ri} for different SUs might be diverse from each other on account of path loss, shadow, etc. When the final decision of the fusion center shows that the channel is available, the transmission consumes energy E_t. Hence, the total energy consumed is given by

$$E = \sum_i I(i)\left(E_s + E_{r_i}\right) + E_t \tag{2.9}$$

where $I(i)$ indicates whether the *i*-th SU reports the sensing information to the fusion center.

When the SU transmits with a stable transmission rate, the transmitting power of SU could decrease by reducing the total false alarm probability Q_F so that the energy consumption during transmission could also decrease. However, choosing the set of SUs to obtain the minimal Q_F might not make sure that the total energy is minimal as well, because for some SUs the false alarm probabilities are low enough but the reporting procedure consumes quite large energy.

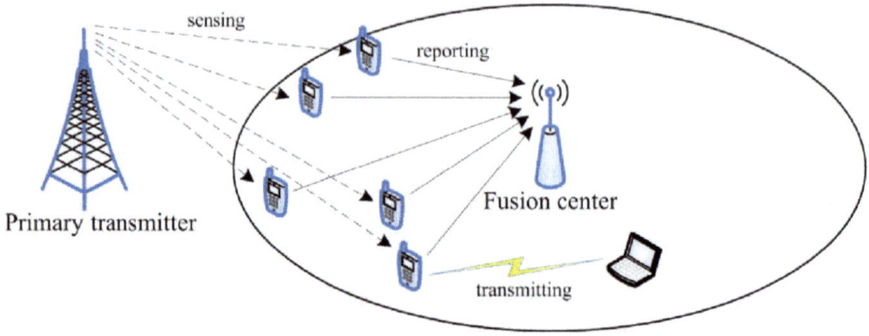

Fig. 2.3 Network scenario

To treat all SUs equally, let all of them achieve the same detection probability P_d for protecting PUs. Because P_d decreases when the number of cooperative users increases, the threshold λ_i for user i can be decreased, which also reduces the false alarm probability P_{f_i} and increases the available spectrum opportunities for transmission. Choosing SUs with low enough false alarm probabilities may reduce Q_F and promotes the reduction of energy consumption. However, some of those SUs might have to consume a relatively large amount of energy to report their sensing decision.

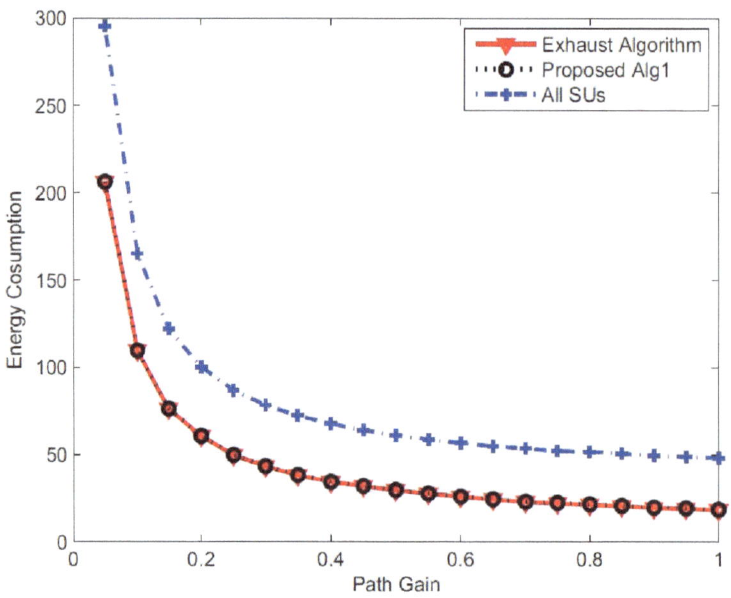

Fig. 2.4 Performance comparison for difference balance strategies of the sensing-transmission tradeoff [26]

To find which SUs should cooperate for spectrum sensing, a heuristic algorithm is proposed in [26] for a sub-optimal solution. The basic idea of the heuristic algorithm is selecting the cooperative SUs one by one. When one more SU is added, the SU corresponding to the minimal total energy consumption is selected. This energy consumption is compared to that without the new SU. The procedure of calculating the energy consumption and adding another SU is repeated till the value of energy consumption can not be reduced.

This suboptimal heuristic algorithm with low complexity chooses the SUs considering both low false alarm probability and low reporting energy. The proposed algorithm achieves almost the same performance as the optimal exhaustive algorithm, as shown in Fig. 2.4. Moreover, compared with the conventional cooperative sensing with all SUs participating in cooperative sensing, the proposed algorithm has a significant gain on energy efficiency, which indicates the importance of the tradeoff between energy and performance.

2.5 Technical Challenges

The energy efficient strategies have been investigated by lots of researchers as introduced above. However, there are still more technical challenges left in pursuing the green wireless communication. In this section, we present some of those challenges with additional implementation issues.

- **Unified evaluation standard for green wireless networks.** In order to decrease the energy consumption of wireless networks, the primary problem is establishing the metrics to evaluate the energy efficiency for different key techniques. For different purposes, the energy efficiency should be analyzed using their corresponding metrics respectively. An appropriate general metrics for energy efficiency is needed considering different purposes and different energy sources for the hybrid traffic cases.

  Lots of green metrics and evaluation methods for wireless networks have been proposed by different organizations and corporations recently. A unified green standard can provide an aim for all modules to make the entire networks greener. Adopting the common aim can avoid the conflict between different modules in the networks or different processes during the wireless transmission.
- **Energy efficiency optimization protocols for green wireless networks.** To design a protocol for evaluating and optimizing the energy efficiency in wireless networks, the primary issue is compatibility. The protocol design should not only increase the performance of energy efficiency decisions but also change existed protocols as less as possible for avoiding the conflict to other mechanisms in the current network protocols.

  More generally, the energy consumption during equipment production is also considered. The energy efficiency of communication equipment is optimized by considering both production and utilization during its whole service life of the equipment. It is possible that the advanced equipments which provide better energy efficiency performance during communications may need different producing method with more energy consumption. Therefore, it is reasonable to optimize the energy efficiency considering both production and utilization.

- **Tradeoff between energy and performance for practical wireless networks.** Some fundamental researches [27][28] on the tradeoff between energy and performance provide the state-of-the-art methods to balance the tradeoff, but the scenarios are simple compared with practical wireless networks. More research efforts are still needed for more practical networks, e.g. multi-cell cellular networks, hierarchical networks, multi-hop wireless networks.

  By sensing the transmission environment, the equipments can obtain more information about the transmission channel condition and traffic requirement, which benefits the energy efficiency improvement. In addition, the sensing process for obtaining the related information also causes energy consumption. With different system overhead and energy consumption, the correctness and precision of the information from sensing are different. Therefore, the energy consumption during both sensing and transmission should be considered when optimizing the energy efficiency.
- **Green emerging wireless networks.** In cognitive radio networks [29], the spectrum dynamics is a major issue. For spectrum sensing, the energy can be saved by avoiding unnecessary sensing. In addition, for the transmission environment with dynamic spectrum, the new energy-erformance tradeoff is also necessary to investigate.

  In hierarchical cellular networks [30], the energy consumption is affected by the power control strategies of different networks. Joint power control between networks is a feasible method to improve the energy efficiency. The base stations which have the overlapping coverage area can also sleep in turn to save energy.

  In wireless Internet of Things (IoT) [31], the large amount of equipments is the main challenge for energy consumption. The protocol design for IDLE state has higher energy saving requirement for intelligent IoT equipments, especially for the equipments which do not work in most of the time. The interference and conflict between equipments also need to be avoided in a distributed manner for decreasing the energy waste.

2.6 Conclusions

We provide an overview on green wireless communication in this chapter. Three key issues of green wireless communications are presented. First, several energy efficiency metrics are introduced. Based on these existing metrics, we discuss some further considerations, including the carbon emission concern and the practical implementation. Second, the current energy efficiency optimization methods are categorized into two types. One is decreasing the equipment usage, and the other is adjusting the resource allocation. Third, the fundamental energy-performance tradeoff is discussed. Following this, we present one of our research works on balancing the energy-optimal sensing-transmission tradeoff for cognitive radio networks as an example. Finally, we discuss some open research problems on metrics, optimization and tradeoff toward green wireless communications. In addition, the technical challenges for several green emerging wireless networks are also discussed for future study.

Acknowledgments

The authors would like to thank Qiufang Wang at Zhejiang University for providing some of the research results on energy-optimal sensing-transmission tradeoff. This work is supported by National Natural Science Foundation of China under grant No. 61001098, Fundamental Research Funds for the Central Universities under grant No. 2010QNA5011 and Scientific Research Fund of Zhejiang Provincial Education Department under grant No. Y200909796.

References

[1] S. Udani, S., Smith, J.: Power Management in mobile computing (a survey), Technical report, University of Pennsylvania (1996)
[2] Jones, C.E., Sivalingam, K.M., Agrawal, P., Chen, Y.C.: A survey of energy efficient network protocols for wireless networks. Wireless Networks 7(4), 373–385 (2001)
[3] Chen, T., Kim, H., Yang, Y.: Energy efficiency metrics for green wireless communications. In: WCSP (October 2010)
[4] IPMVP commitee, International Performance Measurement and Verification Protocol, http://www.nrel.gov/docs/fy02osti/31505.pdf
[5] Verizon, Verizon NEBS Compliance: TEEER Metric Quantification (January 2009), http://www.verizonnebs.com/TPRs/VZ-TPR-9207.pdf
[6] ECR Initiative, Network and Telecom Equipment-Energy and Performance Assessment (2010), http://www.ecrinitiative.org/pdfs/ECR211.pdf
[7] ATIS, Energy Efficiency for Telecommunication Equipment: Methodology for Measurement and Reporting-General Requirements (2009)
[8] 3GPP TR 36.913 Requirements for further advancements for Evolved Universal Terrestrial Radio Access (E-UTRA), http://www.3gpp.org/ftp/Specs/archive/36_series/36.913/36913-800.zip
[9] 3GPP R2-101824, Energy saving techniques for LTE, Huawei, http://www.3gpp.org/ftp/tsg_ran/WG2_RL2/TSGR2_69/Docs/R2-101824.zip
[10] Cardei, M., Du, D.: Improving wireless sensor network lifetime through power aware organization. ACM Journal of Wireless Networks (May 2005)
[11] IEEE 802.15.4: Wireless Medium Access Control (MAC) and Physical layer (PHY) specifications for Low-Rate Wireless Personal Area Networks (LR-WPANs). IEEE Computer Society, LAN/MAN Standards Committee (October 2003)
[12] Rajendran, V., Obraczka, K., Garcia-Luna-Aceves, J.J.: Energy-efficient,collision-free medium access control for wireless sensor networks. In: Sensys 2003 (November 03, 2003)

[13] Minet, P., Mahfoudh, S.: Performance evaluation of the SERENA algorithm to SchEdule RoutEr Nodes Activity in wireless ad hoc and sensor networks. In: AINA 2008 (March 2008)
[14] Kwon, S., Shroff, N.B.: Energy-Efficient Interference-Based Routing for Multi-hop Wireless Networks. In: IEEE INFOCOM 2006 (April 2006)
[15] Hassanein, H., Luo, J.: Reliable Energy Aware Routing in WirelessSensor networks. In: IEEE DSSNS Workshop (April 2006)
[16] Shresta, N.: Reception Awarness for Energy Conservation in Ad Hoc Networks, PhD Thesis, Macquarie University, Sydney, Australia (November 2006)
[17] El Gamal, A., Nair, C., Prabhakar, B., Uysal-Biyikoglu, E., Zahedi, S.: Energy-efficient Scheduling of Packet Transmission over a Wireless Link. IEEE Infocom (2002)
[18] Heinzelman, W., Chandrakasan, A., Balakrishnan, H.: Energy-efficient communication protocol for wireless microsensor networks. In: HICSS 2000, January 2000, pp. 3005–3014 (January 2000)
[19] Kalpakis, K., Dasgupta, K., Namjoshi, P.: Maximum Lifetime Data Gathering and Aggregation in Wireless Sensor Networks. IEEE Networks 2002 (August 2002)
[20] Dai, F., Wu, J.: An extended localized algorithm for connected dominating set formation in ad hoc wireless networks. IEEE Trans. on Parallel and Distributed Systems 15(10) (2004)
[21] Laneman, J.N., Tse, D.N.C., Wornell, G.W.: Cooperative diversity in wireless networks: Efficient protocols and outage behavior. IEEE Trans. Info. Theory 50(12), 3062–3080 (2004)
[22] Dohler, M., Lefranc, E., Aghvami, H.: Space-time block codes for virtual antenna arrays. IEEE PIMRC 2002, 414–417 (September 2002)
[23] Liang, Y.C., Zeng, Y.H., Peh, E.C.Y., Hoang, A.T.: Sensing-Throughput Tradeoff for Cognitive Radio Networks. IEEE Trans. Wireless Commun. 7(4), 1326–1337 (2008)
[24] Maleki, S., Pandharipande, A., Leus, G.: Energy-Efficient Distributed Spectrum Sensing for Cognitive Sensor Networks. IEEE Sensors Journal (June 2010)
[25] Wei, J., Zhang, X.: Energy-Efficient Distributed Spectrum Sensing for Wireless Cognitive Radio Networks. In: IEEE INFOCOM 2010 Workshop (March 2010)
[26] Wang, Q., Wang, W., Zhang, Z.: Energy-optimal cooperative strategy for spectrum sensing. Submitted to IEEE PIMRC (2011)
[27] Berry, R., Gallager, R.: Communication over fading channels with delay constraints. IEEE Trans. Info. Theory 48(5), 1135–1149 (2002)
[28] Neely, M.J.: Optimal energy and delay tradeoff for multiuser wireless downlinks. IEEE Trans. Info. Theory 53(9), 3095–3113 (2007)
[29] Mitola, J., Maguire, G.Q.: Femtocell networks: A survey. IEEE Personal Communications 6(4), 13–18 (1999)
[30] Chandrasekhar, V., Andrews, J., Gatherer, A.: Femtocell networks: A survey. IEEE Commun. Mag. 46(9), 59–67 (2008)
[31] Gershenfeld, N., Krikorian, R., Cohen, D.: The internet of things. Scientific American 291(4), 76–81 (2004)

Chapter 3
Energy-Aware Link Adaptation for a MIMO Enabled Cognitive System

Eren Eraslan, Babak Daneshrad, Chung-Yu Lou, and Chao-Yi Wang

Abstract. Circuit designers have put substantial effort to reduce the energy consumption of specific blocks in wireless communication devices. However, much more significant increase in energy efficiency can be achieved at the system level by proper choice of transmission parameters. Choosing the best mode to transmit given the channel conditions in a wireless link is referred to as link adaptation. In this chapter, we present the state-of-the-art on link adaptation protocols for emerging Multiple-Input-Multiple-Output (MIMO) systems, and we provide a novel energy-aware fast link adaptation protocol which provides orders of magnitude gain in energy efficiency of the communication link.

3.1 Introduction

A MIMO enabled Cognitive System is a very potent concept. Cognitive systems are aware of their surroundings, and can be opportunistic in their transmission. MIMO based communication systems provide significant improvement in capacity by increasing the robustness of the link (space-time block codes - STBC), and/or by improving the spectrum efficiency (spatial multiplexing - SM). Hence, MIMO techniques found their way in every high-speed wireless communication standards, such as IEEE802.11n, IEEE802.16e to name a few. Most of the laptops nowadays have more than one antenna, and cellular base stations are equipped with even more than that. A secondary impact of integrating MIMO into a radio is that it has a multiplicative impact on the number of modes available to close the link, i.e. satisfying the throughput and delay requirements of the application. In order to optimize an objective function, a MIMO radio can change the system parameters such as: number of spatial streams, number of transmitter/receiver antennas, modulation, code rate, and transmit power. Hence, it has a large selection of modes to choose from. A mode-rich cognitive radio can thus sense its

Eren Eraslan · Babak Daneshrad · Chung-Yu Lou · Chao-Yi Wang
UCLA
e-mail: babak@ee.ucla.edu

surroundings and choose the best mode for a given environmental condition to optimize an objective function. This problem is generally referred to as the link adaptation problem.

Past work dealing with the link adaptation problem has mostly focused on maximizing the throughput or decreasing the error rates for robustness without considering the energy consumption of the link. We contend, however, that minimizing the energy consumption of a mobile wireless link is as critical, if not more critical than maximizing its throughput.

The issue of energy saving is very significant in wireless communications since most of the devices operate on batteries, and hence the total number of bits that can be transmitted/received by a wireless link is finite. Power consumption of wireless communication components is approximately 50% of the total power consumption in modern smart-phones [1]. However, battery technology has not evolved as fast as needed to meet the ever increasing energy needs of today's mobile applications. Maximizing the energy efficiency of the link will directly increase the up-time of the battery or decrease the battery capacity needed for similar up-time performance. Given that hundreds of millions of laptops and cell phones are sold each year, a reduction in the battery requirement would also have a huge environmental impact.

In this chapter, we will present energy-aware link adaptation strategies which aim at minimizing the total energy consumed to successfully transfer information bits (Joule/bit) subject to Quality of Service (QoS) constraints such as packet error rate and minimum throughput. We will show that the energy-aware link adaptation strategy can deliver orders of magnitude reduction in the total energy consumption of a link compared to static strategy by choosing the most energy efficient mode (in terms of Joules/bit) given the channel conditions. In arriving at the energy signature of a wireless node we take into account both the versatility provided by the MIMO processing, modulation, code rate, receiver algorithms, bandwidth, data converter power, and transmit power. Orders of magnitude reduction in the total energy draw of a radio without impacting the QoS constraints of the overarching application will go a long way towards reducing the energy footprint of our ever increasing appetite for wireless data communications.

The organization of the chapter is as follows. We first present the state-of-the-art in the link adaptation solutions both for maximizing throughput and minimizing energy consumption. Then, we present energy consumption models for the transmitter/receiver blocks, and novel methods for predicting the performance of the link. With accurate modeling of the packet error rate performance and the energy consumption of the link, we then formulate the energy-aware link adaptation problem and provide a highly practical fast responsive algorithm. Simulation results on the performance of the algorithms are presented for a realistic channel scenario to show that significant gain in energy efficiency can be achieved by a well designed link adaptation protocol which takes the energy consumption into account.

3.2 Existing Work on Link Adaptation

Traditional Single-Input-Single-Output (SISO) radios have only a limited number of modes. Therefore, link adaptation has rather trivial search space and the potential gain associated is limited. There is a rich body of work addressing the link adaptation problem for maximizing the throughput in SISO systems [2] – [5]. On the other hand, link adaptation for MIMO systems is more challenging since they have a large number of operating modes. Most of the research done for MIMO link adaptation has focused on techniques for maximizing the throughput or decreasing error rates. MIMO techniques however come with additional power consumption due to the complex baseband algorithms and duplication of tx/rx chains. Hence, there exists a trade-off between the throughput performance and the energy consumption of the radio. It is desirable to analyze this trade-off and develop an energy-efficient link adaptation algorithm that maximizes the number of successfully transmitted bits per unit energy (bit/J).

Link adaptation papers in the literature can be classified into 4 groups:

- Switching between spatial multiplexing and diversity to decrease Bit-Error-Rate (BER) for fixed rate MIMO systems.
- Medium Access Control (MAC) layer based polling techniques for maximizing the throughput.
- Physical (PHY) layer metric based techniques for maximizing the throughput.
- Limited work on maximizing the energy efficiency for SISO and MIMO

We will elaborate upon each of these groups in the ensuing subsections.

3.2.1 Switching between Spatial Multiplexing and Diversity to Decrease Bit-Error-Rate (BER) for Fixed Rate MIMO Systems

Link adaptation can be used for decreasing the error rates for robustness. In [6], Heath and Love presented an adaptive antenna selection scheme to improve the BER performance of a MIMO system. In [7], Heath and Paulraj proposed a simple way for switching between spatial multiplexing (SM) and transmit diversity for a fixed rate system. They showed that the Demmel Condition number of the channel matrix is a sufficient metric to decide for the crossover point where spatial multiplexing starts outperforming the diversity. This metric essentially characterizes the suitability of the MIMO channel for SM. It was proven that the BER can be decreased significantly by switching between SM and diversity. However, this work was limited to fixed rate uncoded narrow-band MIMO systems.

3.2.2 Medium Access Control (MAC) Layer Based Polling Techniques for Maximizing the Throughput

When the aim is to maximize the throughput of a link, one approach to link adaptation is to do the adaptation at the MAC Layer based on observed PER statistics. The AutoRate Fallback (ARF) link adaptation protocol [8] and its variants [9] – [12] have been widely used in legacy WLANs for maximizing throughput. In the basic form of ARF, all the possible modes are sorted in terms of their rates and the radio automatically switches to a lower rate after two consecutive packet errors and switches to a higher rate after a number of successful packet transmissions (typical number is 10). Some modifications are made on this basic algorithm in order to make it more robust. For example, in [11], a number of neighbor modes (neighborhood is defined in terms of their rates) are probed in a window, and the link adaptation algorithm switches to the higher performance mode based on the PER statistics of the probed modes. These algorithms are very simple to implement, however they require probing of modes and hence they are inherently slow in converging to the optimum mode especially in highly dynamic channels.

3.2.3 Physical (PHY) Layer Metric Based Techniques for Maximizing the Throughput

In order to make the adaptation faster, PHY layer metrics have been considered. In [13], Muquet *et al.* simulated the WiMAX system with a specific channel model (pedestrian A) for various modes and determined the SNR thresholds for switching from one mode to another. However, SNR information alone is not sufficient for determining the performance of MIMO modes. In [14], the authors proposed an SNR table lookup method with an additional dimension which considers the determinant of the channel matrix for switching between the space time coding and spatial multiplexing modes. Similarly, a channel condition number assisted SNR thresholding method is used in [15] where the MAC Layer statistics are also employed to update the switching thresholds in response to the channel changes. A major problem with these types of look-up table methods is that they require extensive simulations in order to characterize the performance of the system. It is not practical to simulate all channel models for all possible modes and packet sizes. In addition, the radio performance in real life might be quite different from its simulated performance.

Given that it is neither practical nor accurate to tabulate the performance of the modes, it is desirable to mathematically model the performance of the system. However, there is no accurate closed form solution for PER in MIMO-OFDM systems for an arbitrary mode, packet size and channel realization. As a consequence, some recent works have focused on the accurate PER prediction problem [16] – [18]. For a linear receiver (i.e. MMSE), the post-processing SNR (PPSNR) can be leveraged to arrive at accurate estimates of the system's uncoded BER. However, determining the PER of a coded system from the uncoded-BER is not straightforward, and is an open area of research. In [16], 4 different link quality metrics

(MMIBM, MIESM, EESM, Raw-BER) are investigated and it is shown via simulations that they have a strong relationship to the PER in a convolutional coded MIMO-OFDM system. All four methods use PPSNRs of individual subcarriers and spatial streams, and combine these PPSNRs in different ways. The authors then obtain the mapping from their respective metrics to the PER using simulations and curve fitting techniques for a specific packet size. In [17], the uncoded-BERs of all subcarriers and spatial streams are averaged to calculate the overall uncoded-BER. This is very similar to the Raw-BER metric used in [16].

Relationship between uncoded-BER and coded-BER in convolutional coded systems is analyzed in [17], and it is shown via simulations that the logarithm of the coded BER is a linear function of the logarithm of the uncoded-BER. Parameters of this linear mapping are found via simulations and tabulated for different code rates. It was shown that this relationship is very accurate in terms of predicting the coded-BER. They also provide direct mapping from uncoded-BER to PER; however it is again for a specific packet size. A different approach for calculating the PER is proposed for 802.11n in [18] based on the assumption that number of bit errors per error event is approximately equal to the free distance of the code. It is still an open problem to find a better closed form mapping from either uncoded-BER or coded-BER to the PER for all packet sizes in a coded MIMO-OFDM system.

3.2.4 Limited Work on Maximizing the Energy Efficiency for SISO and MIMO

There are only few papers that address the energy-aware link adaptation problem for MIMO. None of the papers that are mentioned above considers the energy consumption of the system. They focus solely on maximizing the link throughput or decreasing the error rates. However, error rate prediction methods in these works are still valid for energy-aware link adaptation.

In their pioneering work [20], Cui and Goldsmith formulated the link adaptation problem for minimizing the energy required for transmitting certain amount of information. They modeled the energy consumption of the baseband and radio frequency (RF) circuits as well as the transmit energy which is consumed by the power amplifier (PA). A SISO, single carrier, narrowband, coded/uncoded M-QAM or M-frequency shift keying system operating in AWGN channel was assumed. It is shown that that MQAM is more energy efficient than MFSK at short ranges and the performance difference is more pronounced with coding. Interestingly, for short ranges, uncoded MFSK outperforms coded MFSK in terms of energy efficiency. The results of this paper might not be applicable to MIMO and more realistic channel scenarios. However, it is an important work in terms of its energy consumption models.

The authors extended the energy-aware link adaptation problem to MIMO systems in [19]. However, the underlying system was still a single carrier, narrowband system with one or two antennas at either the transmitter or receiver. Contrary to the traditional belief that MIMO systems are more energy-efficient than SISO, it was shown that in short range-fixed rate applications a SISO system can beat both 2x1 and 2x2 Alamouti based MIMO systems as far as the energy efficiency is concerned. This result is due to the fact that circuit energy dominates at short distances. They don't provide any practical link adaptation algorithm for real systems and all analysis and simulations were done for the average performance in a Rayleigh channel.

Bougard *et al.* in [21] proposed link adaptation for WLANs to minimize the energy consumption of the broadband link. The energy consumption model focused on the power amplifier and ignored other sources of energy draw associated with the baseband digital and radio frequency (RF) components. The optimization problem was formulated based on an idealistic channel capacity expression. First the PPSNRs were calculated for all spatial streams and subcarriers. Then, the AWGN capacity of the system was calculated. The assumption was that the PER is equal to 1 if the capacity of the channel is less than a threshold, and assumed to be 0 otherwise. These thresholds were determined via simulations.

In [22], Kim and Daneshrad formulated the energy-efficient link adaptation problem as a convex optimization problem. In this work, the BER expressions are derived for Maximum Likelihood (ML) and Zero Forcing (ZF) detectors assuming an independent frequency selective channel. Both STBC and SM modes were included in the search space. Additionally, bandwidth and transmit power are included as optimization parameters. The BER expressions are derived for the average performance in fading channels. However, it is desirable to respond to the Rayleigh fading channel fast enough so that the link chooses the most suitable mode for a given realization of the channel. In [22], trends and guidelines are provided for short and long range applications. These trends give very useful insights for a first order guideline. However, the underlying system is uncoded and MAC is not considered.

3.3 Energy-Aware Fast Link Adaptation

In the previous section, we presented the existing work on the topic of link adaptation, both for maximizing throughput and maximizing energy efficiency. In this section, we present an energy-aware link adaptation protocol which picks the optimum transmission mode for maximizing the energy efficiency of the link given the instantaneous channel conditions and QoS constraints. We first describe the system model, and then provide an error rate prediction method for a commonly used transmission method. A complete energy consumption model of a wireless link is provided based on state-of-the-art implementations of individual blocks both at the transmitter and the receiver side. Finally, we present a fast energy-aware link adaptation protocol.

3.3.1 System Description

We consider a generic MIMO-OFDM system (Fig. 3.1) where the transmitter is equipped with N_T antennas, and the receiver uses N_R antennas. For each subcarrier, the $N_R \times 1$ received signal can be expressed as

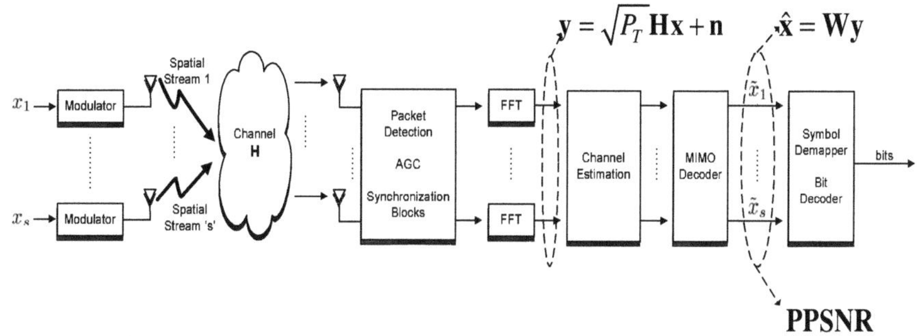

Fig. 3.1 A generic MIMO-OFDM system

$$\mathbf{y} = \sqrt{P_T}\,\mathbf{Hx} + \mathbf{n} \tag{3.1}$$

where $\mathbf{x}$ is the $N_T \times 1$ transmitted signal vector, $\mathbf{H}$ is the $N_R \times N_T$ channel matrix, and $\mathbf{n}$ is the $N_R \times 1$ additive Gaussian noise vector with zero mean and covariance matrix $N_0\,\mathbf{I}$. Time and subcarrier indices are omitted for simplicity. P_T is the transmit power per antenna. Energy of the transmitted signal vector is normalized to unity, i.e. $E\left[\mathbf{xx}^*\right] = \mathbf{I}$.

The receiver can estimate the transmitted signal vector $\mathbf{x}$ by applying Minimum-Mean-Square-Error (MMSE) filter to the received signal vector $\mathbf{y}$, $\hat{\mathbf{x}} = \mathbf{Wy}$. Using the orthogonality principle, $E\left[(\mathbf{Wy} - \mathbf{x})\mathbf{y}^*\right] = \mathbf{0}$, The MMSE solution can be derived as

$$\mathbf{W} = \frac{1}{\sqrt{P_T}}\left(\mathbf{H}^*\mathbf{H} + \frac{N_0}{P_T}\mathbf{I}\right)^{-1}\mathbf{H}^* \tag{3.2}$$

At the output of the MMSE decoder, the post-processing SNR (PPSNR) ζ_k for the kth spatial stream for each subcarrier is calculated as

$$\zeta_k = \frac{P_T\left|(\mathbf{WH})_{k,k}\right|^2}{P_T\sum_{l=1,l\neq k}^{N_T}\left|(\mathbf{WH})_{k,l}\right|^2 + N_0\left(\mathbf{WW}^*\right)_{k,k}} \tag{3.3}$$

where the $(\ldots)_{k,l}$ denotes the $(k,l)^{th}$ entry of the matrix.

The PPSNR simply indicates the ratio of the signal power to the residual interference + noise power after MMSE decoder. This residual noise + interference signal is well approximated as Gaussian [17] for MMSE decoders; therefore the uncoded-BER can be calculated using PPSNR values as it will be explained in next section.

3.3.2 Error Rate Prediction for BICM-MIMO-OFDM Systems

In the latest high-throughput wireless communication standards, such as IEEE802.11a/b/g/n (WLANs) or IEEE802.16 (WMANs), the dominant transmission technique is OFDM with bit interleaved coded modulation (BICM). In BICM-OFDM systems, information bits are encoded (convolutionally), punctured, interleaved and modulated using OFDM modulation. The main advantage of OFDM over single-carrier schemes is its ability to cope with channel dispersions induced by multipath propagations in the channel. Channel equalization for OFDM can be very efficiently performed by simply inserting sufficient guard intervals between OFDM symbols without costly equalization filters.

As it is mentioned earlier, estimation of the instantaneous PER of the link is the most critical task in the link adaptation process, since the optimum mode is chosen solely based on the estimated PER. However, there is no exact closed form PER expression for BICM-MIMO-OFDM systems. In this section, we present PPSNR based PER estimator, which can predict the PER quite accurately.

It was shown that the output of MMSE decoder can be approximated as Gaussian [17], therefore the uncoded-BER for a spatial stream at a subcarrier can be calculated using the AWGN channel BER expressions as follows.

The uncoded-BER of the kth spatial stream at the nth subcarrier is [23]

$$P_{b-uncoded}^{k,n} \approx \alpha Q\left(\sqrt{\beta \zeta_{k,n}}\right) \tag{3.4}$$

where $\zeta_{k,n}$ is the post-processing SNR for the kth spatial stream and the nth subcarrier. α and β are the modulation dependent parameters and can be found in Table 3.1.

Table 3.1 Parameters for calculating uncoded-BER

Modulation	BPSK	QPSK	16QAM	64QAM
α	1	1	3/4	7/12
β	2	1	1/5	1/27

To compute the overall uncoded-BER, we can directly average the uncoded-BERs of individual subcarriers and spatial streams. The overall uncoded-BER is expressed as

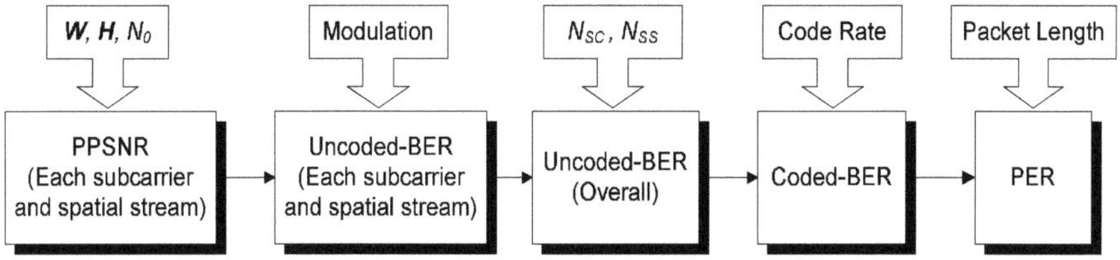

Fig. 3.2 Packet Error Rate prediction process

$$P_{b-uncoded} = \frac{1}{N_{SS} N_{SC}} \sum_{k=1}^{N_{SS}} \sum_{n=1}^{N_{SC}} P_{b-uncoded}^{k,n} \tag{3.5}$$

where N_{SS} is number of spatial streams, and N_{SC} is the number of data subcarriers.

Coded-BER is defined as the BER at the output of Viterbi decoder, and is a linear function of the uncoded-BER in the log domain [17],

$$\ln\left(P_{b-coded}\right) = u \ln\left(P_{b-uncoded}\right) + v \tag{3.6}$$

where u and v are listed in Table 3.2 as a function of the code rate.

Table 3.2 Coefficients of log-linear approximation for calculating coded-BER

Code Rate	1/2	2/3	3/4	5/6
u	9	6.4	5	4
v	15	12	10	8.5

Finally, the relationship $PER = 1-\left(1-P_{b\ coded}\right)^L$ is used to calculate the PER for a given mode, where L is the packet length in number of bits. Figure 3.2 shows all the steps involved in the PER estimation process.

3.3.3 Energy Consumption Model

Accurate modeling of the energy consumption of the radio has a significant impact on the energy savings achieved by the energy-aware link adaptation. In [24] – [26], the authors focused on minimizing the transmission energy without taking other sources of RF and baseband energy consumption into account. Minimizing only the transmission energy (power amplifier energy consumption) might be reasonable in long range applications, since it is the dominant source of energy consumption for long transmission ranges. However, in many indoor and short-range scenarios, the RF circuit and baseband energy consumption can become comparable, or even dominate the transmission energy. In this section, we present more inclusive and accurate energy consumption models for RF and baseband portions of a candidate MIMO-OFDM radio.

The total energy consumption of a link includes both the transmitter and receiver energy consumptions which consist of three major components (see Fig. 3.3):

- RF energy consumption.
- MIMO decoder energy consumption.
- Baseband energy consumption.

For the RF energy consumption modeling, a direct conversion RF transceiver is assumed whose power consumption is a function of P_T, N_T, N_R and bandwidth (BW) according to the model presented in [20], [22].

$$E_{RF} = \begin{pmatrix} p_{RF} \cdot N_T \cdot P_T + \left(b_{RF\_T} \cdot BW + c_{RF\_T}\right) \cdot N_T \\ +\left(b_{RF\_R} \cdot BW + c_{RF\_R}\right) \cdot N_R + c_{RF} \end{pmatrix} \cdot T_{oper} \tag{3.7}$$

E_{RF} in above equation is the total transceiver RF energy consumption where $p_{RF} \cdot N_T \cdot P_T$ represents the total power consumption of the power amplifier (PA), $\left(b_{RF\_T} \cdot BW + c_{RF\_T}\right) \cdot N_T + \left(b_{RF\_R} \cdot BW + c_{RF\_R}\right) \cdot N_R + c_{RF}$ represents the total power consumption of the other RF components, and is proportional to N_T, N_R and BW. The coefficient values for the RF power consumption model are summarized in Table 3.3. The coefficients in Table 3.3 were obtained from already published link adaptation works [20], [22] or based on the actual implementation data [27], [28]. T_{oper} denotes the total time needed to transmit L information bits, i.e. the packet duration, which includes both the data and the preamble portions of the packet.

The MIMO detector energy consumption, E_{MIMO}, is assumed to be proportional to the number of arithmetic operations, O_{MIMO}, involved in detecting L bits (Eq. 3.8). Table 3.3 provides O_{MIMO} for the MMSE algorithm. The coefficient e_{MIMO} is obtained from the results of an ASIC [27].

$$E_{MIMO} = e_{MIMO} \cdot O_{MIMO} \tag{3.8}$$

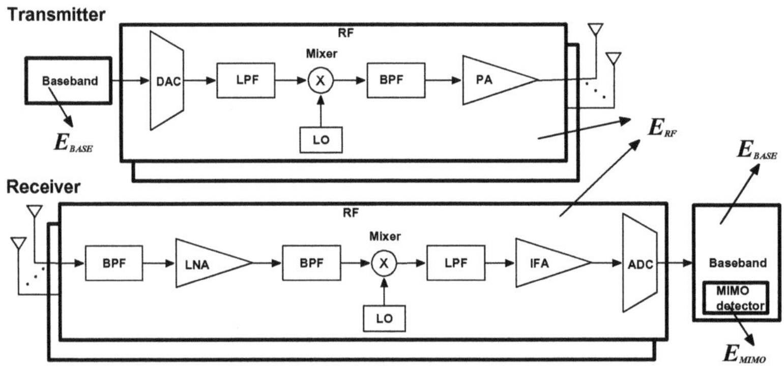

Fig. 3.3 Transceiver block diagram for energy consumption modeling

Table 3.3 Coefficients for energy consumption model

	Value	Description
p_{RF}	$\eta^{-1}\cdot$PAPR	Coefficient related to power amplifier (PA) power consumption. η is the PA efficiency. PAPR is the peak-to-average power ratio. η =0.35, PAPR = 9.51, 9.76, 9.8, 9.83dB for BPSK, QPSK, 16QAM, 64QAM respectively.
b_{RF_T}	1.3×10^{-8}	Coefficient related to transmitter RF components whose power consumption is proportional to BW$\times N_T$. (for example, DAC)
b_{RF_R}	1.8×10^{-8}	Coefficient related to receiver RF components whose power consumption is proportional to BW$\times N_R$. (for example, ADC)
c_{RF}	0.100	Coefficient related to RF components, e.g. low pass filter, whose power consumption is constant
c_{RF_T}	0.042	Coefficient related to transmitter RF components whose power consumption is proportional to N_T, e.g. mixer at the transmitter
c_{RF_R}	0.061	Coefficient related to transmitter RF components whose power consumption is proportional to N_R, e.g. mixer at receiver
b_{BASE_T}	4.09×10^{-9}	Coefficient related to transmitter baseband signal processing power consumption
b_{BASE_R}	1.62×10^{-9}	Coefficient related to receiver baseband signal processing power consumption excluding MIMO detector power.
e_{MIMO}	8.1 μJ / million instructions	Coefficient related to MIMO detection power consumption.
O_{MIMO}	$5L\cdot N_R\cdot q^{-1}+2N_{SC}\cdot N_R\cdot N_T$ $+4N_{SC}\cdot N_T^2+9N_{SC}\cdot N_R\cdot N_T^2$ $+2N_{SC}\cdot N_T^3$	Number of arithmetic operations for MIMO detector.

The baseband signal processing energy, E_{BASE}, at the transmitter and receiver *excluding* the MIMO detector energy consumption is assumed to be linearly proportional to the bandwidth (Eq. 3.9) [22] since the operating clock frequency scales linearly with bandwidth. Finally, the total energy consumption, E_{TOTAL}, is obtained by summing the energy consumptions of the RF, baseband and the MIMO detector.

$$E_{BASE}=\left(b_{BASE\_T}\cdot N_T+b_{BASE\_R}\cdot N_R\right)\cdot BW\cdot T_{oper} \tag{3.9}$$

$$E_{TOTAL}=E_{RF}+E_{BASE}+E_{MIMO} \tag{3.10}$$

The energy consumption model we presented here is of a candidate radio; however the link adaptation protocol that we develop in the next section is capable of working with any valid energy consumption model.

3.3.4 PPSNR Based Fast Link Adaptation Protocol

Past work dealing with the link adaptation problem has mostly focused on maximizing the link throughput subject to QoS constraints. This throughput maximization problem can be formally written as

$$\begin{array}{ll} \text{maximize} & (1-PER)L/T_{oper} \\ \text{subject to} & PER \le PER_{target} \end{array} \tag{3.11}$$

where PER_{target} is the target PER which is usually determined by the application.

When our objective is to communicate L bits across a channel, then maximization of throughput will reduce the total transmission time, but it won't necessarily maximize the energy efficiency for the transmitted bits. The objective of energy-aware link adaptation on the other hand, is to maximize the number of successfully received bits normalized by the total energy consumption of the link, which is equivalent to minimizing the total energy consumed in the link per successfully received bit. Energy-efficiency maximization is formulated as

$$\begin{array}{ll} \text{maximize} & (1-PER)L/E_{TOTAL} \text{ (bit/J)} \\ \text{subject to} & PER \le PER_{target} \\ & TH \ge TH_{target} \end{array} \tag{3.12}$$

where TH indicates the instantaneous throughput and TH_{target} is the target throughput. Energy Efficiency is defined as the number of successfully transmitted bits per unit energy consumption:

$$Energy \cdot Efficiency = (1-PER)L/E_{TOTAL} \tag{3.13}$$

The wireless channel changes rapidly due to the mobility of both the transmitter/receiver and the surrounding objects. As a result, the channel observes small-scale fading (fast-fading) in addition to the path loss and shadowing effects. Some link adaptation work [22] in the literature aims at finding the optimal mode based on the average error rate performance discarding the fast variation in the channel. However, significant performance improvements can be achieved if the link adaptation algorithms are designed to take into account the instantaneous (or short term) channel conditions rather than making decision based on the average behavior of the channel in the long term. We call the latter approach *fast link adaptation*.

The PPSNR based fast link adaptation protocol for MIMO-OFDM systems is summarized as follows (See Fig. 3.4):

- Send a channel sounding packet to estimate the N_R x N_T channel matrix, **H**, for each subcarrier. The sounding packet is sent using all transmit antennas with the highest possible transmit power, and all the antennas are enabled for reception at the receiver side. This enables us to estimate all possible channels between each transmitter/receiver antenna pairs. It should be noted that we need only the preamble portion of this packet to get the channel estimates. The data portion of the packet can either be empty or BPSK modulated with lowest code rate to ensure reception.
- Based on the estimated channel matrices, the PER's and the resulting throughputs for all available modes are calculated using the method presented in the previous section.
- The modes that do not satisfy the target throughput and/or target PER QoS constraints are removed from the search space.
- Energy efficiencies for the remaining handful of modes are calculated based on the predicted PER values using Eq. 3.13. Among the modes that satisfy the QoS constrains, we choose the one that has the maximum predicted energy efficiency number.
- If none of the available modes satisfy the QoS constraints, the algorithm chooses the mode that has the highest predicted throughput value. If all the predicted PER's are equal to 1, which means that the channel capacity is below the application's requirement, the algorithm forces both the transmitter and the receiver to enter to sleep mode to save energy. A second implementation choice in such a case could be letting the application to decide on whether relaxing the QoS constraints or stopping the transmission.
- Transmitter uses the chosen mode by the link adaptation algorithm during the sounding period. The sounding period is defined as the duration between two sounding packets.
- Whole process is repeated when the next sounding packet is transmitted at the end of a sounding period.

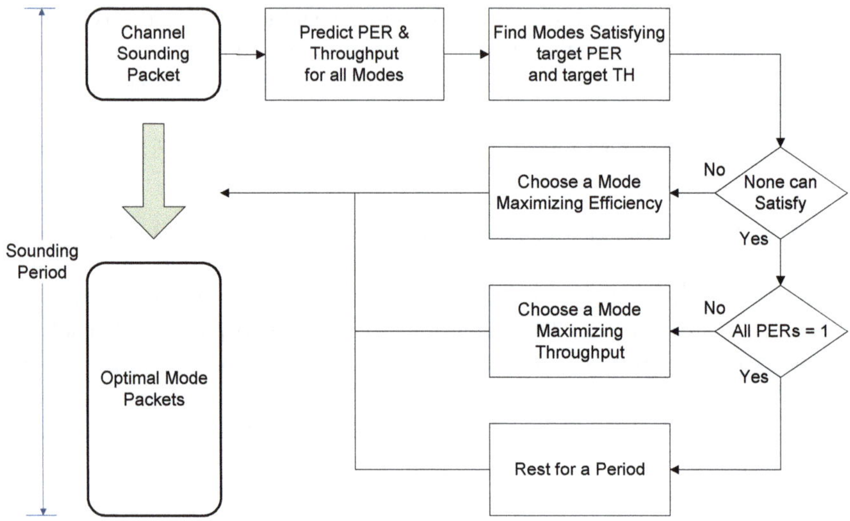

Fig. 3.4 Energy-Aware Fast Link Adaptation Protocol

It is important to note that due to ambient background noise and transceiver non-idealities, the calculated PPSNR values might be noisy. As such, the PER prediction will inherently be off by some random amount. Due to the rapid fall in the BER curve with increasing SNR, it is thus important to stay on the conservative side by simply introducing a few dBs of negative bias to the calculated PPSNR values. In addition, the channel sounding period should be adjusted according to the mobility in the channel (coherence time of the channel) so that the channel does not change significantly between two sounding packets.

3.4 Performance Evaluation

3.4.1 Simulation Setup

Channel Model: The transmitted signal is affected mainly by path loss, shadowing and fast fading in a wireless channel (See Fig. 3.5). Path loss and shadowing effects are known as large-scale effects and they change slowly over time. However, fast fading is due to the mobility in the environment and changes dramatically in a very short time.

We considered the simplified path loss model in [23], and log-normal shadowing in our simulations. For fast fading, we used Jakes' Doppler model in [29]. Each channel between tx-rx antenna pair assumed to have independent fast fading.

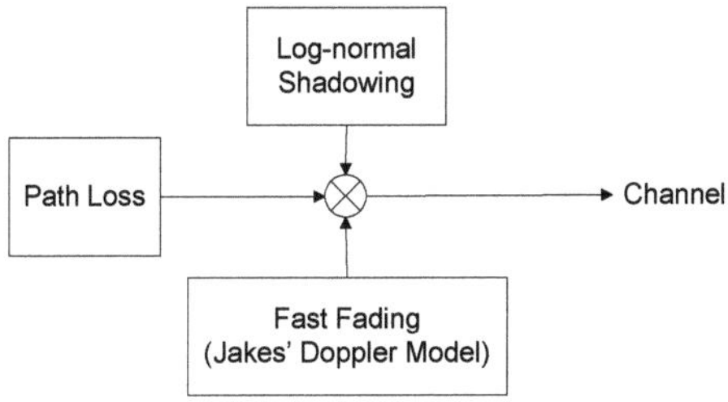

Fig. 3.5 Channel Model used in simulations

Simulation Parameters: The system parameters that are used in the simulation are listed in Table 3.4. Most of the modulation related values are similar to those used in the 802.11n standard.

Table 3.4 Simulation Parameters

Parameter	Value	Description
N_T	1,2,3,4	Number of transmitter antennas
N_R	1,2,3,4	Number of receiver antennas
N_{SS}	1,2,3,4	Number of spatial streams
q	1,2,4,6	Constellation size q=1 : BPSK, q=2 : QPSK, q=4 : 16QAM, q=6 : 64QAM
N_{SC}	52	Number of data subcarriers
BW	20 MHz	Bandwidth
P_T	5 – 20 dBm	Transmit power (in steps of 3 dB)
L	1 KByte	Packet size (information bits)
f_c	2.4 GHz	Carrier frequency
f_D	6 Hz	Doppler frequency (indoor mobility)
σ_ψ^2	4 dB	Log-normal shadowing variance
T_{sound}	1/24 sec	Sounding period
$T_{silence}$	200 μsec	Silence period between the packets (due to MAC Layer Tasks)

We simulated a File Transfer application, which required all the lost packets to be retransmitted until successful reception. We define a target time for transmission of the whole file. The link adaptation algorithm calculates the target throughput based on the remaining file size and the remaining time to transmit. We assumed that the file size is 8Mbytes, and the total target time to finish the transmission is set to 11 seconds.

3.4.2 Results

Here we present simulation results of a realistic link scenario. We assumed that the receiver node goes away from the transmitter for the first 3 seconds, and then it approaches to the transmitter.

Four different protocols were simulated as shown in Table 3.5, and the energy efficiency is plotted over time in Fig. 3.6.

Table 3.5 Simulated Protocols and Results

Protocol Name	Description	Total energy consumed to transmit 8MB	Average throughput	QoS Outage
MAC Based LA – max TH	Uses PER statistics. Tries to maximize the throughput.	32.48 Joules	13.6 Mbps	58.3%
PHY Based LA – max TH	PPSNR based fast link adaptation with the objective of maximizing throughput.	19.13 Joules	20.6 Mbps	2.70%
PHY Based LA – max Efficiency	PPSNR based fast link adaptation with the objective of maximizing energy efficiency.	11.54 Joules	13.9 Mbps	5.53%
4x4 QPSK, Rate ½ - Fixed mode	A fixed mode. No link adaptation. Uses 4x4 QPSK, Rate ½ code, highest possible P_T (20dBm)	43.02 Joules	13.8 Mbps	0%

The MAC based protocol that we simulated is very similar to the one described in [11]. We define 6 neighbor modes, 3 have lower rate than the current mode and 3 have higher rate, and all neighbor modes are probed along with the currently used mode. In each cycle, 60 packets are sent with the neighbor modes (10 for each mode) and 240 packets are sent with the currently used mode. For the next cycle, the MAC based link adaptation algorithm switches to the higher performance mode based on the PER statistics of the probed modes in the current cycle.

The second protocol is the PPSNR based fast link adaptation protocol that we described in the previous section. Its objective is to maximize the throughput. First thing to note here is, if we compare these two protocols PPSNR based protocol achieves 51% throughput improvement over the MAC based protocol. This is because the PHY based protocol responds very quickly to changes in channel, since it sounds the channel and predict the performance beforehand. On the other hand, a MAC based protocol is inherently slow since it needs to build PER statistics. The MAC based protocol suffers from packet losses due to rapid changes in the channel, and thus it does not satisfy the target PER constraint 58% of the time.

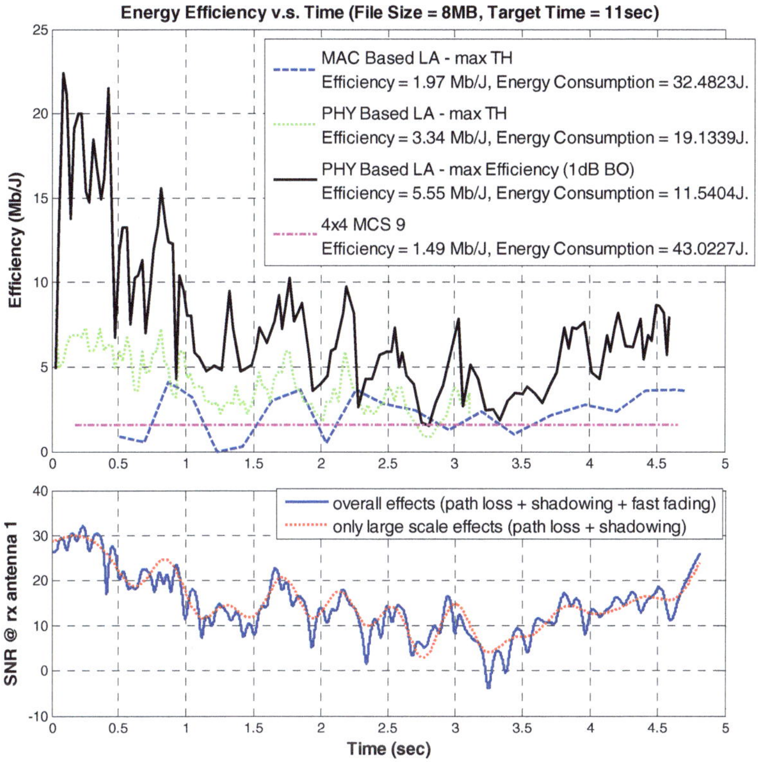

Fig. 3.6 Energy Efficiency vs Time (upper plot), and received SNR for antenna 1 vs Time (lower plot). Lower plot shows the received SNR @ antenna 1 when all 4 transmit antennas are used with maximum transmit power.

Third protocol is the same PPSNR based protocol, but with the objective of maximizing the energy efficiency. Energy efficiency (Mb/Joule) is plotted over time in Fig. 3.6. As can be seen, the PHY based energy-aware link adaptation algorithm consumes 11.5 Joules to complete the transmission of an 8 MB file. It is 4x more energy efficient than the fixed 4x4 QPSK Rate ½ mode. Here we show the performance of a reasonably well performing fixed mode for representation simplicity, however we observe that the PPSNR based energy-aware protocol achieves orders of magnitude improvement in energy efficiency compared to poorly chosen fixed modes while satisfying the QoS constraints. It satisfies the QoS constrains (target PER < 5%) 94% of the time.

It is also important to note here that the PHY based protocol with the objective of maximizing the energy efficiency is 66% more efficient than the one with the objective of maximizing the throughput. This is due to the fact that the latter protocol does not consider energy consumption of the link, and hence chooses energy-inefficient modes while focusing on maximizing the throughput.

3.5 Conclusion

In this chapter, we first presented a comprehensive overview of the existing link adaptation works. Then, a detailed energy consumption model of a wireless link is provided based on state-of-the-art implementations of functional transceiver blocks. We analyzed the error rate prediction of MIMO-OFDM systems, and proposed a link adaptation protocol to maximize the energy efficiency of future MIMO-OFDM radios.

References

[1] Prabhu, R.: Performance Analysis of Energy-Efficient Adaptive Modulation. PhD Dissertation, University of California, Los Angeles (2010)

[2] Qiao, D., Choi, S., Shin, K.G.: Goodput Analysis and Link Adaptation for IEEE 802.11a Wireless LANs. IEEE Transactions on Mobile Computing 1(4) (October-December 2002)

[3] Qiu, X., Chuang, J.: Link Adaptation in Wireless Data Networks for Throughput Maximization under Retransmissions. In: IEEE International Conference on Communications (ICC), vol. 2, pp. 1272–1277 (June1999)

[4] Chevillat, P., Jelitto, J., Barreto, A.N., Truong, H.L.: A Dynamic Link Adaptation Algorithm for IEEE 802.11a Wireless LANs. In: IEEE ICC 2003, vol. 2, pp. 1141–1145 (May 2003)

[5] Pavon, J.P., Choi, S.: Link Adaptation Strategy for IEEE 802.11 WLAN via Received Signal Strength Measurement. In: IEEE ICC 2003, vol. 2, pp. 1108–1113 (May 2003)

[6] Heath, R., Love, D.: Multimode Antenna Selection for Spatial Multiplexing Systems with Linear Receivers. IEEE Transactions on Signal Processing 53(8) (May 2005)

[7] Heath, R., Paulraj, A.: Switching between diversity and multiplexing in MIMO systems. IEEE Transactions on Communications 53(6), 962–968 (June 2005)

[8] Kamerman, A., Monteban, L.: WaveLAN-II: a high-performance wireless LAN for the unlicensed band. Bell Labs Technical Journal 2(3), 118–133 (1997)

[9] Lacage, M., Manshaei, M.H., Turletti, T.: IEEE 802.11 rate adaptation: a practical approach. In: Proc. ACM MSWIN 2004, pp. 126–134 (2004)

[10] Qiao, D., Choi, S.: Fast-responsive link adaptation for IEEE 802.11 WLANs. In: Proc. IEEE International Conference on Communications, vol. 5, pp. 3583–3588 (2005)

[11] Wong, S.H.Y., Yang, H., Lu, S., Bharghavan, V.: Robust rate adaptation for 802.11 wireless networks. In: Proc. MobiCom, pp. 146–157 (2006)

[12] Kim, J., Kim, S., Choi, S., Qiao, D.: CARA: Collision-aware rate adaptation for IEEE 802.11 WLANs. In: Proc. IEEE INFOCOM, pp. 1–11 (2006)

[13] Muquet, B., Biglieri, E., Sari, H.: MIMO Link Adaptation in Mobile WiMAX Systems. In: Proc. IEEE Wireless Communications and Networking Conference (2007)

[14] Han, C., et al.: Link Adaptation Performance Evaluation for a MIMO-OFDM Physical Layer in a Realistic Outdoor Environment. In: IEEE 64th Vehicular Technology Conf. (2006)

[15] Chan, T., Hamdi, M., Cheung, C.Y., Ma, M.: A Link Adaptation Algorithm in MIMO-based WiMAX systems. Journal of Communications 2(5) (August 2007)

[16] Jensen, T., Kant, S., Wehinger, J., Fleur, B.: Fast Link Adaptation for MIMO-OFDM. IEEE Transactions on Vehicular Technology 59(8), 3766–3778 (2010)
[17] Peng, F., Zhang, J., Ryan, W.: Adaptive Modulation and Coding for IEEE 802.11n. In: Proc. IEEE Wireless Communications and Networking Conference, pp. 656–661 (March 2007)
[18] Martorell, G., Palou, F.R., Femenias, G.: Cross-Layer Link Adaptation for IEEE 802.11n. In: Proc. 2nd International Workshop on Cross Layer Design, pp. 1–5 (June 2009)
[19] Cui, S., Goldsmith, A.J., Bahai, A.: Energy-Efficiency of MIMO and Cooperative MIMO Techniques in Sensor Networks. IEEE Journal on Selected Areas in Communications 22(6), 1089–1098 (2004)
[20] Cui, S., Goldsmith, A.J., Bahai, A.: Energy-Constrained Modulation Optimization. IEEE Transactions of Wireless Communication 4(5), 2349–2360 (2005)
[21] Bougard, B., et al.: SmartMIMO: An Energy-Aware Adaptive MIMO-OFDM Radio Link Control for Next-Generation Wireless Local Area Networks. EURASIP Journal on Wireless Communications and Networking, (article ID.98186) (2007)
[22] Kim, H.S., Daneshrad, B.: Energy-Constrained Link Adaptation for MIMO OFDM Wireless Communication Systems. IEEE Transactions on Wireless Communications 9(9), 2820–2832 (2010)
[23] Goldsmith, A.: Wireless Communications. Cambridge U. Press, Cambridge (2005)
[24] Verdu, S.: Spectral efficiency in the wideband regime. IEEE Transactions on Information Theory 48(6), 1319–1343 (2002)
[25] Gamal, A.E., et al.: Energy-efficient scheduling of packet transmissions over wireless networks. In: Proc. IEEE Information Communications (INFOCOM), NY, pp. 1773–1782 (2002)
[26] Schurgers, C., Srivastava, M.B.: Energy efficient wireless scheduling: Adaptive loading intime. In: Wireless Communications and Networking Conf., pp. 706–711 (March 2002)
[27] Garrett, D., Davis, L., ten Brink, S., Hochwald, B., Knagge, G.: Silicon Complexity for Maximum Likelihood MIMO Detection Using Spherical Decoding. IEEE Journal of Solid-State Circuits 39(9), 1544–1552 (2004)
[28] Thomson, J., Baas, B., Cooper, E.M., et al.: An Integrated 802.11a Baseband and MAC Processor. In: IEEE International Solid-State Circuits Conference (2002)
[29] Zheng, Y.R., Xiao, C.: Improved models for the generation of multiple uncorrelated Rayleigh fading waveforms. IEEE Communications Letters 6, 256–258 (2002)

Chapter 4
Energy Efficient MAC

Tae Rim Park and Myung J. Lee

Abstract. In short range and low-rate wireless networks, energy saving has been one of the hottest issues. Compared to high-rate networks designed for multimedia data streaming, the low-rate networks mainly focus on monitoring and control applications. In most of the applications, nodes are expected to operate on battery. For saving energy in those types of networks, medium access control (MAC) protocols have been considered as one of the most essential and actively researched areas. In this chapter, we investigate the energy saving MAC issues and protocols of short range and low rate wireless networks.

4.1 Energy Saving in Wireless Networks

Prospecting the future might be very difficult challenge for almost everyone. However, it is quite persuasive that wireless devices will be spread and change our ambient life style continually. We can easily see similar anticipations from various media around us. For example, Wireless World Research Forum consisting of more than 100 engineers representing industry and research community forecasted that 7 trillion wireless devices would be serving 7 billion people by the year 2020 [1]. The fulfillment of the vision is expected to come from fast market growth of short range and low rate wireless networks usually called sensor networks.

However, it would be impossible to be realized without technological advancement. Especially, energy saving has been considered one of the prominent hurdle in wireless communication and networking community, today. Almost all wireless devices changing our life style such as small environmental and healthcare sensors, smart phones and even tablet PCs are operated on battery, and spend the energy to communicate with the world. It is natural that users of the devices want to use their devices as long as possible without charging or changing the battery. The trouble comes mainly from the slow advancement of battery technology. While most technologies for the devices are evolving very rapidly, the energy density of batteries crawls merely a factor of 3 over the past 15 years [2]. In some

Tae Rim Park · Myung J. Lee
Samsung Adv. Inst of Tech.
e-mail: taerim.park@samsung.com, lee@ccny.cuny.edu

small devices such as environmental sensors, replacing or recharging battery is extremely difficult. In addition, the cost and form factor of small devices make it difficult to adopt a large capacity battery. This might be a major factor of confining lifetime of an application. Moreover, if the batteries are replaced, the waste brings about environmental pollution. According to US Environmental Protection Agency, people in US purchase nearly 3 billion dry-cell batteries every year to power radios, toys, cellular phones, watches, laptop computers, and portable power tools [3]. This is the reason why green technologies are required for low rate and short range wireless communications.

A good approach to showing the energy consumption of the low rate wireless communication is to refer data of IEEE 802.15.4 devices. IEEE 802.15.4, the first international standard designed for low-power and low-rate wireless communication has been widely used not only for researches but for various real applications [4][5]. For example, a transceiver cc2420, currently the market leader supporting the standard drains 17.4 mA when it is in the transmission mode, but drains more in the receive mode [6]. Even when it is not actively receiving a frame, it consumes 19.7 mA only by sampling wireless channel for a possible reception. If a device is operated on two AA batteries of 1600 mAh, the life time of the device might be only 3.4 days even without considering energy consumption of other modules such as a micro controller and sensors [7]. If the devices are deployed with this condition in a large structure or a hostile territory, the life span of the sensor network may be a key failing factor of the project. Recently developed Bluetooth low energy (LE) has the similar order of power consumption. Including the Bluetooth LE transceiver, detailed data for power consumption of competing technologies are summarized in [8].

For the life time of nodes, energy harvesting or scavenging from ambient environment is a proactive way of solving the problem. Recently, researches to improve the efficiency of energy harvesting have been actively conducted in the fields of thermal, motion, vibration and electromagnetic radiation. Energy sources and their corresponding energy densities of up-to-date technologies are introduced in [2]. In most cases, however, these harvesting technologies can only be used in limited situations because of the time and location constraints the number of usable sources. Moreover, not only the efficiency of the harvesting modules but energy source itself except outdoor ambient light[1] may not be enough for today's wireless devices. Although harvested energy can be used for data processing and communication after accumulated for an extended time period, this also curtails application space.

Another approach to extending device life time may lean on ultra low power circuit technologies to reduce energy consumption in the devices [9]. The first step for this is to understand energy expenditures in components of communication technology. Most of all, the wireless signal requires certain energy to be transmitted over the transmission medium. The signal energy is composed of two

[1] Source energy density of outdoor ambient light is roughly 100 mW/cm^2. When the efficiency is 10%, 10mW/cm^2 is harvested form ambient light. Power of indoor light is three orders of magnitude smaller than that of outdoor light [2].

parameters: transmission power and duration. The transmission power is determined by the selected value for a power amplifier of the transmitter. The other parameter, transmission duration, is controlled by defined modulation and coding. The selected modulation and coding determines a nominal bit rate R (bit/s). Thus, the energy to transmit a frame which is L bits of data is a function of three parameters: data rate R, power P_{amp} of a signal amplifier, and frame length L. In Fig 4.1, the energy of a frame is presented where P_0 is the attenuated signal power at a given distance from a transmitter and t_0 is the transmission time determined by R and L.

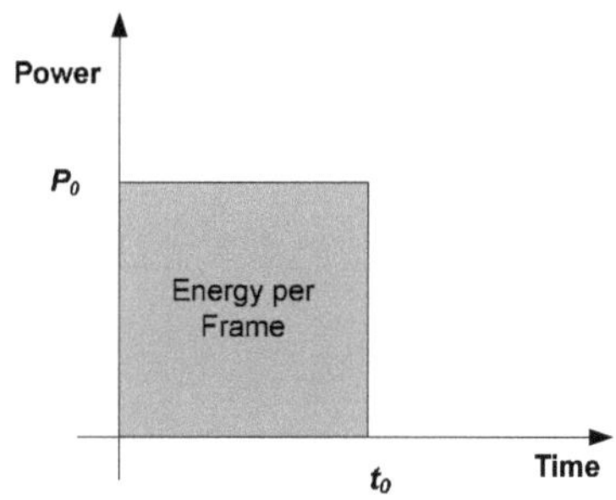

Fig. 4.1 Energy per frame

In order to successfully detect a frame at the receiver, the energy of each bit composing the frame should be higher than the required level at the receiver. With the attenuated signal power P_0, the energy per bit in the received signal is defined

$$E_b = \frac{P_0}{R}$$

The required power to attain a desired bit error rate is derived as a function of E_b and N_0, where N_0 is the noise power spectral density in W/Hz. In order to control the required energy, a transmitter has three options: modulation, coding and transmission power control. Although the consumed energies (areas in Fig.4.1) are equal, controlling modulation and coding requires careful consideration of energy expenditures resulting from selected algorithms. In Fig 4.2, we compare BERs of 4 modulation algorithms.

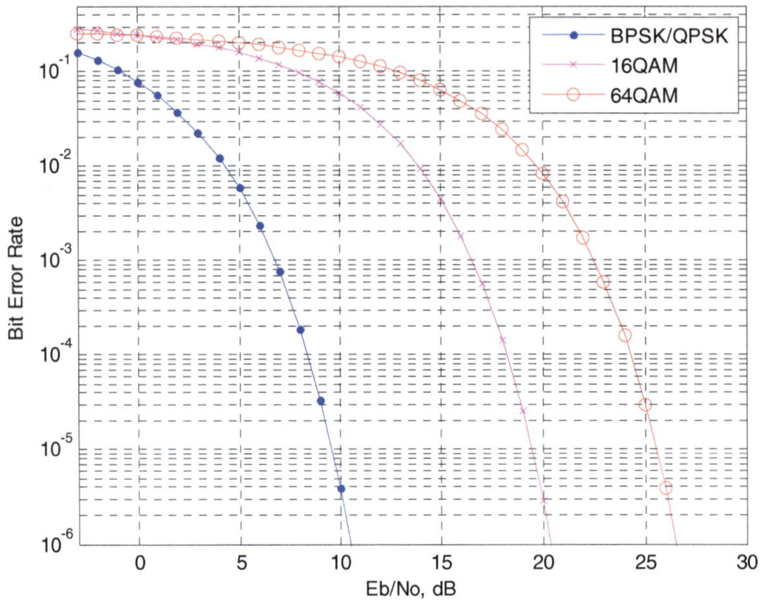

Fig. 4.2 Bit error rate versus Eb/N0

If only modulation and coding are controllable while transmission power is fixed, the best option is to use the highest modulation and coding algorithms satisfying given BER requirement. For example, assume that the required BER is $10^{-5,}$ and if E_b/N_0 of a received signal using QPSK is 30 dB then, the expected BER is much smaller than 10^{-5}. Assume also adaptive modulation is possible, and the transmission power is fixed. If the transmitter changes the modulation to 16QAM, then E_b/N_0 would be reduced to 27dB, still meeting the BER requirement (refer to Fig.4 2)[2]. Adopting 64 QAM, E_b/N_0 becomes 25.3 dB, so it cannot fulfill the BER requirement. Thus, under the assumed fixed power budget, 16QAM is the most energy efficient option in this example.

However, if the transmission power can be controlled, the situation permits different solutions. In this case, using BPSK is the most efficient method in view of only signal energy. This result may be different when energy consumption of other components in communication devices is counted. Also, relatively low channel utilization of wireless channel is another issue.

Unfortunately, in many simple devices, adaptive modulation and coding are not supported [4][6][10][11]. In practice, power control is used only in limited cases since power control may change network topology, which in turn will have impact on higher layer performances like routing and transport. A concern is also on the

[2] When transmitting the same information, if the signal powers of QPSK and 16 QAM are equal, the bit duration of QPSK is twice longer than that of 16QAM.

reduction of the signal margin, which may hamper the signal quality in dynamic noisy wireless environments.

For power consumption, other components than power amplifier deserve some explanations. Beside the issue of efficient channel utilization, transmit and receive circuitry spend energy for modulation/demodulation and coding/decoding. An internal architecture of a network device is presented in [12], which shows that power consumption of modulation and coding block is 151mW and that of demodulation and decoding block is 279mW. The detailed discussion about designing protocol regarding modulation and coding is presented in [13].

The topics we have discussed so far focus on the energy consumption when the transmission and reception occur. However, in reality, the wireless channel is shared by more than one network devices, and a frame is transmitted when an event in the device happen. Since the event happen randomly in each node, the exact transmission times of other devices are hard to estimate. If the transmission is not properly coordinated, the data frames may be transmitted at the same time from more than one node. Consequently, the frame cannot be decoded correctly at the intended receiver, and retransmission is required. Moreover, if a node is not expected to receive a frame for a certain time period, it may be able to turn off receiving circuits to save energy. Thus, in addition to determining the proper modulation, coding, and power level, transmission should be coordinated among network nodes within their radio coverage. Medium access control layer takes charge of those operations in the protocol stack. That is the reason why MAC has taken much attention for energy saving. In the next chapter, we investigate MAC functions and related energy saving issues.

4.2 Medium Access Control

MAC is located as a sublayer in the second layer, data link layer (DLL) when a network protocol stack is classified into 7 layers following the open system interconnection (OSI) model. The DLL defines functions to reliably exchange data between two nodes. Strictly, MAC defines methods to coordinate access order among network nodes. In practice, however, the term MAC is frequently used as representing name of the second layer. One possible reason is that the access coordination is the most important issue in networks using shared medium. Also, explaining the other second layer functions in the view of MAC algorithm is acceptable since the MAC function is woven with the other functions. Thus, we also refer DLL as MAC unless otherwise mentioned in this chapter. In order to discern the original MAC functions, we use the term *access control.*

As introduced briefly in the previous section, functions in the MAC more or less affect energy consumption of a network node. When the assessment of functions is necessary for implementation, it is beneficial to divide the functions into low level operational blocks [14]. Time synchronization might be a good example. If the function is designed as an independent component block, it might be easily accessed and implemented. However, if the purpose of classification is for discussion of the energy saving algorithms, high level conceptual blocks are more helpful. We divide MAC into three high level conceptual blocks: link control, access control and power management. Among them, we focus on the access control and power management. The link control contains functions of error control (automatic repeat request), flow control, modulation control, code control and power control. In the networks we have been focused, the impact of the error control and flow control are relative low because of low data rates. Also, as mentioned in the previous section, modulation, code and power control are seldom used in those networks.

4.2.1 Access Control

Access control has been studied extensively over the years, and a plethora of protocols have been devised. Based on the fundamental coordinating domain for individual access, they are classified into TDMA, FDMA, CDMA and SDMA. Time division multiple access (TDMA) protocols divide the time axis and determine accessing order by schedule or competition. Thus, whole sharing frequency bandwidth among nodes is utilized by one node at a time. Frequency division multiple access (FDMA) protocols divides the available frequency bandwidth into a number of sub-channels. Then, it allows nodes individual access by assigning the sub-channels. Code division multiple access (CDMA) protocols spread signals over the frequency bandwidth. It enables nodes individual access by assigning different codes. Space division multiple access (SDMA) protocol utilize spatial signal differences created from multiple antennas. However, we only focus TDMA since other access protocols require relatively complex hardware for handling narrow bandwidth channels or signal processing.

One of the traditional ways of classifying TDMA is classifying them into controlled access and random access. In controlled access usually called TDMA, nodes pre-negotiate the transmission time before the transmission. Thus, usually, a centralized controller in the network coordinates the access. Guaranteed time slot service in IEEE 802.15.4 is a good example. Fixed time duration is announced by a beacon from a coordinator, and exclusive access for only one node is agreed among nodes in the network. Point Coordinate Function in IEEE 802.11 uses different type of a control method [15]. In the protocol, a node called station (STA) can transmit a data frame only on the polling from a central coordinator, access point (AP)[3]. These control access protocols are free from collision because of the nature of the control used. Thus, these are favored especially when applications require low packet error rate and low latency such as in VOIP and automatic control.

[3] IEEE 802.11 working group provides more advanced methods. HCCA (Hybrid coordination function Controlled Channel Access) in IEEE 802.11e enables to change the random access mode to the controlled access mode similar PCF on the request from AP. Power Save Multi-Poll (PSMP) reduces polling overhead with one aggregated polling frame [16].

Random access protocols allow fully distributed operation. On the random packet arrival, a node estimates network and channel condition and determines the access time. ALOHA protocol is the first protocol in this category. In the pure ALOHA, a node immediately transmits a frame when it comes from the upper layer. Since there is no coordination, the latency of waiting for the transmission time is minimized, and transceiver architecture becomes extremely simple. IEEE 802.11 and IEEE 802.15.4 adopt an advanced version of the algorithm, carrier sense multiple access and collision avoidance (CSMA/CA). Although two standard protocols are slightly different, CSMA/CA is most widely used random access protocol. Nodes adopting CSMA/CA regulate channel access probability by a backoff mechanism in which they randomly select transmission time slots within bounded number of slots. Especially, the probability of access is changed adaptively by extending the time slot bound. In addition, nodes sense carrier on the medium before transmission not to intervene on-going transmission of other nodes.

However, carrier sensing over the wireless medium is not always complete because the transmitted signal attenuates as a function of distance. For example, if two transmitters are located at the outside of sensing distance of each other, and there is a receiver in the transmission distance of both transmitters, one of the transmitters becomes a hidden terminal of the other transmitter. In this case, sensing the carrier before transmission is useless. Especially, when size of a frame is large, the collision probability increases; consequently, more energy is spent for transmitting and retransmitting the frame. IEEE 802.11 adopts virtual carrier sensing in which RTS and CTS frames are exchanged before data frame transmission. Request to send (RTS) and clear to send (CTS) are short control frames. They have intended receiving nodes respectively, but every node overhears the frames and takes information on the frames without filtering. A node having data to transmit fills in RTS frame the time duration required to finish the data transmission. The intended receiver of the RTS responds with CTS after writing the same time duration down on that. Thus, the nodes that have received RTS or CTS know about on-going channel activity although it is not physically sensible. However, even with this RTS/CTS handshake collision still exists. When a CTS is transmitted, another node which has not received the first RTS may initiate its own data transmission by transmitting another RTS. Then, two data transaction can proceed independently and be collided. Another problem of RTS/CTS handshake is that the overhead for control frame exchange is considerable. Especially, when sizes of data frames are small, relying on retransmission after possible collision without the RTS/CTS control frames is a better choice.

The selection of proper access control depends on the required performance of applications. Interestingly, performance measures such as throughput, packet error rate, latency and energy consumption affect each other. Especially, energy consumption in low rate applications has a trade off relation with latency. Thus, in order to select an access control method, it is essential to estimate the energy consumption of access methods in consideration under a certain level of latency bound. The Energy consumption can be a function of traffic, a number of competing nodes and protocol parameters such as beacon interval and back off times. Usually, random access protocols have better energy consumption and latency when the traffic and the number of nodes are small because controlled access protocols have extra energy consumption for time synchronization and additional latency to wait for assigned access orders. However, as the traffic and the number of nodes increase, the performances of the random access protocols quickly drop because of collision and retransmission. Refer to [19] for the selection of a CSMA/CA protocol and a fixed access time protocol.

4.2.2 Power Management

Power management function is designed to minimize energy consumption of activities for transmission and reception in a wireless device. It resides in the management plane. Different from access control functions and link control functions residing in the data plane, a power management algorithm is not independently operated, but operated with other functions. In some protocols where the access time is controlled, the power management functions are tightly woven with access control functions, and it may be difficult to clearly discern the power management functions.

The first discussion of power management is to investigate energy consumption for communication activities. In order to identify the exact sources of inefficient energy expenditure, actual energy consumption pattern is analyzed in [18]. The authors presented four sources of power consumption; namely, collision, overhearing, control packet overhearing, and idle listening. Among those, they reported that idle listening, which is channel sensing activity to be ready for a possible incoming frame, is the major power drainer. As shown in [19], 90% of time is used for idle listening in many applications. The power level for idle listening is same as that of frame reception or similar in many low power devices [6][10][11]. Thus, the frame transmission and reception methods should be designed in energy efficient manners to reduce idle listening time.

A possible approach to solving this problem is defining a specific role of a network device. For example, a small sensor may be designed to transmit frames to a coordinator located within one hop range without receiving any data frame. In this case, the device does not need to spend energy for monitoring channel (idle listening) for possible incoming frames. IEEE 802.15.4 supports this type of devices with the name of a Reduced Function Device (RFD) [4]. The RFD joins the network as a member but does not support the functions of a network coordinator. Its communication is allowed only with a Full Function Device (FFD), mostly a coordinator. The CSMA/CA algorithm of IEEE 802.15.4 requires monitoring the channel only at the last slot of chosen backoff slots. If a RFD sensor device is designed for transmission only, it can minimize energy consumption even for transmission by turning off the radio while decreasing the backoff counter until the last slot. However, this type of approach is hardly generalized. Even small sensor devices are required to have the receiving function to reconfigure control parameters or to request on-demand data transmission. Moreover, in multi-hop sensor networks, frame reception is an essential function for relaying frames.

Another interesting approach to solving the idle listening is to use low power radio only for monitoring communication channel [9] [19] [20] [21]. The method, usually called a wake-up radio reasons that the high quality signal and high power processing are not required to simply detect incoming frames[4]. In the system, a network node adopts two radios: a low power wake-up radio and a high power main radio. The main radio supports high speed data transmission and reception. It is more energy efficient than the wake-up radio when it is used for large data exchange. But, it consumes much larger energy for monitoring channel. Thus, the system attempts the main radio to sleep when it is not involved in any frame exchange. Usually, the wake-up radio consumes extremely small energy (in the order of uW) and is used only to transmit binary information or the destination address to trigger the main radio of a designated node. Nodes wait for incoming frame only with the wake-up radio. If a node has a frame to transmit, it transmits the wake-up request with the wake-up channel first then, transmits the data frame through the main radio.

One of the problems is the additional cost. For many sensor applications where low hardware cost and small form factor are the key success indicator, the increased cost can become a main hurdle of the method. The second problem is the technology for the low power wake-up radio itself. Beside the problem how to align coverage of the two radios, making a stable ultra low power wake-up radio is still challenging. The state of the art researches are introduced in [9][20].

The most common approach is designing a power management function in a MAC protocol with a single radio. The function controls when to turn on or off the radio circuitry in a device. The goal of the power management algorithm is to assign as much sleeping time (turning off the radio) as possible while meeting the application requirements. In the algorithm design, the latency should be handled very carefully because a device cannot receive any frame while it is sleeping, differently from the algorithms using a wake-up radio. Therefore, the design of algorithms and protocols for power management differs in order to meet application requirements and characteristics. In the next section, we present details of power managements in MAC.

4.3 MAC Protocol Classification with Power Management Functions

As we have presented in previous sections, energy saving is one of most important issues in low rate and short range networks. For that, the MAC layer, specifically power management function takes a critical role in energy saving. An optimal power management function is to minimize idle listening by turning on the receiver only when it is directly related to packet transmission or reception. However, in the distributed environment, packet transmission and reception occurs in a random manner. Significant number of MAC protocols has been proposed to address the optimal usage with network specific characteristics such as traffic

[4] Only energy detection is considered because waveform detection is defeating the purpose of using ultra-low power wake-up radio.

patterns, topology and latency bound. In [22], more than 70 sensor network MACs are already listed. Although different techniques can contribute to better performance under specific traffic and environmental condition, some MAC protocols support similar functions. Thus, categorizing and analyzing the MAC protocols are essential to understand, adopt or even design a new energy efficient MAC for a new application. For this purpose, we classify energy saving MAC protocols into two groups based on the characteristics of target networks: MACs for asymmetric single-hop networks and MACs for symmetric multi-hop networks.

Asymmetric single-hop networks are basic forms of wireless networks. Representative MAC examples of this type of networks are IEEE 802.11 power saving mode and IEEE 802.15.4 beacon mode. Usually, this type of networks consists of a coordinator and devices located within one hop communication range of the coordinator. The coordinator is a network controller having more processing power and energy.

On the other hand, symmetric multi-hop networks are an extended form of networks. When an application requires the coverage greater than one hop range of a node, it is natural to use multi-hop network technology. For the purpose, the first candidate may be using multiple asymmetric single-hop networks and connecting those clusters with wireless or wired links. The other method is to design a symmetric multi-hop network. The latter is the common approach for low power and low rate sensor networks with small symmetric sensor devices. Several example MACs are found in [18][19][23-26]. The networks usually consist of devices without a specific coordinator having more resources than other devices, although there may exist different roles such as sink and source in view of applications.

For both groups of MAC protocols for asymmetric single hop networks or symmetric multi-hop networks, a key of MAC layer energy saving is how to minimize unnecessary energy consumption. Turning on the radio circuitry only when it involves frame transaction activity is the best policy. Therefore, an energy saving MAC algorithm should support a function to buffer frames for sleeping receivers at the transmitter and a function to retrieve the buffered frames at the receiver. Although the retrieving function may be performed non-periodic manner for some applications, this function is usually performed periodically in the consideration of latency bound. We call this period a wake-up interval.

4.3.1 MAC for Asymmetric Single-Hop Networks

For asymmetric MACs, it can be assumed that a coordinator is supported by a power supply or a large capacity battery, and turns on the radio all the time for incoming frames. Unless the channel time is scheduled for specific devices, a device may transmit a frame without any concern when a frame is generated. Thus, the energy saving for asymmetric MACs focuses on finding what the best strategy of receiving a buffered frame is. The asymmetric single-hop MAC protocols are classified again into synchronous MACs and asynchronous MACs based on the methods of timer management in a device. The synchronous MACs further sub-divide into *automatic delivery* and *transmitter notification*.

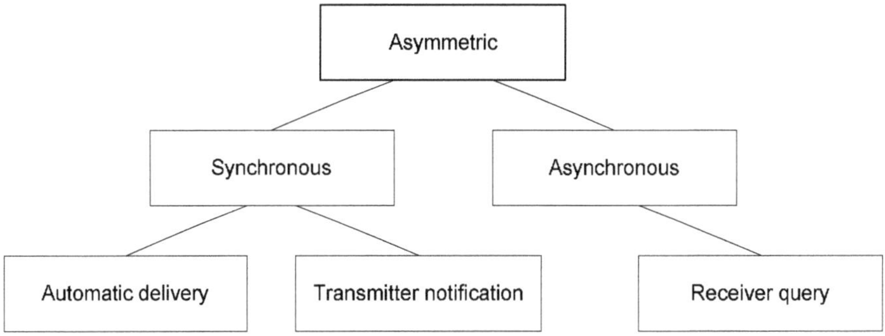

Fig. 4 3 Asymmetric protocol classification

4.3.1.1 Automatic Delivery

Automatic delivery is adopting the merits of fixed time access control algorithms for energy saving purpose. In those algorithms, a transmitter and a receiver synchronize their clocks, and then assign a specific time duration between them. In asymmetric network, the synchronization process is relatively simple since all nodes can hear a broadcast frame from a coordinator and adjust their timers. The frame transmission happens only in the assigned time duration. Thus, a device can turn off the radio circuitry and save energy at other times.

A good example of *automatic delivery* is found in IEEE 802.11. In the standard, an access point (AP), a controller and gateway of a network, provides network services to nodes named stations (STAs). It turns on its radio all the time and waits for the incoming frames. For a downlink frame, the transmission depends on whether a STA wants power saving support or not. If a STA is in the active mode where a STA always turns on its receiver circuitry, the AP directly transmits the frame to the STA. But, if a STA is in power saving mode, then it buffers the frame until the transmission is allowed. Currently, IEEE 802.11 supports three power saving modes: power saving mode (PSM), unscheduled-automatic power save delivery (U-APSD), and scheduled-automatic power save delivery (S-APSD). Among them, S-APSD belongs to the synchronous *automatic delivery*. The example time line of S-APSD is presented in Fig.4. In S-APSD, a schedule is agreed

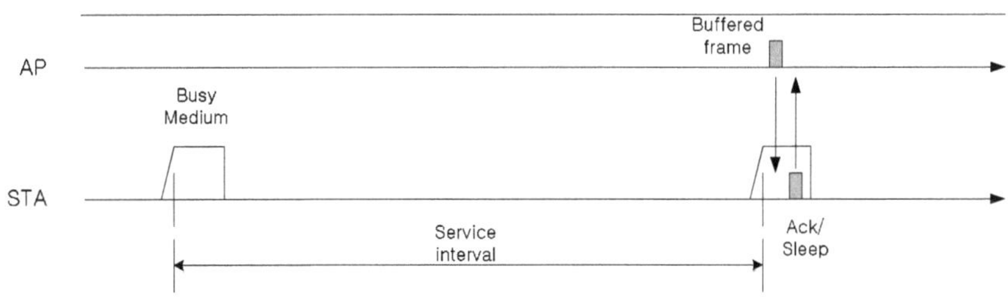

Fig. 4 4 Example timeline of scheduled automatic power saved delivery

upon between an AP and a STA. Then, the STA wakes up at every pre-scheduled time. The AP buffers a frame destined to the STA if a packet is arrived before the scheduled time. When the time comes, it transmits the buffered frame assuming the STA is awake. A random access method should be used during the defined time duration since the independent access right is not supported in the S-APSD.

Another example of *automatic delivery* can be found in IEEE 802.15.4. In the beacon mode of the protocol, a coordinator provides a synchronization service by periodically broadcasting a beacon frame. Thus, all devices have a common time line, and the time line is divided into fixed time intervals, also known as beacon intervals. The beacon interval is divided into two time periods: an active period and an optional inactive period. The active period is divided further into a Contention Access Period (CAP) and an optional Contention Free Period (CFP). This time structure called the superframe is illustrated in Fig 4.5.

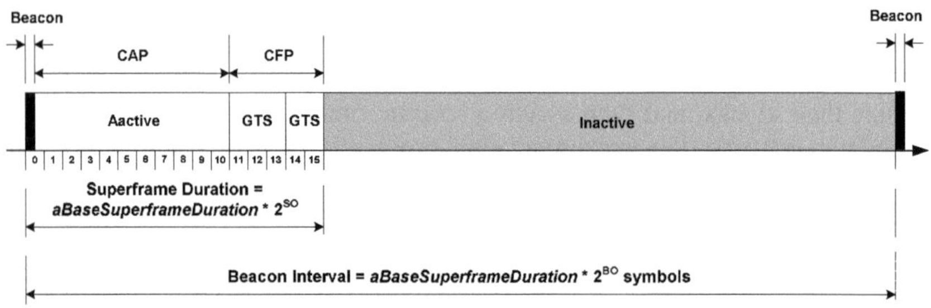

Fig. 4.5 Superframe structure in the beacon mode of IEEE 802.15.4.

The *automatic delivery* happens at the CFP. In order to use the time slot, a device requests or is assigned Guaranteed Time Slot (GTS) by exchanging the control frames during the CAP period. Then, at the assigned time slot, a frame is transmitted only by the device requested the slot.

Ideally, *automatic delivery* is the best energy saving algorithm because it does not have any control frame exchange in the scheduled active duration. However, one of the fundamental problems of *automatic delivery* is from the inherent feature of synchronous protocols. If the clocks of the transmitter and the receiver are not synchronized by external devices such as GPS, the devices have to spend extra energy for synchronization. For example, in order to synchronize the scheduled time for S-APSD or GTS, a STA or a device in the networks needs to be periodically resynchronized, so that it periodically receives a beacon and re-adjusts its timer. In addition, each device has to wake-up slightly earlier than the scheduled time for a margin of tolerance. In practice, oscillators exhibit a slight random deviation from their normal operating frequency. This phenomenon is called clock drift or clock skew, and it is due to impure crystals and several environmental conditions like temperature and pressure. The clock drift in sensor networks is reported in the range between 1 and 100 PPM [27][28].

4.3.1.2 Transmitter Notification

The second category of the synchronous MAC protocols is *transmitter notification*. In these MAC protocols, a common active duration among a coordinator and all devices is agreed upon. At the beginning of each active time, the coordinator notifies the existence of buffered frames for all devices. Then, a device upon receiving a buffered frame notification, requests the frame transmission to the coordinator by transmitting a request frame. If a device received empty buffer notification explicitly or implicitly, it turns off the radio and saves energy.

This notification may be used in a slightly different way. If a device wants a long sleep, the time drift will be considerably large, and the device may not be able to get a notification mostly payload in a beacon frame in the expected time. In this case, the device may turn on the receiver until it receives a beacon without keeping synchronized time information. Although this type of operation is defined in IEEE 802.15.4 as a non-beacon tracking, we do not consider this method since this asynchronous communication requires fairly large energy consumption at each attempt for reception. Therefore the usage is very limited.

An example of *transmitter notification* MACs is power save mode (PSM) in IEEE 802.11. In the protocol, a STA requests power management service to an AP. Then, the AP buffers all frames toward the STA. An AP of IEEE 802.11 periodically broadcasts a beacon to announce information about its capabilities, configuration, and security information. In PSM, in addition to these information, the AP notifies the STA whether a frame is buffered or not using the beacon. If a STA receives the beacon indicating a buffered frame, it transmits PS-Poll frame to request the buffered frame. After receiving the poll frame, the AP assumes that the STA is ready, and transmits the buffered frame. Also in IEEE 802.15.4, the similar operation is possible in the beacon mode named indirect transmission. Fig.4.6 illustrates IEEE 802.11 PSM.

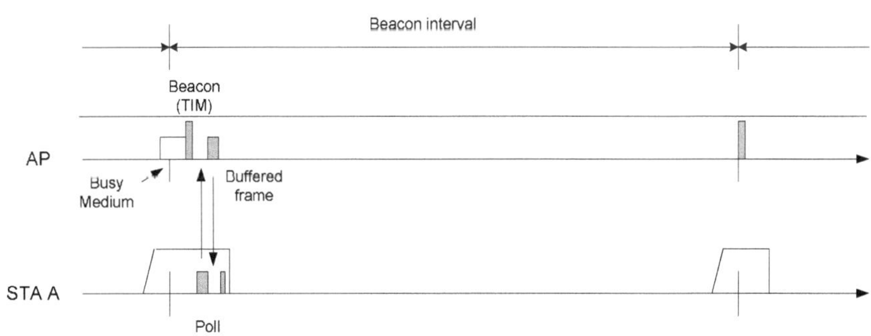

Fig. 4.6 IEEE 802.11 Power Save Mode

This notification benefits in two ways. Most of all, a device can resynchronize without extra efforts. As mentioned before, *automatic delivery* MAC requires additional efforts to be synchronized. The other benefit is that a device is sure of the buffer status and can go to sleep without waiting for possible frames. The coordinator is also sure about status of a receiver since it is requested explicitly. This reduces channel time waste that can be caused by transmitting frames to a device sleeping or moving. However, comparing to *automatic delivery*, it has the overhead for notification (beacon in PSM) and request (PS-Poll in PSM). Other issues are from the feature that the notification is announced only by the AP. First, in order to contain buffer status of all associated nodes, the required length of beacon becomes problematic. In practice, a beacon in IEEE 802.11 has a bit map (TIM element) for associated STAs. Second, each STA cannot have its own optimized schedule (wake-up interval). Last, the common schedule may increase contention rate when after the beacon especially if the number of buffered frames increase.

4.3.1.3 Receiver Query

Different from previous two categories of MAC protocols, *receiver query* is a receiver oriented asynchronous algorithm. Without any agreed schedule, a device just announces to a coordinator that it is in the power saving mode repeating waking-up and sleeping. Then, the coordinator buffers any frame to the device. The device usually periodically wakes up and inquires of the coordinator about a buffered frame. If there exists any buffered frame, the coordinator transmits the frame. Otherwise, a short control frame is transmitted to inform the empty buffer state.

Unscheduled-automatic power save delivery (U-APSD) in IEEE 802.11 is a good example of the *receiver query*. The protocol itself does not define a periodic wake-up interval for the query. However, periodic inquiry is essential because of latency bound. In the protocol, the trigger frame is used to query the buffered frame. Also, any uplink data frame can be utilized as the query. If no data fame exists, a device uses a null frame as a trigger frame. If queried, the AP transmits a buffered frame. If no frame is buffered, a null data frame is transmitted.

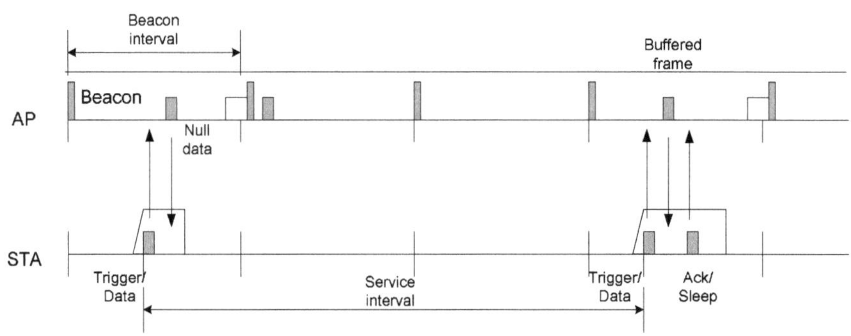

Fig. 4.7 Example time line of unscheduled-automatic power save delivery

The best merit of *receiver notification* is that no agreed schedule between a coordinator and a node is required. This results in no extra effort for synchronization such as periodic beacon frame transmission and reception. In addition, compared to transmitter notification in asymmetric MAC, every device can optimize its own wake-up interval based on traffic characteristic and latency bound. On the other hand, *receiver notification* requires active participation (query transmission) of devices. Thus, if the utilization is low and the number of devices adopting the algorithm is large, it would impose unnecessary channel time.

4.3.2 MAC for Symmetric Multi-hop Networks

In symmetric multi-hop networks, all devices are assumed to be operated on battery power. Thus, all devices repeat waking-up and sleeping to save energy. Here, designing efficient receiving methods becomes important since all devices participate in relaying frames in multi-hop networks. Equally important issue is how to transmit a frame to the device repeating waking-up and sleeping. Similar to asymmetric single-hop MACs, the symmetric multi-hop MAC protocols are classified into synchronous MAC and asynchronous MAC based on the clock management method of a device. However, in the multi-hop networks, *automatic delivery* is not considered because it is impractical for a transmitter to synchronize with all the receivers' different active durations without counting explicit control frame exchanges with all the receivers. Also, *transmitter notification* is performed in different ways because every neighbor can be a transmitter of a power saving receiver. The asynchronous MACs need also different algorithms and they are sub-divided into *transmitter sweep* and *receiver notification* as shown in Fig.4.8.

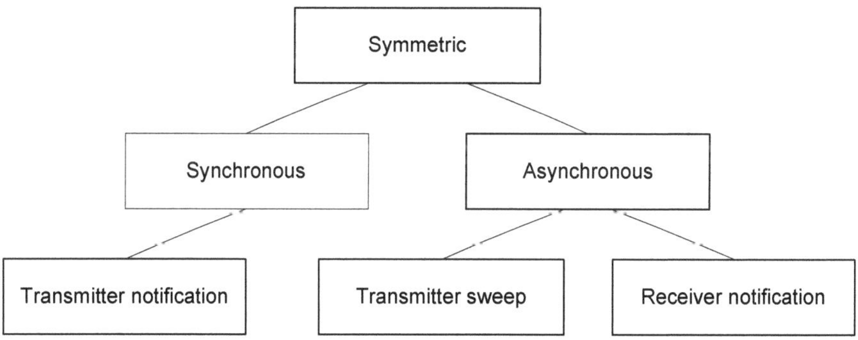

Fig. 4.8 Symmetric protocol classification

4.3.2.1 Transmitter Notification

Transmitter notification enables communications among energy saving nodes in multi-hop networks by globally synchronizing all active durations of the network nodes. During the active duration, a node transmits a beacon to ascertain the existence of active duration and to resynchronize clocks. In the active duration, any device having a frame to transmit notifies this by transmitting a control frame such as RTS. The device receiving the RTS replies with CTS, and the data frame is transmitted right after the CTS or at time duration dedicated to data frame transmission. Differently from asymmetric single-hop networks, every node can contend to transmit a beacon at the beginning of a common active duration. Also, an additional frame such as RTS is required separately to notify the existence of data to a destination since the node transmitting a beacon and having a data frame to transmit may be different.

Sensor MAC (SMAC) is a widely known sensor network protocol adopting *transmitter notification* [18]. In the initialization stage of a network, all devices are in active mode. A device having the smallest value[5] for synchronization broadcasts its own schedule periodically in a sync frame. This divides time lines into periodic blocks consisting of a short active duration and a long inactive duration. A device that has received a sync frame follows the schedule in the received frame. After then, it also competes to broadcast its sync frame at the beginning of the every active duration. And, this schedule is propagated to whole network. The active duration is subdivided into three sub-durations for sync, RTS/CTS, and data. In the duration for sync, a device broadcasts and receives a sync frame. The transmission attempt is with probability to reduce collision rate. If a device has a data to transmit, it first exchanges RTS and CTS frames, which enables the devices involved in this transaction to stay on to exchange data in the active duration for data. On the other hand, the other devices turn off the radio to save energy until the next time for a sync frame. The example time of SMAC is presented in Fig.4.9.

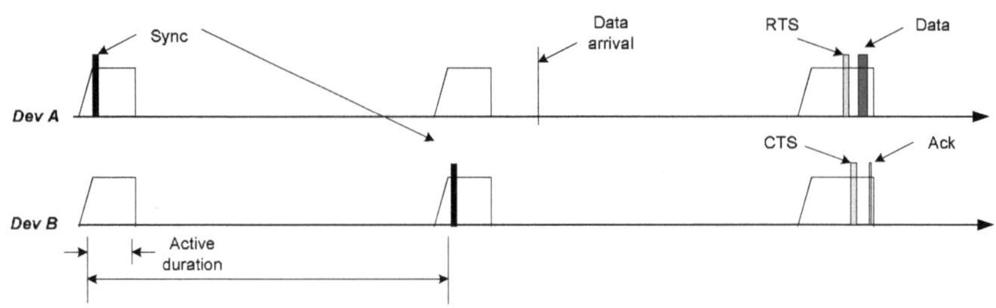

Fig. 4.9 Example timeline of SMAC

A merit of *transmitter notification* is that the active duration can be used as the same way of non-power saving mode. If the active duration is scheduled for sufficient time and the reservation with RTS/CTS is not used, frames can be relayed to several hops within the same active duration. Also, broadcast can be implemented very easily. However, it is a difficult task to have a common schedule among all network devices. Especially when more than one device start transmitting a sync in the initialization stage, it makes nodes belong to different schedule groups. In [18], border nodes of two different schedules solve the problem by having both periodic active durations of the two groups. Another problem of *transmitter notification* is that the time for beacon transmission and the margin for synchronization are required compared to asynchronous protocols.

[5] The timer value might be set with randomly generated one or predetermined one based on the policy of an application.

4.3.2.2 Transmitter Sweep

MAC protocols classified into *Transmitter sweep* enable communications among network devices adopting energy saving algorithm without synchronization. What motivated this class of protocols is the considerable overhead for synchronization and the very low traffic rate. That is, if the number of transmissions is very small compared to the number of periodic wake-ups, it will be more efficient to spend more resources (i.e., overhead) for transmission than for periodic wake-up and channel probing. In *transmitter sweep*, devices periodically wake up and check whether there is any transmission activity on the channel or not. If any activity is recognized, a device stays awake until the data is received. On the other hand, when a device has a frame to transmit, it notifies this with long preamble or a stream of control frames to wake the destination node up. If this notification occupies the channel longer than one wake-up interval, all the devices within a hop transmission range will wake up and be ready to receive a frame from the transmitter.

The first protocol in this category is BMAC [23]. In the protocol, a transmitter broadcasts a preamble longer than one wake-up interval, and the data transmission follows. To receive the data, devices periodically wake up and check whether there is on-going preamble or not. If the preamble is detected, devices keep receiving until the transmission is completed with data. This ensures devices to have the minimum periodic active duration. In this way, the life time of devices becomes maximized when the traffic is very low. However, this approach has also some drawbacks. The preamble wakes up all neighbors though they are not the intended destinations. In addition, although the destination already recognizes preamble transmission at the beginning, the transmitter and the receiver have to transmit and receive the long preamble for the entire wake-up interval.

XMAC is designed to resolve the problem [24]. In the protocol, short control frames named short preamble are transmitted until the destination replies with an early Ack. If the early Ack is received, a data frame is transmitted. Since the short preamble has the destination address, other devices can turn off the radio circuitry while the transmitter attempts to wake the destination up. However, the time duration to detect short control frame is longer than that of preamble. We presented the cxamplc timc linc of XMAC in Fig.4.10.

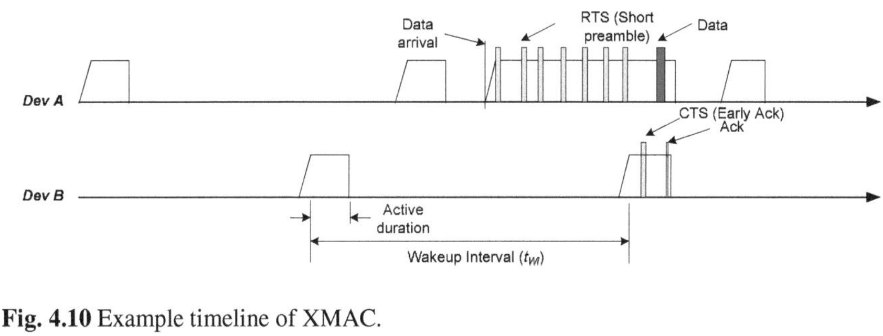

Fig. 4.10 Example timeline of XMAC.

By minimizing the active duration, *transmitter sweep* maintains better energy efficiency than *transmitter notification* when the data rate is low. Interestingly, energy efficiency of BMAC is the best if the traffic rate is extremely low [7]. However, by the asynchronous feature of *transmitter notification*, energy consumption for transmission is considerably large. Also, the channel occupancy time reduces channel utilization.

In order to reduce this transmission overhead, local synchronization is proposed in [26][29]. In these protocols named WiseMAC and SCP-MAC respectively, wake-up time information of neighbors is logged in a device's own time line after exchanging a data frame. Then, while keeping its own schedules for receiving, the device starts transmission just before the expected active time of the receiver. This local synchronization reduces transmission overhead considerably, but it requires high complexity of MAC HW and SW. In addition, if a node is expected to receive frames from a number of nodes, the overhead for potential contention should be tamed gently.

Another interesting approach considering channel contending overhead is proposed in [30]. In the protocol named Z-MAC, a timeline is divided into slots when traffics in the network increase. Then, by the defined rule, a privilege to access the channel earlier than other nodes is assigned to one node similar to controlled access protocols. Since Z-MAC is defined on top of BMAC, the energy consumption of Z-MAC is comparable to BMAC, but the throughput is higher.

4.3.2.3 Receiver Notification

Receiver notification is another class of asynchronous MAC protocols to efficiently use the wireless channel. Compared to the *transmitter sweep* which is initiated by a transmitter, the transmission of *receiver notification* is initiated by a receiver. In *receiver notification*, each device notifies its schedule whenever it enters its periodic active duration. A device having data to transmit turns on the radio circuitry and waits for the notification. If the notification is received, it considers that the receiving device is in the active duration, and transmits an RTS frame. If a CTS is received, it transmits the buffered data frame.

Examples of *receiver notification* are IEEE 802.15.5 asynchronous energy saving (AES) mode [31], RI-MAC [32], and RICER [25]. The example timeline of IEEE 802.15.5 AES is presented in Fig.4.11. In AES, each device broadcasts a wake-up notification (WN) frame to notify the length of active duration. A device having data to transmit waits for the WN of destination. If the active duration of the destination announced by WN is long enough, a data frame is transmitted. If not, an extension request (EREQ) is transmitted to request an extension of the active duration. If the request is confirmed by an extension reply (EREP), the device transmits a data.

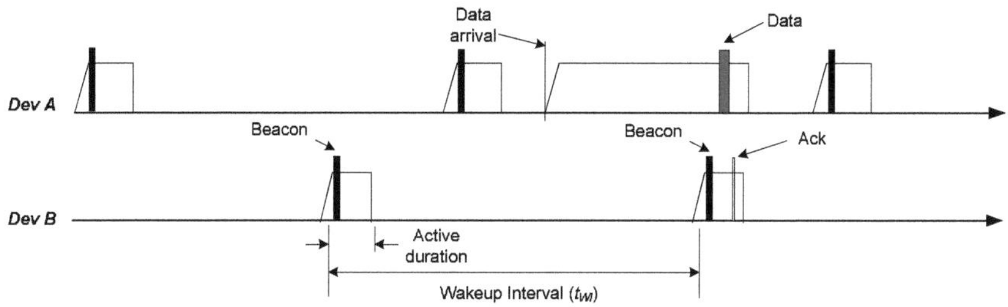

Fig. 4.11 Example time line of IEEE 802.15.5 asynchronous energy saving mode

A merit of *receiver notification* is that channel time can be used efficiently. Compared to *transmitter sweep*, the other nodes can transmit data frames while a device is waiting for a beacon of a destination. In addition, the active duration can be extended flexibly to accommodate traffic burst. However, a problem is that the contention happens inside of the active duration of a receiver. Note that in *transmitter sweep,* the contention is resolved before the active duration of a receiver. Another issue is that the broadcasting is not possible in *receiver notification* since the active durations of devices are not synchronized.

4.3.3 Summary

In this section, we have focused on the power management techniques. Table 4.1 summarizes the discussed protocols. For more detailed performance comparison and other issues related the protocols, we recommend to read [33][34].

Table 4.1 Summary of discussed power management protocols

Asymmetric			Symmetric		
Synchronous		Asynchronous	Synchronous	Asynchronous	
Automatic Deliver	Transmitter Notification	Receiver Query	Transmitter Notification	Transmitter Sweep	Receiver Notification
IEEE 802.11 S-APSD, IEEE 802.15.4 GTS	IEEE 802.11 PSM, IEEE 802.15.4 beacon-indirect	IEEE 802.11 U-APSD	SMAC	BMAC, XMAC, (SCP-MAC, WiseMAC, , Z-MAC[6])	IEEE 802.15.5 AES, RI-MAC, RICER

4.4 Conclusion

In this chapter, we have discussed energy efficiency in wireless communication. Then, we have presented major functional blocks in MAC protocols and explained why MAC protocols take critical roles to achieve energy efficiency. In order to compare the power management algorithms in MAC protocols proposed for energy saving, we classified them into two categories: asymmetric single-hop MAC and symmetric multi-hop MAC. The asymmetric single-hop MAC is sub-divided into *automatic delivery*, *transmitter notification* and *receiver query*, and the symmetric multi-hop MAC is classified again into *transmitter notification*, *transmitter sweep*, and *receiver notification*. Characteristics, merits and demerits of each subcategory are discussed with representative MAC protocols. All protocols are designed to achieve energy efficiency. Under the given network topology, the performance of the protocols is determined by traffics, environmental condition and system characteristics such as clock accuracy. Therefore, it is recommended to select a proper approach based on the understanding of application requirements and system characteristics.

References

[1] David, K., Dixit, D., Jefferies, N.: 2020 Vision. IEEE Vehicular Technology Magazine 5(3), 22–29 (2010)
[2] Vullers, R.J.M., van Schaijk, R., Doms, I., Van Hoof, C., Mertens, R.: Micropower energy harvesting. Solid-State Electronics 53(7), 684–693 (2009)
[3] `http://www.epa.gov/osw/conserve/materials/battery.htm`
[4] IEEE 802.15.4-2006, Part 15.4: Wireless LAN Medium Access Control (MAC) and Physical Layer (PHY) specifications for Low-Rate Wireless Personal Area Networks (LR-WPANs) (2006)
[5] `http://www.zigbee.org`
[6] Chipcon: 2.4GHz IEEE802.15.4ZigBee-ready RF Transceiver datasheet (rev1.2), Chipcon AS, Oslo, Norway (2004)
[7] Park, T.R., Lee, M.J.: Power saving algorithms for wireless sensor networks on IEEE 802.15.4. IEEE Communications Magazine 46(6), 148–155 (2008)
[8] Patel, M., Wang, J.: Applications, challenges, and prospective in emerging body area networking technologies. IEEE Wireless Communications 17(1), 80–88 (2010)
[9] Rabaey, J., Ammer, J., Otis, B., Burghardt, F., Chee, Y.H., Pletcher, N., Sheets, M., Qin, H.: Ultra-low-power design. IEEE Circuits and Devices Magazine 22(4), 23–29 (2006)
[10] Texas Instruments, cc2430 preliminary datasheet (rev2.01), Texas Instruments, Dallas (2006)
[11] Freescale, MC13192/MC13193 2.4 GHz Low Power Transceiver for the IEEE 802.15.4 Standard (Rev 2.9), Freescale Semiconductor (2005)
[12] Min, R., Chandrakasan, A.: A framework for energy-scalable communication in high-density wireless networks. In: ISLPED, pp. 36–41 (2002)
[13] Shih, E., Cho, S.-H., Ickes, N., Min, R., Sinha, A., Wang, A., Chandrakasan, A.: Physical layer driven protocol and algorithm design for energy-efficient wireless sensor networks. In: ACM MobiCom, pp. 272–287 (2001)

[14] Klues, K., Hackmann, G., Chipara, O., Lu, C.: A Component Based Architecture for Power-Efficient Media Access Control in Wireless Sensor Networks. In: ACM SenSys (2007)
[15] IEEE 802.11, Part 11: Wireless LAN Medium Access Control (MAC) and Physical Layer (PHY) Specifications - Revision of IEEE Std 802.11-1999 (2007)
[16] IEEE 802.11, Part 11: Wireless LAN Medium Access Control (MAC) and Physical Layer (PHY) Specifications Amendment 5: Enhancements for Higher Throughput (2009)
[17] Ergen, S.C., Di Marco, P., Fischione, C.: MAC Protocol Engine for Sensor Networks. In: IEEE GLOBECOM (2009)
[18] Ye, W., Heidemann, J., Estrin, D.: An energy-efficient MAC protocol for wireless sensor networks. In: IEEE INFOCOMM, pp. 1567–1576 (2001)
[19] Guo, C., Charlie Zhong, L., Rabaey, J.M.: Low power distributed MAC for Ad hoc sensor radio networks. In: IEEE GlobeCom (November 2001)
[20] Pletcher, N.M., Gambini, S., Rabaey, J.: A 52 w wake-up receiver with 72 dbm sensitivity using an uncertain-if architecture. IEEE Journal of Solid-State Circuits, 44(1), 269–280 (2009)
[21] Stathopoulos, T., McIntire, D., Kaiser, W.J.: The energy endoscope: Real-time detailed energy accounting for wireless sensor nodes. In: IEEE IPSN, pp. 383–394 (2008)
[22] `http://pds.twi.tudelft.nl/~koen/MACsoup`
[23] Polastre, J., Hill, J., Culler, D.: Versatile low power media access for wireless sensor networks. In: ACM SenSys, pp. 95–107 (2004)
[24] Buettner, M., Yee, G.V., Anderson, E., Han, R.: X-MAC: a short preamble MAC protocol for duty-cycled wireless sensor networks. In: ACM SenSys, pp. 307–320 (2006)
[25] Lin, E.-Y.A., Rabaey, J.M., Wolisz, A.: Power-Efficient Rendez-vous Schemes for Dense Wireless Sensor Networks. In: IEEE ICC (2004)
[26] Enz, C.C., et al.: WiseNET: An Ultralow-Power Wireless Sensor Network Solution. IEEE Comp. 37(8) (August 2004)
[27] Elson, J., Girod, L., Estrin, D.: Fine-Grained Time Synchronization using Reference Broadcasts. In: The Fifth Symposium on Operating Systems Design and Implementation (OSDI 2002), Boston, MA (December 2002)
[28] Sivrikaya, F., Yener, B.: Time synchronization in sensor networks:A survey. IEEE Network 18, 45–50 (2004)
[29] Ye, W., Silva, F., Heidemann, J.: Ultra-low duty cycle MAC with scheduled channel polling. In: ACM SenSys, pp. 321–334 (2006)
[30] Rhee, I., Warrier, A., Aia, M., Min, J.: Z-mac: a hybrid MAC for wireless sensor networks. In: ACM SenSys, pp. 90–101 (2005)
[31] IEEE 802.15.5, Part 15.5: Mesh Topology Capability in Wireless Personal Area Networks (WPANs) (2009)
[32] Sun, Y., Gurewitz, O., Johnson, D.B.: RI-MAC: a receiver-initiated asynchronous duty cycle MAC protocol for dynamic traffic loads in wireless sensor networks. In: ACM SenSys, pp. 1–14 (2008)
[33] Karl, H., Will, A.: Protocols and Architectures for Wireless Sensor. John Wiley & Sons, Chichester (2005)
[34] Demirkol, I., Ersoy, C., Alag, F.: Mac protocols for wireless sensor networks: a survey. IEEE Communication Magazine, 115–121 (2006)

Tae Rim Park received B.S and M.S degrees from In-ha University, and Ph.D degree in electrical engineering and computer science from Seoul National University, Korea in 2005. He is currently working as a senior research engineer at signals and systems lab. in Samsung Advanced Institute of Technology. At the time of this work, he was an assistant research professor with the Electrical Engineering Department of City University of New York (CUNY). He has been doing researches on embedded systems and wireless networks. He is an active contributor to IEEE 802.15.4e, IEEE 802.15 and ZigBee standard.

Myung Jong Lee received B.S and M.S degrees from Seoul National University, Korea, and a Ph.D degree in electrical engineering from Columbia University, New York. He is currently a professor in the department of Electrical Engineering of City University of New York. His research interests include wireless sensor networks, ad hoc networks, and CR networks. He publishes extensively in these areas and holds 25 U.S and International Patents (pending included). He actively participates in international standard activities (the chair of IEEE 802.15.5 TG and former Vice Chair of ZigBee NWK WG). His group also contributed NS-2 module for IEEE 802.15.4, a standard NS-2 distribution.

Chapter 5
Energy-Efficient Routing in Delay-Tolerant Networks*

Zygmunt J. Haas, Seung-Keun Yoon, and Jae H. Kim

Abstract. The lifetime of a wireless network is significantly affected by the energy consumed on data transmission. One approach which allows reduction of the transmission energy consumption is the new networking paradigm - the *Delay Tolerant Networks* (*DTN*s). The topology of DTNs consists of nodes with short transmission range, thus allowing the reduction of energy consumption. However, this short transmission range leads to sparse network topologies, raising the challenge of an efficient routing protocol. *Epidemic Routing Protocol* (*ERP*), in which data packets are replicated on nodes that come in contact, is one of such DTN routing protocols. The basic ERP exhibits the shortest delay in packet delivery, but this short delay comes at the expense of large energy consumption. In our past publications, we have proposed a number of new variants of ERP for DTN - the *Restricted Epidemic Routing* (*RER*) protocols - which allow to efficiently tradeoff between the energy consumption of a single packet and the packet delivery delay. In this chapter, we extend our study to determine and to compare the overall lifetime of a network when the various RER protocols are used.

5.1 Introduction

In an intermittently connected mobile ad-hoc network, packets are relayed from the source node to the destination node, while relying on the mobility of nodes and

Zygmunt J. Haas · Seung-Keun Yoon
School of Electrical and Computer Engineering;
Wireless Networks Laboratory (WNL); Cornell University; Ithaca, NY 14853
e-mail: {zhaas,sy255}@cornell.edu

Jae H. Kim
Boeing Research & Technology; The Boeing Company; Seattle, WA 98124
e-mail: jae.h.kim@boeing.com

* This article is based on "Tradeoff between Energy Consumption and Lifetime in Delay-Tolerant Mobile Networks", by S-K. Yoon, Z.J. Haas, and J.H. Kim, which appeared in IEEE MILCOM 2008, San Diego, CA, November 17-19, 2008.

on the future contingency of encounters among the network nodes. This, in general, results in long delays. *Delay Tolerant Network* (*DTN*) is an intermittently connected network that can tolerate some degree of packet delivery delay, while using a short-range transmission to conserve energy. Several protocols have been proposed for DTNs. *Epidemic Routing* [1] uses packet flooding method which involves packet replication and packet propagation to increase the number of copies of the packet that is destined to the destination node. When there are multiple copies in the system, the probability for the destination node to receive one of the copies increases, and hence the average delivery time decreases. However, Epidemic Routing also results in excessive expense of resources such as the energy consumption for multiple packet transmissions and the network capacity.

There have been several publications proposing different ways to overcome the drawback of the Epidemic Routing protocol [2 – 9]. For example, *SWIM* [2, 3] uses an anti-packet which is created by the destination node and propagated throughout the system using Epidemic Routing. When a node receives an anti-packet, it is notified that the destination node has received the data packet, and the node deletes the local copy of the data packet, thus preventing unnecessary future transmissions. The *Spray and Wait* routing scheme [4] employs a different way to reduce the total number of copies in the system. Since it is impossible for the nodes to know the number of copies in the system, the packet could contain the information of how many times it can be transmitted. In our past work [5] we proposed and evaluated several schemes using different methods of restricting the *Epidemic Routing* protocol. We derived the analytical models for these *Restricted Epidemic Routing* (*RER*) schemes using various Markov chain models. For these schemes, we were able to find the most efficient tradeoff between energy consumption and delivery time delay. These works, however, were focused only on conserving node energy for a single routing attempt.

Node energy can be conserved by reducing the number of copies at the expense of longer delivery delay [5 – 7]. When the battery capacities of the nodes are limited, after several attempts of Epidemic Routing, some of the nodes' batteries may become depleted faster than others. This reduces the number of active nodes and the average number of copies in the system. Hence the probability of the destination node to receive a copy decreases as well. This could be improved by using the residual battery energy information, a concept that is similar to the energy-aware routing in ad hoc networks, where nodes are forced to consume energy evenly [10 – 12].

There have been other approaches to improve the efficiency of energy consumption as well. The gossip based algorithm [13, 14] controls the packet flow by transmitting packets to only a fraction of encountered nodes based on a certain probability. Utility-based replication controls the packet flow by transmitting replicated packets only to selected nodes. The selective transmissions could be based on the history of past encounters [15, 16], on current nodes' resource [17, 18, 19], or on the anticipated destination location [20].

Coding-based protocols also reduce energy consumption during packet routing in DTNs. For example, with erasure-coding [21], to reduce energy consumption during a packet transmission, packets carry only partial information. Network-coding [22] was used to combine multiple different packets to reduce the number of transmissions. In such coding-based and network-coding protocols, the destination node has to receive more than one packet to recover the whole data.

The remainder of this chapter is organized as follows. First, we analyze the Epidemic Routing protocol and its performance, and we define the lifetime of a network that uses Epidemic Routing in Section 5.2. In Section 5.3, we consider several Restricted Epidemic Routing schemes and determine how much these schemes can extend the network lifetime. In Section 5.4, we propose a scheme that uses residual battery information. Our simulation and evaluation results are discussed in Section 5.5. We conclude our work in Section 5.6.

5.2 The Lifetime of the Epidemic Routing Protocol

Nodes in a DTN are usually powered by non-rechargeable batteries, thus each node has only limited amount of energy before it becomes inactive. Typically, during multiple attempts of Epidemic Routing, before the last node's battery becomes depleted, there will be a time period when the number of active nodes decreases (gradually, in practical scenarios) to zero. In this period, the packet delivery probability will also decrease gradually to zero. As the usefulness of a network is limited by some minimal delivery probability, hence, practically, the lifetime of a network should be defined somewhere in this "gradually decreasing" time period.

5.2.1 The Unrestricted Epidemic Routing Scheme

First, we analyze the *Unrestricted Epidemic Routing* (*UER*) protocol in order to calculate the average number of copies in the system and the packet delivery probability as a function of time. We represent the network as N mobile nodes, in addition to a stationary destination node. We characterize the encounter between any two particular nodes in the network as a random process that occurs with the rate λ [events/sec]. We further assume that the time between each encounter is an exponentially distributed random variable T with parameter λ. Since we modeled the encounter process as Poisson arrivals, no more than one encounter can occur at the same time. Using these assumptions, we represent the UER scheme as a Markov chain, shown in Figure 5.1.

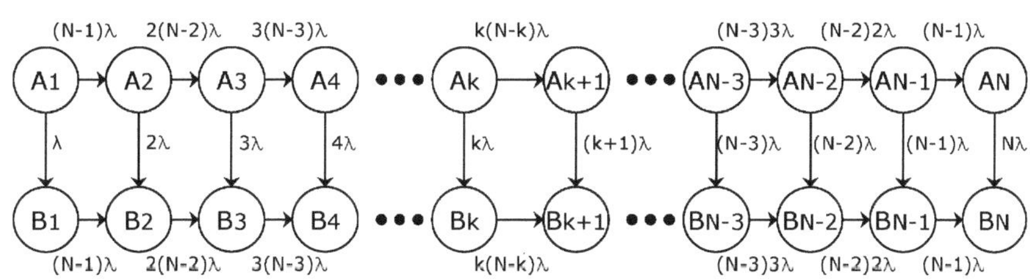

Fig. 5.1 The transition diagram of the UER Markov chain model for number of packet copies

In Figure 5.1, a state A_k represents that there are k copies in the system, but none of them has yet reached the destination node. A state B_k indicates that there are k copies in the system, and at least one of them has reached the destination node. Suppose that there are k copies in the system, then the transition rate from a state A_k to a state B_k is $k\lambda$. Since there are $(N-k)$ nodes that have not received a copy of the packet, the transitional rate from state a k to a state $(k+1)$ is $k(N-k)\lambda$ for both states, A_k and B_k. Using the above argument, we calculate the probability of the system being in a state k state as:

$$\begin{cases} P_k(t) = \int_0^t P_{k-1}(x) \cdot (k-1)\{N-(k-1)\}\lambda \cdot e^{-k(N-k)\lambda(t-x)} dx \quad , \quad (2 \le k \le N) \\ P_1(t) = e^{-(N-1)\lambda \cdot t} \end{cases} \tag{5.1}$$

$$\begin{cases} P_{A,k}(t) = \int_0^t P_{A,k-1}(x) \cdot (k-1)\{N-(k-1)\}\lambda \cdot e^{-k(N-k+1)\lambda(t-x)} dx \quad , \quad (2 \le k \le N) \\ P_{A,1}(t) = e^{-N\lambda \cdot t} \end{cases} \tag{5.2}$$

$$P_{B,k}(t) = P_k(t) - P_{A,k}(t) \tag{5.3}$$

Using these probabilities, we calculate the expected number of copies of the packet in the system and the cumulative distribution function of the time by which the destination node received a copy of the packet.

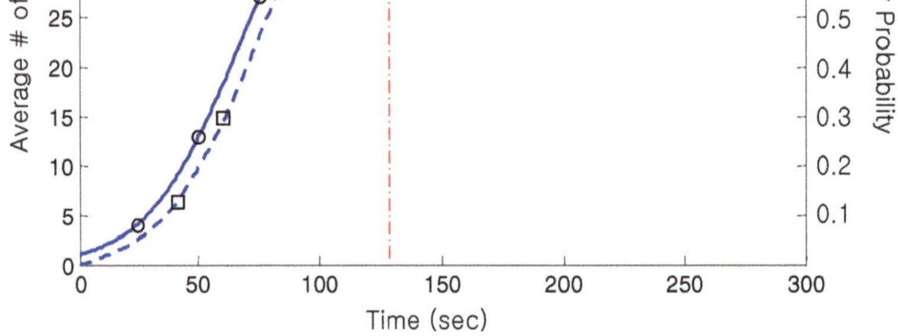

Fig. 5.2 The average number of copies and the delivery probability, versus time

Figure 5.2 shows an example result of the average number of copies in the system and the probability of the destination node having received the packet, versus time. The total number of nodes in the system is $N = 50$, and the total area is a 1000[m] by 1000[m] closed (torus-shape) with the encounter rate λ = 0.4043[events/sec]. After 130[sec] from the time of a packet creation, the average number of total copies of the packet in the system reaches 47 and the average packet delivery probability by this time is 90%.

5.2.2 *The Variance of Energy Consumption in UER*

We measure energy in *Energy Units* [*EU*], and we suppose that the total capacity of a battery of each node is 40[EU], where a single packet transmission requires 1[EU]. Then, with *Time-To-Live* (*TTL*) of a packet set to 130[sec], the average amount of energy used to transmit 47 copies in the system plus the energy to transmit a copy to the destination node is 47[EU]. Suppose a packet is created at an arbitrarily chosen node every 150[sec]. Then, the maximal (ideal) lifetime of the network would be approximately 106[min] (50·40[EU] ·150[sec] / 47[EU] ≈ 6383[sec]).

However, this ideal lifetime holds only if the batteries of all the nodes become depleted at the same time. It is obvious that a source node transmits more copies than other nodes, since it carries the packet longer than any other node. After the source node (referred to here as the "1st node") creates a packet, the second copy is created when the source node encounters another node (2nd node) and transmits a copy of the packet. Now, the third copy can be transmitted either by the 1st node or the 2nd node. Since either node is equally likely to transmit the third copy, the expected number of transmission by both, the 1st node and the 2nd node, is 1/2. Following the same argument, the nth copy of the packet can be transmitted by any of the (n-1) nodes and, thus, the expected numbers of transmissions of each of the nodes equals 1/(n-1). Hence, when there are total of C copies in the system, we can derive the expected value of the total number of transmissions of the nth node (TR_n) as:

$$TR_n = \sum_{j=n}^{C-1} 1/j \qquad (1 \leq n \leq C\text{-}1). \tag{5.4}$$

Eq. (5.4) shows that the energies consumed by the nodes are not equal. During multiple attempts of UER, some nodes will end up having their batteries depleted sooner than other nodes. When a node's battery becomes depleted, there is one less active node for the next UER routing and, thus, all the rates in the Markov chain decrease. Consequently, the packet delivery probability at the destination node decreases as well. In summary, as time goes by, the delivery probability decreases.

5.2.3 Defining the Lifetime of an Epidemic-Routed Network

Figure 5.3 shows the simulation results of the delivery probability during multiple routing attempts of the UER scheme. Included in the figure is also the "ideal expectation" curve, which corresponds to the case when the batteries of all the nodes are equally depleted. In this latter case, all the nodes die simultaneously. The network consists of 50 nodes in a 1000[m] by 1000[m] closed (torus-like) area. The encounter rate is set at $\lambda = 0.004043$[events/sec], TTL is 130[sec], and a packet is created at an arbitrarily chosen node every 150[sec]. The simulation results are averaged over 1000 runs of 100 trials each (for the total of 100,000 trials), and all confidence intervals are 95%. As we have seen in Figure 5.2, with theses parameters, the delivery probability is 90% when all the nodes are active. As the nodes consume their battery energy, the number of active nodes decreases gradually, and after 39 routing attempts the delivery probability at the destination node drops below 80%. The ideal lifetime of the network is 43 routing attempts. (50·40[EU] / 47[EU] = 43), which is equivalent to about 107.5[min] (43·150[sec] = 6450[sec] $\approx$ 107.5[min]). At this time, the UER delivery probability at the destination node is only approximately 60%, which typically would be considered too small to be useful. In UER, even after the ideal lifetime, there are some (active) nodes which are left with some residual energy. However, since a network with so low packet delivery probability would typically be of no benefits for most applications, this residual energy cannot be used and is effectively lost.

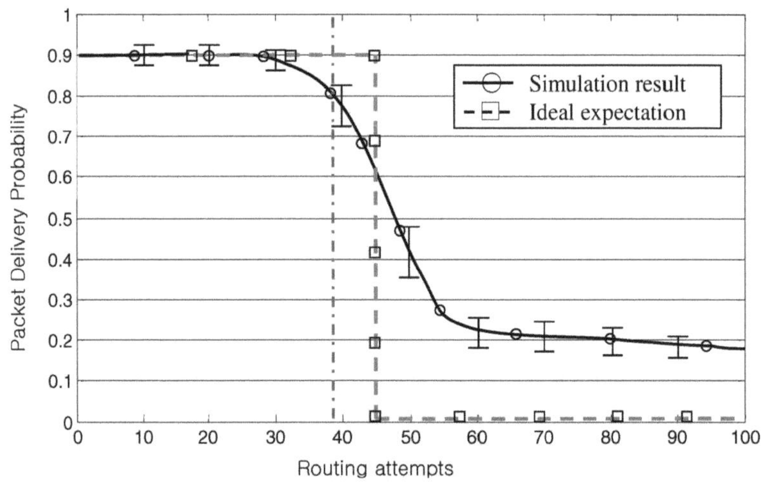

Fig. 5.3 The delivery probability for multiple routing attempts of the UER scheme

In order to define a lifetime of an ER network, we first need to set two delivery probabilities. The first one is the Target Delivery Probability (TDP), which is defined as the average delivery probability (by some time, TTL) when all the nodes are active. The second one is the Minimum Delivery Probability (MDP), which is defined as the threshold of the packet delivery probability below which the network ceases to be useful. Hence, the lifetime of the network (referred to as the "MDP lifetime") is defined as the time period (in units of time or number of packet routing attempts) until the delivery probability at the destination node drops below the minimum delivery probability. In this chapter, for exemplary purpose, we set the TDP to 90% and the MDP to 80%. Figure 5.3, for instance, shows that the MDP lifetime of the UER scheme is approximately 97.5 min (39 ·150[sec] = 5850[sec]). The difference between the ideal lifetime and the MDP lifetime of this UER example network is 10 min (600[sec]).

In the next section, we will introduce three different variations of the UER scheme, which we referred to as the *Restricted Epidemic Routing* (*RER*) schemes: the *Exclusion* (*EX* scheme) scheme, the *Limited Time* (*LT* scheme) scheme, and the *Limited Number of Copies* (*LC* scheme) scheme. We will see how these RER schemes allow increasing the ideal lifetime and the MDP lifetime of a DTN.

5.3 Extending the Lifetime with RER Schemes

In our previous work, we derived the tradeoff functions for several RER schemes to evaluate their efficiency in reducing the number of copies of a packet at the expense of an increase in the delivery delay (for a particular level of the delivery probability). By reducing the number of copies, consequently the total energy consumption for a single routing attempt is reduced as well, and hence one could expect the lifetime of the network to extend too.

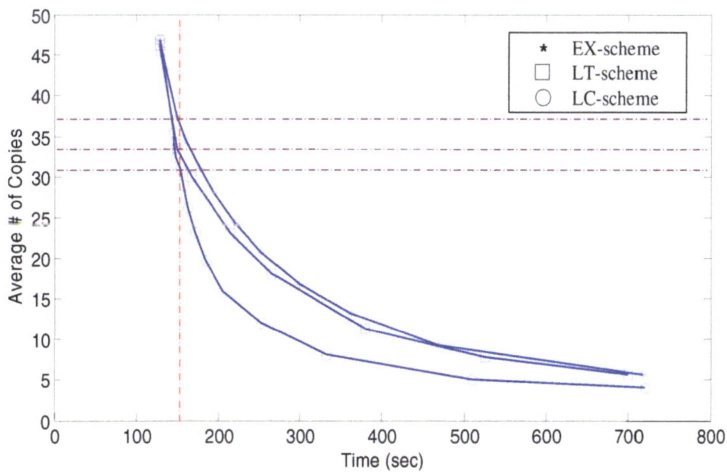

Fig. 5.4 The energy-vs-latency tradeoff of the RER schemes (90% delivery probability)

Three graphs in Figure 5.4 depict for each RER scheme the tradeoff functions between the number of copies and the delivery time delay with 90% delivery probability. From this tradeoff, we can estimate the required number of copies to obtain the required level of TDP (equal to 90% in our exemplary case) within the time of TTL. Suppose that a packet is created every 150[sec] and we do not want to have multiple packets being routed in the system at the same time. Then, the TTL should not exceed 150[sec]. In Figure 5.4 at 150[sec], the points for each scheme where the corresponding vertical lines cross the tradeoff curves are the minimum number of copies required. The required numbers of copies are 31 copies for LC scheme, 33 copies for LT scheme, and 37 copies for EX scheme. These results will be used in the sequel to show how to extend the lifetime for each one of the RER schemes.

5.3.1 The Exclusion Scheme (EX Scheme)

The *Exclusion* (*EX*) scheme excludes some of the nodes from participating in the epidemic routing. Before the source node encounters another node and copies the packet, it decides based on some criteria (e.g., randomly) which nodes are to be excluded from the epidemic routing process. Thus, the EX scheme is merely the UER scheme, except that the total number of nodes being used in the Epidemic Routing is reduced to a certain value of M that is smaller than N. Consequently, the Markov chain model for EX scheme is the same as in Figure 5.1, except that the transition rates from the states and the total number of states decrease.

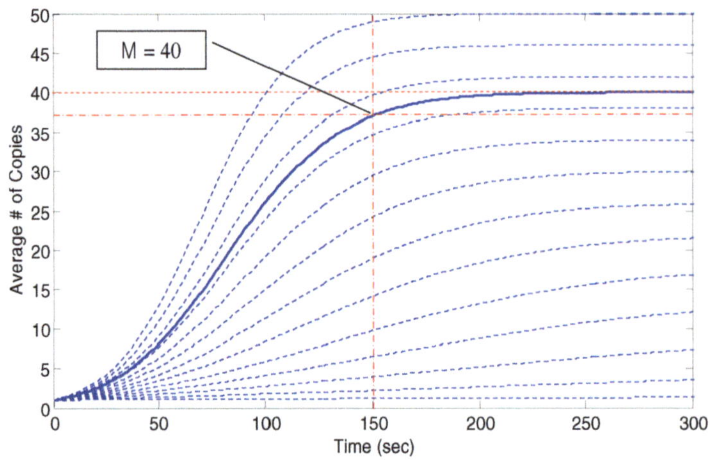

Fig. 5.5 Average number of copies for EX scheme with different values of M

In order to determine the required number of nodes to be used (M) in the EX scheme (as to satisfy the condition that TDP=90%), first we use the Markov chain model for the EX scheme to evaluate the number of copies in the system at

150[sec]. As we have seen at the beginning of this Section, the EX scheme with 37 copies in the system at 150[sec] yields the required TDP level.

Next, Figure 5.5 shows that when the total number of nodes being used is limited to $M = 40$, the average number of copies in the system grows to 37 at 150[sec]. (Note that, as opposed to Figure 5.4, Figure 5.5 shows the dynamic nature of the epidemic routing process.) Hence, in order for the EX scheme to satisfy the TDP=90% requirement, the total number of nodes to be used during the epidemic routing should be set $M = 40$. Assuming that every node has initial energy capacity of 40[EU], since the average number of copies at 150[sec] for the EX scheme is 37, the ideal lifetime of the EX scheme is approximately 135 min ($50 \cdot 40$[EU]$\cdot$150[sec] / 37[EU] $\approx$8108[sec]).

5.3.2 *The Limited Time Scheme (LT Scheme)*

The *Limited-Time* (*LT*) scheme has two different timers. One is the same TTL as in the other RER schemes, and the other (referred to as the *Propagation Time Limit* (*PTL*)) is set to the time until when the nodes that carry a packet are allowed to copy it onto other nodes. Typically, TTL > PTL. After PTL expires, all the nodes just carry the packet until TTL. Of course, a node will transmit its copy of the packet to the destination node at any time that it encounters the destination node. The Markov chain model for LT scheme is exactly the same as in Figure 5.1. Hence, until the time of PTL, the average number of copies in the system will be also the same as in the UER scheme. Between PTL and TTL, the number of copies does not change.

The behavior of the LT scheme makes it simple to find the value of PTL for the LT scheme that satisfies the TDP (=90%) requirement. As discussed at the beginning of the Section 5.3, the LT scheme requires 33 copies in order to satisfy the TDP = 90% requirement. The corresponding PTL is the time when the average number of copies in the system reaches 33, as shown in Figure 5.6.

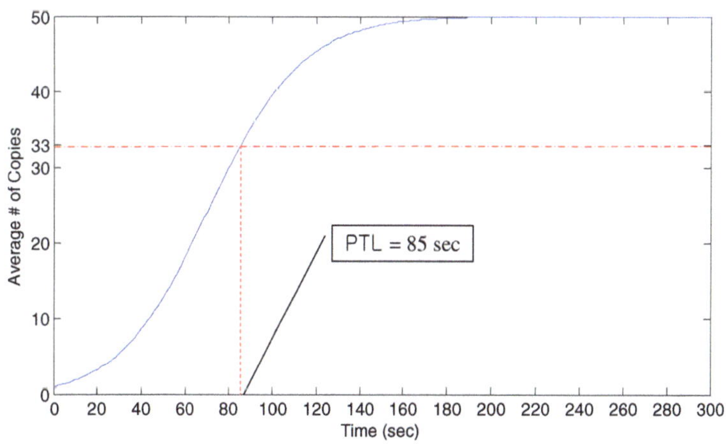

Fig. 5.6 The average number of copies for the LT scheme as a function of time

As one can see in Figure 5.6, the average number of copies in the system becomes 33 at 85[sec]. Therefore, the PTL should be set to 85[sec]. Assuming that the initial energy capacity of every node in the network is 40[EU], since the average number of copies at 150[sec] for the LT scheme is 33, the ideal lifetime of the LT scheme is approximately 151.5[min] (50·40[EU]·150[sec] / 33[EU] ≈9091[sec]).

5.3.3 The Limited Number of Copies Scheme (LC Scheme)

The LC scheme restricts the Epidemic Routing by limiting the total number of copies that can be transmitted to other nodes in the system. In order to limit the total number of copies, each copy of a packet is associated with a number, which we refer to as permits. A "permit" is a right to transmit a copy of the packet to another node and each transmission deletes one permit from the transmitting node. Permits can be forwarded with a packet to a receiving node. Thus, if the source node sets the number of permits to m, then upon encounter of another node, the source can transmit a copy of the packet (thus decreasing the number of remaining permits to m-1), in addition to forwarding some, all, or none of the remaining m-1 permits to the other node.

For expositional purposes, it is easier to refer to permits as copies of the packet. Thus, when a node forwards p permits to another node, we will refer to as the node transmitting p (additional) copies of the packet onto the other node. However, it is understood that a packet is transmitted only once. Thus, for example, if a node with k copies encounters another node, it transmits a copy of the packet onto the other node, thus reducing the number of its copies to k-1. The transmitting node then can forward onto the other node any number between 0 and k-1 additional copies, which the receiving node can use to transmit copies to other encounter nodes, or to forward some of the "unused" copies to nodes onto which it had transmitted a copy of the packet already.

The LC scheme is substantially different from the UER scheme and, consequently, another approach is needed to analyze the LC scheme. We use a 2-dimensional Markov chain to model the number of nodes in the system which received (at least one) copy of the packet. Figure 5.7 is an example of such a 2-dimensional Markov chain for the LC in which the total number of copies in the system is limited to 12. This is accomplished by initially setting the parameter m to 12.

In Figure 5.7, the x- and the y-axis indicate, respectively, the number of nodes in the system which received a copy (i.e., at least one copy) and the number of nodes that are still able to transmit a copy of the of the packet onto another node (i.e., such a node needs to have at least two copies of the packet). A state in the Markov chain in Figure 5.7 is labeled with a sequence of numbers, separated by the signs "/". Each position in the sequence indicates a node which carries at least one copy of the packet, and the actual number in the position indicates the number of copies that such a node carries. For instance, in the state [6/2/2/1/1], there are 5 nodes carrying at least one copy of the packet: one of the nodes has 6 copies, which means it still can propagate 5 copies onto other nodes; two other nodes, each with 2 copies, which means that each of these nodes can copy the packet onto one other node each; and two nodes with just one copy, meaning that these two nodes cannot transmit onto other nodes anymore. The next state is determined by the number of nodes carrying more than one copy and the assumption that each node has the same probability of encountering another node. Using the probability of being in a particular state, we can derive the average number of nodes that can still transmit onto other nodes (n_k), when there are k nodes with copies in the system. Eventually, we translate this 2-dimensional Markov chain into a Markov chain model similar to the one in Figure 5.1 (with two rows of m states each). The transitional rates from a state k to a state (k+1) now becomes $n_k (N\text{-}k)\lambda$ for both states, A_k and B_k.

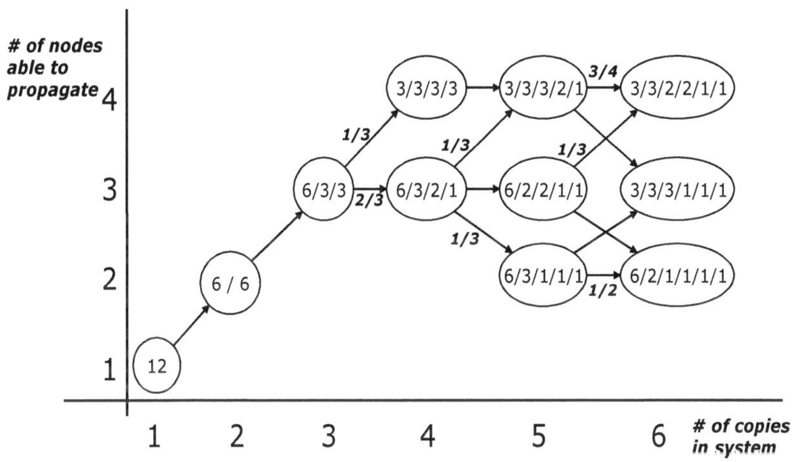

Fig. 5.7 The 2-D transition diagram of Markov chain model for LC scheme (m=12)

In order to determine the required value of the parameter m that limits the total number of copies in the LC scheme, while satisfying the TDP =90% condition, we use a similar method to the one we used in determining the value of M for the EX scheme. As stated at the beginning of Section 5.3, this TDP condition is met for the LC scheme when there are 31 copies at 150[sec] in the LC scheme (this value is derived with the assistance of the above 2-D Markov chain model).

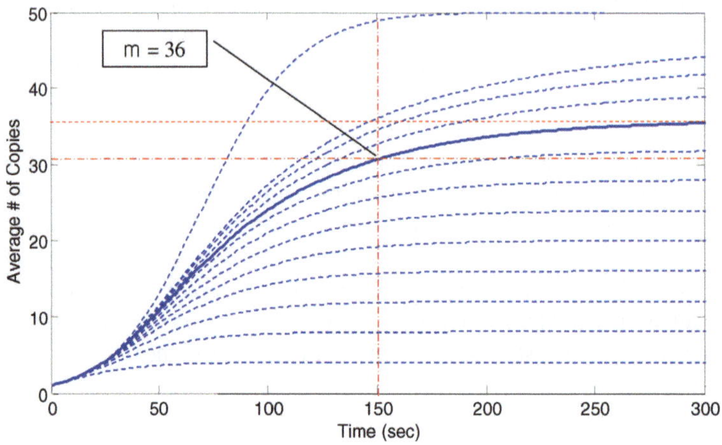

Fig. 5.8 The average number of copies for the LC scheme with different values of *m*

Figure 5.8 shows that when the total number of copies is set to m = 36, the average number of copies in the LC scheme becomes 31 at 150[sec]. In summary, in order for the LC scheme to satisfy the TDP (=90%) condition, the total number of copies that can be propagated in the system should be limited to 36 copies. Assuming that the initial energy of the batteries of the nodes is 40[EU], and since the average number of copies at 150[sec] in the LC scheme is 31, the ideal lifetime of this LC scheme is approximately 161[min] (50·40[EU]·150[sec] / 31[EU] ≈9677[sec]).

5.3.4 The MDP Lifetime of the RER Schemes

So far, we have considered how the RER schemes can extend the ideal lifetime of a DTN. Figure 5.9 depicts the simulation result of the MDP lifetime of the LC scheme, together with the scheme's ideal lifetime for comparison. The simulation results were averaged over 1000 runs of 100 trials each (for the total of 100,000 trials), and all confidence intervals are 95%.

According to Figure 5.9, the MDP lifetime of LC scheme is approximately 130[min] (52·150[sec] = 7800[sec]), which is 32.5[min] longer than the MDP lifetime of the UER scheme. We have seen that the ideal lifetime of the LC scheme is approximately 161[min], which is about 55[min] longer than the ideal lifetime of the UER scheme. Clearly, the LC scheme extended both the ideal lifetime and the MDP lifetime, as compared with the UER scheme. Nevertheless, the gap between the ideal lifetime and the MDP lifetime of the LC scheme increased by 22.5 min (from 55[min] to 32.5[min]). The LT scheme and the EX scheme also show a similar trend; i.e., both RER schemes extend the ideal lifetime and the MDPE lifetime as compared with the UER scheme, although the differences between the ideal lifetime and the MDP lifetime increase as well.

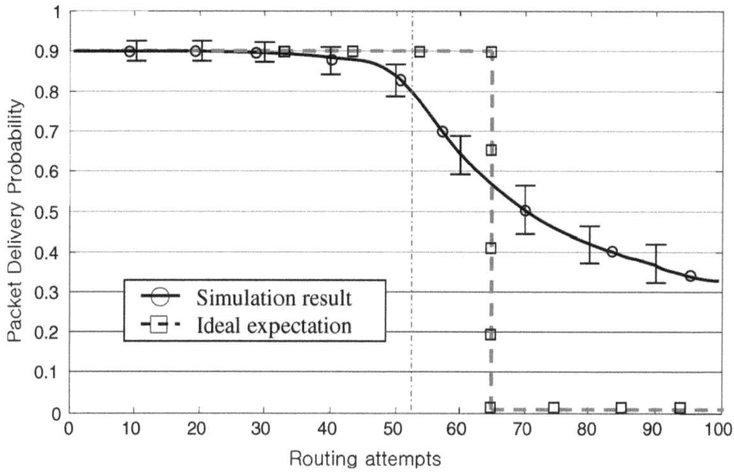

Fig. 5.9 The delivery probability for multiple-routing attempts for the LC scheme

5.4 Exploiting Residual Battery Information

In general, the performance of Epidemic Routing depends mostly on the number of active nodes in the system. Hence, decrease in the number of active nodes results in poorer performance, which means a drop in the delivery probability. The ideal lifetime of a routing scheme is possible only when all nodes' batteries become depleted at the same time, which practically is impossible to achieve. However, it is possible to control the nodes' energy consumption to some degree, as to come closer to this goal.

Suppose two nodes encounter and one node transmits its packet to the other node. By sharing the information regarding their residual battery energy, the nodes can decide how many packets should be exchanged. Moreover, when the nodes transmit not only the packets, but also permits to replicate the packet in the future, it would make sense for the node with larger residual battery energy to accept more permits. Thus, the residual energy information could serve as an indicator for current and for future energy use. For example, in the LC scheme, when a node transmits a packet copy to another node, these two nodes can control the number of permits that should be transferred from the transmitting to the receiving node. Indeed, the LC scheme can easily exploit the residual energy information in its operation, allowing overall improvement in the network lifetime. It is worth noticing that it is more difficult to use the residual energy information in the other RER schemes.

5.4.1 The LC Scheme with Residual Battery Information

Normally in LC scheme, when a node transmits a copy of a packet to another node, it divides its permit load, passing half of the permit load onto the receiving node. In that way, the two nodes end up with the same number of permits to replicate the packet in the future. However, if the nodes are aware of their residual battery energy information, they can divide the number of permits according to their residual battery energy, instead. A simple way to divide the permit load is to split the number of permits in proportion to the residual battery energy. As a result, the node with more residual battery energy will end up generating and transmitting more copies of the packet. We refer to this scheme as the *Limited number of copies with Battery information* (*LCB*) scheme.

Since the LCB scheme is a variant of the LC scheme, in order to find the value of the parameter m that limits the total number of copies for a single-packet routing, we use a very similar method that we used for calculation of the parameter m in the LC scheme. Hence, to achieve the TDP=90%, the total number of copies of a packet that can be generated in the system should be set to 36 copies, where the average number of copies at TTL (=150[sec]) is 31. Similarly to the LC scheme, the ideal lifetime of the LCB scheme is also approximately 161[min] (50·40[EU]·150[sec] / 31[EU] ≈9677[sec]).

5.5 Simulation Results and Performance Comparison

The simulations were performed for a 1000[m] by 1000[m] closed (torus-like) area, with N=50 mobile nodes plus one destination node. The destination node was stationary and placed in the middle of the area. The transmission range of all the nodes was set to 25[m]. The direction and the velocity of each mobile node were uniformly distributed random variables, with the direction being distributed from 0° to 360° and the velocity from 20[m/s] to 50[m/s]. These settings resulted in the encounter rate λ of 0.004043. A node was chosen arbitrarily to create a new packet every 150[sec]. The lifetime of the network was defined by setting TDP to 90% and MDP set to 80%. In order to satisfy the TDP, the TTL parameter of the packets in the UER scheme was set to 130[sec], while for the other RER schemes the TTL parameter was set to 150[sec]. The initial battery capacity of every node was 40 [EU] at the start of every simulation run. The restricting parameters for the RER schemes were set based on the derivations in Section 5.3.

5.5.1 The MDP Lifetime

Figure 5.10 shows the simulation results of the packet delivery probability as a function of the number of routing attempts. The red vertical lines indicate the corresponding MDP lifetimes of each scheme. The packet delivery probabilities for all the schemes are initially at the TDP level of 90% from 0 [sec] to 3000 [sec]. After 4500[sec], the packet delivery probability of the UER scheme starts to decrease and the rate drops below the MDP at approximately 5850[sec]. As can be observed, the EX scheme has the shortest MDP lifetime from among the RER schemes, followed by the LT scheme, and then the LC scheme. The LCB scheme has the longest MDP lifetime.

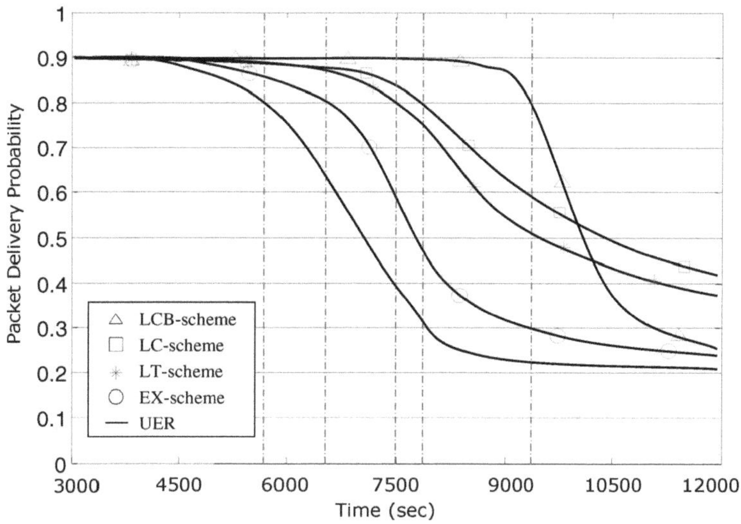

Fig. 5.10 The lifetimes of the UER and the RER schemes

The LC scheme has longer MDP lifetime than the LT scheme and the EX scheme since it transmits fewer copies per packet routing. Interestingly, even though the LCB scheme is very similar to the LC scheme, the LCB scheme results in much longer MDP lifetime than the LC scheme. Another finding is that the delivery probability of the LCB scheme decreases sharply after it drops below the MDP, and that then the delivery probability becomes lower than the other schemes. In this sense, the LCB scheme is a better approximation to the concept of ideal lifetime than the other schemes. Indeed, for the LCB scheme, the nodes consume almost equivalent amount of energy, most of the batteries are depleted close to the MDP lifetime, and after this time most of the nodes become inactive.

5.5.2 Evaluation of Lifetime for Each Scheme

The MDP lifetimes are listed in Table 5.1 and compared with the ideal lifetime for each scheme. The difference between these two lifetimes for each scheme is also shown in the table. E(n) indicates the average number of transmission per node, and σ(n) indicates the standard deviation of the number of transmission per node. The last column is the coefficient of variation values calculated as σ(n) / E(n).

The results in the table demonstrate that the LCB scheme has the longest MDP lifetime, and that its MDP lifetime is closer to its ideal lifetime than any other scheme. With the exception of the LCB scheme, although the rest of the RER schemes have longer lifetime than the lifetime of the UER scheme, the difference between the ideal lifetime and the MDP lifetime of each one of these RER schemes is larger than the corresponding difference of the UER case. Results for E(n) shows that small number of transmission per node increases the lifetime due to smaller consumption of battery energy. However, the results in the table also

Table 5.1 The ideal lifetimes and MDP lifetimes for the considered routing schemes

Scheme	Ideal lifetime (α)	MDP lifetime (β)	α-β	E(n)	σ(n)	σ(n) /E(n)
UER	6383[sec]	5850[sec]	533[s]	0.93	1.360	1.462
EX	8108[sec]	6450[sec]	1658[s]	0.71	1.206	1.699
LT	9091[sec]	7500[sec]	1591[s]	0.66	1.187	1.798
LC	9677[sec]	7800[sec]	1877[s]	0.61	1.213	1.988
LCB	9677[sec]	9250[sec]	427[s]	0.61	0.904	1.482

demonstrate that reducing the coefficient of variation of the number of transmission per node tends to bring the MDP lifetime closer to the corresponding ideal lifetime.

As a final remark, we point out that we expect the above trends to be significantly stronger when the network parameters, such as for example the mobility patterns of the nodes, are more heterogeneous than in the cases that we simulated here.

5.6 Conclusion

In this chapter, we first discussed various studies on energy-efficient routing in Delay-Tolerant Networks (DTNs), pointing out that although the Epidemic Routing scheme allows for fastest delivery of a packet to its destination, it is also the most wasteful in energy. We then introduced various Restricted Epidemic Routing (RER) schemes, whose purpose is to trade off latency for energy. We modeled a DTN as a Markov Chain and evaluated its performance as a function of time. This transitional solution allows us to derive the limiting parameters for the RER schemes and to calculate the respective lifetimes of those schemes.

We pointed out that most of the studies on energy-efficient DTN routing focused on single-packet routing, following the understanding that minimizing the energy required for a single-packet routing would correspond to overall energy-efficient routing as well and, thus, extend the overall network lifetime. However, our main observation has been that the lifetime of a network is also strongly influenced by the way that the nodes deplete their batteries. For example, if the energy is unevenly depleted from the network nodes, some nodes will die quickly, resulting in premature inability of the network to route new packets, and thus in a short network lifetime. Therefore, optimizing the energy of single-packet routing would not, in general, result in maximal network lifetime.

To study the effect of multi-packet routing on the network lifetime, we introduced the Minimum Delivery Probability (MDP) parameter, which represents the lower bound on the acceptable performance of the network, below which the network could be considered useless. We then computed the MDP lifetimes of the various RER schemes, which we defined as the time duration until when the delivery probability edges below the MDP.

We also defined the notion of the Ideal Lifetime, which is the maximum lifetime of a network running a particular routing protocol. The Ideal Lifetime occurs when all the nodes run out of energy at the same time.

In an attempt to "emulate" the ideal lifetime, we modified the Limited Number of Copies scheme to include a provision by which nodes with larger amount of residual energy transmit more often than nodes with lesser residual energy. We termed this scheme the LCB (Limited number of Copies with Battery information) scheme.

An interesting result was that the Exclusion scheme, the Limited Time scheme, and Limited Number of Copies scheme did not extend the MDP lifetime as much as they extended their corresponding ideal lifetimes. The difference between the MDP lifetime and the ideal lifetime was even larger than that of the Unrestricted Epidemic Routing scheme. However, the LCB scheme outperformed all the other schemes in terms of both, the ideal lifetime and the MDP lifetime.

Based on these results, we postulate that the average number of transmissions alone is not the only factor in extending the network lifetime. Rather, we speculate that decreasing the variance of the number of transmissions across the network nodes and across multiple routings may play a crucial role in maximizing the network lifetime. Indeed, use of the information of residual battery energy is one such a method that decreases this variance. A future direction of this work could be to study the extension of the network lifetime by examining how the variance of the number of transmission could be minimized through other approaches and for other routing schemes that were not discussed in this chapter.

Acknowledgement

This work has been supported in part by the grants from the National Science Foundation numbers ANI-0329905 and CNS-0626751, by the AFOSR contract number FA9550-09-10121, and by a grant from The Boeing Company.

References

[1] Vahdat, A., Becker, D.: Epidemic Routing for Partially- Connected Ad Hoc Networks. Technical Report, Duke University (April 2000)

[2] Small, T., Haas, Z.J.: The Shared Wireless Infostation Model: A New Ad Hoc Networking Paradigm (or Where there is a Whale there is a Way). In: MobiHoc, Annapolis, Maryland (June 1-3, 2003)

[3] Haas, Z.J., Small, T.: A New Networking Model for Biological Applications of Ad Hoc Sensor Networks. IEEE/ACM Trasnactions on Networking 14(1) (February 2006)

[4] Spyropoulos, T., Psounis, K., Raghavendra, C.S.: Spray and Wait: An Efficient Routing Scheme for Intermittently Connected Mobile Networks. In: ACM SIGCOMM Workshop on Delay Tolerant Networking, Philadelphia, PA, (August 22-26, 2005)

[5] Yoon, S., Haas, Z.J.: Efficient Tradeoff of Restricted Epidemic Routing in Mobile Ad Hoc Networks. In: IEEE MILCOM, Orlando, Florida (October 29-31, 2007)

[6] Small, T., Haas, Z.J.: Resource and Performance Tradeoffs. In: ACM SIGCOMM, Philadelphia, Pennsylvania (August 26, 2005)
[7] Neglia, G., Zhang, X.: Optimal Delay-Power Tradeoff in Sparse Delay Tolerant Networks: a preliminary study. In: ACM SIGCOMM Workshop on Challenged Networks (CHANTS), Pisa, Italy (September 15,2006)
[8] Spyropoulos, T., Psounis, K., Raghavendra, C.S.: Spray and Focus: Efficient Mobility-Assisted Routing for Heterogeneous and Correlated Mobility. In: PerCom Workshop on Intermittently Connected Mobile Ad Hoc Networks (ICMAN), White Plains, NY (March 2007)
[9] Ramanathan, R., Hansen, R., Basu, P., Rosales-Hain, R., Krishnan, R.: Prioritized Epidemic Routing for Opportunistic Networks. In: ACM/SIGMOBILE First International Workshop on Mobile Opportunistic Networking (MobiOpp), Puerto Rico (June 11, 2007)
[10] Shah, R.C., Rabaey, J.M.: Energy Aware Routing for Low Energy Ad Hoc Sensor Networks. In: IEEE Wireless Communications and Networking Conference, Orlando, Florida, March 17-21 (2002)
[11] Chang, J., Tassiulas, L.: Energy Conserving Routing in Wireless Ad-hoc Networks. In: INFOCOM, Conference on Computer Communications, Tel Aviv, Israel, March 26-30 (2000)
[12] Maleki, M., Dantu, K., Pedram, M.: Power-aware Source Routing Protocol for Mobile Ad Hoc Networks. In: International Symposium on Low Power Electronics and Design, Monterey, California, August 12-14 (2002)
[13] Haas, Z.J., Halpern, J.Y., Li, L.: Gossip-based Ad Hoc Routing. In: IEEE INFOCOM, New York, NY (June 23, 2002)
[14] Ramanathan, P., Singh, A.: Delay differentiated gossiping in delay tolerant networks. In: Proceedings of International Conference on Communications, pp. 3291–3295 (2008)
[15] Lindgren, A., Doria, A., Scheln, O.: Probabilistic routing in intermittently connected networks. In: The Fourth ACM International Symposium on Mobile Ad Hoc Networking and Computing (2003)
[16] Nelson, S.C., Bakht, M., Kravets, R.: Encounter-based routing in DTNs. In: Proceedings of the IEEE INFOCOM, Rio de Janerio, Brazil (2009)
[17] Yoon, S.K., Haas, Z., Kim, J.H.: Tradeoff between Energy Consumption and Lifetime in Delay Tolerant Mobile Network. In: IEEE Milcom, San Diego, CA (November 2008)
[18] Balasubramanian, A., Levine, B.N., Venkataramani, A.: Replication routing in DTNs: A resource allocation approach. IEEE/ACM Transactions on Networking 18(2), 596–609 (2010)
[19] Banerjee, N., Corner, M.D., Levine, B.N.: Design and field experimentation of an energy-efficient architecture for DTN throwboxes. IEEE/ACM Transactions on Networking 18(2), 554–567 (2010)
[20] Leguay, J., Conan, V., Friedman, T.: Evaluating MobySpace-based routing strategies in delay-tolerant networks. Wireless Communications and Mobile Computing (May 2007)
[21] Wang, Y., Jain, S., Martonosi, M., Fall, K.: Erasure-Coding Based for Opportunistic Networks. In: ACM SIGCOMM, Workshop on Delay Tolerant Networking and Related Topics, WDTN 2005 (August 2005)
[22] Lin, Y., Liang, B., Li, B.: Performance Modeling of Network Coding in Epidemic Routing. In: 1st International MobiSys Workshop on Mobile Opportunistic Networking, San Juan, Puerto Rico (June 11, 2007)

Chapter 6
Relay Selection Strategies for Green Communications

Nicholas Bonello

Abstract. This chapter presents a microeconomic-based 'green' radio framework, in which a network operator contracts with one of its inactive end-users, with the specific intention of reducing the transmitted power of its source nodes and the costs incurred in providing the service. The proposed approach bears comparison with a tender process, where each inactive end-user submits a tender to the operator, describing the energy- and cost-savings that may be provided if selected to relay data to the intended destination. The operator will then associate additional measures to each submitted tender that quantify the downsides which may result from cooperating with that specific end-user. Finally, there is the tender evaluation and selection procedure, which may be compared with solving a multi-objective optimization problem. Our results disclose that the proposed framework can attain up to 68% energy savings and up to 65% cost savings over the direct transmission benchmarker.

6.1 Introduction

The global concern on environmental issues has lately triggered a growing interest in perceiving environmentally conscious solutions for the wireless sector, which must materialize in a reality dominated by an ever-increasing drive for substantial cost-reductions and demands for high transmission-rate services. Along these lines, the focus of this chapter lies on a perfectly competitive commercial firm, which operates a network to provide diverse wireless communications services to its subscribers. It is assumed that at a specific instant, an end-user d requests a service necessitating R bps/Hz. With the ever-increasing mobile penetration rates evidenced in our current times, it is realistic to conjecture that at the same instant, a significant proportion of the operator's clientele is inactive and owns a somewhat similar

Nicholas Bonello
University of Sheffield, United Kingdom
e-mail: `NBonello1@sheffield.ac.uk`

technology to that maintained by the intended destination node d. The network operator is considering shifting some of its responsibilities towards its customer d to one of its inactive end-users, with the specific intention to reduce its carbon footprint by minimizing the transmit power at its source nodes, thus reducing the energy consumption, and to minimize the costs incurred in delivering the service. An inactive end-user may strategically be at a better position to offer the same type of service than the operator itself, due to reasons such as a superior channel quality, closer proximity to the destination node, etc.

The concept of 'shifting responsibilities' may be compared to that of cooperation, wherein a relaying end-user is allocated part of the responsibilities of the network operator; i.e., the transfer of data. Recent work such as [1, 2] investigated the use of cooperation between mobile terminals in the deployment of energy-efficient wireless communication systems, which have lately been grouped under the umbrella of the so-called 'green' radio frameworks [3]. Power-minimization techniques have also attracted the attention of many researchers, for instance in the context of wireless sensor networks, thereby forming a body of literature that is certainly too large to be reviewed thoroughly in a single paragraph. As it was mentioned in [4], the distinctive feature of this prior work lies in its emphasis on the uplink communications, thus addressing the designs for energy-efficient schemes for the battery-powered nodes whilst ignoring the more resource abundant central node(s). On the other hand, in 'green' radio frameworks, it is the network operator with its high energy-consuming source nodes that steal the limelight, which essentially forces a shift in our focus from the uplink to the downlink communications. Power-minimization in our 'green' radio framework is discussed in Section 6.3, following a brief description of the underlying transmission schemes in Section 6.2.

In this chapter, cooperation between the source nodes and the end-users is treated as being an instance of outsourcing [5, 6]. The process of outsourcing derives its origin from the classical economic principle of 'division of labour', whereby an organization contracts with a third-party service provider to fulfill some of its responsibilities. However, an investigation of practical outsourcing mechanisms for 'green' communications certainly cannot stop on the energy-savings issues but must also address the inherent economic implications. In all respects, one may hypothesize that the operator's concerns on environmental issues are undoubtedly not wholly altruistic, since energy use effectively translates into costs [7]. It is therefore understandable that any revenue-seeking network operator will be reluctant to adopt any 'green' strategy if this entails a substantial increase in the costs incurred. On the other hand, one cannot expect a node to relay data to the intended destination without receiving any financial incentive. After all, the relay node under contract will undoubtedly have to bear certain expenses to provide its services to the network operator. These issues will be tackled in Section 6.4, by employing techniques similar to those used in cost-minimization problems in microeconomics.

Our approach assumes that each potential relay node will be submitting a tender to the operator, describing the energy- and cost-savings (i.e., the accrued benefits) if contracted. To each submitted tender, the operator will associate additional measures that quantify the downsides that may result from outsourcing to that specific node. Our attention will be focussed on two potential disadvantages. Firstly, it is argued that in contrast to direct transmission, the operator is now dependent on a third-party which may or may not deliver the savings pledged in the submitted tender. Accordingly, Section 6.6 derives a criterion that quantifies the risk undertaken by the operator in trusting a particular node. Secondly, outsourcing to a node may inflict interference on other nearby active end-users as discussed in Section 6.7. Subsequently, Section 6.8 describes the tender evaluation and selection procedure, by which the operator will determine the winning tender—from the set of Pareto-optimal solutions—that strikes the best compromise between the benefits and the potential drawbacks of outsourcing. The results presented in Section 6.9 disclose up to 68% energy savings and up to 65% cost savings over the direct transmission scenario (i.e., no outsourcing), whilst minimizing the risk taken as well as the interference inflicted on other active end-users in the network.

6.2 System Model and Transmission Schemes

The wireless network is modeled by the standard directed graph model (cf. [8], amongst others) described by $G = \{N, L\}$, containing a finite nonempty set $N = S \cup M$ representing the nodes (corresponding to the vertices of G), and a collection L of ordered pairs of distinct nodes describing the directed links (identifying the edges of G). The set of nodes N can be further decomposed into two separate finite and nonempty sets: the set $S \subset N$ of stationary source nodes, which are operated and maintained by a revenue-seeking network operator, and the set $M \subset N$ of mobile terminals exploited by the network subscribers. A directed link $(a, b) \in L$ between a pair of adjacent nodes $\{a, b\} \in N$ describes the outgoing link from node a or equivalently, the incoming link to node b. The parameter $h_{(a,b)}$ denotes the channel gain associated with each directed link $(a, b) \in L$.

It is assumed that at a specific instant, an end-user operating node $d \in M$ requests a specific service from a source node $i \in S$. Assuming direct transmission of $P_{(i,d)}$ watts on link $(i, d) \in L$, the spectral efficiency (in bps/Hz) experienced by the intended destination node d is given by [9]

$$R_{(i,d)} = \log_2 \left(1 + \frac{h_{(i,d)} P_{(i,d)}}{\Gamma N_{(i,d)}}\right), \tag{6.1}$$

where Γ represents the signal-to-noise ratio (SNR) gap to capacity, which depends on the specific modulation scheme and the desired bit error ratio (BER), whilst $N_{(i,d)}$ represents the total interference and the additive white Gaussian noise (AWGN) on link (i, d). The term $h_{(i,d)}/\Gamma N_{(i,d)}$ will be replaced by $\alpha_{(i,d)}$, in order to simplify the notation. Alternatively, source node i may opt to involve one of the inactive end-users $j \in M$, $j \neq d$, in the provision of the same service of $R_{(i,d)}$ bps/Hz to the intended destination node d. A standard two-phase cooperative protocol [10] will be considered, wherein the first phase, the source node selectively multicasts data to a previously inactive end-user j and to the intended destination node d. The spectral efficiencies for nodes $\{j, d\} \in M$ during this first phase of the cooperation are formulated by $\tilde{R}_{(i,j)|d \in M} = \log_2(1 + \alpha_{(i,j)}\tilde{P}_{(i,d)})/2$ and $\tilde{R}_{(i,d)} = \log_2(1 + \alpha_{(i,d)}\tilde{P}_{(i,d)})/2$, where $\tilde{P}_{(i,d)}$ denotes the transmitted power by the source node during the first phase of cooperation. Node j will then decode the data received from source node i during the first cooperative phase and then forwards the data to the original destination node d during the second cooperative phase. Subsequently, node d will combine the two data streams received during the two ensuing cooperative phases using maximum ratio combining, thus attaining

$$R_{(i,d)|j \in M} = \frac{1}{2} \log_2 \left(1 + \alpha_{(i,d)}\tilde{P}_{(i,d)} + \alpha_{(j,d)}P_{(j,d)}\right), \tag{6.2}$$

where $P_{(j,d)}$ is the transmit power used by the relaying end-user during the second cooperative phase. The one-half multiplicative term for $\tilde{R}_{(i,j)|d \in M}$, $\tilde{R}_{(i,d)}$ and $R_{(i,d)|j \in M}$ results from the fact that the spectral efficiencies in question are realized after two time slots, where each time slot corresponds to one phase of the cooperative protocol. For notational convenience, subscripts will be omitted and the spectral efficiency experienced by the end-user d will be denoted by R, which can either be achieved by cooperating with another end-user j—thus obtaining $R = R_{(i,d)|j \in M}$—or else via direct transmission, in which case $R = R_{(i,d)}$.

6.3 The 'Green' Solution

This section addresses one facet of the proposed 'green' radio framework, which is that of power (and thus energy) minimization. In this regard, the aim of the network operator is to contract that particular relay node j—from its pool of inactive end-users—that minimizes the power vector $\mathbf{P} = [\tilde{P}_{(i,d)}, P_{(j,d)}]$, $\mathbf{P} \in \mathbb{R}^2_{++}$, where $\mathbb{R}_{++}$ denotes the set of positive real numbers. With the aid of the equations for $\tilde{R}_{(i,j)|d \in M}$, $\tilde{R}_{(i,d)}$ and $R_{(i,d)|j \in M}$, one can formulate a linear program as follows:

$$\begin{aligned} \min_{\{\mathbf{P}\}} \quad & \tilde{P}_{(i,d)} + P_{(j,d)} \\ \text{s.t.} \quad & \frac{\tilde{P}_{(i,d)} + P_{(j,d)}}{2} < \frac{2^R - 1}{\alpha_{(i,d)}}, \\ & \tilde{P}_{(i,d)} \geq \frac{2^{2R} - 1}{\alpha_{(i,j)}}, \\ & P^{\min}_{(j,d)} \leq P_{(j,d)} \leq P^{\max}_{(j,d)}, \\ & \alpha_{(i,d)}\tilde{P}_{(i,d)} + \alpha_{(j,d)}P_{(j,d)} = 2^{2R} - 1, \end{aligned} \tag{6.3}$$

where $P^{\min}_{(j,d)}$ and $P^{\max}_{(j,d)}$ represent the minimum and maximum power for the relaying node, respectively. Let $f_0(\mathbf{P})$ denote the objective function of (6.3) with optimal value $f_0(\mathbf{P}^*)$, which will be referred to as the 'green' solution in the forthcoming discourse. Furthermore, it is assumed that each of the currently inactive node $j \in M$ can be uniquely and unambiguously identified by the pair $(\alpha_{(i,j)}, \alpha_{(j,d)})$. Clearly, the network operator would not like to expend any efforts on those currently inactive nodes that cannot provide a feasible solution $\mathbf{P} \in D_0$, where D_0 represents our feasible region. This necessitates the identification of a number of constraints to be imposed on $\alpha_{(i,j)}$ and $\alpha_{(j,d)}$—constituting our so-called initial relay selection rules—which enable the quick exclusion of those nodes that cannot contribute to any energy-savings over the scenario without outsourcing. These rules can be derived by first reformulating the equality constraint of (6.3) to

$$\tilde{P}_{(i,d)} = \frac{2^{2R} - 1}{\alpha_{(i,d)}} - \frac{\alpha_{(j,d)}}{\alpha_{(i,d)}} P_{(j,d)}. \tag{6.4}$$

Subsequently, by comparing (6.4) with the first constraint of (6.3) and then replacing $P_{(j,d)}$ with $P^{\max}_{(j,d)}$, we obtain

$$\frac{2^{2R} - 1 - \alpha_{(j,d)} P^{\max}_{(j,d)}}{\alpha_{(i,d)}} < \frac{2(2^R - 1)}{\alpha_{(i,d)}} - P^{\max}_{(j,d)}. \tag{6.5}$$

Therefore, given R and a source-to-destination channel gain value[1] $\alpha_{(i,d)}$, one can eliminate all currently inactive nodes that violate the constraint

$$\alpha_{(j,d)} > \frac{2^R(2^R - 2) + 1 + \alpha_{(i,d)} P^{\max}_{(j,d)}}{P^{\max}_{(j,d)}}. \tag{6.6}$$

Examining the intersection point of (6.4) with the first and second constraints of (6.3), one can derive a strict lower bound value on $\alpha_{(i,j)}$, which is formulated by

$$\alpha_{(i,j)} > \frac{\alpha_{(i,d)}(2^{2R} - 1)(\alpha_{(j,d)} - \alpha_{(i,d)})}{2\alpha_{(j,d)}(2^R - 1) - \alpha_{(i,d)}(2^{2R} - 1)}, \tag{6.7}$$

given that the corresponding $\alpha_{(j,d)}$ satisfies $\alpha_{(j,d)} > 2\alpha_{(i,d)}(2^R - 1)/(2^{2R} - 1)$. If $\alpha_{(j,d)}$ is strictly smaller than the right-hand side of the latter inequality, then the inequality represented in (6.7) will provide a strict upper bound value on $\alpha_{(i,j)}$ (due to the sense reversal). Additionally, replacing $P_{(j,d)}$ in (6.4) by $P^{\min}_{(j,d)}$ and exploiting the second inequality constraint of (6.3) gives

$$\alpha_{(i,j)} \geq \frac{(2^{2R} - 1)\alpha_{(i,d)}}{2^{2R} - 1 - \alpha_{(j,d)} P^{\min}_{(j,d)}}, \tag{6.8}$$

for all corresponding values of

[1] We recall that the channel gain value $\alpha_{(a,b)}, \forall\, (a, b) \in L$, subsumed the term $\Gamma N_{(i,d)}$ (cf. Section 6.2).

$$\alpha_{(j,d)} \in \left(\frac{2^R(2^R-2)+1+\alpha_{(i,d)}P^{\min}_{(j,d)}}{P^{\min}_{(j,d)}}, \frac{(2^{2R}-1)}{P^{\min}_{(j,d)}} \right), \tag{6.9}$$

where $(\cdot\,,\cdot)$ identifies an open interval. Consequently, if source node $i \in S$ fails to locate a single inactive node $j \in M$ with a pair of $(\alpha_{(i,j)}, \alpha_{(j,d)})$ values conforming to the initial relay selection rules laid out by the constraints represented in (6.6), (6.7) and (6.8), then problem (6.3) is infeasible (i.e., $f_0(\mathbf{P}^*) = +\infty$) and thus the network operator cannot outsource. In the forthcoming discourse, it will be assumed that the set $M' \subset M$ denotes the set of nodes that fulfill the initial relay selection rules. Any node $j \in M'$ will be referred to as being an 'eligible' or a 'potential' relay node; i.e., a node that is eligible to out-service/relay or equivalently, a potential contractor.

6.3.1 Closed-Form Solutions

The closed-form solution of the optimization problem represented in (6.3) can be determined by first formulating the Lagrangian:

$$\begin{aligned}\mathscr{L}(\boldsymbol{P},\boldsymbol{\mu},\boldsymbol{s}) = f_0(\boldsymbol{P}) &+ \mu_0\left(\tilde{P}_{(i,d)} + P_{(j,d)} + \frac{2-2^{R+1}}{\alpha_{(i,d)}} + s_0^2\right)\\ +\mu_1\left(\frac{2^{2R}-1}{\alpha_{(i,j)}} - \tilde{P}_{(i,d)} + s_1^2\right) &+ \mu_2\left(P_{(j,d)} - P^{\max}_{(j,d)} + s_2^2\right)\\ &+\mu_3\left(\alpha_{(i,d)}\tilde{P}_{(i,d)} + \alpha_{(j,d)}P_{(j,d)} + 1 - 2^{2R}\right),\end{aligned} \tag{6.10}$$

with the Lagrange multiplier vector $\boldsymbol{\mu} \in \mathbb{R}^4$ and vector $\mathbf{s} = [s_0^2, s_1^2, s_2^2]$ representing the slack variables. We are here using s_k^2 rather than s_k in order to signify that the slack is positive and in particular, we note that s_0^2 is strictly positive since the corresponding constraint is never binding. The feasible region of (6.3) is a convex set formed by linear constraints and so, the constraint qualification will inevitably be met [11]. Consequently, the necessary and sufficient conditions for $\boldsymbol{P}^*$ to solve the minimization problem of (6.3) is that $\exists\, \boldsymbol{\mu}^*$ such that the Kuhn-Tucker first order conditions (FOCs) are satisfied [12, 13]. These conditions effectively translate into

$$\frac{\partial \mathscr{L}(\boldsymbol{P}^*,\boldsymbol{\mu}^*,\boldsymbol{s})}{\partial \tilde{P}_{(i,d)}} = 1 + \mu_0^* - \mu_1^* + \mu_3^*\alpha_{(i,d)} = 0, \tag{6.11}$$

$$\frac{\partial \mathscr{L}(\boldsymbol{P}^*,\boldsymbol{\mu}^*,\boldsymbol{s})}{\partial P_{(j,d)}} = 1 + \mu_0^* + \mu_2^* + \mu_3^*\alpha_{(j,d)} = 0, \tag{6.12}$$

$$\frac{\partial \mathscr{L}(\boldsymbol{P}^*,\boldsymbol{\mu}^*,\boldsymbol{s})}{\partial s_k} = 2s_k\mu_k^* = 0,\ \forall\, k \in [0,2], \tag{6.13}$$

together with the constraints on the primal variables formulated in (6.3). The marginal FOCs represented in (6.13) are commonly referred to as the complimentary slackness conditions, which are satisfied when either: (a) $\mu_k = 0$ and $s_k > 0$, which implies that the constraint associated with μ_k is inactive; or (b) $\mu_k \neq 0$ and $s_k = 0$ and so the corresponding constraint is binding; or else (c) $\mu_k = s_k = 0$, thus implying a liminal constraint [14].

The optimization problem represented in (6.3) has two solutions. The first solution occurs when $\mu_k^* = 0$ and $s_k^2 \neq 0$, $\forall\, k \in \{0,2\}$, whilst $s_1^2 = 0$. Using the first and second marginal FOCs represented in (6.11) and (6.12), we obtain $\mu_1^* = \left(\alpha_{(j,d)} - \alpha_{(i,d)}\right)/\alpha_{(j,d)}$ and $\mu_3^* = -1/\alpha_{(j,d)}$. The optimal transmit power vector $\boldsymbol{P}^*$ can be calculated by taking into account that $s_1^2 = 0$ and then applying the equality constraint represented in (6.4), thus yielding

$$\tilde{P}^*_{(i,d)} = \left(2^{2R} - 1\right)/\alpha_{(i,j)} \tag{6.14}$$

and

$$P^*_{(j,d)} = \frac{\left(2^{2R} - 1\right)\left(\alpha_{(i,j)} - \alpha_{(i,d)}\right)}{\alpha_{(i,j)}\alpha_{(j,d)}}, \tag{6.15}$$

where $P^*_{(j,d)} \geq P^{\min}_{(j,d)}$. It is also worth pointing out that the constraint formulated in (6.8) is binding when $P^*_{(j,d)} = P^{\min}_{(j,d)}$. The second solution occurs when $\mu_k^* = 0$ and $s_k^2 \neq 0$, $\forall\, k \in \{0,1\}$, whilst $s_2^2 = 0$. By applying the Kuhn-Tucker FOCs, we obtain $\mu_2^* = (\alpha_{(j,d)} - \alpha_{(i,d)})/\alpha_{(i,d)}$, $\mu_3^* = -1/\alpha_{(i,d)}$ and the allocated powers of $P^*_{(j,d)} = P^{\max}_{(j,d)}$ and $\tilde{P}^*_{(i,d)} = (2^{2R} - 1 - \alpha_{(j,d)}P^{\max}_{(j,d)})/\alpha_{(i,d)}$. The derived Lagrange multipliers will be further exploited in Section 6.6.

6.3.2 Partitions in the Feasible Region

Summarizing the analysis presented in Section 6.3.1, the minimum value function $f_0(\boldsymbol{P}^*)$ obtained for the previously described first and second solutions of (6.3) is represented by

$$\mathcal{M}_1 := \frac{2^{2R} - 1}{\alpha_{(i,j)}\alpha_{(j,d)}}\left[\alpha_{(i,j)} + \alpha_{(j,d)} - \alpha_{(i,d)}\right] \tag{6.16}$$

and

$$\mathcal{M}_2 := \left[2^{2R} - 1 + (\alpha_{(i,d)} - \alpha_{(j,d)})P^{\max}_{(j,d)}\right]/\alpha_{(i,d)}, \tag{6.17}$$

respectively. Let us proceed by defining $\mathcal{M}_{\min} := \min_{\{\boldsymbol{P}^* \in \mathcal{P}^*\}} f_0(\boldsymbol{P}^*)$, where $\mathcal{P}^* \in \mathbb{R}^2_{++}$ represents the set of optimal solutions of (6.3) for any pair of $(\alpha_{(i,j)}, \alpha_{(j,d)})$ in the feasible region, as determined by the previously derived constraints represented in (6.6), (6.7) and (6.8). The function $\mathcal{M}_{\min}$ is realized at the intersection point of $\mathcal{M}_1$ and $\mathcal{M}_2$. Equating (6.16) with (6.17) will in fact show that the end-user $j \in M'$ capable of providing the most energy-efficient solution to the network operator is characterized by having $(\alpha_{(i,j)}, \alpha_{(j,d)})$ satisfying

$$\alpha_{(j,d)} = \frac{(2^{2R} - 1)\left(\alpha_{(i,j)} - \alpha_{(i,d)}\right)}{\alpha_{(i,j)} P^{\max}_{(j,d)}} \tag{6.18}$$

for any given value of $\alpha_{(i,d)}$. From (6.18), one can proceed to define the function

$$\Omega(\alpha_{(i,j)}, \alpha_{(j,d)}) := (2^{2R} - 1)\left(\alpha_{(i,j)} - \alpha_{(i,d)}\right) - \alpha_{(i,j)}\alpha_{(j,d)} P^{\max}_{(j,d)}, \tag{6.19}$$

which essentially divides our feasible region D_0 into two partitions; where each partition contains one of the aforementioned solutions to (6.3). In our forthcoming discourse, we will refer to an eligible relay node $j \in M'$, which is characterized by a pair of channel gain values $(\alpha_{(i,j)}, \alpha_{(j,d)})$ that leads to $\Omega(\alpha_{(i,j)}, \alpha_{(j,d)}) < 0$, as being 'located in the first partition' (of the feasible region D_0). Comparably, an eligible relay node associated with $\Omega(\alpha_{(i,j)}, \alpha_{(j,d)}) > 0$ is said to be 'located in the second partition'. It will become evident from our forthcoming discussions that the location of an eligible relay node with respect to the two partitions of D_0 does not only influence the amount of energy-savings granted to the operator but will also affect the per-watt reimbursements given (cf. Section 6.4.2), the predisposition of a potential relay node towards cooperation (cf. Section 6.5) as well as the robustness of the underlying algorithm (cf. Section 6.6). We can summarize the value of $f_0(\boldsymbol{P}^*), \forall\, \boldsymbol{P}^* \in \mathcal{P}^*$, by

$$f_0(\boldsymbol{P}^*) = \begin{cases} \mathcal{M}_1 \text{ for } \Omega(\alpha_{(i,j)}, \alpha_{(j,d)}) < 0, \\ \mathcal{M}_{\min} \text{ for } \Omega(\alpha_{(i,j)}, \alpha_{(j,d)}) = 0, \\ \mathcal{M}_2 \text{ for } \Omega(\alpha_{(i,j)}, \alpha_{(j,d)}) > 0. \end{cases} \tag{6.20}$$

For the sake of simplifying our analysis, let us consider a straightforward numerical example by modeling the channel gain for any generic link $(a, b) \in L$ by $\alpha_{(a,b)} = K/d^3_{(a,b)}$, where parameter $d_{(a,b)}$ denotes the distance between two nodes $\{a, b\} \in N$ in meters whilst constant $K = 1 \times 10^6$. An end-user $d \in M$, located 500 m from the source node $i \in S$, is requesting a service from the network operator with $R = 0.04$ bps/Hz. It is further assumed that there is a pool of inactive end-users, each identified with a pair $(\alpha_{(i,j)}, \alpha_{(j,d)})$, and located on the line-of-sight between the source and destination nodes. Fig. 6.1 portrays the variation of $P_{(j,d)}$ and $f_0(\boldsymbol{P}^*)$ versus $d_{(j,d)}$ in this specific scenario. The source node will first employ the initial relay selection rules formulated in (6.6), (6.7) and (6.8) in order to identify which of the currently inactive nodes will be eligible to participate in the ensuing cooperation by providing a feasible solution to (6.3). It can be observed from Fig. 6.1 that any end-user that is approximately located between 43 m to 471 m from the source node can contribute to a feasible solution $\mathbf{P} \in D_0$. Subsequently, source node i can locate that particular intermediate node j which can contribute to the most energy-efficient solution over the direct transmission scenario by applying (6.20), thus locating that node with $f_0(\boldsymbol{P}^*) = \mathcal{M}_{\min}$ (if available) or with the closest $f_0(\boldsymbol{P}^*)$ to $\mathcal{M}_{\min}$.

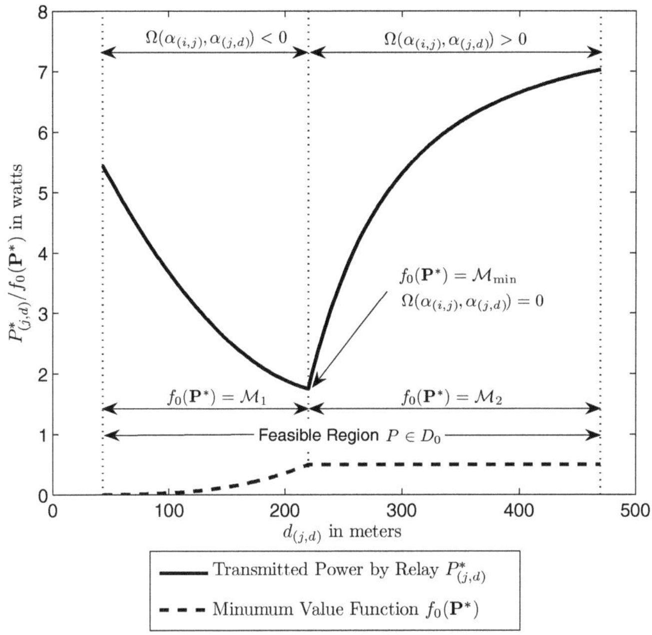

Fig. 6.1 The variation of the transmitted power by the relay $P^*_{(j,d)}$ and the minimum value function $f_0(\boldsymbol{P}^*)$ versus the relay-to-destination distance $d_{(j,d)}$. Here, it is assumed that: (i) the intended destination is located 500 m from the source node and $R = 0.04$ bps/Hz and (ii) there is a pool of inactive end-users located on the line-of-sight between the source and destination.

A central question to our forthcoming discussion is the following: How to ensure that the now-derived 'green' solution is also cost-minimizing? This question will be answered by applying techniques similar to those used in cost-minimization problems in microeconomics, as detailed in the next section.

6.4 The Least-Cost Solution

Simplistically speaking, a relay node under contract will incur two types of costs. Firstly, there are the variable (and thus nonsunk) costs, which constitute those output sensitive costs resulting from the utilization of energy expended during relaying. Additionally, one must not overlook the opportunity costs; i.e., the forward-looking costs that may result from the sacrificed opportunities by the node whilst offering its relaying services or from the potential transmission or reception delays for its own data. In view of these arguments, we will introduce a mechanism where an inactive end-user is allowed to ask for a nonnegotiable reimbursement of $c_{(j,d)}$ monetary units (m.u.) per-watt of transmitted power, if selected to relay data to the intended destination d. In practice, these reimbursements may be realized in the form of degressive charging; i.e., the more an end-user offers its services as a relay, the less that end-user pays for its own data transfer. Subsequently, one can formulate the total costs[2] incurred by the network operator by

$$C_{(i,j,d)} = c_{(i,d)}\tilde{P}_{(i,d)} + c_{(j,d)}P_{(j,d)}, \tag{6.21}$$

where $c_{(i,d)}$ represents the total costs to operate and maintain the source node, per unit watt of transmitted power. Subsequently, one can proceed to formulate the cost-minimization problem and employ similar techniques to those presented in Section 6.3, in order to identify that solution $\boldsymbol{P}$ which minimizes the new objective $C_{(i,j,d)}$ as well as satisfies all the constraints formulated in (6.3), given the fixed prices of $c_{(i,d)}$ and $c_{(j,d)}$. However, we are not intrigued by the actual cost-minimizing solution per se; we would be more concerned with the restrictions that must be imposed on $c_{(j,d)}$ to guarantee that the least-cost solution and the 'green' solution are one and the same. In other words, the least $C_{(i,j,d)}$ value in (6.21) should ideally be attained when the transmit powers $\tilde{P}_{(i,d)}$ and $P_{(j,d)}$ in (6.21) correspond to the previously derived 'green' solution. This problem will be tackled from a microeconomic perspective, to which we lay the foundations in Section 6.4.1. Then, in Section 6.4.2, we derive strict upper bounds on the value of reimbursement $c_{(j,d)}$ as a function of $c_{(i,d)}$. It is assumed that the value of $c_{(j,d)}$ is determined by the relay node; however, any node asking for a reimbursement that is higher than the aforementioned upper bound value will effectively be out-pricing itself compared to its competitors and thus will be treated as being ineligible to out-service.

6.4.1 Microeconomic Setting

For the sake of simplifying our discourse, let us first consider a simple preliminary example to introduce the underlying concepts and related terminology. Consider a revenue-seeking firm with a two-input one-output technology described by the commonly employed Cobb-Douglas production function [15], formulated by $Q = f(K,L) = \sqrt{KL}$, where Q, K and L denote the output quantity produced by the firm and the input factor demands of capital and labor. Fig. 6.2 illustrates an isoquant and isocost map, which conveniently displays a family of isoquants for output quantities $Q_2 > Q_1 > Q_0$. We also show the so-called isocost function dictating the total costs incurred, hereby formulated by $TC = rK + wL$, where TC, r and w denote the total costs and the given fixed input prices of capital and labor respectively. It can be observed from Fig. 6.2 that points corresponding to the input combinations A and C may be deemed to be technically efficient for producing Q_0 units of output, however they are not cost-minimizing since they correspond to a higher level of cost than the isocost line passing through the least-cost solution point B. In other words, by moving from point A to B or from C to B, the firm can produce

[2] Strictly speaking, the terminology of 'total costs' refers to the total *power dependent* costs; i.e., we are excluding the fixed costs as well as those variable costs that bear no dependence on the transmitted power.

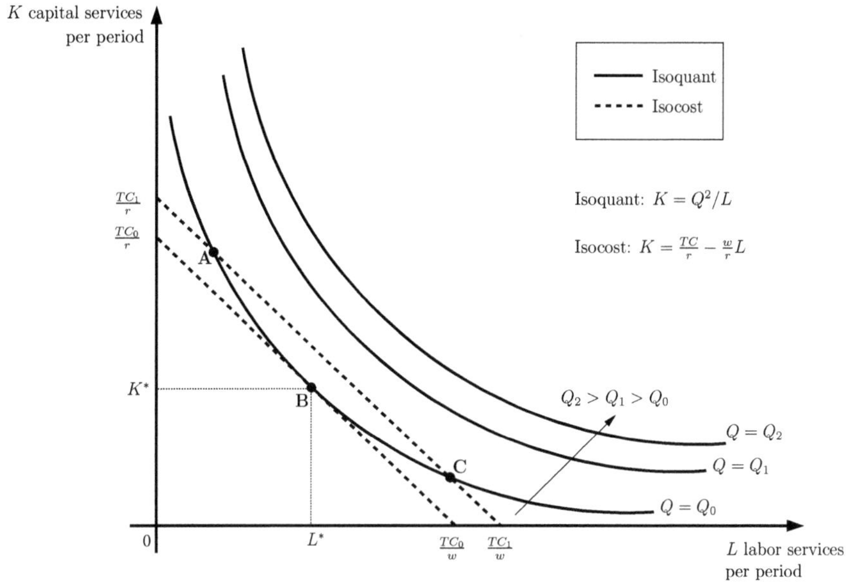

Fig. 6.2 An isoquant and isocost map for the preliminary example introduced in Section 6.4.1, in which we consider a revenue-seeking firm with a two-input, one-output technology described by the Cobb-Douglas production function [15]. This map shows a family of isoquant curves for output quantities $Q_2 > Q_1 > Q_0$ as well as the isocost lines, which are formulated by $TC = rK + wL$, where TC represents the total costs incurred, whilst r and w denote the given fixed input prices of capital and labor respectively. The least-cost solution for producing Q_0 units of output corresponds to the input combination point B.

the same output of Q_0 units whilst reducing the total costs from TC_1 to TC_0. It can be easily shown that the least-cost solution B is located at that point where the slope of the isocost line is equal to the slope of the isoquant. Furthermore, the negative of the slope of the isoquant is equal to the marginal rate of technological substitution of labor for capital [16]; i.e.,

$$MRTS_{\{L,K\}} := \frac{MPP_{\{L\}}}{MPP_{\{K\}}}, \tag{6.22}$$

where

$$MPP_{\{L\}} := \frac{\partial Q}{\partial L} \quad \text{and} \quad MPP_{\{K\}} := \frac{\partial Q}{\partial K} \tag{6.23}$$

denote the marginal physical product (MPP) of labor and of capital respectively. In this example, the slope of the isocost line is equal to $-w/r$ and consequently, the least-cost solution is located at that point where the ratio of marginal physical products is equal to the ratio of input prices.

Against this background, a network operator may also be considered to be a firm with two input factor demands, corresponding to the transmission power used by the source node and the transmission power outsourced to the selected contractor. Similarly to the simple example described above, we can also identify the production function for the network operator's two input, one output technology, which formulates the relation between the output of the firm (i.e., the spectral efficiency experienced by the intended destination d) with respect to the input factor demands. In effect, the production function of the network operator is represented in (6.2), whilst the reformulation of (6.4) corresponds to the isoquant line, from which we observe that

$$MRTS_{\{P_{(j,d)},\tilde{P}_{(i,d)}\}} := \frac{MPP_{\{P_{(j,d)}\}}}{MPP_{\{\tilde{P}_{(i,d)}\}}} = \frac{\alpha_{(j,d)}}{\alpha_{(i,d)}}. \tag{6.24}$$

Up to this point, the only noticeable difference between the scenario in question and the previously described example is that our isoquant is a linear function; i.e., $\Delta MRTS_{\{P_{(j,d)},\tilde{P}_{(i,d)}\}} = 0$. Using technical microeconomic phraseology, this condition would imply that the network operator can *perfectly* substitute one of the input factor demands for the other. In simpler terms, if for the sake of the argument we have $MRTS_{\{P_{(j,d)},\tilde{P}_{(i,d)}\}} = 2.5$, this would imply that the network operator can *perfectly* replace 1 watt of outsourced power to the relay with 2.5 watts of power at its source node—given that $\boldsymbol{P} \in D_0$—and the output R will remain unchanged. In this context, we speak about a 'perfect' substitution since in this case, the elasticity of substitution ϕ, formulated by

$$\phi := \frac{\Delta(\tilde{P}_{(i,d)}/P_{(j,d)})}{\Delta MRTS_{\{P_{(j,d)},\tilde{P}_{(i,d)}\}}} \tag{6.25}$$

is infinite, since $\Delta MRTS_{\{P_{(j,d)},\tilde{P}_{(i,d)}\}} = 0$. We can also reformulate the total output sensitive costs for the network operator delineated by (6.21) to

$$\tilde{P}_{(i,d)} = \frac{C_{(i,j,d)}}{c_{(i,d)}} - \frac{c_{(j,d)}}{c_{(i,d)}} P_{(j,d)}, \tag{6.26}$$

which corresponds to the isocost line. Both the isocost and the isoquant lines are depicted in Fig. 6.3. Table 6.1 summarizes the microeconomic terms used in this chapter, together with their definitions and corresponding equations.

6.4.2 Strict Upper Bounds on the Value of Compensation

As a direct consequence of the linearity of our isoquant, the least-cost solution in this case is a corner point solution rather than an interior solution, such as the input combination point B (cf. Fig. 6.2) for the previously described example. More explicitly, we should aim for the *lower* corner point solution such that the least-cost solution coincides with the 'green' solution as desired. Consequently, the slope of our

Table 6.1 Microeconomic terminology used in this chapter

Terminology	**Definition** [13, 16]	**Equation**
Production Function	Specifies the maximum output quantity produced by a firm (i.e., the network operator) given its input quantities (i.e., transmit powers).	(2)
Isoquant Line/Curve	Displays all input combinations $(P_{(j,d)}, \tilde{P}_{(i,d)})$ producing a given level of output R by the firm.	(4)
Isocost Line/Curve	Displays the set of input combinations yielding the same total costs $C_{(i,j,d)}$ for the firm.	(26)
Marginal Physical Product (MPP)	The extra amount of output that can be produced by a firm when utilizing one additional unit of a specific input whilst keeping the levels of the other input(s) constant.	(23)
Marginal Rate of Technological Substitution (MRTS)	The rate at which the available technology allows for the substitution of one input by another.	(24)
Elasticity of Substitution (ϕ)	A measure of how quickly the $MRTS_{\{P_{(j,d)}, \tilde{P}_{(i,d)}\}}$ changes whilst moving along the isoquant.	(25)

isoquant line must be always less (i.e., more positive) than the slope of our isocost line. The specific condition translates to a constraint on the price of reimbursement, which is formulated by

$$c_{(j,d)} < \frac{\alpha_{(j,d)}}{\alpha_{(i,d)}} c_{(i,d)}. \tag{6.27}$$

However, we note that this constraint does not necessarily guarantee that the least-cost solution is also lower than the costs incurred for direct transmission, which constitute our budget constraint. In order to ensure the latter, it is important to also compare intercepts; more specifically, the intercept of the isocost line with the intercept of the first inequality constraint of (6.3); i.e.,

$$\frac{C_{(i,j,d)}}{c_{(i,d)}} < 2\left(\frac{2^R - 1}{\alpha_{(i,d)}}\right). \tag{6.28}$$

By replacing $C_{(i,j,d)}$ with (6.21), with $\tilde{P}_{(i,d)}$ and $P_{(j,d)}$ substituted by their respective optimal[3] solutions, one obtains the following strict upper bound on the price of reimbursement

[3] Given that constraint (6.27) is satisfied, the adjective 'optimal' alludes to both the 'green' solution and to the least-cost solution, since both solutions would be identical.

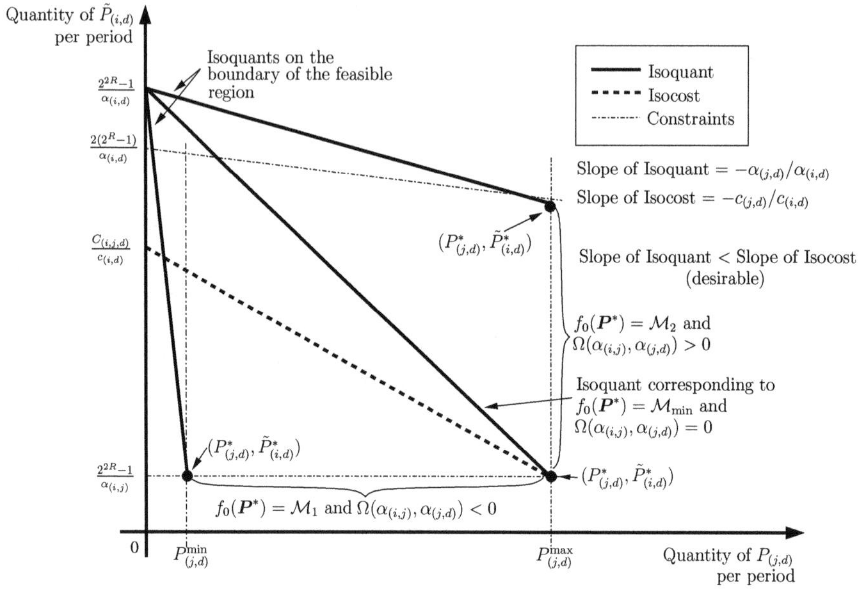

Fig. 6.3 The isoquant map (not to scale) for the network operator, displaying three isoquant lines for producing the same output of R bps/Hz, each assuming different channel gain values $\alpha_{(j,d)}$. Two isoquant lines correspond to solutions $\boldsymbol{P}^*$ of (6.3) that are located on the boundary of the feasible region D_0. The other isoquant line shown corresponds to that solution of (6.3) leading to $\mathcal{M}_{\min}$. The figure also shows the isocost line, which is formulated in (6.26). We remark that the constraints shown in the figure are those formulated in (6.3). The quantities of transmitted power shown on the axis are measured per 'period', where one period corresponds to the time duration of one phase of the cooperation protocol described in Section 6.2.

$$c_{(j,d)} < \left(\frac{2(2^R - 1) - \alpha_{(i,d)}\tilde{P}^*_{(i,d)}}{\alpha_{(i,d)}P^*_{(j,d)}} \right) c_{(i,d)}. \tag{6.29}$$

We recall from Section 6.3.2 that the feasible region D_0 is divided in two partitions and so, the power-minimization problem and the cost-minimization problem will both have two separate solutions, one for each partition. Replacing $\tilde{P}^*_{(i,d)}$ and $P^*_{(j,d)}$ in (6.29) by their respective values for each partition will lead to the following strict upper bounds on the per-watt compensation provided to a relaying node:

$$c_{(j,d)} < \underbrace{\left(\frac{2\alpha_{(i,j)}(2^R - 1) - (2^{2R} - 1)\alpha_{(i,d)}}{\alpha_{(i,d)}[\alpha_{(i,j)} - \alpha_{(i,d)}](2^{2R} - 1)} \right) \alpha_{(j,d)}\, c_{(i,d)}}_{k_1(\alpha_{(i,j)}, \alpha_{(j,d)})} \tag{6.30}$$

for $\Omega(\alpha_{(i,j)}, \alpha_{(j,d)}) < 0$ and

$$c_{(j,d)} < \underbrace{\left(\frac{2^R(2-2^R) - 1 + \alpha_{(j,d)} P^{\max}_{(j,d)}}{\alpha_{(i,d)} P^{\max}_{(j,d)}} \right)}_{k_2\left(\alpha_{(j,d)}\right)} c_{(i,d)}, \tag{6.31}$$

for $\Omega(\alpha_{(i,j)}, \alpha_{(j,d)}) > 0$ and a given $\alpha_{(i,d)}$ and $P^{\max}_{(j,d)}$, where $\Omega(\alpha_{(i,j)}, \alpha_{(j,d)})$ is formulated in (6.19). We further note that the strict upper bound value for the per-watt compensation for that node with $f_0(\boldsymbol{P}^*) = \mathcal{M}_{\min}$ can be calculated using either (6.30) or (6.31), given that $\left(\alpha_{(i,j)}, \alpha_{(j,d)}\right)$ satisfies the relationship formulated in (6.18). It is also worth mentioning that the strict upper bounds formulated in (6.30) or (6.31) are slightly tighter than those of (6.27). This is due to the fact that intercept of the isoquant line represented in (6.4) is higher than the intercept of the first constraint of the power-minimization problem in (6.3). Consequently, one can still formulate an isocost line with a gradient that satisfies (6.27) but with an intercept that is higher than that of the first constraint of the power-minimization problem. We remark that the intercept of the latter prescribes the costs incurred for the direct transmission scenario whilst the intercept of the isocost line dictates the total costs incurred when outsourcing (cf. Figs. 6.2 and 6.3).

6.4.3 *Further Ramifications*

Albeit the simplicity of the upper bounds formulated in (6.27), (6.30) and (6.31), there are implicated some rather interesting ramifications. We commence by investigating the case when the per-watt compensation demanded by an eligible relay violates the aforementioned upper bounds. This context is depicted by the isoquant and isocost map of Fig. 6.4, in which the input combination points A and C identify the least-cost and the 'green' solution respectively.[4] It can be observed that by moving along the isoquant from point A to C, the network operator will be adopting a 'greener' strategy whilst incurring higher costs, which is clearly undesirable. Let us now proceed with the more desirable case when $c_{(j,d)}$ satisfies the aforementioned upper bounds. Firstly, we observe that the slope of the isoquant given in (6.24) has an absolute value that is always higher than unity, as transpires from constraint (6.6). Secondly, it can readily be demonstrated that both $k_1\left(\alpha_{(i,j)}, \alpha_{(j,d)}\right)$ in (6.30) and $k_2\left(\alpha_{(j,d)}\right)$ in (6.31) are also larger than unity, for all values of $\left(\alpha_{(i,j)}, \alpha_{(j,d)}\right)$ satisfying the initial relay selection rules of (6.6), (6.7) and (6.8). As a consequence, the network operator can still afford paying $c_{(j,d)} > c_{(i,d)}$ and attain coincident least-cost and 'green' solutions as well as cost-savings over the direct transmission scenario. The reason behind this somewhat surprising statement revolves around the fact that $MRTS_{\{P_{(j,d)}, \tilde{P}_{(i,d)}\}} > 1$. Based on this rationale, the transmit power from a relay node *may* be valued higher than the transmit power from a source node, since the network operator can perfectly substitute a quantity of x_1 watts of power at its source node with x_2 watts of outsourced power to the relay, where $x_2 < x_1$ and $[x_1, x_2] \in D_0$.

[4] Points A and C do not match due to the violation of (6.27).

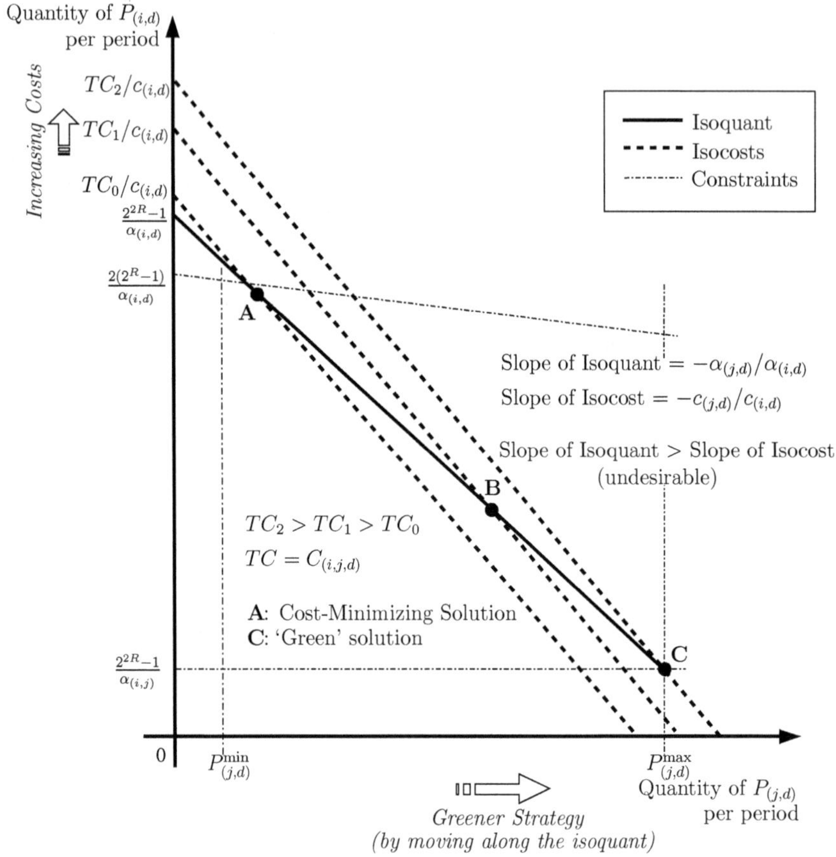

Fig. 6.4 The isoquant and isocost map (not to scale) for the network operator, where the slope of the isoquant line is greater than the slope of the isocost line, which is clearly undesirable since the least-cost solution will not coincide with the 'green' solution. In fact, the input combination point A is the least-cost input bundle whilst the input combination point C is the green solution. By moving along the isoquant line, the network operator will be adopting a 'greener' strategy at the expense of higher costs.

For a more complete picture, one must also investigate the strict upper bound value on the reimbursements provided to a relay; i.e., $c^+_{(j,d)}P^*_{(j,d)}$ m.u., where $c^+_{(j,d)}$ denotes the strict upper bound value on the per-watt compensation as determined by (6.30) and (6.31). We will revisit the simple example introduced in Section 6.3.2, where we consider an end-user d requesting a service of $R = 0.04$ bps/Hz. Without loss of generality, let us assume that $c_{(i,d)} = 1$ m.u. per-watt of transmitted power. In fact, it can be observed from (6.27), (6.30) and (6.31) that $c_{(i,d)}$ can effectively be treated as a scaling factor and consequently, our results may be scaled accordingly for any other value of $c_{(i,d)}$. Fig. 6.5 illustrates the value of $c^+_{(j,d)}P^*_{(j,d)}$ for any

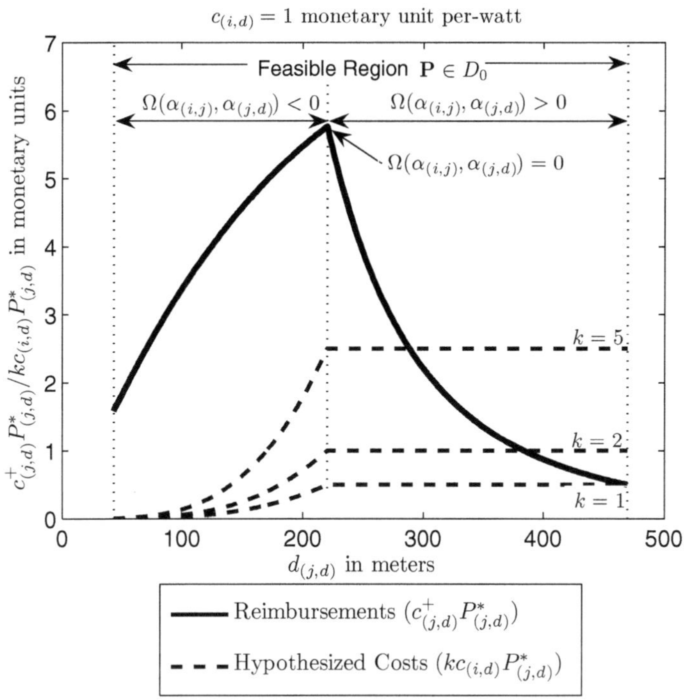

Fig. 6.5 The strict upper bound value on the reimbursements provided to a relay; i.e., $c^{+}_{(j,d)} P^{*}_{(j,d)}$ m.u., where $c^{+}_{(j,d)}$ represents the strict upper bound value on the per-watt compensation as determined by (6.30) and (6.31). We also show a hypothetical case of having variable costs equal to, twice and five times the variable costs of the source node. We recall from our discussion of Section 6.4.2 that the per-watt compensation provided to a relay *may* be higher than the per-watt costs of the source node.

potential contractor that can contribute to a feasible solution $\mathbf{P} \in D_0$. We remark that this figure must be interpreted in conjunction with the minimum value function $f_0(\mathbf{P}^*)$ shown in Fig. 6.1, in order to determine whether the financial reward being offered to a cooperating end-user really reflects the aforementioned benefits (cf. Section 6.1) gained by the network operator from outsourcing. Accordingly, it can be observed that maximum compensation is being provided to the relay node leading to $f_0(\mathbf{P}^*) = \mathcal{M}_{\min}$ and that the higher the energy-savings relished, the higher is the compensation paid to the relaying node. Consequently, it may be argued that the strict upper bounds formulated in (6.27), (6.30) and (6.31) constrain that node interested in offering its services to adopt a value-based pricing strategy; i.e., the price asked by an eligible relay node will very much depend on the value that the network operator places on receiving the ensuing relaying service.

Comparing $P^{*}_{(j,d)}$ in Fig. 6.1 with the corresponding compensation value $c^{+}_{(j,d)} P^{*}_{(j,d)}$ in Fig. 6.5 will also shed some light on the predisposition of the inactive nodes towards offering their service as a relay. For the sake of simplifying our arguments better, let us consider two eligible relay nodes: one located in the region corresponding to $\Omega(\alpha_{(i,j)}, \alpha_{(j,d)}) = 0$ whilst the other located in the region $\Omega(\alpha_{(i,j)}, \alpha_{(j,d)}) > 0$. Moreover, let us assume that these two nodes have the same opportunity costs. By observing $P^*_{(j,d)}$ in Fig. 6.1, one can also deduce that these two nodes will also have the same variable costs, since the transmit power used by these nodes—if selected to cooperate—will be the same; i.e., $P^*_{(j,d)} = P^{\max}_{(j,d)}$. Consequently, one may say that these two nodes in question will incur the same costs to offer their services as a relay. However, the node located in the partition $\Omega(\alpha_{(i,j)}, \alpha_{(j,d)}) > 0$ will receive a lower reimbursement value than the node with $\Omega(\alpha_{(i,j)}, \alpha_{(j,d)}) = 0$. Clearly, a node in the second partition $\Omega(\alpha_{(i,j)}, \alpha_{(j,d)}) > 0$ cannot expect to receive the same compensation as a node with $\Omega(\alpha_{(i,j)}, \alpha_{(j,d)}) = 0$, even if they both have to compensate for the same costs, since the latter node is granting much more energy-savings to the network operator than the former one. In other words, the former node will not be pricing its service competitively if it asks for the same compensation as a node with $\Omega(\alpha_{(i,j)}, \alpha_{(j,d)}) = 0$. Let us elaborate slightly further and hypothesize that the total costs incurred by a node to relay data for the intended destination d are equal to $kc_{(i,d)}P^*_{(j,d)}$, where $k = 1, 2$ and 5. Superimposing $kc_{(i,d)}P^*_{(j,d)}$ on $c^+_{(j,d)}P^*_{(j,d)}$ in Fig. 6.5 clearly shows that the reimbursement provided to some of the eligible relays located in the region with $\Omega(\alpha_{(i,j)}, \alpha_{(j,d)}) > 0$ is not even enough to offset the costs. In this regard, it is plausible to suggest that some of the eligible relays located in the second partition of our feasible region may be more reluctant to cooperate that the nodes in the first partition of the feasible region.

6.5 Modeling Pricing Strategies

As we briefly alluded to in Section 6.4, eligible relay nodes are allowed to determine the per-watt compensation $c_{(j,d)}$ that they would like to receive, if contracted to offer their relaying services. Against this backdrop, the per-watt compensation requested by a potential relay node j to relay data to destination d is modeled by

$$c_{(j,d)} = c^+_{(j,d)} - \rho - \epsilon_j, \ \forall\, j \in M', \tag{6.32}$$

where the slack variable $\rho \in \mathbb{R}_{++}$ is a time-invariant parameter determined by the network operator, which essentially determines the amount of cost-savings generated by the network operator when outsourcing to relay node j over the direct transmission scenario.[5] In light of the discussions of Section 6.4.2, we remark that if the per-watt compensation $c_{(j,d)}$ of (6.32) requested by node $j \in M'$ violates the constraints formulated in (6.30) or (6.31), then that specific node j will automatically be disqualified from further consideration. We recall from Section 6.4.2, that if node

[5] Here, it is implicitly assumed that the network operator is aiming for a 'green' strategy that is also cheaper than direct transmission, hence ρ is strictly positive. It is also assumed that the value of ρ has been disclosed to all potential contractors.

$j \in M'$ demands a per-watt compensation $c_{(j,d)}$ that violates the constraints formulated in (6.30) or (6.31), then cooperating with this relay is even more costly than direct transmission. If the requested $c_{(j,d)}$ value also violates the constraint represented in (6.27), then the cost-minimizing input combination is not a 'green' solution. The parameter $c^{+}_{(j,d)}$ in (6.32) retains the same definition as that described in Section 6.4.3, which can be calculated locally by the eligible relay node j. Then, variable ϵ_j in (6.32) denotes an independent and identically distributed random variable (r.v.) that caters for the variability of the opportunity costs for our pool of eligible relay nodes M'. The latter costs may for example depend on the remaining battery power level of the node and/or specific user-defined settings. In this light, one can interpret (6.32) as being a value-based pricing strategy (i.e., $c^{+}_{(j,d)} - \rho$), together with a r.v. ϵ_j modeling the variability of the end-users.

As it was also alluded to in Section 6.4.2, it is anticipated that eligible relay nodes in the second partition of the feasible region may show reluctance to cooperate, whilst potential relays located in the first partition will be willing (or at most indifferent) to offer their services. We emulate this disparate behavior exhibited by the eligible relay nodes across the two partitions by sampling ϵ_j from two different distributions. More explicitly, we let

$$\epsilon_j \sim \begin{cases} \mathcal{N}(0,\sigma^2) & \text{for } \Omega(\alpha_{(i,j)}, \alpha_{(j,d)}) \leq 0, \\ \mathcal{SN}(0,\omega^2,\psi) & \text{for } \Omega(\alpha_{(i,j)}, \alpha_{(j,d)}) > 0, \end{cases} \tag{6.33}$$

where $j \in M'$. The distribution $\mathcal{N}(0,\sigma^2)$ refers to the standard Normal distribution with zero mean and variance σ^2, whilst $\mathcal{SN}(0,\omega^2,\psi)$ corresponds to the so-called skew-Normal distribution [17], with zero location, scale parameter $\omega \in \mathbb{R}_{++}$ and shape parameter $\psi \in \mathbb{R}$. The expected value of a r.v. $Z \sim \mathcal{SN}(0,\omega^2,\psi)$ is then given by [17]

$$\mathrm{E}\{Z\} = \omega\,\lambda\sqrt{\frac{2}{\pi}}, \tag{6.34}$$

whilst the third and fourth standardized moments of Z are formulated by [17]

$$\gamma_1\{Z\} = \frac{4-\pi}{2}\,\lambda^3\left(\frac{2}{\pi - 2\lambda^2}\right)^{\frac{3}{2}} \tag{6.35}$$

and

$$\gamma_2\{Z\} = \frac{8(\pi-3)\lambda^4}{(\pi - 2\lambda^2)^2}, \tag{6.36}$$

where

$$\lambda = \frac{\psi}{\sqrt{1+\psi^2}}. \tag{6.37}$$

It can be readily demonstrated that the skewness $\gamma_1\{Z\}$ is a monotonically increasing function of ψ whilst the kurtosis $\gamma_2\{Z\}$ is a strictly increasing function of $|\psi|$, where $|\cdot|$ denotes the modulus operator. Additionally, it can be observed from (6.34) and (6.37) that increasing the shape parameter ψ, also increases the expected value $\mathrm{E}\{Z\}$. Consequently, the aforementioned reluctance manifested by potential contractors in the second partition can be modeled by an asymmetrical distribution; more explicitly, by a positively-skewed and heavy-tailed (i.e., high kurtosis) distribution, which can be achieved by *appropriately* modifying the shape parameter to $\psi \in \mathbb{R}_{++}$. Elaborating slightly further, it can also be inferred from Fig. 6.5 that not all potential contractors in the second partition will exhibit the same degree of reluctance towards cooperation. In view of the preceding discussions presented in Section 6.4.2, it can be argued that eligible relays with $\Omega(\alpha_{(i,j)}, \alpha_{(j,d)}) > 0$ that are closer to the source node will be more reluctant to cooperate than those potential relays in the same partition that are closer to the node with $\Omega(\alpha_{(i,j)}, \alpha_{(j,d)}) = 0$. Inevitably, we must seek for a parameter which can quantify—at least to some extent—the amount of reluctance towards cooperation exhibited by an eligible relay node located in the second partition. Both $\Omega(\alpha_{(i,j)}, \alpha_{(j,d)})$ of (6.19) and $k_2\left(\alpha_{(j,d)}\right)$ of (6.31) may serve this purpose. Here, we will focus on the latter by first noticing from (6.6) and (6.31) that

$$\lim_{\alpha_{(j,d)} \to \alpha_{(j,d)}^{-}} k_2\left(\alpha_{(j,d)}\right) = 1, \tag{6.38}$$

where $\alpha_{(j,d)}^{-}$ denotes the strict lower bound value on $\alpha_{(j,d)}$ as identified by (6.6). Therefore, the eligible relay node that manifests the most reluctance to cooperate is located in the region $\Omega(\alpha_{(i,j)}, \alpha_{(j,d)}) > 0$ and having a channel gain value $\alpha_{(j,d)}$ leading to $k_2\left(\alpha_{(j,d)}\right) \approx 1$. Subsequently, we can model deviations in the characteristic behavior of potential relay nodes by modifying the shape parameter ψ for the skew-Normal distribution of (6.33) to $\psi^{\max}/k_2\left(\alpha_{(j,d)}\right)$, where the shape parameter $\psi^{\max} > 1$ is chosen to model high reluctance by virtue of the high skewness and kurtosis. On the other hand, an eligible relay node $j \in M'$ in the second partition and closer to that node with $\Omega(\alpha_{(i,j)}, \alpha_{(j,d)}) = 0$, will be characterized by a higher value of $k_2\left(\alpha_{(j,d)}\right)$ and consequently, it will be associated with a lower value of the shape parameter ψ, which will subsequently lead to a lower value of $\mathrm{E}\{\epsilon_j\}$ and a more symmetrical distribution with lower skewness $\gamma_1\{\epsilon_j\}$ and kurtosis $\gamma_2\{\epsilon_j\}$. At this stage, it is worth adding that when $\psi = 0$, the skew-Normal distribution turns into the standard Normal distribution with variance ω^2. In this light, one may view the skew-Normal distribution as a generalization of the standard Normal distribution, which caters for non-zero skewness [17].

6.6 Sensitivity to Perturbations

Let us assume that all potential contractors would have submitted their tender for the provision of their relaying service at time t. The tender submitted by a potential relay node $j \in M'$ will specify its pair of channel gain values $(\alpha_{(i,j)}, \alpha_{(j,d)})$ as well as the demanded per-watt compensation $c_{(j,d)}$, which respectively quantify the energy and cost-savings that will accrue from outsourcing to that specific node $j \in M'$. Let us further assume the network operator would have selected as well as initiated cooperation with the winning bidder by the time $t + \delta t$. Up to this stage of our analysis, we are implicitly assuming that for any generic link $(a, b) \in L$, the channel gain $\alpha_{(a,b)}$ at time $t + \delta t$ is exactly the same as its estimate at time t, where the latter will be denoted by $\hat{\alpha}_{(a,b)}$. For this specific reason, we have completely disregarded the notion of $\hat{\alpha}_{(a,b)}$ in our previous discourse, since it was implicitly assumed that $\alpha_{(a,b)} = \hat{\alpha}_{(a,b)}$. In this section, we will relax this assumption and consider a more realistic scenario of having some mismatch (although small) between $\alpha_{(a,b)}$ and $\hat{\alpha}_{(a,b)}$; i.e., $\hat{\alpha}_{(a,b)} = \alpha_{(a,b)} + \delta\alpha_{(a,b)}$, where

$$|\delta\alpha_{(a,b)}| \ll \min\{\alpha_{(a,b)}, \hat{\alpha}_{(a,b)}\}, \ \delta\alpha_{(a,b)} \in \mathbb{R}. \tag{6.39}$$

The mismatch $\delta\alpha_{(a,b)}$ may result from any geographical displacement by nodes taking place between time t and $t + \delta t$, from fluctuations on the channel or simply due to imperfect estimation of the channel. In practice, this new reality of having $\hat{\alpha}_{(a,b)} \neq \alpha_{(a,b)}$ implies that the energy and cost-savings promised on a submitted tender with timestamp t, might not accrue after the ensuing cooperation commencing at time $t + \delta t$. In this section, we will develop a criterion, referred to as the sensitivity criterion, which can be calculated by the network operator in order to determine how much trust to put in an eligible relay node to deliver what it has promised in the submitted tender.

When $\hat{\alpha}_{(a,b)} \neq \alpha_{(a,b)}$, our optimization problem of (6.3) is said to be perturbed and consequently, a perturbation may be added to the underlying constraints as follows:

$$\tilde{P}_{(i,d)} + P_{(j,d)} + \frac{2 - 2^{R+1}}{\alpha_{(i,d)}} + s_0^2 = \xi_0, \tag{6.40}$$

$$\frac{2^{2R} - 1}{\alpha_{(i,j)}} - \tilde{P}_{(i,d)} + s_1^2 = \xi_1, \tag{6.41}$$

$$P_{(j,d)} - P_{(j,d)}^{\max} + s_2^2 = \xi_2, \tag{6.42}$$

$$\alpha_{(i,d)}\tilde{P}_{(i,d)} + \alpha_{(j,d)}P_{(j,d)} + 1 - 2^{2R} = \xi_3. \tag{6.43}$$

We define $f_0(\boldsymbol{P}^*, \boldsymbol{\Xi})$ to be the minimum value function of the optimization problem of (6.3), subject to the perturbation vector $\boldsymbol{\Xi} = [\xi_0, \xi_1, \ldots, \xi_3]$. The optimal value of $f_0(\boldsymbol{P}^*, \mathbf{0}) = f_0(\boldsymbol{P}^*)$ corresponds to the previously considered scenario with $\alpha_{(a,b)} = \hat{\alpha}_{(a,b)}$. We know that $f_0(\boldsymbol{P}^*, \mathbf{0})$ is differentiable and that strong duality holds.[6] Consequently, the optimal dual variables μ_k^* derived in Section 6.3.1 can shed light on the sensitivity of the optimal value with respect to perturbations by applying the following result, which states that [18]

$$\mu_k^* = -\frac{\partial f_0(\boldsymbol{P}^*, \boldsymbol{0})}{\partial \xi_k}, \ \forall\, k \in [0, 3]. \tag{6.44}$$

This result suggests that a small positive (negative) perturbation ξ_k would relax (tighten) the corresponding constraint represented in (6.40) to (6.43), which would yield to a decrease (increase) in the optimal value $f_0(\boldsymbol{P}^*)$ by $\mu_k^* \xi_k$ [18]. We will derive the aforementioned sensitivity criterion starting from Δf^*, which is defined by

$$\Delta f^* := f_0(\boldsymbol{P}^*, \boldsymbol{\Xi}) - f_0(\boldsymbol{P}^*). \tag{6.45}$$

In this light, Δf^* quantifies how sensitive (or robust) the optimal value $f_0(\boldsymbol{P}^*)$ is to perturbations.

Let us proceed by calculating Δf^* for each of the two solutions of the perturbed version of (6.3). From our analysis of Section 6.3.1, we observe that constraints (6.41) and (6.43) will be binding for the first solution (i.e., when $\Omega(\hat{\alpha}_{(i,j)}, \hat{\alpha}_{(j,d)}) < 0$). Using (6.39), (6.44) and the values for the Lagrange multipliers derived in Section 6.3.1, we obtain

$$\begin{aligned} \Delta f^*_{\Omega<0} &= -\mu_1^* \xi_1 - \mu_3^* \xi_3, \\ &\approx \frac{\left(\hat{\alpha}_{(i,d)} - \hat{\alpha}_{(j,d)}\right) \xi_1 + \xi_3}{\hat{\alpha}_{(j,d)}} =: \widehat{\Delta f}^*_{\Omega<0}. \end{aligned} \tag{6.46}$$

The perturbation values ξ_1 and ξ_3 can be calculated by comparing the perturbed and unperturbed versions of their respective binding constraints at the optimal solution $\boldsymbol{P}^*$. For instance, the unperturbed version of (6.41) would imply that

$$(2^{2R} - 1) - \hat{\alpha}_{(i,j)} \tilde{P}^*_{(i,d)} = 0, \tag{6.47}$$

where $\hat{\alpha}_{(i,j)} = \alpha_{(i,j)}$. Introducing fluctuations on the channel gain value in (6.47) by allowing $\hat{\alpha}_{(i,j)} = \alpha_{(i,j)} + \delta\alpha_{(i,j)}$ and then comparing (6.47) with (6.41) yields $\xi_1 = \delta\alpha_{(i,j)} \tilde{P}^*_{(i,d)}$. The perturbation ξ_3 can be calculated using a similar method, thus giving $\xi_3 = -(\delta\alpha_{(i,d)} \tilde{P}^*_{(i,d)} + \delta\alpha_{(j,d)} P^*_{(j,d)})$. Replacing the values of ξ_1 and ξ_3 in (6.46) leads to

$$\widehat{\Delta f}^*_{\Omega<0} = \frac{\left(\hat{\alpha}_{(i,d)} - \hat{\alpha}_{(j,d)}\right) \delta\alpha_{(i,j)} \tilde{P}^*_{(i,d)} - \delta\alpha_{(i,d)} \tilde{P}^*_{(i,d)} - \delta\alpha_{(j,d)} P^*_{(j,d)}}{\hat{\alpha}_{(j,d)}}. \tag{6.48}$$

[6] Strong duality holds when the optimal primal objective is equal to the optimal dual objective, thus resulting in a zero duality gap.

We observe that (6.48) still depends on unknown mismatch values and thus cannot be calculated by the network operator. Let us tackle this issue by first focusing on the polarity of the mismatch values. Firstly, we note that $\widehat{\Delta f}^*_{\Omega<0} > 0$ indicates that the previously considered case with $\boldsymbol{\Xi} = \mathbf{0}$ has presented an optimistic opinion as to what energy and cost savings can be supplied by an eligible relay $j \in M'$. On the other hand, $\widehat{\Delta f}^*_{\Omega<0} < 0$ suggests that a potential relay under consideration may contribute to even more energy and cost savings at time $t + \delta t$ than the ones calculated at time t.[7] The latter scenario may appear to be more desirable than the former; however, one can hypothesize situations in which the latter scenario will also be detrimental. For instance, an eligible relay associated with $\widehat{\Delta f}^*_{\Omega<0} < 0$ may be considered to be inadmissible because there can be bids submitted by other eligible relays which may *appear* to be more competitive (when in fact they aren't). As we will discuss in more detail in Section 6.8, the winning tender will be selected from one of the solutions located on the Pareto-optimal frontier. In this light, an eligible relay associated with $\widehat{\Delta f}^*_{\Omega<0} < 0$ might not be deemed to be a Pareto-optimal solution (even though it may be), and thus is excluded from further consideration. In this light, we will treat both scenarios as being equally undesirable and for this specific reason, we are more interested in the absolute value of $\widehat{\Delta f}^*_{\Omega<0}$. Subsequently, using (6.48) and applying the triangle inequality, we obtain

$$\left|\widehat{\Delta f}^*_{\Omega<0}\right| \leq \frac{\left[|\hat{\alpha}_{(i,d)} - \hat{\alpha}_{(j,d)}||\delta\alpha_{(i,j)}| + |\delta\alpha_{(i,d)}|\right] \tilde{P}^*_{(i,d)} + |\delta\alpha_{(j,d)}| P^*_{(j,d)}}{\hat{\alpha}_{(j,d)}}. \quad (6.49)$$

Let us denote the strict upper bound value of (6.49) by $\left|\widehat{\Delta f}^*_{\Omega<0}\right|^+$. Then, if we assume that the fluctuations of each channel gain value do vary much from each other such as they can be effectively treated as a constant, we obtain

$$\left|\widehat{\Delta f}^*_{\Omega<0}\right|^+ \propto \frac{\left[|\hat{\alpha}_{(i,d)} - \hat{\alpha}_{(j,d)}| + 1\right] \tilde{P}^*_{(i,d)} + P^*_{(j,d)}}{\hat{\alpha}_{(j,d)}}. \quad (6.50)$$

Similarly, we can calculate Δf^* for the second solution of the perturbed version of (6.3), which gives

$$\Delta f^*_{\Omega\geq 0} = -\mu_2^*\xi_2 - \mu_3^*\xi_3 \approx \frac{\left(\hat{\alpha}_{(i,d)} - \hat{\alpha}_{(j,d)}\right)\xi_2 + \xi_3}{\hat{\alpha}_{(i,d)}}. \quad (6.51)$$

The perturbation ξ_2 in (6.51) is always equal to zero since constraint (6.42) is independent from the channel gain parameters; thus we obtain

$$\widehat{\Delta f}^*_{\Omega\geq 0} = \frac{\xi_3}{\hat{\alpha}_{(i,d)}} = -\frac{\left(\delta\alpha_{(i,d)}\tilde{P}^*_{(i,d)} + \delta\alpha_{(j,d)}P^*_{(j,d)}\right)}{\hat{\alpha}_{(i,d)}} =: \widehat{\Delta f}^*_{\Omega\geq 0}. \quad (6.52)$$

[7] The scenario leading to $\widehat{\Delta f}^*_{\Omega<0} = 0$ would then correspond to the case with almost no fluctuations on the channel gain parameters; i.e., $\boldsymbol{\Xi} \approx \mathbf{0}$.

Taking the absolute value of (6.52) and subsequently applying the triangle inequality yields to

$$|\widehat{\Delta f}^*_{\Omega\geq 0}| \leq \frac{|\delta\alpha_{(i,d)}|\tilde{P}^*_{(i,d)} + |\delta\alpha_{(j,d)}|P^*_{(j,d)}}{\hat{\alpha}_{(i,d)}} =: |\widehat{\Delta f}^*_{\Omega\geq 0}|^+. \tag{6.53}$$

By assuming constant mismatch terms, we obtain $|\widehat{\Delta f}^*_{\Omega\geq 0}|^+ \propto f_0(\boldsymbol{P})/\hat{\alpha}_{(i,d)}$. Summarizing the two results for each solution, we define the aforementioned sensitivity criterion for the tender submitted by an eligible relay node $j \in M'$ by

$$\mathcal{S}_j := \begin{cases} \left(\left[|\hat{\alpha}_{(i,d)} - \hat{\alpha}_{(j,d)}| + 1\right] \tilde{P}^*_{(i,d)} + P^*_{(j,d)}\right)/\hat{\alpha}_{(j,d)} & \text{for } \Omega < 0, \\ \left(\tilde{P}^*_{(i,d)} + P^*_{(j,d)}\right)/\hat{\alpha}_{(i,d)} & \text{for } \Omega \geq 0. \end{cases} \tag{6.54}$$

If a relatively high $\mathcal{S}_j$ value is associated with a tender submitted by an eligible relay node j, this would indicate that the previously calculated value of $f_0(\boldsymbol{P})$ *may* be highly sensitive to any perturbation; which would subsequently imply a higher probability that the energy and cost-savings predicted by the network operator at time t *might* not accrue at time $t + \delta t$. We on purposely emphasize the words 'may' and 'might' for two main reasons: (i) $\mathcal{S}_j$ is based upon the upper bound value of the $\widehat{\Delta f}^*$, which would shed light on the *worst-case* sensitivity to perturbations and (ii) the absolute values of the mismatch terms are assumed to be comparable, which may not always be the case in reality.

6.7 Interference Criterion

The network operator would also like to ensure that collaboration with relay node j does not disturb other currently active end-users, which may be also generating revenue for the operator by exploiting one of its offered services. We will assume that the network operator owns the technology to identify and locate these users, which we will refer to as vulnerable users (VUs). This generic framework can also be extended to include VUs associated with other independent network operators, sharing the same frequency bands in the same geographical area. In this context, it may be assumed that all network operators share the location information of their respective VUs for their mutual benefit. Subsequently, the operator will also take into account the interference inflicted on the VUs by an eligible relay j, by associating another criterion $\mathcal{I}_j$, defined by $\mathcal{I}_j := P^*_{(j,d)}/\overline{d}^2_{(j,v)}$, $\overline{d}_{(j,v)} = \frac{1}{V}\sum_{v=1}^{V} d_{(j,v)}$, to each submitted tender. The criterion $\mathcal{I}_j$ will be referred to as the interference criterion, whilst V and $d_{(j,v)}$ denote the number of VUs and the distance (in meters) between the VU with index $v \in [1, V]$ and a potential contractor j.

6.8 Compromise Programming

Let each submitted tender be represented by vector $\boldsymbol{T}_j = \left[f_0(\boldsymbol{P}^*), C_{(i,j,d)}, \mathcal{S}_j, \mathcal{I}_j\right]$, $\boldsymbol{T}_j \in \mathcal{T}$, where $\mathcal{T}$ denotes the set of all tenders received by the network operator. The next step in the development of our framework would be to evaluate the tenders submitted by all potential contractors willing to provide their relaying services to the operator, subject to adequate reimbursements. In this regard, the network operator must seek that potential relay node which: (a) is associated with the lowest optimal value $f_0(\boldsymbol{P}^*)$, thus leading to the 'greenest' transmission strategy; (b) minimizes the total costs $C_{(i,j,d)}$ incurred;[8] (c) minimizes the sensitivity of $f_0(\boldsymbol{P}^*)$ and $C_{(i,j,d)}$ on potential perturbations, which implies that the submitted tender enjoys the highest level of trust amongst all candidates; as well as (d) minimizes the interference inflicted on the aforementioned VUs. Ideally, the winning tender would correspond to that tender $\boldsymbol{z}$ that minimizes all the four objectives (aka the utopia point). However, it is highly unlikely that the network operator will receive such a tender; especially when considering that some of these objectives *may* actually be conflicting to each other. Note our emphasis on the word 'may' since we cannot determine a-priori those criteria that will be *always* conflicting, since some of our objectives are dependent on randomly generated parameters. A case in point is $C_{(i,j,d)}$, which depends on $c_{(j,d)}$, where the latter will be modeled on (6.32) and thus dependent on the r.v. ϵ_j. Another example is the interference criterion $\mathcal{I}_j$, which is dependent on $d_{(j,v)}$, which may be treated as another r.v. since we will later assume a random topology for the position of nodes (cf. Section 6.9).

We will treat this problem as a multi-criterion optimization problem, which will be solved by the following two steps. Firstly, we determine the Pareto-optimal set containing all those solutions (i.e., tenders)—within the decision space $\mathcal{T}$—whose corresponding objective vector components in $\boldsymbol{T}_j$ cannot be all simultaneously improved.[9] We will denote the entire set of Pareto-optimal solutions by $\mathcal{T}^*$. The winning tender would then correspond to that solution $\boldsymbol{T}_j \in \mathcal{T}^*$ that is the closest to $\boldsymbol{z}$, which is determined by minimizing some distance metric. In our case, we have opted for the commonly-used Tchebycheff metric (aka the L_∞-metric) defined by:

$$L_\infty(\boldsymbol{T}_j, \boldsymbol{z}) := \max_{k=1}^{4} \frac{|t_k - z_k|}{t_k^{\max} - z_k}, \tag{6.55}$$

where t_k and z_k denote the kth objective in $\boldsymbol{T}_j \in \mathcal{T}^*$ and $\boldsymbol{z}$, respectively, whilst $t_k^{\max}$ represents the worst value obtainable for the kth objective in any tender $\boldsymbol{T}_j \in \mathcal{T}$. This method is generally know as compromise programming [19, 20] or the method of global criterion, and would correspond to the well-known min-max method when used with the distance metric of (6.55).

[8] The total costs would depend on $\boldsymbol{P}^*$ and on the demanded per-watt compensation $c_{(j,d)}$, as shown in (6.21). Note that we can only substitute $\boldsymbol{P}^*$ in (6.21), if and only if $c_{(j,d)}$ satisfies certain constraints. We have detailed this specific point in Section 6.4.2.

[9] The Pareto-optimal solutions are sometimes referred to as being non-inferior, admissible or efficient and their corresponding objective vector $\boldsymbol{T}_j$ is said to be non-dominated.

6.9 Simulation Results

We consider scenarios where the end-users are uniformly distributed across a square having an area of 200 $\times$ 200 meters squared. This area is serviced by one source node $i \in S$ positioned on one of the corners of the square. We have considered downlink spectral efficiencies of $R = 0.04$ bps/Hz, 0.20 bps/Hz and 0.50 bps/Hz. The per-watt compensation value $c_{(j,d)}$ demanded by a node was modeled according to (6.32) as detailed in Section 6.5, with parameters $\rho = 10^{-6}$, $\sigma^2 = \omega^2 = 0.12/\rho$ and $\psi^{\max} = 8$. The total number of VUs was set to $V = 20$. All the reported results have been averaged over half a million simulation runs.

Fig. 6.6 displays the average percentage energy-savings provided by the proposed 'green' radio framework in comparison to the scenario without outsourcing. It can be observed that our framework will be providing approximately 68% energy savings (on average, over the scenario without outsourcing) when $R = 0.04$ bps/Hz and $N_u = 250$ inactive end-users. It must be emphasized that not all the N_u-number of inactive end-users will be eligible to participate in the ensuing cooperation. For example, some of the inactive end-users may be located behind the intended destination, others may violate the initial relay selection rules of (6.6), (6.7) and (6.8). Other inactive end-users may be reluctant to cooperate or else may have high opportunity costs. These end-users will be asking for a per-watt compensation value $c_{(j,d)}$ that violates the strict upper bounds derived in Section 6.4.2 and

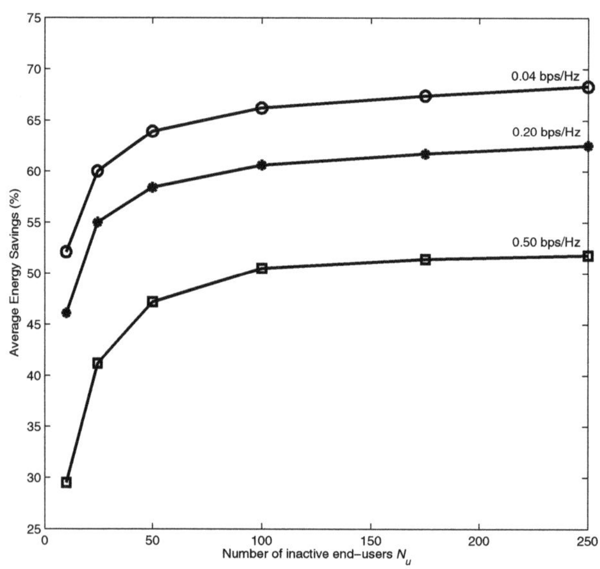

Fig. 6.6 The average percentage energy savings (in comparison with the scenario without outsourcing) that will accrue from the deployment of our framework, versus the number of inactive end-users N_u.

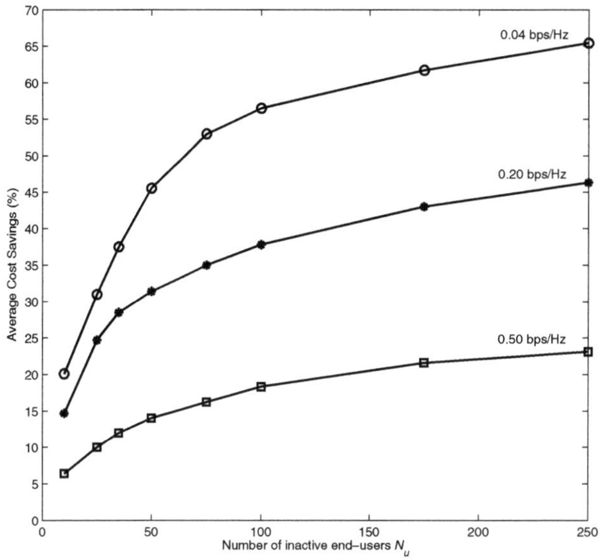

Fig. 6.7 The average percentage cost savings (in comparison with the scenario without outsourcing) versus the number of inactive end-users. The total number of VUs was set to $V = 20$.

thus their submitted tender will be subsequently disqualified. We also report that the average number of nodes submitting a competitive tender was equal to 50, 47 and 31 nodes when $R = 0.04$ bps/Hz, 0.20 bps/Hz and 0.50 bps/Hz respectively, and with $N_u = 250$. Fig. 6.7 illustrates the average percentage cost savings versus the number of inactive end-users. It can be observed that outsourcing with $N_u = 250$ inactive end users will yield cost savings of approximately 65%, 46% and 23% when $R = 0.04$ bps/Hz, 0.20 bps/Hz and 0.50 bps/Hz respectively.

All our results reported in Figs. 6.6 and 6.7 were determined via compromise programming with respect to all four objectives (i.e., power/energy, cost, sensitivity and interference). Fig. 6.8 illustrates the average energy and cost savings obtained after tackling the multi-criterion optimization problem with respect to the energy and cost objectives only. It can be verified from Fig. 6.8 that these savings are higher than the ones reported in Figs. 6.6 and 6.7. For example, outsourcing with $N_u = 250$ inactive end-users will yield approximately 70% energy savings and 65% cost savings when $R = 0.20$ bps/Hz, which represent a respective increase of approximately 8 and 20 percentage points over the energy and cost savings shown in Figs. 6.6 and 6.7. However, the selected bidder will (on average) be inflicting more interference on the VUs and its tender will be less robust to potential perturbations. For example, it was observed that when $R = 0.50$ bps/Hz and $N_u = 250$ inactive end-users, the average sensitivity and interference criteria were respectively increased by approximately 70 and 75 percentage points over the same criteria attained after optimizing with

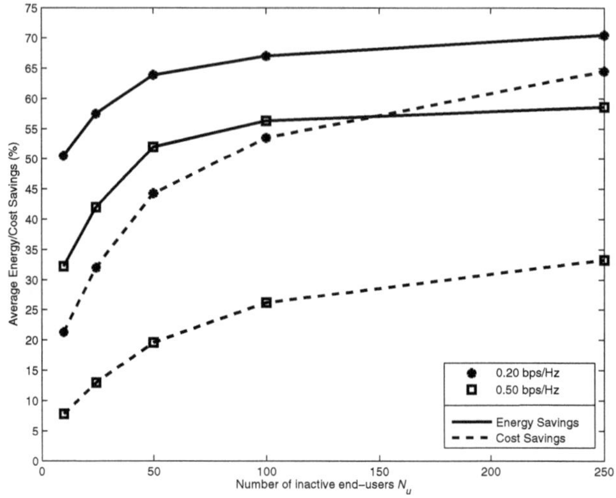

Fig. 6.8 The figure shows the average percentage energy and cost savings (in comparison with the scenario without outsourcing) versus the number of inactive end-users N_u. It can be observed that the savings are higher than the ones reported in Figs. 6.6 and 6.7, in which we have considered all four objectives (i.e., transmit power, cost, sensitivity and interference).

respect to all four objectives. The latter point implies that our predicaments depicted in Fig. 6.8 are less likely to materialize (in practical scenarios with nonzero perturbations on the constraints) than the energy and cost savings reported in Figs. 6.6 and 6.7. In this regard, the network operator may opt for higher savings by trusting less predictable nodes, which inflicts more interference on the VUs or else, select nodes that offer a better tradeoff between the four objectives.

6.10 Conclusions

In this chapter, we have exploited outsourcing for the provision of an energy- and cost-efficient solution for the so-called 'green' radio frameworks. Our approach may be compared to a tender evaluation and selection procedure, whereby a network operator solves a multi-objective optimization problem to determine the winning tender that strikes the best compromise between the benefits (energy and cost savings) and the drawbacks (potential broken promises by contractors and added interference) of outsourcing. Our results show that the proposed framework can attain up to 68% energy savings and up to 65% cost savings over the scenario without outsourcing, whilst still managing to keep the aforementioned potential downsides to a minimum.

References

1. Wang, R., Thompson, J.S., Haas, H.: A new framework for designing power-efficient resource allocation under rate constraints. In: Proceedings of the IEEE Vehicular Technology Conference Fall, Anchorage, Alaska, USA, pp. 1–5 (2009)
2. Raman, C., Foschini, G., Valenzuela, R., Yates, R., Mandayam, N.: Relaying in downlink cellular systems. In: Proc. 44 Ann. Conf. Inf. Sciences and Systems, Princeton, NJ, USA, March17-19, pp. 1–6 (2010)
3. Grant, P., McLaughlin, S., Aghvami, H., Fletcher, S.: Green Radio–Towards Sustainable Wireless Networks (May 2009), `http://www.ee.princeton.edu/seminars/iss/Spring2009/slides/grant.pdf`
4. Comaniciu, C., Mandayam, N.B., Poor, H.V.: Radio resource management for green wireless networks. In: Proc. IEEE Veh. Tech. Conf., Anchorage, AK, USA, pp. 1–5 (2009)
5. Mclvor, R.: The Outsourcing Process, Strategies for Evaluation and Management. Cambridge University Press, Cambridge (2005)
6. Burkholder, N.C.: Outsourcing: The Definitive View, Applications, and Implications. John Wiley & Sons, Chichester (2006)
7. Akhtman, J., Hanzo, L.: Power versus bandwidth efficiency in wireless communication: The economic perspective. In: Proc. IEEE Veh. Tech. Conf., Anchorage, AK, USA, September 20-23, pp. 1–5 (2009)
8. Bertsekas, D.P., Gallager, R.: Data Networks. Prentice-Hall, Englewood Cliffs (1992)
9. Goldsmith, A.: Wireless Communications. Cambridge University Press, Cambridge (2004)
10. Sendonaris, A., Erkip, E., Aazhang, B.: User cooperation diversity – part I: System description. IEEE Trans. Commun. 51, 1927–1938 (2003)
11. Chiang, A.C., Wainwright, K.: Fundamental Methods of Mathematical Economics. McGraw Hill Higher Education, New York (2005)
12. Mas-Colell, A., Whinston, M.D., Green, J.R.: Microeconomic Theory. Oxford University Press, Oxford (1995)
13. Kreps, D.M.: A Course in Microeconomic Theory. Harvester Wheatsheaf (1990)
14. Horsley, A., Wrob, A.J.: Liminal inequality constraints and second-order optimality conditions. CDAM Research Report Series 2003 LSE-CDAM-2003-16, London School of Economics (December 2003), `http://www.cdam.lse.ac.uk/Reports/Files/cdam.lse.ac.uk/Reports/Files/cdam-2003-16.pdf`
15. Cobb, C.W., Douglas, P.H.: A theory of production. Amer. Econ. Rev. 18, 139–165 (1928)
16. Morgan, W., Katz, M.L., Rosen, H.S.: Microeconomics. McGraw Hill Higher Education, New York (2009)
17. Azzalini, A.: A class of distributions which includes the normal ones. Scand. J. Statist. 12, 171–178 (1985)
18. Boyd, S., Vandenberghe, L.: Convex Optimization. Cambridge University Press, Cambridge (2004)
19. Zeleny, M.: Compromise programming. In: Cochrane, J.L., Zeleny, M. (eds.) Multiple Criteria Decision Making, pp. 262–301. Univeristy of South Carolina Press (1973)
20. Yu, P.L.: A class of solutions for group decision problems. Management Science 19, 936–946 (1973)

Chapter 7
Cooperative Relay Scheduling in Energy Harvesting Sensor Networks

Huijiang Li, Neeraj Jaggi, and Biplab Sikdar

Abstract. This chapter studies relay scheduling in wireless sensor networks with energy harvesting and cooperative communications capabilities. Sensor networks are increasingly deployed in inaccessible and remote regions for applications such as environmental monitoring, relief operations, and defense. In such networks, energy harvesting and cooperative communication paradigms are used simultaneously to design energy-efficient relay scheduling strategies. The following scheduling problem is formulated to maximize the network utility: Given an estimate of the current network state, should a source sensor node transmit its data directly to the destination sensor node, or should it use a relay sensor node to help with the transmission? An upper bound is obtained on the performance of an arbitrary scheduler. Scheduling policies are then developed to choose the appropriate transmission mode depending upon the available energy at the sensors as well as the states of their energy harvesting and data generation processes. Two separate scenarios are considered where the state of the relay node is either fully or partially observable at the source node, and the scenarios are modeled using a Markov Decision Process (MDP) and a Partially Observable Markov Decision Process (POMDP) respectively. It is shown that the POMDP can be transformed into an equivalent MDP. Optimal scheduling strategies are evaluated using value iteration algorithm, and various insights towards optimal relay scheduling are discussed. Simulation results are used to show the performance of the proposed strategies.

Huijiang Li
Rensselaer Polytechnic Institute, 110 8th ST, Troy NY 12180, USA
e-mail: `lih4@rpi.edu`

Neeraj Jaggi
Wichita State University, Wichita KS 67260, USA
e-mail: `neeraj.jaggi@wichita.edu`

Biplab Sikdar
Rensselaer Polytechnic Institute, 110 8th ST, Troy NY 12180, USA
e-mail: `bsikdar@ecse.rpi.edu`

7.1 Introduction

Technological advances in the design of low-cost, low-power sensor devices with sensing, processing and wireless communication capabilities, usually on a single chip, has facilitated the wide-spread deployment of wireless sensor networks (WSNs). WSNs have found wide application in a variety of fields including but not limited to environmental and infrastructure monitoring, military and emergency applications, and medical applications such as diagnostic techniques, health and stress monitoring, management of chronic diseases, patient rehabilitation etc. [1, 2, 3, 4]. Most of these applications require small and low-cost devices with long operating lifetimes. In addition, the major hurdle for the wide adoption of the Body Sensor Network (BSN) technology is the limited energy supply.

Compared to computing and communication technologies, the relatively slow rate of progress in the battery technology does not promise battery driven WSN nodes in near future [5]. Batteries are often seen to be the heaviest part of modern handheld electronic devices, as a lithium-ion battery could take up to 36% of the total weight of a Motorola V180 cell phone [6]. The lack of suitable electrode materials and electrolytes, and difficulties in mastering the interface between them are the main reasons that slow down the advance of battery technology [7]. The lithium-ion battery is the most popular type of battery for portable devices. Compared to Ni-MH and Lead-acid batteries, lithium-ion cell has higher energy density, power rate and cycle life, and is relatively green since recycling is feasible although costly. However, to support the advances in microelectronics, miniature power sources such as solid-state, lithium-based, thin-film batteries, are required. Many wireless sensor network applications require the network to operate for several months to years without battery replacement. Such requirements are difficult to meet with current batteries which have low capacity and current, resulting from the limited volume of active materials and low energy efficiency [6]. Other battery quality issues such as self-discharging and performance degradation subject to extreme temperature conditions also impact wireless sensor network applications. In addition, battery manufacturing often involves unsustainable resources with safety being a potential concern. For example, currently compounds of lithium-ion batteries such as $LiCoO_2$ and $LiMn_2O_4$ are from ores, which are not renewable energy resources [7].

Thus, design of sustainable and green batteries suitable for long-term wireless sensor networks operation is a challenging research area requiring collaboration among a range of disciplines [7]. Meanwhile, energy harvesting from external sources, which has been around in large scale applications such as harvesting energy from the water wheel and windmill, is becoming a promising way to realize energy efficient sensor nodes. However, energy harvesting leads to new challenges such as time-varying and correlated harvesting rates. Cooperative communications can be used to design energy efficient communication strategies in such energy constrained networks. It has been shown that cooperative diversity gains can be achieved in distributed networks where nodes help each other by relaying transmissions [8], resulting in either higher network capacity or lower energy consumption with the same capacity. In this chapter, energy harvesting and cooperative communications are used simultaneously to design energy-efficient relay scheduling strategies in sensor networks. The rest of the chapter is organized as following. Section 7.2 presents an overview of the state-of-the-art energy harvesting options and background on cooperative communication background. Section 7.3 formulates the problem of energy efficient scheduling in sensor networks with both energy harvesting and cooperative communication capabilities and Section 7.4 concludes the chapter.

7.2 Energy Harvesting and Cooperative Communication

7.2.1 Energy Harvesting

The process of deriving energy from renewable energy sources, converting it into consumable electrical energy is called energy harvesting or energy scavenging. In wireless sensor networks, energy can be harvested to power small, low-power MEMS devices. Most common sources of energy harvesting are mechanical vibration, solar and thermal energy [9].

7.2.1.1 Vibration

The harvesters transduce the vibrational energy into electrical energy using one of three methods: electromagnetic (inductive), electrostatic (capacitive) and piezoelectric.

Electromagnetic energy harvesting [9, 10] follows the principle of electromagnetic induction, wherein voltage is generated across a conductor when it moves through a magnetic field. Electrostatic generators transform mechanical work into electric energy by manual or other power. Detail introduction and review of electrostatic energy harvester can be found in [11]. Electromagnetic systems require no input voltage source to initiate power generation but may require transformers since typical voltage generated is between 0.1 and 0.2 V. Wafer-scale fabrication techniques limit the implementation of MEMS electromagnetic devices [6]. Electrostatic devices could be integrated into microsystems but require an additional voltage source to initially charge the capacitor.

Piezoelectric transduction has received the most attention. Piezoelectric material is capable of generating electrical charges in response to mechanical pressure, force or vibration. Review articles [12, 6] show that piezoelectric energy harvesting, with comparably large power densities, is the simplest way due to the inherent ability of piezoelectric devices to detect vibrations and convert to voltage directly. [13] builds a system by using piezoelectric harvesters built into a shoe to collect energy from heel strikes. One simple implementation is to tap power from pressure by a unimorph strip made from piezoceramic composite material. The other is to use a stave made from a multilayer laminate of PVDF (polyvinyli-dineflouride) foil. The stave is very thin and the shape of it is chosen to conform to the footprint and bending distribution of a standard shoe sole. The performance of piezoelectric energy transduction is also compared [13] with a system where vibration energy is harvested through a standard electromagnetic generator. It shows that although the power generated by the piezoelectric systems is limited, it could be integrated in the least invasive way and is able to serve as power supply for applications with low-power and low-duty-cycle such as a 12-bit RFID system integrated onto a jogging sneaker. Various methods of converting energy from different vibrations with piezoelectric harvesters have been investigated. For example, for unmanned air vehicles, energy harvesting from aeroelastic vibrations is involved. [14] presents a time-domain piezoaeroelastic modeling of a generator wing with embedded piezoceramics. With an aim to reduce the computational cost, frequency-domain piezoaeroelastic analysis of a generator wing with continuous electrodes is also studied in [14].

Piezoelectric energy harvesting from ambient vibration energy can potentially supply 10-100's of μW. Low power output challenges both the logic circuit design and efficient power delivery circuits. [15] proposes a bias-flip rectifier circuit to improve the power extraction capability. For most wireless sensor network applications with piezoelectric energy harvesting, the storage of harvested energy is necessary. [12] provides a review of such storage architectures. [16] proposes that the energy generated could be stored with capacitors or rechargeable batteries. The usage of capacitors is discussed in [17, 18]. The use of nickel metal hydride batteries is considered in [19] and is shown to be preferable than capacitors. Lithium-based batteries are shown to be ideal in energy harvesting systems [20, 21, 22]. The thin-film lithium-based batteries could be directly embedded into an energy harvesting package with piezoelectric materials creating a self-charging, multifunctional device [22, 23].

The capacitive, inductive or piezoelectric energy conversion from vibration to electricity is mainly based on linear oscillators where the mechanical oscillators are designed to be resonantly tuned to the dominant ambient frequency. In many applications the ambient vibrations have energy distributed over a wide frequency spectrum. Devices designed with single resonant frequency may not be robust to variations in the excitation frequency. [19] looks at a random vibration scenario in an automobile compressor, and experimentally investigates the possible power generated. Wide band excitation with harmonic excitation is considered in [24, 25]. Energy harvesting under unknown or random excitations is formulated first with linear random vibration theory to obtain closed-form expressions in [26]. An electromechanical system with stationary Gaussian white noise based excitation is formulated in [27]. The characteristics of vibration energy's frequency spectrum may not always allow for frequency tuning. [28] proposes a non-resonant oscillator approach which could achieve potentially 400% power harvesting gain compared to linear model. A bistable toy-model oscillator is realized using an inverted pendulum made of a four-layer piezoelectric beam.

7.2.1.2 Solar Energy

Solar photovoltaics (PV) is one of the fastest growing power-generation technologies in the world. Solar cell or photovoltaic cell converts solar light directly into electrical energy using photovoltaic effects. When photons striking a PV cell are absorbed, electricity is generated. Solar energy harvesting achieves the highest power density according to [29]. However, the solar energy supply could be location dependent and is highly time varying. A solar energy harvesting system involves components with different characteristics including the solar panel and the energy storage modules. A solar energy harvesting system implementation should take into account factors such as availability of light source, characteristic of the photovoltaic cell used, the chemistry and capacity of the batteries if used as energy storage, the power supply requirements, and different application behaviors [29, 9]. Novel solar cell comparisons can be found in [6].

To integrate the solar energy harvesting function into MEMS system, improving the energy conversion efficiency is important. In June 2010, a Silicon Valley based company, SunPower, claimed the new solar cell efficiency of record 24.2%. [30] presents that for day-integrated, above-bandgap direct sunlight, the Si wire arrays can achieve an absorption efficiency as high as 85%. It also shows that the light absorbed can be collected with a peak external quantum efficiency of 0.89. Although the absorption rate is not equivalent directly to sunlight-to-electricity conversion efficiency, the absorption enhancement and collection efficiency enable smaller size cells and may potentially increase the PV efficiency owing to an effective optical concentration of up to 20 times.

As the solar panel size is of concern, micro-solar power system has recently received more research attention since the energy harvesting is more challenging under limited energy storage capacity and energy utilization efficiency compared to well-addressed macro-solar power system [31]. Several micro-solar power systems have been developed by different research institutes. Heliomote [32] from University of California, LA, is a wireless sensor node with solar energy harvesting module where two AA sized NiMH rechargeable batteries are used to store the energy. A harvesting aware protocol is demonstrated and graphical interface is available to monitor the network operation. Long lifetime is one of the primary objectives of many solar energy harvesting systems and the battery life usually limits the system due to frequent charging and discharging. University of California, Berkeley, presents Prometheus (Perpetual Self Sustaining Telos Mote) [33], which adopts a two-stage energy storage system to slow down the deterioration of the battery. In the design of Prometheus, both a supercapacitor and a Lithium rechargeable battery are used and the supercapacitor serves as the primary buffer for energy charging and discharging. Trio [34], a large-scale outdoor test bed with solar energy harvesting which integrates Prometheus is also from UC Berkeley. The comparison and analysis of Heliomote and Trio can be found in [31].

To operate with high conversion efficiency, maximum power point tracking (MPPT) is desired. The maximum power point (MPP) is the point on the I-V curve where the power output at the given level of source light intensity is maximized.

MPPT is used to adjust the operating point of the solar panel to achieve optimal output. In micro-scale energy harvesting system, the maximal energy drawn from PV modules is very limited (i.e., hundreds of milliwatts) [35] due to the small size of the cells which constraints the power consumption of MPPT to a few milliwatts.

[36] considers a system with approximate MPPT which does not require a microcontroller unit to reduce the overhead. [37] is a solar energy harvesting system with built-in MPPT using only one supercapacitor for energy storage. The system is shown to achieve higher efficiency than Heliomote and Prometheus, but the control unit to run MPPT might potentially consume more power. [35] proposes a solar energy harvester using a pilot-cell as the reference for MPPT which does not consume extra power but is subject to static characterization of the panel. DuraCap [38] is a solar-powered energy harvesting system which aims to maximize the MPPT over a variety of solar panels. More discussion on solar energy harvesting system with MPPT can be found in [39].

7.2.1.3 Thermal Energy

Thermoelectric generators are solid-state power sources which convert temperature differences (gradients) directly into electrical energy using Seebeck effect. Thomas Johann Seebeck found that a voltage is produced across a circuit made from two dissimilar metals, with junctions at different temperatures. Solid-state thermoelectric generators have the advantage of long operation time, low maintenance and high reliability [9]. However, low energy conversion efficiency and operational temperature of the material in thermocouple modules limit the commercialization of thermoelectric generators. Thus, most research on thermal energy harvesting has focused on materials [40, 41, 42]. Thermoelectric devices comparison is further discussed in [6].

Although thermoelectric generators are the primary way for harvesting energy from heat, the requirement of large temperature gradient may result in limited performance in MEMS device. Instead of temperature gradient, temperature variation can also be converted to energy by using pyroelectric effect. [43] compares the pros and cons of using thermoelectric and pyroelectric generators.

7.2.2 Cooperative Communication

The idea of cooperative communication was first brought out and studied using a discrete memoryless three-terminal relay network by van der Meulen [44] and Cover and El Gamal [45]. The channel capacity between the the source and the destination, the strategies on the relay, energy efficiency and distribution among the network, and other aspects of relay networks have been the focus of intensive research [46, 47, 48, 49, 50]. The most commonly studied cooperative strategies are amplify-and-forward (AF) and decode-and-forward (DF). With AF, the relay node amplifies the source's transmission and retransmits it to the destination. The amplifier gain might depend on the estimated channel coefficient between the source and the relay.

The destination can decode the information by combining both the received signals. With the DF strategy, the relay node receives and decodes the source's transmission, then re-encodes and transmits it to the destination. The destination then combines the reception from the sender and the reception from the relay to decode the data.

Cooperative communication as virtual MIMO in which individual sensor nodes with single antenna cooperate to form multi-antenna transmitters or receivers has also been actively studied. [51] investigates the effect of propagation parameters on energy and delay efficiency of virtual MIMO scheme and develops a semi-analytical approach measuring the energy consumption of both virtual MIMO and SISO based sensor networks taking into account the extra training overhead. [52] proposes a threshold-based distributed MIMO system to achieve the maximum throughput and maintain stable queues at the same time by dynamically deciding the size of co-operative clusters. Cooperative communication introduces many opportunities for cross-layer design and optimization. Addressing the throughput, delay and energy efficiency of a wireless network with cooperative communication, [53] proposes a Medium Access Control (MAC) protocol, CoopMAC. The design of [53] is to be compatible with IEEE 802.11 MAC layer. Additional control signals are exchanged for each station to store the information about the relay candidates. The RTS/CTS mode defined in IEEE 802.11 is extended to include a Helper ready To Send (HTS). [54] is CSMA/CA based, in which two-hop paths are first searched as a non-cooperative network and then cooperative-routing is proposed to integrate the cooperative gain into the route. Besides RTS/CTS, Relay-Start (RS) and Relay-Acknowledgement (RA) are also broadcast to set up the link. [55], based also on IEEE 802.11 MAC layer DCF algorithm, chooses the relay with the best link quality and four-way handshake among the source, the relay and the destination is proposed with additional Cooperative RTS (C-RTS) and Cooperative CTS (C-CTS) defined.

Energy efficiency of cooperative communication has been investigated. [56] proposes a relay selection scheme to minimize the total energy consumption of the transmission process. In [56], multiple relays acquire the channel state information (CSI) when the source is broadcasting information. It shows that the energy consumption for data transmission can be decreased by using more relays, however the overhead of obtaining CSI increases as the number of relays increases. [57] shows the energy saving possibility with virtual MIMO. [58] studies the energy efficiency of single-relay cooperative communication networks and proposes a relay selection strategy in which the relay selected is with the least distributively evaluated signalling overhead . [59] assumes an information fusion center in densely deployed sensor networks. Evaluating the trade-off between energy efficiency and system reliability, the architecture of the fusion center is analyzed and the problem of whether sensors should cooperatively communicate with the fusion center or not is studied.

Scheduling of relays has been studied from different aspects. [60] focuses on the relay selection schemes to achieve the sub-optimal signal-to-noise ratio (SNR) but with complexity linear in the number of relays. [61] considers the optimal relay assignment problem in an ad-hoc network where multiple source-destination pairs compete for relays. Polynomial time algorithm has been developed to find relay for every pair and the minimum capacity among all pairs is maximized. [62] considers the scheduling problem in a basic three-node cooperative network. The scheduler proposed decides the role of one node as either a source or a destination or a relay under the criteria of fairness, real-time channel adaption and computational complexity. It shows that the combination of selecting the source in a round-robin way and deciding the relay according to normalized instantaneous signal-to-nose ratio can be a practical solution.

Wireless sensor networks integrating cooperative communication and energy harvesting capability is promising, which addresses the energy supply limitation concern of wireless sensor networks and utilizes the spatial diversity and environment energy. [63] considers that multiple sensors capable of energy harvesting voluntarily serve as cooperative amplify-and-forward relays. The fundamental constraint is that only the harvested and stored energy at those nodes can be used to relay. The number of relays needed to achieved the maximum spatial diversity is studied as well as the transmission power relays should use to achieve certain level of reliability. The energy harvesting at a sensor is assumed to be a stationary and ergodic process with mean specified in the unit of J/sec. The performance of the system is studied in terms of symbol-error-rate (SER). [64] designs data link automatic repeat request (ARQ) protocol with cooperative communications in energy harvesting sensor networks. One or multiple relays are requested to retransmit the data to the receiver when the information first transmitted from the source is not successfully received by the receiver. Relays selection conforms to the rule that at each sensor the average energy consumption rate does not exceed the energy generation rate.

7.3 Optimal Scheduling for Relay Usage in Sensor Networks with Energy Harvesting

In this section, wireless sensor networks with energy harvesting capability is considered and the problem of scheduling cooperative, relay based communications is addressed. Time-slotted source-relay-destination system is considered, where a sensor (the source) has the option to have another sensor (the relay) help to transmit its data to the destination. All sensor nodes under consideration are equipped with energy harvesting capability.

From an energy efficient perspective, the source may achieve the same bit error rate (BER) for a lower transmission power if it uses a relay, as compared to a direct transmission. However, this increases the power consumed by the relay and as a result, the relay sensor may not have energy to report its own data in the future. The energy harvesting capability at sensor node also adds a new dimension to the power distribution in the network. Thus, at any given instant of time, the problem of interest is for the source to determine whether to transmit data on its own or cooperatively with the relay in order to maximize the long term ratio of the data that is successfully delivered, to the total data that is generated.

In order to deduce an energy efficiency optimal scheduler, in addition to its own state information (e.g. current battery level, battery recharge and event occurrences), the source also needs to know the state information at the relay. An upper bound on the performance of an arbitrary scheduler is first derived. Then two scenarios with different system observabilities are considered. In the ideal, fully observable system, it is assumed that the source can always obtain complete knowledge about the current status at both the source and the relay. The transmission scheduling problem is then formulated as a Markov Decision Process (MDP). However, this is unrealistic and in a more practical system it is reasonable to assume that when a relay transmits or relays data, the headers of the packets may include the relay's state information. In periods without data transmission at the relay, conveying the state information of the relay in real time represents a significant overhead. Thus, in a partially observable system, where the source has to base its decision on stale state information, the scheduling problem to determine the optimal decision is formulated as a Partially Observable Markov Decision Process (POMDP). It will be shown that the POMDP can be decomposed into an equivalent MDP, the solution of which also gives the optimal solution for the POMDP.

7.3.1 Basic Metric and Methodology

7.3.1.1 Quality of Coverage

Activation of sensors varies the topology of the wireless sensor network. Thus, coverage and connectivity are common measurement of the performance of wireless sensor networks scheduler. [65] investigate wireless networks where sensors have long sleep time. The network performance is measured as the probability that any given location is not monitored by any active sensor. The performance of the network is also measured as the probability that the length of the time interval during which any location that is not covered by any active sensor is limited to some extent. Taking into account that the sensing range and the transmission range of sensors could be different, [66] designs a scheduling algorithm for sensor activation such that ensure the area is covered to a given percentage and all active sensors are connected. [67] measures the time-average utility to find an near-optimal activation scheme in a sensor network with energy harvesting to provide better quality of coverage. The scheduler in [68] does not active or de-active sensor but considers that a sensor goes into de-active mode if it does not have enough energy to detect and handle events and quality of coverage is used as the metric. For the scenario we consider in the following sections, we do not discriminate the transmission range and sensing range and consider similar Quality of Coverage as in [68].

7.3.1.2 MDP

Due to the random nature of both the recharging processes and the event occurrence process, the scheduling question outlined above is a stochastic decision problem. Markov decision process models have gained notable reputation over diverse fields including communications engineering. The reader is referred to [69] for detailed study of MDP models. A MDP model consists of decision epochs, states, actions, rewards for actions chosen at different epochs, and transition probabilities which determine the next state based on current state and the action taken [69].

[70] applied MDP model to study the activation problem in wireless sensor networks with energy-harvesting sensors. In the scenario [70] considered, the sensor has the flexibility to stay alive or asleep. Staying alive the sensor can observe any new event but might waste the energy when there is no event for a while. Since the recharging and discharging process is a stochastic process, to maximize the overall events reported, an optimal activation problem is modeled as a MDP and analyzed.

[68] considers transmission strategies in Body Sensor Networks (BSNs) which consists of sensors located around human body and equipped with energy harvesting capability. Transmission strategies with different energy consumption and correspondingly different packet error rates are available in the system considered in [68]. The goal is to minimize the probability that the sensor has no enough energy to detect and report an event while maximized the chance of correct transmission. The problem is addressed as a Markov decision process in which the recharging process and event process are modeled as stochastic processes, an instant time decision on transmission strategy costs different energy consumption determining the next system state and the reward is the successful transmission probability.

In our case, time is divided into slots also referred to as decision epochs. The decision of whether to use a relay or not at a particular slot, together with the recharging processes, determines the available battery levels at both the source and the relay at the next slot. Time-averaged quality of coverage is the objective we are optimizing, and the reward is defined to reflect instant successful event reports. Since the decision is made frequently (every time slot), the average reward optimality criterion is preferred [69] and value iteration is adopted. The MDP formulation of the problem addressed is in Section 7.3.3.

7.3.2 An Upper Bound on the Performance

In the perspective of this chapter, a discrete time model is assumed where time is slotted in intervals of unit length. Each slot is long enough so that a source node and a relay node can either cooperatively transmit one data packet for the source, or both can transmit one of their own packets. At most one data packet is generated at a node in a slot. Each sensor has a rechargeable battery and an energy harvesting device. The energy generation process at each sensor is modeled by a correlated, two-state process with parameters (q_{on}, q_{off}). In the *on* state (i.e. when ambient conditions are conducive to energy harvesting), the sensor generates energy at a constant rate of c units per time slot. In the *off* state, no energy is generated. If the sensor harvested energy in the current slot, it harvests energy in the next slot with probability q_{on} and no energy is harvested with probability $1 - q_{on}$. On the other hand, if no energy was harvested in the current slot, no energy is harvested in the next slot with probability q_{off}, and energy is harvested with probability $1 - q_{off}$, $0.5 < q_{on}, q_{off} < 1$.

It is assumed that the energy generated during a recharge event is available at the end of the slot. The two-state model can account for harvesting methods such as vibration-based event-driven harvesting, as well as solar harvesters. More complicated harvesting models can be accommodated by increasing the number of states in the Markov model.

The data packets that the sensors report to a sink are also generated according to a correlated, two-state process with parameters (p_{on}, p_{off}) with $0.5 < p_{on}, p_{off} < 1$ where in the *on* state an event (i.e. data packet) is generated in each slot, and no events are generated in the *off* state. It is assumed that an event is generated and detected at the beginning of a time slot. The average duration of a period of continuous events, $E[N]$, is

$$E[N] = \sum_{i=1}^{\infty} i(p_{on})^{i-1}(1-p_{on}) = \frac{1}{1-p_{on}}, \tag{7.1}$$

and the steady-state probability of event occurrence, π_{on}, is

$$\pi_{on} = \frac{1-p_{off}}{2-p_{on}-p_{off}}. \tag{7.2}$$

Similarly, the average length of a period without events and its steady-state probability are $\frac{1}{1-p_{off}}$ and $\pi_{off} = 1-\pi_{on}$, respectively. Also, the average length of a period with energy harvesting and the steady-state probability of such events are $\frac{1}{1-q_{on}}$ and $\mu_{on} = \frac{1-q_{off}}{2-q_{on}-q_{off}}$, respectively. Finally, the expected length of periods without recharging and its steady-state probability are $\frac{1}{1-q_{off}}$ and $\mu_{off} = 1-\mu_{on}$, respectively. The parameters corresponding to the source and relay nodes are denoted with a superscript of s and r, respectively (e.g. p^s_{on}).

To develop our models, a three-node network with one source, one relay and one destination node is considered. A source sensor has two transmission modes: 1) direct mode in which a packet is transmitted directly from the source to the destination and δ^s_1 units of energy is consumed at the source; 2) relay mode, consuming δ^s_2 units of energy, in which the packet is transmitted by the source and relayed by the relay sensor. A relay sensor also has two transmission modes: own-traffic mode and relay mode. In the own-traffic mode, the relay sensor transmits its own packet to the destination consuming δ^r_1 units of energy while in the relay mode the relay sensor's own traffic is discarded if any and δ^r_2 units of energy is consumed to relay another sensor node's packet. Let $\delta_1 > \delta_2$ where the superscript is dropped (s or r to indicate source and relay sensors, respectively) to indicate that the relation holds for both source and relay sensors. Assume that the sensors are working in real-time monitoring scenarios. Thus no retransmission is attempted for errors and packets are not buffered. Also, a sensor is considered available for operation if it has enough energy to transmit or relay a packet.

The communication strategy of a sensor pair {source, designated relay} is governed by a policy Π that decides on the transmission mode to be used for reporting events. The action taken by the sensor pair in time slot t is denoted by a_t with $a_t \in \{0,1,2,3,4\}$ denoting {no transmission, no transmission}, {direct, no transmission}, {relay, relay}, {direct, own-traffic}, and {no transmission, own-traffic}, respectively. The basic objective of the decision policy Π is to maximize quality of coverage, defined as follows. Let $\mathscr{E}_o(T)$ denote the number of events that occurred in the sensing region of a sensor over a period of T slots in the interval $[0,T]$. Also, let $\mathscr{E}_d(T)$ denote the total number of events that are detected and correctly reported by a sensor over the same period under policy Π. The time average of the fraction of events detected and correctly reported by both the source and relay nodes represents the quality of coverage and is given by

$$U(\Pi) = \lim_{T\to\infty} \frac{\mathscr{E}_d^s(T) + \mathscr{E}_d^r(T)}{\mathscr{E}_o^s(T) + \mathscr{E}_o^r(T)}. \tag{7.3}$$

A transmission action can be successfully taken only if the corresponding sensor has enough power (δ_1 for direct/own-traffic mode and δ_2 for relay mode) and an event occurs at the beginning of the slot. Let T_i with $i \in \{0,1,2,3,4\}$ denote the number of time slots in which action i is successfully taken over the period $[0,T]$, under the optimal policy Π_{OPT}. Then events reported by source and relay nodes will be $\mathscr{E}_d^s(T) = T_1 + T_2 + T_3$ and $\mathscr{E}_d^r(T) = T_3 + T_4$ respectively. As $T \to \infty$, the number of events occurring in the interval $[0,T]$ satisfies

$$\lim_{T\to\infty} \frac{\mathscr{E}_o^s(T)}{T} = \pi_{on}^s = \frac{1 - p_{off}^s}{2 - p_{on}^s - p_{off}^s} \tag{7.4}$$

$$\lim_{T\to\infty} \frac{\mathscr{E}_o^r(T)}{T} = \pi_{on}^r = \frac{1 - p_{off}^r}{2 - p_{on}^r - p_{off}^r}. \tag{7.5}$$

Let the available energy at the sensor at the beginning of slot t be L_t and assume that the initial energy level is L_0. The expected energy level of the source sensor and the relay sensor at time T are given by

$$E[L_T^s] = L_0^s - (T_1 + T_3)\delta_1^s - T_2\delta_2^s + T\mu_{on}^s c^s, \tag{7.6}$$

$$E[L_T^r] = L_0^r - (T_3 + T_4)\delta_1^r - T_2\delta_2^r + T\mu_{on}^r c^r. \tag{7.7}$$

Using the fact that $E[L_T^s] \geq 0$ and $\delta_1^s > \delta_2^s$

$$\lim_{T\to\infty} \frac{T_1 + T_3 + T_2}{T} \leq \frac{\mu_{on}^s c^s}{\delta_2^s}. \tag{7.8}$$

Additionally, since $\frac{T_1+T_2+T_3}{T} \leq \frac{\mathscr{E}_o^s(T)}{T}$,

$$\lim_{T\to\infty} \frac{T_1 + T_3 + T_2}{T} \leq \min\{\frac{\mu_{on}^s c^s}{\delta_2^s}, \pi_{on}^s\}. \tag{7.9}$$

Using the fact that $E[L^r_T] \geq 0$, $T_2 \geq 0$ and $\frac{T_3+T_4}{T} \leq \frac{\mathscr{E}^r_o(T)}{T}$, it can also be shown that

$$\lim_{T\to\infty} \frac{T_3+T_4}{T} \leq \min\{\frac{\mu^r_{on}c^r}{\delta^r_1}, \pi^r_{on}\}. \tag{7.10}$$

Finally, combining Eqns. (7.9) and (7.10), the performance of the optimal policy and thus any arbitrary policy is bounded by

$$\begin{aligned} U(\Pi_{OPT}) &= \lim_{T\to\infty} \frac{\mathscr{E}^s_d(T)+\mathscr{E}^r_d(T)}{\mathscr{E}^s_o(T)+\mathscr{E}^r_o(T)} \\ &\leq \frac{\min\{\frac{\mu^s_{on}c^s}{\delta^s_2}, \pi^s_{on}\} + \min\{\frac{\mu^r_{on}c^r}{\delta^r_1}, \pi^r_{on}\}}{\pi^s_{on}+\pi^r_{on}}. \end{aligned} \tag{7.11}$$

7.3.3 MDP Formulation

The problem of developing the optimal scheduling problem is fairly challenging due to the number of variables involved and their complex interactions. In this section it will show that the problem may be modeled as a MDP, thereby the optimal policy can be obtained[1].

Denote the system state at time t by $X_t = (L^s_t, E^s_t, C^s_t, L^r_t, E^r_t, C^r_t)$ where $L^s_t, L^r_t \in \{0,1,2,\cdots,K\}$ represents the energy available in the sensors at time t, and $E^s_t, E^r_t \in \{0,1\}$ equals one if an event to be reported during time interval $[t,t+1)$ occurred at time t and zero otherwise. Also, $C^s_t, C^r_t \in \{0,1\}$ equals one if the sensor recharged during time interval $[t-1,t)$ and zero otherwise. (i.e. it is assumed that the sensor does not know at time t if it will recharge during interval $[t,t+1)$). The battery capacity of the sensor is assumed to be K. The action taken at time t is denoted by $a_t \in \{0,1,2,3,4\}$ as described in Section 7.3.2. The next state of the system depends only on the current state and the action taken. Thus the system constitutes a Markov Decision Process [69].

Let θ^s and θ^r denote the reward gained by the system for each source sensor and relay sensor event that is successfully reported. The values of θ^s and θ^r may be chosen to reflect the importance of the observations of each sensor. Alternatively, θ^s and θ^r may also be made equal to the probability that a transmitted packet is received without errors, in order to account for channel errors. The reward function $R(X_t, a_t)$ is then given by,

$$R(X_t,a_t) = \begin{cases} \theta^s & \text{if } a_t = 1 \text{ or } 2, E^s_t = 1 \\ \theta^s+\theta^r & \text{if } a_t = 3, E^s_t = E^r_t = 1 \\ \theta^r & \text{if } a_t = 4, E^r_t = 1 \\ 0 & \text{otherwise.} \end{cases} \tag{7.12}$$

[1] Readers are referred to [69] for an introduction to MDPs.

Let g_t^s and g_t^r be the amount of energy gained by the source and relay sensors in the interval $[t, t+1)$ respectively. Then,

$$g_t^s = \begin{cases} c^s & \text{w.p. } C_t^s q_{on}^s + (1 - C_t^s)(1 - q_{off}^s) \\ 0 & \text{otherwise} \end{cases} \tag{7.13}$$

$$g_t^r = \begin{cases} c^r & \text{w.p. } C_t^r q_{on}^r + (1 - C_t^r)(1 - q_{off}^r) \\ 0 & \text{otherwise} \end{cases} \tag{7.14}$$

where *w.p.* stands for "with probability". Let l_t^s and l_t^r be the amount of energy spent by the source and relay sensors in the interval $[t, t+1)$ respectively. Then,

$$l_t^s = \begin{cases} \delta_1^s & \text{if } a_t = 1 \text{ or } a_t = 3 \\ \delta_2^s & \text{if } a_t = 2 \\ 0 & \text{otherwise} \end{cases} \tag{7.15}$$

$$l_t^r = \begin{cases} \delta_1^r & \text{if } a_t = 3 \text{ or } a_t = 4 \\ \delta_2^r & \text{if } a_t = 2 \\ 0 & \text{otherwise.} \end{cases} \tag{7.16}$$

To complete the MDP formulation, the system state at time $t+1$ is given by

$$X_{t+1} = (L_{t+1}^s, E_{t+1}^s, C_{t+1}^s, L_{t+1}^r, E_{t+1}^r, C_{t+1}^r), \tag{7.17}$$

where

$$L_{t+1}^s = \max\{\min\{L_t^s + g_t^s - l_t^s, 0\}, K\} \tag{7.18}$$

$$L_{t+1}^r = \max\{\min\{L_t^r + g_t^r - l_t^r, 0\}, K\} \tag{7.19}$$

$$E_{t+1}^s = \begin{cases} 1 & \text{w.p. } E_t^s p_{on}^s + (1 - E_t^s)(1 - p_{off}^s) \\ 0 & \text{otherwise} \end{cases} \tag{7.20}$$

$$E_{t+1}^r = \begin{cases} 1 & \text{w.p. } E_t^r p_{on}^r + (1 - E_t^r)(1 - p_{off}^r) \\ 0 & \text{otherwise} \end{cases} \tag{7.21}$$

$$C_{t+1}^s = \begin{cases} 1 & \text{w.p. } C_t^s q_{on}^s + (1 - C_t^s)(1 - q_{off}^s) \\ 0 & \text{otherwise} \end{cases} \tag{7.22}$$

$$C_{t+1}^r = \begin{cases} 1 & \text{w.p. } C_t^r q_{on}^r + (1 - C_t^r)(1 - q_{off}^r) \\ 0 & \text{otherwise.} \end{cases} \tag{7.23}$$

The objective is to maximize the average reward criteria over an infinite horizon. The optimal solution can be computed by using value iteration [69]. Since the induced Markov chain is unichain, from Theorem 8.5.2 of [69], there exists a deterministic, Markov, stationary optimal policy Π_{MD} which also leads to a steady-state transition probability matrix. Considering the average expected reward criteria, the optimality equations are given by [71]

$$\lambda^* + h^*(X) = \max_{a\in\{0,1,2,3,4\}} \left[R(X,a) + \sum_{X'=(0,0,0,0,0,0)}^{(K,1,1,K,1,1)} p_{X,X'}(a) h^*(X') \right],$$
$$\forall X \in \{(0,0,0,0,0,0), \cdots, (K,1,1,K,1,1)\} \qquad (7.24)$$

where $p_{X,X'}(a)$ represents the transition probability from state X to X' when action a is taken, λ^* is the optimal average reward and $h^*(i)$ are the optimal rewards when starting at state $i = (0,0,0,0,0,0)$, $\cdots$, $(K,1,1,K,1,1)$. For the purpose of evaluation, the relative value iteration technique [71] is used to solve Eqn. (7.24).

7.3.4 POMDP Formulation

For the partially observable system, the decision problem is first formulated as a POMDP, and then the equivalent MDP formulation is presented, the detailed process can be found in [74].

7.3.4.1 System States and Observations

The system state at time t is denoted by $X_t = (L_t^s, E_t^s, C_t^s, L_t^r, E_t^r, C_t^r)$, as in Section 7.3.3, except that the variable $E_t^r \in \{0,1\}$ is defined similarly for the relay but equals one if the event process is *on* during time interval $[t-1,t)$. Note that while E_t^r, C_t^s and C_t^r are based on the interval $[t-1,t)$, E_t^s is based on the interval $[t,t+1)$. The state of the relay at time t is defined in terms of the previous slot since that is the latest information the source may have about the relay. Assume that the battery at a sensor has a finite capacity K. Denote the set of actions described in Section 7.3.2 by $\mathscr{A} = \{0,1,2,3,4\}$. The action taken at time t is denoted by $a_t \in \mathscr{A}$.

The *system observation* at time t at the source sensor is denoted by Y_t. The source is assumed to always have full information about itself. If the action taken at time $t-1$ is 2,3 or 4, then the relay was active, and the observation matches the state and equals X_t. However, if the action taken was 0 or 1, the relay was inactive. Thus the state of the event and energy generation processes at the relay are not known, along with the energy level at the relay (due to the possibility of recharging). Thus the observation Y_t is characterized by,

$$Y_t = \begin{cases} X_t & \text{if } a_{t-1} \in \{2,3,4\} \\ (L_t^s, E_t^s, C_t^s, \phi_L, \phi_E, \phi_C) & \text{if } a_{t-1} \in \{0,1\} \end{cases}$$

where ϕ_ω denotes that a variable ω is unknown.

7.3.4.2 POMDP Transformation

In the presence of only partial observations, the optimal action depends on the current and past observations, and on past actions. Existing work has shown that a

POMDP may be formulated as a completely observable MDP with the same finite action set [71, 72, 73]. The state space for the equivalent MDP comprises of the space of probability distributions on the original state space. Thus in the general case, the state space of the equivalent MDP may become uncountable or infinite. In this case, the structure of the POMDP leads to a countable state space for the equivalent MDP, guaranteeing the existence of an optimal solution to the average cost (reward) optimality equation [73]. As a result, the solution to the equivalent MDP with complete state information provides the optimal actions to take in the POMDP, and with the optimal reward.

Denote the state space of the equivalent MDP as Δ, and its state at time t as Z_t. Then $Z_t \in \Delta$ is a information vector of length $|\mathscr{X}|$, whose i-th component is given by,

$$Z_t^{(i)} = Pr[X_t = i | y_t, \cdots, y_1; a_{t-1}, \cdots, a_0], \quad i \in \mathscr{X}. \tag{7.25}$$

Z_t satisfies $\mathscr{I}' Z_t = 1$, where $\mathscr{I}'$ denotes a row vector of length $|\mathscr{X}|$ with all elements equal to 1, and $\mathscr{X}$ denotes the state space of system state X_t.

It can be shown [74] that the state space of the equivalent MDP is countable and that the state at time t, Z_t, can be represented in the form $Z_t = (L_t^s, E_t^s, C_t^s, L^r, E^r, C^r, i)$, representing the following: (a) the relay had no transmissions in the past i slots; (b) the state of the relay when it last transmitted was (L^r, E^r, C^r); (c) the current state at the source is (L_t^s, E_t^s, C_t^s).

The POMDP is then transformed to an equivalent MDP with state space Δ and the optimality equations for this MDP are given by [73]:

$$\Gamma^* + h^*(Z) = \max_{a \in \mathscr{A}} \left[\bar{R}(Z,a) + \sum_{y \in \mathscr{Y}} V(y,Z,a) h^*\left(\frac{\bar{T}(y,Z,a)}{V(y,Z,a)} \right) \right], \qquad \forall Z \in \Delta \tag{7.26}$$

where $h^*(Z)$ is the optimal reward when starting at state Z, $\bar{T}(y, Z_t, a_t)$ represents the probability of event $\{X_{t+1} = i, Y_{t+1} = y\}$ given past actions and observations, $V(y, Z_t, a_t)$ is interpreted as the probability of $Y_{t+1} = y$ given the past actions and observations, and $\bar{R}(Z,a) = Z'[R(i,a)]_{i \in \mathscr{X}}$ is the reward function which will be discussed in Section 7.3.4.3. These equations can be solved using the relative value iteration algorithm [71], however, exact closed-form expressions for h^* and Γ^* may not exist, particularly for finite K.

7.3.4.3 Equivalent MDP Reward Function

Recall that θ^s and θ^r denote the rewards gained by the system for each source sensor and relay sensor event that is successfully reported, respectively. For the partially observable system, the reward associated with the states $Z \in \Delta$ of the equivalent MDP, denoted as $\bar{R}(Z,a)$, is the same as that of the optimal reward for the original

POMDP [73]. Then, the reward function of the equivalent MDP at time t is given by,

$$\bar{R}(Z,a) = \begin{cases} \theta^s & \text{if } a_t=1, E_t^s=1, L_t^s \geq \delta_1^s \\ \theta^s Pr[L_t^r \geq \delta_2^r | L_{t-i}^r = L^r] & \text{if } a_t=2, E_t^s=1, L_t^s \geq \delta_2^s \\ \theta^s + \theta^r Pr[L_t^r \geq \delta_1^r, E_t^r=1 | L_{t-i}^r = L^r, E_{t-i}^r = E^r] & \text{if } a_t=3, E_t^s=1, L_t^s \geq \delta_1^s \\ \theta^r Pr[L_t^r \geq \delta_1^r, E_t^r=1 | L_{t-i}^r = L^r, E_{t-i}^r = E^r] & \text{if } a_t=4 \\ 0 & \text{otherwise.} \end{cases} \quad (7.27)$$

7.3.5 Simulation Results

This section explores the impact of various parameters on the performance of the proposed schedulers using simulations presenting the results when only a single three-node-group (i.e. source, relay and destination) is present in the network. Networks with multiple groups are further considered in [75]. All simulations were run for a duration of 5000000 time units and physical layer aspects such as bit errors were not considered. All figures show the quality of coverage (unless noted otherwise) defined as the ratio of the number of events successfully reported to the total number of events generated.

Figure 7.1 demonstrates the effect of the event generation process on the performance of a fully observable system with MDP formulated policy, along with the theoretical upper bound from Section 7.3.2, and all parameters are specified in the caption. In all the four cases, the recharge process parameters and transmission energies are the same for both the source and the relay sensors. Since p_{on} and p_{off} are constrained in the range (0.5,1.0), the four choices of (0.6,0.6), (0.6,0.9), (0.9,0.6) and (0.9,0.9) in Figure 7.1 give an indication of the performance in diverse settings of low-low, low-high, high-low and high-high correlation probabilities at the relay. For each of the four cases, the parameters p_{on}^s and p_{off}^s of the event generation process of the source node are varied from 0.55 to 0.95. The quality of coverage decreases as p_{on}^s increases, since an increase in p_{on}^s increases both π_{on}^s and the average length of periods with continuous packets, while decreasing π_{off}^s and the interval between bursts of traffic. With a relatively higher (close to 1) p_{on}^r and lower (close to 0.5) p_{off}^r, as in Figure 7.1(c), the event generation rate at the relay node increases, thereby reducing the energy available at the relay for helping the source sensor. Thus the quality of coverage degrades in this case, and in contrast, improves with a lower p_{on}^r and higher p_{off}^r (Figure 7.1(b)).

In Figure 7.1, the theoretical bound is tighter when p_{on}^s is low and p_{off}^s is high. For an intuition behind this observation, it is noted that the bound in Eqn. (7.11) uses two approximations: (a) to achieve the first term in its numerator, the term $\frac{T_1+T_3}{T}\frac{\delta_1^s-\delta_2^s}{\delta_2^s}$ is omitted during derivation, and, (b) the second term in the numerator of Eqn. (7.11) neglects the term $\frac{T_2}{T}\frac{\delta_2^r}{\delta_1^r}$ during derivation. When p_{on}^s is low and p_{off}^s is high, $\pi_{on}^s < \frac{\mu_{on}^s c^s}{\delta_2^s}$. The source sensor has enough energy and tends to transmit the packet directly and the relay spends most of its energy and time slots on its own

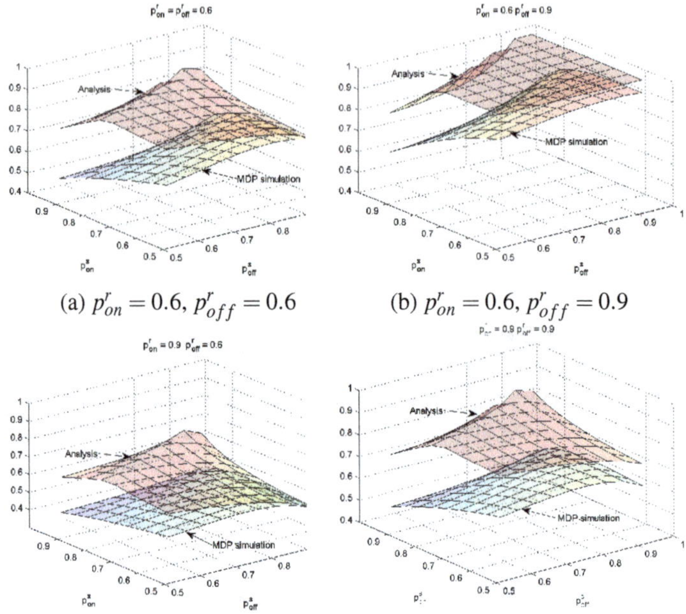

(a) $p^r_{on} = 0.6$, $p^r_{off} = 0.6$ (b) $p^r_{on} = 0.6$, $p^r_{off} = 0.9$

(c) $p^r_{on} = 0.9$, $p^r_{off} = 0.6$ (d) $p^r_{on} = 0.9$, $p^r_{off} = 0.9$

Fig. 7.1 Effect of p_{on} and p_{off} on the quality of coverage (z-axis) of a fully observable system and the theoretical upper bound. Parameters used: $q^s_{on} = q^r_{on} = 0.85$, $q^s_{off} = q^r_{off} = 0.7$, $c^s = c^r = 1$, $\delta^s_1 = \delta^r_1 = 2$, $\delta^s_2 = \delta^r_2 = 1$, $\theta^s = \theta^r = 1$, $K = 20$.

traffic. Consequently, T_2 is small and the approximation error from the second term is very small, and the bound is thus tight. On the other hand, as π^s_{on} increases, the first term of the numerator of Eqn. (7.11) is determined by $\frac{\mu^s_{on} c^s}{\delta^s_2}$ which introduces an error corresponding to the term $\frac{T_1+T_3}{T} \frac{\delta^s_1-\delta^s_2}{\delta^s_2}$ in Eqn. (7.8). Also, when π^s_{on} is large, the packet generation rate at the source is high, resulting in a low battery level at the source and a higher use of the relay to transmit the source's traffic. Thus the fraction $\frac{T_2}{T}$ increases and the error introduced by the second term in the numerator of Eqn. (7.11) also increases, and the bound becomes looser.

Similar set of results are evaluated to explore the impact of the recharge process on the performance. For those results, the event generation process parameters and the transmission energies were the same for both the source and relay sensors. It was observed that the quality of coverage increases as q^s_{on} increases and q^s_{off} decreases, i.e. when the steady state probability of recharging and the average length of continuous recharging slots of the source node increase. The rate of increase in the quality of coverage with q^s_{on} is approximately linear when q^s_{off} is close to 0.5 and approximately exponential when it is close to 1. Also, as q^r_{on} increases and q^r_{off} decreases, the quality of coverage increases in general. The theoretical bound is tighter when both q^s_{on} is small (close to 0.5) and q^s_{off} is large (close to 1). The intuition behind the results is similar to that given for Figure 7.1.

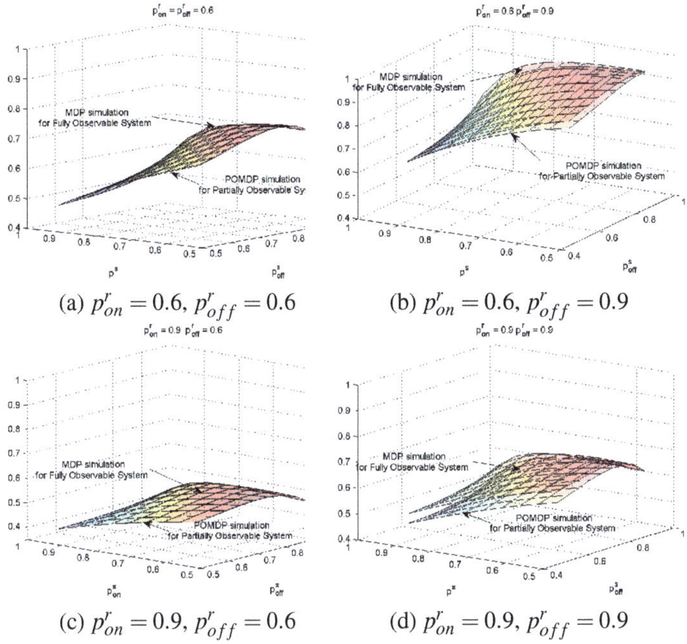

(a) $p^r_{on} = 0.6$, $p^r_{off} = 0.6$ (b) $p^r_{on} = 0.6$, $p^r_{off} = 0.9$

(c) $p^r_{on} = 0.9$, $p^r_{off} = 0.6$ (d) $p^r_{on} = 0.9$, $p^r_{off} = 0.9$

Fig. 7.2 Effect of p_{on} and p_{off} on the quality of coverage (z-axis) of a partially observable system and a fully observable system. Parameters used: $q^s_{on} = q^r_{on} = 0.85$, $q^s_{off} = q^r_{off} = 0.7$, $c^s = c^r = 1$, $\delta^s_1 = \delta^r_1 = 2$, $\delta^s_2 = \delta^r_2 = 1$, $\theta^s = \theta^r = 1$, $K = 20$.

Figure 7.2 demonstrates the effect of the event generation process on the performance for a partially observable system formulated as POMDP, compared with results of a fully observable system. All parameters are specified in the caption. In all the four cases, the recharge process parameters and the transmission energies are the same for both the source and the relay sensors. The four choices of $(0.6,0.6)$, $(0.6,0.9)$, $(0.9,0.6)$ and $(0.9,0.9)$ give an indication of the performance in diverse settings of low-low, low-high, high-low and high-high correlation probabilities at the relay. For each of the four cases, the parameters p^s_{on} and p^s_{off} of the event generation process at the source node are varied from 0.55 to 0.95. Overall, the quality of coverage is higher when the relay has a lower p^r_{on} and higher p^r_{off}. The percentage of relay usage in the four cases are shown in Table 7.1. The table shows the overall quality of coverage (QoC) defined earlier along with the individual QoCs defined as the following: *Source QoC* is the ratio of the number of packets transmitted to the number of packets generated by the source and *Relay QoC* is the ratio of the number of packets transmitted (its own) to the number of packets generated by the relay. *Relay usage* is defined as the ratio of the number of source packets transmitted using the relay to the total number of packets transmitted by the source. For cases $(0.6,0.6)$ and $(0.9,0.9)$, the steady-state probabilities of the event occurrence at the relay (π^r_{on}) are the same (1/2), but as the length of continuous events at the relay ($E(N)$, defined in Eqn. (7.1)) increases, the source tends to transmit the traffic

Table 7.1 Relay usage summary. (Parameters used: $q^s_{on} = q^r_{on} = 0.85$, $q^s_{off} = q^r_{off} = 0.7$, $p^s_{on} = 0.85$, $p^s_{off} = 0.7$, $c^s = c^r = 1$, $\delta^s_1 = \delta^r_1 = 2$, $\delta^s_2 = \delta^r_2 = 1$, $\theta^s = \theta^r = 1$, $K = 20$)

(p^r_{on}, p^r_{off})	(0.6,0.6)	(0.6,0.9)	(0.9,0.6)	(0.9,0.9)
QoC	0.5662	0.7309	0.4510	0.5179
Source QoC	0.5945	0.7765	0.4934	0.4932
Relay usage	0.3372	0.7477	0.0000	0.0000
Relay QoC	0.5284	0.5793	0.4157	0.5507

directly as long as it has enough energy. When both $E(N)$ and π^r_{on} are low ((0.6, 0.9) case), the relay is used intensively by the source. The reason behind this is that when the reward for transmitting one source packet equals that of transmitting one relay packet, from Eqn. (7.27) The system reward is maximized when the relay has enough energy such that the relay and the source could transmit their own traffic, respectively, in the same slot. Thus, intuitively, as the traffic rate at the relay increases, the relay tends to use its energy for its own traffic. The QoC of a fully observable system is slightly higher than that of the partially observable system. The difference is larger when p^r_{off} is higher. This is so because when p^r_{off} is higher, the relay has more energy available most of the times, and thus can be used more often for its own transmissions or for relaying transmissions. In the MDP case, the sensor has complete information about the relay's state and can utilize the relay fully, whereas in the POMDP case, the source is not able to utilize the relay fully due to partial state information availability, resulting in larger performance difference.

7.4 Conclusion

Harvesting ambient energy sources has been a promising and green way to address the limited onboard battery capacity which constrains many wireless sensor network applications in remote or inaccessible areas. Cooperative communication has been shown to increase energy efficiency. In this chapter, energy-efficient relay scheduling strategies are designed for energy harvesting sensor networks. Given the knowledge of the current network state, in order to maximize the long-term utility of the network, the problem of whether a source sensor should transmit its data directly or cooperatively with a relay has been formulated as a MDP or POMDP respectively, depending on whether the system is fully or partially observable. An upper bound on the performance of arbitrary strategy is obtained. Simulation results are used to show the performance of the proposed strategies, and various insights towards optimal relay scheduling are discussed.

References

1. Akyildiz, I., Su, W., Sankarasubramaniam, Y., Cayirci, E.: Wireless Sensor Networks and Applications: a Survey. Int. J. Comp. Sci. and Network Security 7(3), 264–273 (2007)
2. Aziz, O., Lo, B., King, R., Darzi, A., et al.: Pervasive body sensor network: an approach to monitoring the post-operative surgical patient. In: Proc. International Workshop on Wearable and Implantable Body Sensor Networks, BSN 2006 (2006)
3. Falck, T., Espina, J., Ebert, J.P., Dietterle, D.: BASUMA - the sixth sense for chronically ill patients. In: Proc. International Workshop on Wearable and Implantable Body Sensor Networks, BSN 2006 (2006)
4. Jovanov, E., O'Donnell Lords, A., Raskovic, D., et al.: Stress monitoring using a distributed wireless intelligent sensor system. IEEE Engineering in Medicine and Biology Magazine 22(3), 49–55 (2003)
5. Paradiso, J.A., Starner, T.: Energy scavenging for mobile and wireless electronics. IEEE Pervasive Computing 4(1), 18–27 (2005)
6. Cook-Chennault, K.A., Thambi, N., Sastry, A.M.: Powering MEMS Portable Devices-A Review of Non-Regenerative and Regenerative Power Supply Systems With Emphasis on Piezoelectric Energy Harvesting Systems. Smart Mater. Struct. (2008), doi:10.1088/0964-1726/17/4/043001
7. Armand, M., Tarascon, J.-M.: Building better batteries. Nature (2008), doi:10.1038/451652a
8. Laneman, J.N., Tse, D., Wornell, G.W.: Cooperative diversity in wireless networks: Efficient protocols and outage behavior. IEEE Trans. Inf. Theory 50(12), 3062–3080 (2004)
9. Chalasani, S., Conrad, J.M.: A Survey of Energy Harvesting Sources for Embedded Systems. IEEE South east con (2008)
10. Sasaki, K., Osaki, Y., Okazaki, J., Hosaka, H., Itao, K.: Vibration-based automatic power-generation system. Technol.-Micro Nanosyst.-Information Storage Processing Syst. 11, 965–969 (2005)
11. Beeby, S.P., Tudor, M.J., White, N.M.: Energy harvesting vibration sources for Microsystems applications. J. Measurement Science and Technology 17, 175–195 (2006)
12. Sodano, H.A., Inman, D.J., Park, G.: A Review of Power Harvesting from Vibration Using Piezoelectric Materials. The Shock and Vibration Digest 36, 197–205 (2004)
13. Kymissis, J., Kendall, C., Paradiso, J., Gershenfeld, N.: Parasitic Power Harvesting in Shoes. In: Proc. of the Second IEEE International Conference on Wearable Computing (ISWC). IEEE Computer, Los Alamitos (1998)
14. De Marqui Jr. C., Vieira, W.G.R., Erturk, A., Inman, D.J.: Piezoaeroelastic Modeling and Analysis of a Generator Wing With Continuous and Segmented Electrodes. J. Intell. Mater. Syst. Struct. 21, 983–993 (2010)
15. Ramadass, Y.K., Chandrakasan, A.P.: An Efficient Piezoelectric Energy harvesting Interface Circuit Using a Bias-Flip Rectifier and Shared Inductor. IEEE Journal of Solid-State Circuits 45(1), 189–204 (2010)
16. Starner, T.: Human-Powered Wearable Computing. IBM Systems Journal 35, 618–629 (1996)
17. Umeda, M., Nakamura, K., Ueha, S.: Energy Storage Characteristics of a Piezo-Generator Using Impact Induced Vibration. Japanese Journal of Applied Physics (1997), doi:10.1143/JJAP.36.3146
18. Kimura: M, Piezoelectric Generation Device. United States Patent Number 5801475 (1998)
19. Sodano, H., Inman, D., Park, G.: Generation and storage of electricity from power harvesting devices. J. Intell. Mater. Syst. Struct. 16, 67–75 (2005)

20. Pereira, T., Scaffaro, R.: Guo et al Performance of thin-film lithium energy cells under uniaxial pressure. Advanced Engineering Materials 10(4), 393–399 (2008)
21. Pereira, T., et al.: The performance of thin-film Li-ion batteries under flexural deflection. Journal of Micromechanics and Microengineering 16(12), 2714–2721 (2006)
22. Pereira, T., Guo, Z., Nieh, S., et al.: Embedding thin-film lithium energy cells in structural composites. Composites Science and Technology 68(7-8), 1935–1941 (2008)
23. Anton, S.R., Erturk, A., Inman, D.J.: Piezoelectric energy harvesting from multifunctional wing spars for UAVs: Part 2. Experiments and storage applications. In: Proc. SPIE, vol. 7288 (2009), doi:10.1117/12.815799
24. Soliman, M.S.M., Abdel-Rahman, E.M., El-Saadany, E.F., Mansour, R.R.: A wideband vibration-based energy harvester. J. Micromech. Microeng. 115021 (2008), doi:10.1088/0960-1317/18/11/115021
25. Liu, F., Phipps, A., Horowitz, S., et al.: Acoustic energy harvesting using an electromechanical Helmholtz resonator. J. Acoust. Soc. Am. 123, 1983–1990 (2008)
26. Halvorsen, E.: Energy harvesters driven by broadband random vibrations. J. Microelectromech. Syst. 17, 1061–1071 (2008)
27. Adhikari, S., Friswell, M.I., Inman, D.J.: Piezoelectric energy harvesting from broadband random vibrations. Smart Mater. Struct. (2009), doi:10.1088/0964-1726/18/11/115005
28. Cottone, F., Vocca, H., Gammaitoni, L.: Nonlinear Energy Harvesting. Phys. Rev. Lett. (2009), doi:10.1103/PhysRevLett.102.080601
29. Raghunathan, V., Kansal, A., Hsu, J., et al.: Design considerations for solar energy harvesting wireless embedded systems. In: IEEE IPSN, pp. 457–462 (2005)
30. Kelzenberg, M.D., et al.: Enhanced absorption and carrier collection in Si wire arrays for photovoltaic applications. Nature Materials 9, 239–244 (2010)
31. Jeong, J., Jiang, X., Culler, D.: Design and Analysis of MicroSolar Power Systems for Wireless Sensor Networks, In: 5th International Conference on Networked Sensing Systems, INSS 2008, pp. 181–188 (2008)
32. Lin, K., et al.: Heliomote: enabling long-lived sensor networks through solar energy harvesting. In: 2005: Proceedings of the 3rd International Conference on Embedded Networked Sensor Systems (2005)
33. Jiang, X., Polastre, J., Culler, D.: Perpetual environmentally powered sensor networks. In: Information Processing in Sensor Networks, vol. 2005, pp. 463–468 (2005)
34. Dutta, P., et al.: Trio: Enabling Sustainable and Scalable Outdoor Wireless Sensor Network Deployments. IEEE SPOTS, 407–415 (2006)
35. Brunelli, D., Moser, C., Thiele, L., Benini, L.: Design of a solar-harvesting circuit for batteryless embedded systems. IEEE Transactions on Circuits and Systems I: Regular Papers 56(11), 2519–2528 (2009)
36. Park, C., Chou, P.H.: AmbiMax: Efficient, autonomous energy harvesting system for multiple-supply wireless sensor nodes. In: Proc. Third Annual IEEE Communications Society Conference on Sensor, Mesh, and Ad Hoc Communications and Networks (SECON), pp. 168–177 (2006)
37. Simjee, F., Chou, P.H.: Everlast: Long-life, supercapacitor-operated wireless sensor node. In: Proc. ISLPED, pp. 197–202 (2006)
38. Chen, C.-Y., Chou, P.H.: DuraCap: a Supercapacitor-Based, Power-Bootstrapping, Maximum Power Point Tracking Energy-Harvesting System. In: Proceeding ISLPED 2010 Proceedings of the 16th ACM/IEEE International Symposium on Low Power Electronics and Design (2010)
39. Alippi, C., Galperti, C.: An adaptive system for optimal solar energy harvesting in wireless sensor network nodes. IEEE Trans. Circuits Syst. I, Reg. Papers 55(6), 1742–1750 (2008)

40. Bottner, H., et al.: New thermoelectric components using microsystem technologies. J. Microelectromech. Syst. 13, 414–420 (2004)
41. Hochbaum, A.I., et al.: Enhanced thermoelectric performance of rough silicon nanowires. Nature 451, 163–167 (2008)
42. Snyder, G.J., Toberer, E.S.: Complex thermoelectric materials. Nature Materials 7, 105–114 (2008)
43. Sebald, G., Guyomar, D., Agbossou, A.: On thermoelectric and pyroelectric energy harvesting. Smart Mater. Struct. (2009), doi:10.1088/0964-1726/18/12/125006
44. van der Meulen, E.: Three-terminal communication channels. Adv. Appl. Prob. 3, 120–154 (1971)
45. Cover, T., El Gamal, A.: Capacity theorems for the relay channel. IEEE Transactions on Information Theory 25(5), 572–584 (1979)
46. Host-Madsen, A.: On the Capacity of Wireless Relaying. In: Proc. IEEE VTC (2002)
47. Kramer, G., Gastpar, M., Gupta, P.: Cooperative Strategies and Capacity Theorems for Relay Networks. IEEE Transactions on Information Theory 51(9), 3037–3063 (2005)
48. Hong, Y.-W., Scaglione, A.: Energy-efficient broadcasting with cooperative transmissions in wireless sensor networks. IEEE Trans. Wireless Comm. 5(10), 2844–2855 (2006)
49. Guo, L., Ding, X., Wang, H., Li, Q., Chen, S., Zhang, X.: Cooperative Relay Service in a Wireless LAN. IEEE Journal on Selected Areas in Communications 25(2), 355–368 (2007)
50. Deng, X., Haimovich, A.M.: Power allocation for cooperative relaying in wireless networks. IEEE Communication Letters 9(11), 994–996 (2005)
51. Jayaweera, S.: Virtual MIMO-based cooperative communication for energy-constrained wireless sensor networks. IEEE Trans. Wireless Comm. 5(5), 984–989 (2006)
52. Yang, H., Shen, H.-Y., Sikdar, B., Kalyanaraman, S.: A threshold based MAC protocol for cooperative MIMO transmissions. In: Proceedings of IEEE INFOCOM Minisymposium (2009)
53. Liu, P.: Cooperative wireless communications: a cross-layer approach. IEEE Wireless Communications 13(4), 84–92 (2006)
54. Azgin, A., Altunbasak, Y., AlRegib, G.: Cooperative MAC and routing protocols for wireless ad hoc networks. In: Proceedings of IEEE GLOBECOM, pp. 2854–2859 (2005)
55. Moh, S., et al.: CD-MAC: Cooperative Diversity MAC for Robust Communication in Wireless Ad Hoc Networks. Communications. In: IEEE International Conference on ICC 2007, pp. 3636–3641 (2007)
56. Madan, R., Mehta, N., Molisch, A., et al.: Energy-Efficient Cooperative Relaying over Fading Channels with Simple Relay Selection. IEEE Transactions on Wireless Communications 7(8), 3013–3025 (2008)
57. Cui, S., Goldsmith, A.J., Bahai, A.: Energy-efficiency of MIMO and Cooperative MIMO Techniques in Sensor Networks. IEEE Journal on Selected Areas in Communications 22(6), 1089–1098 (2004)
58. Zhou, Z., Zhou, S., Cui, J.-H., Cui, S.: Energy-Efficient Cooperative Communication Based on Power Control and Selective Single-Relay in Wireless Sensor Networks. IEEE Transactions on Wireless Communications 7(8), 3066–3078 (2008)
59. Quek, T.Q.S., Dardari, D., Win, M.Z.: Energy efficiency of dense wireless sensor networks: to cooperate or not to cooperate. IEEE Journal on Selected Areas in Communications 25(2), 459–470 (2007)
60. Jing, Y., Jafarkhani, H.: Single and Multiple Relay Selection Schemes andtheir Achievable Diversity Orders. IEEE Transactions on Wireless Communications 8(3), 1414–1423 (2009)

61. Shi, Y., Sharma, S., Hou, Y.T., Kompella, S.: Optimal Relay Assignment for Cooperative Communications. In: Proceedings of the 9th ACM International Symposium on Mobile Ad hoc Networking and Computing (2008), doi:10.1145/1374618.1374621
62. Krikidis, I., Belfiore, J.C.: Scheduling for Amplify-and-Forward Cooperative Networks. IEEE Transactions on Vehicular Technology 56(6), 3780–3790 (2007)
63. Medepally, B., Mehta, N.B.: Voluntary Energy Harvesting Relays and Selection in Cooperative Wireless Networks, IEEE Transactions on Wireless Communications 9(11), 3543–3553 (2010)
64. Tacca, M., MontiA, P.: Cooperative and Reliable ARQ Protocols for Energy Harvesting Wireless Sensor Nodes, IEEE Transactions on Wireless Communications 6(7), 1536–1576 (2007)
65. Hsin, C.-F., Liu, M.: Network Coverage Using Low Duty-Cycled Sensors: Random & Coordinated Sleep Algorithms. In: Proceedings of the 3rd International Symposium on Information Processing in Sensor Networks (2004), doi:10.1145/984622.984685
66. Liu, C., et al.: Random Coverage with Guaranteed Connectivity: Joint Scheduling for Wireless Sensor Networks. IEEE Transactions on Parallel and Distributed Systems 17(6), 562–575 (2006)
67. Jaggi, N., Kar, K., Krishnamurthy, A.: Near-Optimal Activation Policies in Rechargeable Sensor Networks under Spatial Correlations. ACM Transactions on Sensor Networks (TOSN) 4(3), 17:1–17:36 (2008)
68. Seyedi, A., Sikdar, B.: Energy Efficient Transmission Strategies for Body Sensor Networks with Energy Harvesting. IEEE Transactions on Communications 58(7), 2116–2126 (2010)
69. Puterman, M.: Markov Decision Processes - Discrete Stochastic Dynammic Programming. John Wiley and Sons, Chichester (1994)
70. Jaggi, N., Kar, K., Krishnamurthy, A.: Rechargeable Sensor Activation under Temporally Correlated Events. ACM/Springer Wireless Networks (WINET) 15(5), 619–635 (2009)
71. Bertsekas, D.: Dynamic Programming and Optimal Control. Athena Scientific, Belmon MA (2000)
72. Cassandra, A., Kaelbling, L., Littman, M.: Acting optimally in partially observable stochastic domains. Proceedings of the 12th National Conference on Artificial Intelligence 2, 1023–1028 (1994)
73. Fernandez-Gaucherand, E., Arapostathis, A., Marcus, S.: On the average cost optimality equation and the structure of optimal policies for partially observable Markov decision processes. Annals of Operations Research 29(1-4), 439–470 (1991)
74. Li, H., Jaggi, N., Sikdar, B.: Cooperative Relay Scheduling under Partial State Information in Energy Harvesting Sensor Networks. In: Proceedings of IEEE GLOBECOM (2010)
75. Li, H., Jaggi, N., Sikdar, B.: Relay Scheduling for Cooperative Communications in Sensor Networks with Energy Harvesting. Accepted for publication in IEEE Transactions on Wireless Communications (2010)

Chapter 8
Energy-Efficient Parallel Packet Forwarding*

Weirong Jiang and Viktor K. Prasanna

Abstract. As the Internet traffic continues growing rapidly, parallel packet forwarding becomes a necessity in Internet infrastructure to meet the throughput requirement. On the other hand, energy/ power consumption has been a critical challenge for Internet infrastructure. It has been shown that two thirds of power dissipation inside a core router is due to packet forwarding. This chapter studies the problem of energy-efficient parallel packet forwarding in Internet Infrastructure. According to whether the data structure is shared or duplicated among multiple engines, two types of parallel packet forwarding systems are discussed. For the system with shared data structure, we study how to partition the data structure and map onto multiple engines, so that the worst-case energy/ power consumption is minimized. For the system with duplicated data structure, we formulate as an optimization problem how to distribute traffic load onto multiple engines to minimize the overall power consumption while satisfying the throughput demand.

8.1 Introduction

The Internet infrastructure consists of routers and switches to achieve interconnectivity. Its primary function is to forward packets, where the header information (such as the destination IP address) extracted from each packet is looked up in the forwarding table inside the routers/ switches. With the network traffic growing rapidly, the throughput requirement becomes difficult to meet by using a single packet forwarding engine. For example, current backbone link rates have been pushed beyond OC-768 (40 Gbps) rate, which requires a throughput of 125 million packets per

Weirong Jiang
Juniper Networks Inc., Sunnvyale, California, USA
e-mail: weirongj@acm.org

Viktor K. Prasanna
University of Southern California, Los Angeles, California, USA
e-mail: prasanna@usc.edu

* This work was supported by U.S. National Science Foundation under grant 1018801.

second (MPPS) for minimum size (40 bytes) packets. Employing multiple packet forwarding engines has been a standard in today's routers/switches. As depicted in Figure 8.1, in such a system, multiple packet forwarding engines process the network traffic in parallel. For each incoming packet, the dispatcher will decide which forwarding engine will be assigned to process the packet. When the traffic assigned to an engine is higher than the processing capacity of the engine, a queue is needed to buffer the packets.

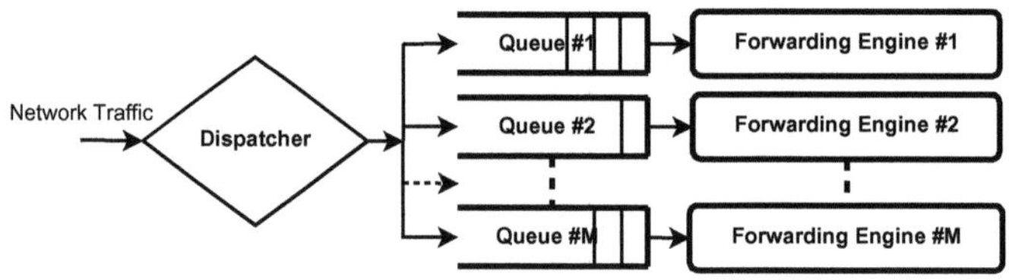

Fig. 8.1 Conceptual model of a parallel packet forwarding system.

8.1.1 Challenges

Most research in parallel packet forwarding is on balancing the traffic among these engines to achieve high throughput [8, 4, 29]. Little of them take power or energy consumption into account. On the other hand, as routers achieve aggregate throughputs of trillions of bits per second, power consumption has become an increasingly critical concern in backbone router design [30, 10]. Some recent investigations [23, 7] show that power dissipation has become the major limiting factor for next generation routers and predicts that expensive liquid cooling may be needed in future routers. As shown in Figure 8.2, the capacity of backbone routers used to double every 18 months until 5 years ago. Today, terabit routers with 10–15 kW are at the limit due to the power density. As a result, a thirty-fold shortfall in capacity by 2015 is foreseen as compared to the historical trend for single rack routers [23]. Recent analysis by researchers from Bell labs [23] reveals that, almost two thirds of power dissipation inside a core router is due to packet forwarding engines.

8.1.2 Related Work

Various techniques including data structure, hardware, system or network-level optimization [10, 13, 12, 7, 26] have been proposed to reduce the power consumption of routers and/or switches.

In [22] clock gating is used to turn off the clock of unneeded processing engines of multi-core network processors to save dynamic power when there is a low traffic workload. A finer-grained clock gating scheme is proposed in [12] to lower the

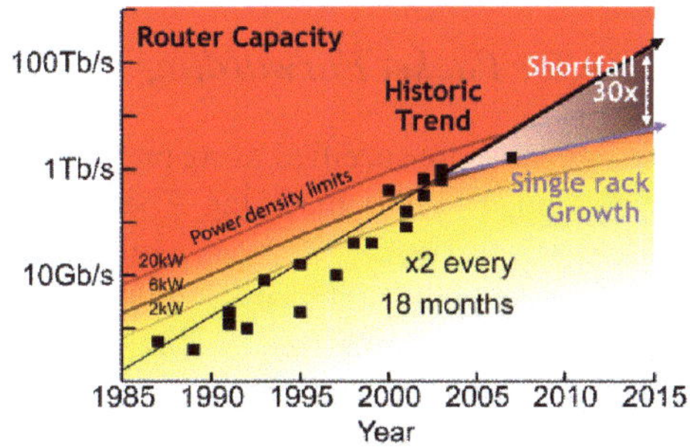

Fig. 8.2 Router capacity limited by power [23].

dynamic power consumption of pipelined forwarding engines. In [18] the more aggressive approach of turning off these processing engines is used to reduce both dynamic and static power consumption. Dynamic frequency and voltage scaling are used in [16] and [25], respectively, to reduce the power consumption of the processing engines.

Chabarek et al. [7] enumerate the power demands of two widely used Cisco routers. The authors further use mixed integer optimization techniques to determine the optimal configuration at each router in their sample network for a given traffic matrix. Nedevschi et al. [26] assume that the underlying hardware in network equipment supports sleeping and dynamic voltage and frequency scaling. The authors propose to shape the traffic into small bursts at edge routers to facilitate sleeping and rate adaptation.

Though reducing the power consumption of the Internet infrastructure has been a topic of significant interest, little of the existing work focuses on the *parallel* packet forwarding systems.

8.1.3 Organization

The rest of this chapter is organized as follows. Section 8.2 introduces the background on parallel packet forwarding systems which can be classified into two types: *heterogeneous* and *homogeneous* systems, based on whether the data structure is shared or duplicated among multiple engines. Section 8.3 considers the heterogeneous system and discusses the solution to partitioning and mapping the data structure onto the multiple pipelined packet forwarding engines to achieve energy efficiency. Section 8.4 focuses on the homogeneous system and develops a theoretical framework to minimize the overall power consumption while satisfying the throughput demand. Section 8.5 summarizes this chapter.

8.2 Background

8.2.1 Data Structure for Packet Forwarding

The entries in the forwarding table are specified using prefixes. The kernel of packet forwarding is IP lookup i.e. longest prefix matching. The most common data structure for IP lookup is some form of trie [28]. A trie is a binary tree, where a prefix is represented by a node. The value of the prefix corresponds to the path from the root of the tree to the node representing the prefix. The branching decisions are made based on the consecutive bits in the prefix. A trie is called a uni-bit trie if only one bit at a time is used to make branching decisions. Figure 8.3 (b) shows the uni-bit trie for the prefix entries in Figure 8.3 (a).

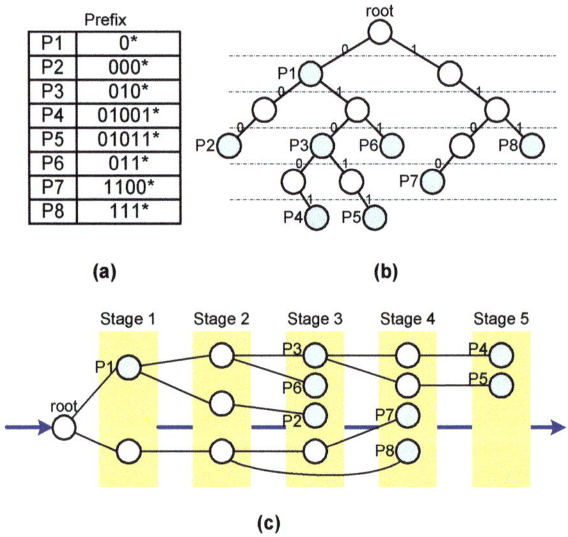

Fig. 8.3 (a) Prefix entries; (b) Uni-bit trie; (c) Fine-grained trie-to-pipeline mapping.

Given a uni-bit trie, IP lookup is performed by traversing the trie according to the bits in the IP address. The information about the longest prefix matched so far, which is updated when meeting a node containing a prefix, is carried along the traversal. Thus, when a leaf is reached, the longest matched prefix along the traversed path is returned. The time to look up a uni-bit trie is equal to the prefix length.

8.2.2 Pipelined Packet Forwarding Engines

Pipelining can dramatically improve the throughput of trie-based solutions. A straightforward way to pipeline a trie is to assign each trie level to a separate stage.

Each stage contains a block of static random addressable memory (SRAM) that stores the trie nodes. As a result, a lookup request can be issued every clock cycle [3]. But such a simple scheme results in unbalanced memory distribution across the pipeline stages. In an unbalanced pipeline, the global clock rate is determined by the access time to the "fattest" stage. Since it is unclear at hardware design time which stage will be the "fattest", we must allocate memory with the maximum size for each stage. Such an over-provisioning results in memory wastage and excessive power consumption [2]. Some proposed solutions [2, 19] balance the memory distribution across stages at the cost of lowering the throughput. In [14], a fine-grained node-to-stage mapping scheme is proposed for linear pipeline architectures. It is based on the heuristic that allows any two nodes on the same level of a trie to be mapped onto different stages of a pipeline. The heuristic is enabled by storing in each node the distance value to its child nodes. When a packet is passed through the pipeline, the distance value is decremented by 1 when it goes through each stage. Only when the distance value becomes 0, the child node's address is used to access the memory in that stage. Balanced memory distribution across pipeline stages is achieved, while a high throughput of one packet per clock cycle is sustained. Figure 8.3 (c) shows the mapping result for the uni-bit trie in Figure 8.3 (b) using the fine-grained mapping scheme.

8.2.3 Parallel Packet Forwarding

A parallel packet forwarding system consists of multiple engines. The number of engines is denoted P. According to whether the data structure is shared or duplicated among the multiple engines, parallel packet forwarding systems can be classified into two types. When the data structure is shared, each engine contains a different subset of the forwarding table. The full forwarding table is partitioned into P subsets and mapped onto the P engines. Any incoming packet is mapped to only one subset, and thus processed by the engine containing that subset. We call such a system to be *heterogeneous*. Essential to the heterogeneous system is how to perform the partitioning and mapping, which guides the construction of the dispatcher. When the data structure is duplicated, each engine contains the same forwarding table. An incoming packet can go through any of these engines. We call such a system to be *homogeneous*. Essential to the homogeneous system is the dispatcher that distributes incoming traffic load to the multiple forwarding engines. The basic goal of the dispatcher is to fully utilize the capacity of the multiple engines to maximize the overall throughput.

8.3 Energy-Efficient Multi-pipeline Architecture

8.3.1 Motivation

Ternary Content Addressable Memories (TCAMs), where a single clock cycle is sufficient to perform one IP lookup, are popular in today's routers. However,

TCAMs do not scale well in terms of clock rate, power consumption, or chip density [14]. Taylor [30] estimates that the power consumption per bit of TCAMs is on the order of 3 micro-Watts, which is 150 times more than for Static Random Access Memories (SRAMs). On the other hand, SRAM-based pipeline architectures have been developed as a promising alternative to TCAMs for high performance packet forwarding engines in next generation routers [3, 14, 6]. However, most SRAM-based solutions still suffer from high power consumption [32], which comes mainly from two sources. First, the size of the memories to store the search structure is large. Second, the number of accesses to these large memories (i.e. the pipeline depth), is large. The overall power consumption for one IP lookup can be expressed as in Equation 8.1:

$$Power_{overall} = \sum_{i=1}^{H} [P_m(S_i, \cdots) + P_l(i)] \tag{8.1}$$

Here, H denotes the number of memory accesses, i.e. the pipeline depth, $P_m(.)$ the function of the power dissipation of a memory access (which usually has a positive correlation with the memory size), S_i the size of the ith memory being accessed, and $P_l(i)$ the power consumption of the logic associated with the ith memory access. Since the logic dissipates much less power than the memory in these architectures [19, 12], our main focus is on reducing the power consumption of the memory accesses. Note that the power consumption of a single memory is affected by many other factors, such as the fabrication technology and sub-bank organization, which are beyond the scope of this paper. According to Equation 8.1, to reduce the worst-case power consumption, we should bound the number of memory accesses, as well as minimize the memory size for each access.

To reduce the memory size, we exploit chip-level parallelism by partitioning the forwarding table and allowing IP lookup be performed on only one of the partitions. We map the partitioned forwarding table onto multiple SRAM-based pipelines to ensure each pipeline requires the same amount of memory. We propose a two-phase partitioning scheme, including an effective method called *height-bounded split*, to partition a routing trie into several height-bounded subtries. As a result, all the IPs traverse the subtries in a bounded number of memory accesses. The partitioning scheme is enabled by using small TCAMs as part of the index table in the dispatcher.

8.3.2 Algorithms

First, we define the following terms.

Definition 1. The *depth* of a trie node is the directed distance from the trie node to the trie root. The depth of a trie refers to the maximum depth of all trie leaves.

Definition 2. The *height* of a trie node is the maximum directed distance from the trie node to a leaf node. The height of a trie refers to the height of the root. In fact, the depth of a trie is equal to its height.

Definition 3. The *size* of a trie is the number of nodes in the trie.

In the worst case, the number of memory accesses needed for an IP lookup is equal to the trie height. We propose a holistic scheme to (1) partition a full routing trie into many height-bounded subtries, and (2) map those subtries onto multiple pipelines so that each pipeline contains equal numbers of trie nodes. By these means, each IP lookup is completed through a bounded number of accesses on small-size memories, so that power efficiency is achieved according to Equation 8.1.

Our scheme consists of two phases: prefix expansion and height-bounded split, as illustrated in Figure 8.4 (a) and (b).

8.3.2.1 Prefix Expansion

Several initial bits are used as the index to partition the trie into many disjoint subtries. The number of initial bits to be used is called the *initial stride*, denoted I. A larger I can result in more subtries of smaller height, which can help balance the memory distribution among pipelines as well as across stages when mapping subtries to pipelines. However, a large I can result in size (prefix) expansion, where the sum of the sizes (number of prefixes) of all subtries is larger than the size (number of prefixes) of the original trie. In core routers, the majority of prefixes has lengths between 16 and 24 [11]. $I < 16$ does not result in much size (prefix) expansion.

8.3.2.2 Height-Bounded Split

After prefix expansion, we obtain $K = 2^I$ subtries, whose heights can vary from 1 to $32 - I$. We must partition these subtries further, to reduce the height of the resultant subtries, with minimum overhead. Meanwhile, we must map those subtries onto multiple pipelines to achieve a balanced memory allocation among those pipelines. In other words, given D pipelines (denoted P_i, $i = 1, \cdots, D$) and K subtries each of which has W_i nodes ($i = 1, 2, \cdots, K$), we must ensure that each pipeline (except the last pipeline) contains $B_P = \left\lceil \frac{\sum_{i=1}^{K} W_i}{D} \right\rceil$ nodes.

The above problem is a variant of packing problem [17], and is NP-hard. We develop a heuristic that allows a subtrie to be split and mapped onto different pipelines. First, we sort those subtries obtained from prefix expansion, in decreasing order of size. Then we traverse those subtries one by one. Each subtrie is traversed in the post-order. Once the height bound, denoted B_H, is reached, a new subtrie is split. After the number of nodes mapped onto a pipeline exceeds the size bound of the pipeline, B_P, we map the rest of nodes onto the next pipeline. Algorithm 1 shows a recursive implementation of the height-bounded split, where *size*(P_i) denotes the number of nodes mapped onto the i th pipeline.

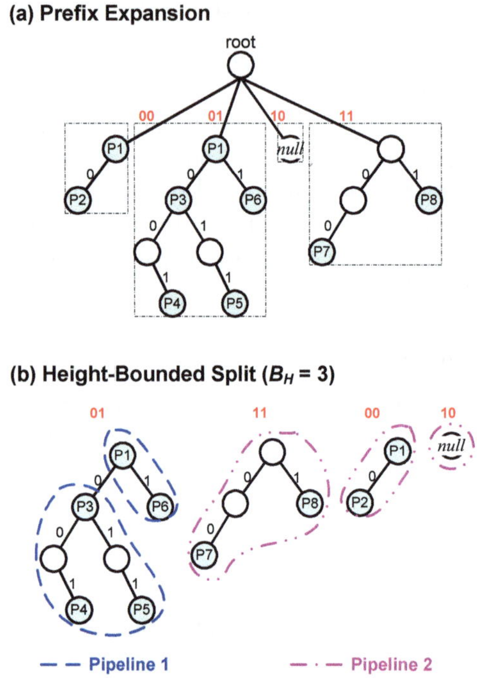

Fig. 8.4 Partition the trie shown in Figure 8.3 (b) and map onto 2 pipelines. The value of the height bound is $B_H = 3$.

8.3.3 Implementation Issues

After height-bounded splitting, the resulting subtries may be rooted at different depths of the original trie. For the subtries rooted at the depth of I, we use an index SRAM (called SRAM_A). For the rest of the subtries, we need an index TCAM and an index SRAM (called SRAM_B). The TCAM stores the prefixes to represent the subtries. The two index SRAMs store the information associated with each subtrie: (1) the mapped pipeline ID, (2) the ID of the stage where the subtrie's root is stored, and (3) the address of the subtrie's root in that stage. Figure 8.5 shows the index table used to achieve the mapping shown in Figure 8.4.

The index table resides in the dispatcher. An arriving input IP searches the index SRAM_A and the index TCAM in parallel. I initial bits of the input IP are used to index the SRAM_A. Meanwhile, the entire input IP searches the index TCAM and obtains the subtrie ID corresponding to the longest matched prefix. Then, the IP uses the subtrie ID to index the SRAM_B to retrieve the associated information. The result obtained from the SRAM_B has a higher priority than that from the SRAM_A. The number of entries in the SRAM_A is 2^I, while the number of entries in the index TCAM and SRAM_B is at most 2^{32-B_H}.

Algorithm 1. Height-bounded split: $HBS(n,i)$

Require: n: a node;
Require: i: the ID of the pipeline for n to be mapped on.
Ensure: i: the ID of the pipeline for the next node to be mapped on.

if $n == null$ **then**
  Return i.
end if
$i = HBS(n.left\_child, i)$
$i = HBS(n.right\_child, i)$
if $size(P_i) < B_P$ **then**
  Map n onto P_i.
else
  Map n onto P_{i+1}.
end if
if $n.height >= B_H$ **then**
  Mark n as a subtrie root.
end if
Return $i+1$.

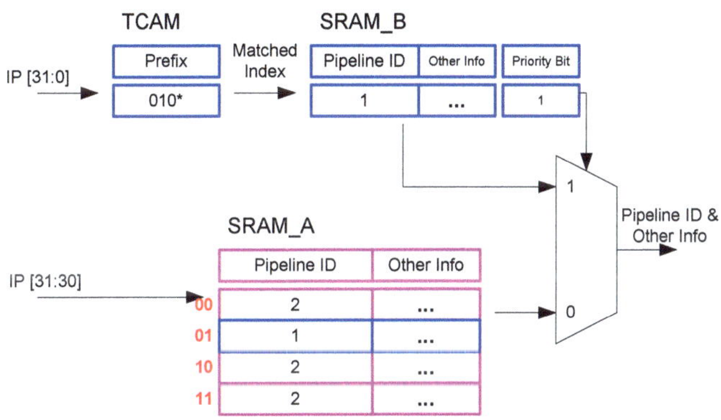

Fig. 8.5 Index table for the mapping shown in Figure 8.4

8.3.4 Performance Evaluation

We conducted simulation experiments on the 8 real-life backbone routing tables collected from [27] on 11/30/2007. Their information is listed in Table 8.1. For ASIC implementation, we used CACTI 5.3 [5] and an accurate TCAM model [1] to evaluate the performance of SRAMs and TCAMs, respectively. The number of pipelines was $D = 4$, and the initial stride for prefix expansion was $I = 12$.

Table 8.1 Representative Routing Tables

Routing table	Location	# of prefixes
RIPE-NCC	Amsterdam, Netherlands	243474
LINX	London, UK	240797
SFINX	Paris, France	238089
NYIIX	New York, USA	238836
DE-CIX	Frankfurt, Germany	243732
MSK-IX	Moscow, Russia	238461
PAIX	Palo Alto, USA	243731
PTTMetro-SP	Sao Paulo, Brazil	243242

8.3.4.1 Selection of Height Bound

How to set an appropriate height bound is an issue worth discussing. Using a small height bound can reduce the number of memory accesses for IP lookup, but it may also result in a large number of subtries, which requires a large table to index those subtires. A subtrie whose height is h may be split into $O(2^{h-B_H})$ subtries, whose heights are bounded by B_H.

We conducted experiments with various values of the height bound to evaluate such a trade-off. As Figure 8.6 shows, a smaller height bound resulted in a larger number of TCAM entries in the index table. The architecture achieved the lowest power consumption[1] when the value of the height bound was 16. In the following experiments, the height bound was $B_H = 16$, and the pipeline depth was $H = B_H + 1 = 17$.

8.3.4.2 Architecture Performance

We mapped the 8 routing tables onto a 4-pipeline 17-stage architecture. As Figure 8.7 shows, our partitioning and mapping scheme achieved a balanced memory allocation among the 4 pipelines. To perform the node-to-stage mapping, we used the same scheme as [14], to achieve balanced node distribution across stages within each pipeline. The results are shown in Figure 8.8. The first several stages of a pipeline could not be balanced, since there were few nodes at the top levels of any subtrie.

According to Figure 8.8, each stage contained fewer than 16K nodes. Thus 14 address bits were enough to index a node in the local memory of a stage. The pipeline depth was 17, and the first stage was dedicated for the subtrie roots. Thus we needed 4 bits to specify the distance for each node. Assuming each node needed 15 bits as the pointer to next-hop information, the total memory needed to store 243732 prefixes from the largest routing table DE-CIX in this architecture was $(13+4+15) \times 2^{14} \times 17 \times 4 \approx 34$ Mb $= 4.25$ MB, where each stage needed 64 KB of memory. According to CACTI 5.3 [5], a 64 KB SRAM using 65 nm technology needed 0.8017 ns to access and dissipated 0.0235 nJ power. For DE-CIX, there were 986 entries in the index TCAM. According to the TCAM model [1], a 986-row 18-bit TCAM using 65 nm technology needed 1.4653 ns to access and

[1] The measurement of the power consumption was similar as in Section 8.3.4.2.

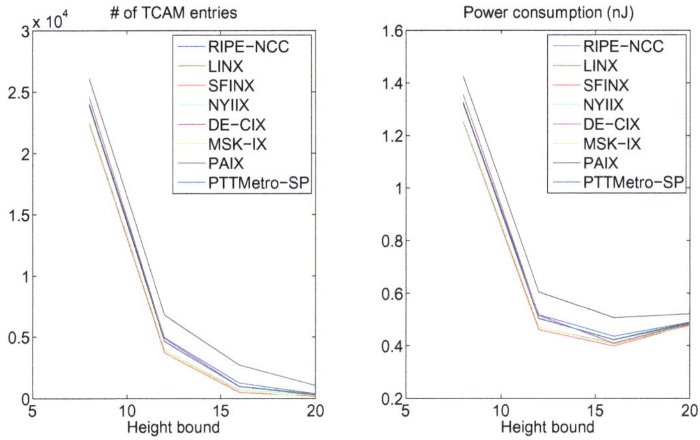

Fig. 8.6 Impact of height bound

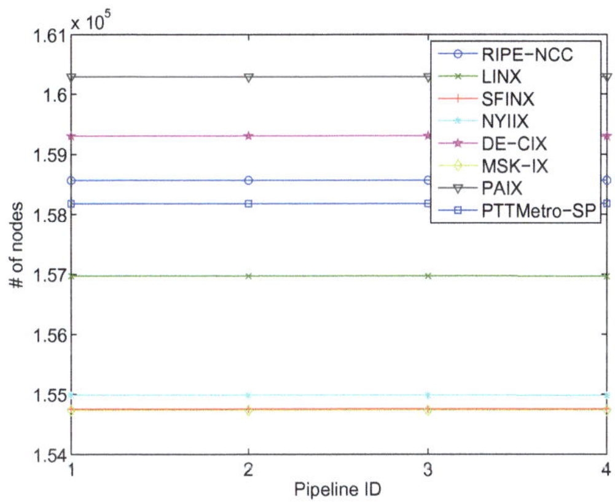

Fig. 8.7 Node distribution over 4 pipelines

dissipated 0.0325 nJ power. The overall clock rate could achieve $\frac{1}{0.8017} = 1.25$ GHz, by using two copies of the index TCAM working in parallel. The throughput was thus 400 Gbps (i.e. 10× OC-768 rate) for minimum size (40 bytes) packets. The energy consumption for one IP lookup was $0.0235 \times 16 + 0.0325 \times 2 = 0.441$ nJ [2].

The results for the 8 routing tables are compared with the 4-partition CoolCAM [32] and IPStash [15] in Table 8.2. The CoolCAM results did not include the power dissipation of the index TCAM. The IPStash results were normalized using the best-case reduction ratio given in [15]. Our architecture achieved up to 7-fold and 3-fold reductions in power consumption over CoolCAMs and IPStash, respectively.

[2] The two index SRAMs, one with $2^{10} \times 10$ bits = 1.25 KB and the other with $2^{12} \times 18$ bits = 9 KB, were so small that their contribution to the overall performance could be ignored.

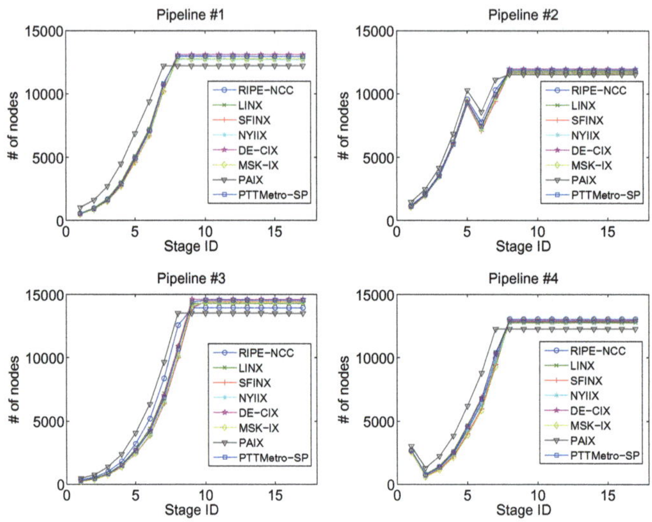

Fig. 8.8 Node distribution over 17 stages in each pipeline.

Table 8.2 Comparison on energy consumption (*nJ*) of different solutions

Routing table	RIPE-NCC	LINX	SFINX	NYIIX	DE-CIX	MSK-IX	PAIX	PTTMetro-SP
4-partition CoolCAM [32]	2.8881	2.8571	2.8246	2.8337	2.8920	2.8286	2.8920	2.8856
IPStash [15] (normalized)	1.1553	1.1428	1.1298	1.1335	1.1568	1.1314	1.1568	1.1543
4-pipeline arch ($B_H = 16$)	0.4353	0.4227	0.3986	0.4095	0.4410	0.4050	0.5055	0.4215

8.4 Power-Aware Load Distribution

In this section, we focus on the parallel packet forwarding system consisting of homogeneous engines. We develop a theoretical framework by modeling the power dissipation of a single engine as the function of the traffic load assigned to that engine. Then the power-aware load distribution in parallel forwarding becomes an optimization problem to minimize the overall power consumption while meeting the throughput requirement. We consider two types of power functions, which are different in terms of whether supporting sleep mode or not.

8.4.1 Optimization Framework

8.4.1.1 Problem Definition

The problem of distributing traffic load onto multiple forwarding engines to minimize the overall power consumption while satisfying the throughput requirement can be defined as follows.

$$\min \quad \sum_{i=1}^{M} P_i(x_i) \tag{8.2}$$

$$\text{s.t.} \quad \sum_{i=1}^{M} x_i = T \tag{8.3}$$

$$0 \leq x_i \leq B_i,\ i = 1, \ldots, M \tag{8.4}$$

In the above definition, M denotes the total number of forwarding engines and T the total throughput requirement. For the ith forwarding engine, $i = 1, \ldots, M$, x_i denotes its traffic load, $P_i(.)$ its power function with respect to its traffic load and B_i its throughput upper bound (i.e. the bandwidth). The bandwidth of each engine is lower than that of the output link attached to the engine.

8.4.1.2 Power Function of Each Engine

To solve the above optimization problem, we should first understand the power function of each engine. The most common are the following two types of power functions.

Sleep-disabled

According to [31, 24], the power consumption of most of today's network devices can be modeled as a linear function with respect to the traffic load, as shown in (8.5):

$$P(x) = ax + b \tag{8.5}$$

where x denotes the traffic load. a is the coefficient for dynamic power consumption while b the static power consumption. This type of power function assumes that the forwarding engine does not have the sleep mode. As a result, even when there is no traffic to be processed, the forwarding engine still dissipates (static) power.

Sleep-enabled

Recent work [26] proposes that next generation network devices should support sleep mode. With sleep mode enabled, the forwarding engine will not dissipate any power when there is no traffic load. Thus we have following power function for such kinds of forwarding engines.

$$P(x) = (ax + b)I_x = ax + bI_x = \begin{cases} 0 & x = 0 \\ ax + b & x > 0 \end{cases} \tag{8.6}$$

where I_x is a unit step function:

$$I_x = \begin{cases} 0 & x = 0 \\ 1 & x > 0 \end{cases} \tag{8.7}$$

8.4.1.3 Specific Case: Property-Homogeneous Engines

We start the discussion from a specific (and simple) case where all the forwarding engines have the same properties including the power function and the bandwidth. In other words, $\forall i$, $i = 1,2,\ldots,M$, $a_i = a$, $b_i = b$, $B_i = B$, where a, b and B are constants. We call these engines to be *property-homogeneous*.

1. Considering sleep-disabled forwarding engines, the objective function (1) will be

$$\sum_{i=1}^{M} P_i(x_i) = \sum_{i=1}^{M} [ax_i + b] = a\sum_{i=1}^{M} x_i + Mb = aT + Mb \tag{8.8}$$

   As a, b, T and M are given constants, there is no optimization to be done to minimize the overall power consumption.
2. Considering sleep-enabled forwarding engines, the objective function (1) will be

$$\sum_{i=1}^{M} P_i(x_i) = \sum_{i=1}^{M} [ax_i + bI_{x_i}] = a\sum_{i=1}^{M} x_i + b\sum_{i=1}^{M} I_{x_i} = aT + b\sum_{i=1}^{M} I_{x_i} \tag{8.9}$$

   Thus the optimal solution is to minimize the number of active forwarding engines while satisfying the throughput requirement. Since all the fowarding engines have the same bandwidth, the optimal solution for meeting throughput of T is to turn on $\lceil \frac{T}{B} \rceil$ forwarding engines while keeping the rest of forwarding engines to sleep. Then the overall power consumption is minimized to be: $\min \sum_{i=1}^{M} P_i(x_i) = aT + b\lceil \frac{T}{B} \rceil$.

8.4.1.4 General Case: Property-Heterogeneous Engines

In most cases, the properties of forwarding engines differ from each other. In other words, $\exists i, j$, $i, j = 1,2,\ldots,M$, $a_i \neq a_j$, $b_i \neq b_j$, and $B_i \neq B_j$. We call these engines to be *property-heterogeneous*.

1. Considering sleep-disabled forwarding engines, the objective function (1) will be

$$\sum_{i=1}^{M} P_i(x_i) = \sum_{i=1}^{M} [a_i x_i + b_i] = \sum_{i=1}^{M} a_i x_i + \sum_{i=1}^{M} b_i \tag{8.10}$$

   Then the optimization problem becomes a linear programming (LP) problem which can be solved in polynomial time. Note that $\sum_{i=1}^{M} b_i$ is constant, which will not affect the decision on traffic load distribution. Hence $\sum_{i=1}^{M} b_i$ can be omitted from (8.10) when solving the linear programming (LP) problem.
2. Considering sleep-enabled forwarding engines, the objective function (1) will be

$$\sum_{i=1}^{M} P_i(x_i) = \sum_{i=1}^{M} [a_i x_i + b_i I_{x_i}] \tag{8.11}$$

The optimization problem then becomes a non-linear programming problem which is hard to be solved. We convert it into a mixed integer programming (MIP) problem[3] by introducing a penalty parameter K to remove the unit step function. Then we have:

$$\min \quad \sum_{i=1}^{M} [a_i x_i + b_i y_i] \tag{8.12}$$

$$\text{s.t.} \quad \sum_{i=1}^{M} x_i = T \tag{8.13}$$

$$0 \leq x_i \leq B_i,\ i = 1, \ldots, M \tag{8.14}$$

$$y_i <= K * x_i,\ i = 1, \ldots, M \tag{8.15}$$

$$y_i >= x_i / K,\ i = 1, \ldots, M \tag{8.16}$$

$$y_i \in \{0, 1\} \tag{8.17}$$

We can prove that when we set $K >> \max_{i=1}^{M} B_i$, the above problem is identical to (8.11) with constraints (8.3)(8.4). A simple proof is as follows. If $x_i = 0$, then y_i must be 0; otherwise, i.e. $x_i > 0$, then y_i must be 1. Hence $y_i = I_{x_i}$. $i = 1, 2, \ldots, M$.

8.4.2 Implementation Issues

While we focus mainly on theoretical analysis, here we discuss briefly the issues in system design and implementation. The kernel of the dispatcher is the linear programming (LP) / mixed integer programming (MIP) solver. To obtain the parameters used in the optimization framework, we need following components in the dispatcher system.

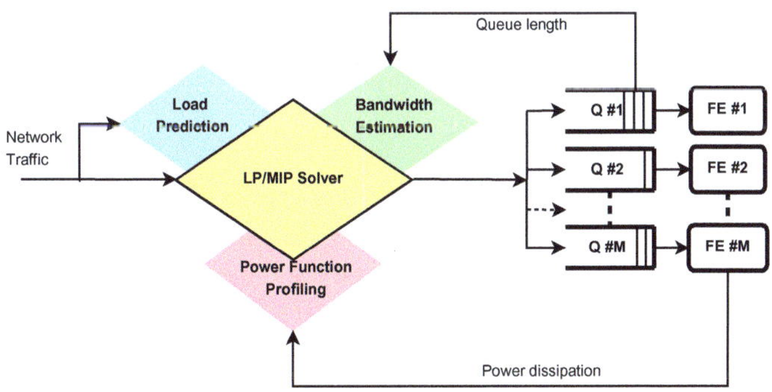

Fig. 8.9 Example of the dispatcher system design.

[3] Although a mixed integer programming (MIP) problem is *NP-hard*, there exist various computation-efficient MIP solvers.

- **Load predication** to predict T in real time.
- **Bandwidth estimation** to estimate B_i $(i = 1, 2, \ldots, M)$ for each forwarding engine if their values are unknown or varying.
- **Power function profiling** to retrieve the parameters a_i, b_i $(i = 1, 2, \ldots, M)$ for each forwarding engine if these parameters cannot be pre-determined.

Then we have the overall architecture as shown in Figure 8.9, where 'Q' and 'FE' are the abbreviations of 'Queue' and 'Forwarding Engine', respectively.

8.4.2.1 Load Predication

The optimal solution is only possible when we can predict future traffic load. Various load prediction techniques have been proposed in literature [9, 21]. One of the accurate prediction algorithms is Auto-Regressive Moving Average (ARMA) adopted in [21]. We use ARMA for load prediction in our experiments.

8.4.2.2 Bandwidth Estimation

In most cases, the bandwidth of each forwarding engine is known apriori. But in case the forwarding bandwidth of some forwarding engine is unknown or varying, we need to perform real-time estimation using other information. For example, we can monitor the queue length of a forwarding engine. Since the dispatcher keeps track of the traffic load distributed to that forwarding engine, we can infer the forwarding bandwidth based on the queue length of that engine.

8.4.2.3 Power Function Profiling

Usually we can pre-determine the power function of each forwarding engine. If we want to model the power function of a forwarding engine on the fly, we need the real-time information of the power dissipation of the engine. Since the dispatcher contains the traffic load records, we can profile the power functions based on power and load information.

8.4.3 Experimental Results

To evaluate how much power/ energy reduction can be achieved by using the proposed optimization framework, we conducted simulation experiments using real-life traffic traces from LBNL/ICSI Enterprise Tracing Project [20]. We downloaded 25 trace files collected on the same day (2005/01/07) and concatenated them into one trace. Its statistics is shown in Table 8.3 where throughput is measured in terms of the number of packets per second (PPS). Figure 8.10 depicts the traffic rate variation during the entire period.

Table 8.3 Statistics of the 18-hour traffic trace

Trace	Date	# of packets	Duration	Max. throughput	Min. throughput
LBNL/ICSI	20050107	26325056	17 hours 33 minutes	11083 PPS	0 PPS

We consider the general case where the system consists of four parallel forwarding engines whose parameters are different from each other. The system configuration for the simulation is summarized in Table 8.4. The parameters were set to comply with the power models of network devices observed in [31, 24]. Among the four forwarding engines, the third one (FE #3) represents a high-end engine and is the most power-hungry, while the fourth one (FE #4) represents a low-end engine with the least power consumption.

In following experiments, we consider two scenarios: sleep-disabled and sleep-enabled forwarding engines, respectively. We compared the performance achieved using our optimal (**power-aware**) solution with that using the traditional (**power-unaware**) load balancing -based parallel forwarding scheme.

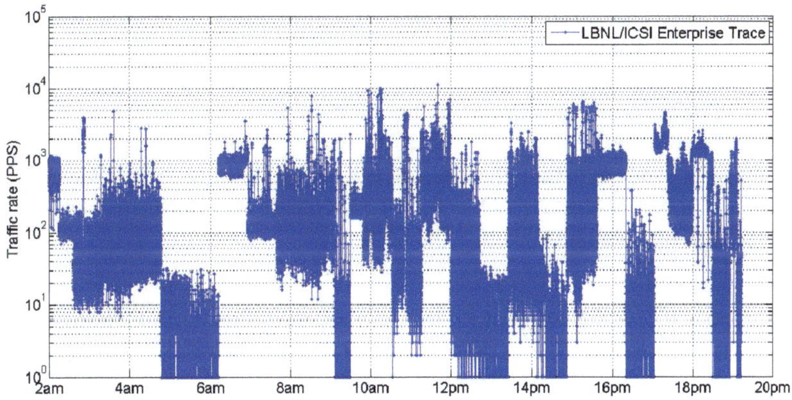

Fig. 8.10 Traffic rate variation of the trace.

Table 8.4 Summary of simulation configurations

Parameter	FE #1	FE #2	FE #3	FE #4
a	0.1	0.05	0.2	0.02
b	300	200	500	100
B	3000	1500	6000	600

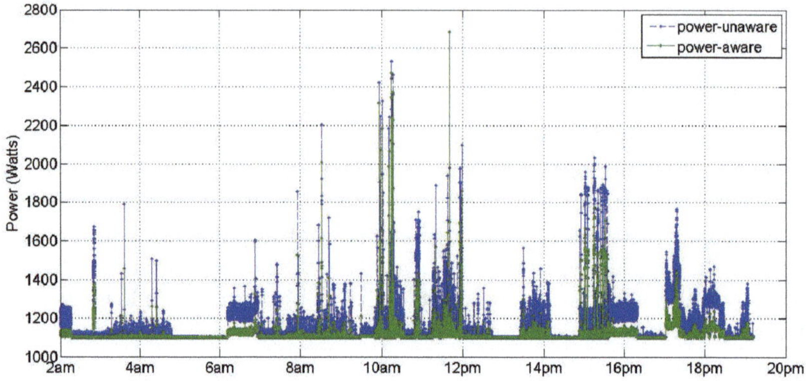

Fig. 8.11 Power consumption of power-aware versus power-unware parallel forwarding schemes (with sleep-disabled forwarding engines).

8.4.3.1 With Sleep-Disabled Engines

Figure 8.11 shows that our power-aware scheme achieved lower power consumption than the traditional power-unaware scheme, especially when the traffic load is low. Since the static power consumption cannot be reduced in sleep-disabled forwarding engines, the overall energy reduction was marginal: our solution achieved 3% lower energy consumption than the traditional scheme. However, if we consider the dynamic part only, our solution achieved **4.1**-fold reduction in energy consumption than the traditional scheme.

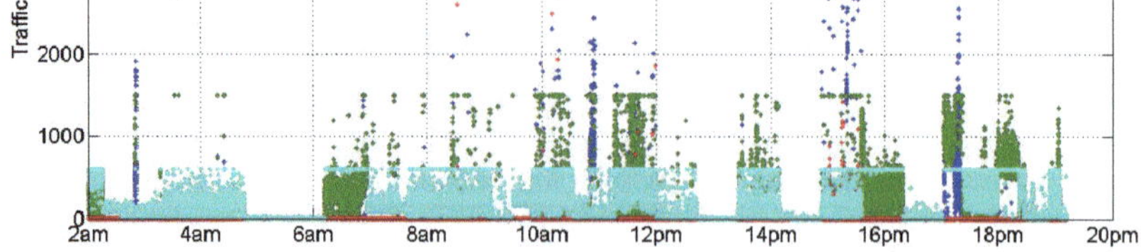

Fig. 8.12 Load distribution on 4 sleep-disabled forwarding engines with power-aware parallel forwarding.

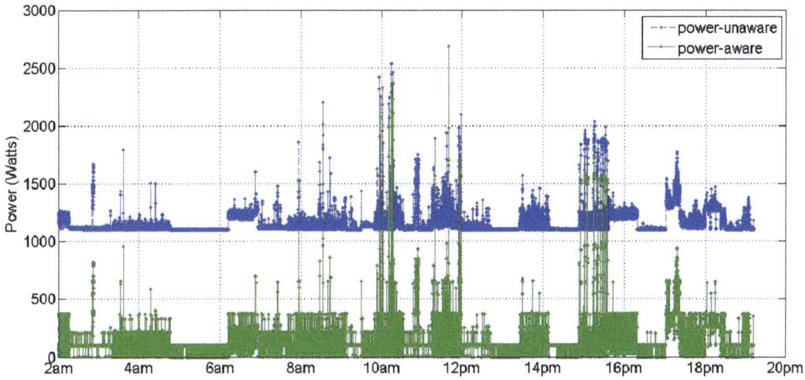

Fig. 8.13 Power consumption of power-aware versus power-unware parallel forwarding schemes (with sleep-enabled forwarding engines).

Figure 8.12 shows the load distribution on the 4 forwarding engines. We can see that, the forwarding engine with the lowest dynamic power consumption usually reached its throughput upper bound (i.e. the bandwidth). The forwarding engines with lower dynamic power consumption tended to receive higher volume of traffic than those with higher dynamic power consumption. As a result, the short-term load distribution among the forwarding engines was not balanced. But the overall throughput requirement was still met.

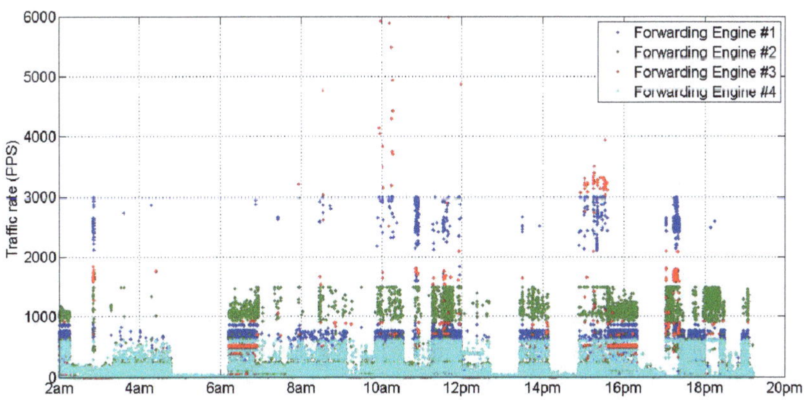

Fig. 8.14 Load distribution on 4 sleep-enabled forwarding engines with power-aware parallel forwarding.

8.4.3.2 With Sleep-Enabled Engines

When forwarding engines have the option to sleep, our power-aware parallel forwarding scheme can achieve significant power/ energy reduction. As shown in Figure 8.13, our power-aware scheme achieved much lower power consumption than the traditional power-unaware scheme. The fundamental reason is that the traditional parallel forwarding scheme does not exploit the sleep mode of forwarding engines to reduce the static power consumption. Our power-aware parallel forwarding scheme turned on only portion of forwarding engines based on the optimization results. As a result, our solution achieved **9.13**-fold reduction in energy (and average power) consumption than the traditional power-unaware scheme.

Figure 8.14 shows the load distribution on the 4 forwarding engines. In most situations, the forwarding engines with lower power consumption tended to receive higher volume of traffic than those with higher power consumption, which is similar as Figure 8.12. When the traffic rate was low, the engine with high power consumption tended to sleep. But when the traffic rate was high, there were some situations where it was better to turn on one engine with high dynamic power consumption than multiple engines with low dynamic power consumption. Thus we observed different load distribution between Figure 8.14 and Figure 8.12. Again, though the short-term load distribution among the forwarding engines was not balanced, the overall throughput requirement was still met.

8.5 Summary

Energy/power consumption has become a major concern in the design of next generation network infrastructure. Reducing the energy/power consumption of packet forwarding systems has been a topic of significant interest [10,7,26]. Most of the existing work focuses on either single forwarding engines or the system- and network-level optimizations. This chapter represents one of the first study on energy/power-aware parallel forwarding in routers/ switches.

First, we considered the system where the forwarding table is partitioned and mapped onto multiple engines. We identified the reason for the high power consumption of SRAM-based pipelined packet forwarding engines to be the unbounded number of accesses on large size memories. A two-phase scheme was proposed to partition a trie into many height-bounded subtries which were then mapped onto multiple pipelines. The partitioning and mapping scheme was carefully designed to balance the memory requirement among the pipelines. The scheme was enabled by combining small TCAM into the index table that worked as the dispatcher. Our experiments using real-life backbone routing tables showed that the proposed 4-pipeline architecture achieved up to 7-fold and 3-fold reductions in energy consumption over the state-of-the-art TCAM-based and SRAM-based solutions, respectively, while sustaining a throughput of 400 Gbps (i.e. $10\times$ OC-768 rate) for minimum size (40 bytes) packets.

Second, we considered the system where each engine contains a copy of the full forwarding table. Given that the power dissipation of each engine can be modeled as a function of traffic load going through that engine, we formulated the optimization problem that minimizes the overall power consumption while satisfying the throughput demand. We discussed two types of power functions, in terms of whether they support sleep mode or not. We solved the two problems via linear programming (LP) and mixed integer programming (MIP), respectively. Our simulation using a 18-hour real-life traffic trace showed that our solution achieved up to **9.13**-fold reduction in energy (and average power) consumption compared with the traditional parallel forwarding scheme based on load balancing. We also discussed the system design issues and identified the challenges for real implementation.

Many other issues remain open. For example, we did not consider the power consumption of the queues / buffers in this chapter. Some applications require the parallel forwarding system preserve intra-flow packet order, which makes the problem more difficult. We hope our initial work can motivate more follow-up research in this area. We also believe the ideas proposed in this chapter can be applied to other parallel computing systems with high power density, such as data centers, server farms, or clusters.

References

1. Agrawal, B., Sherwood, T.: Ternary CAM power and delay model: Extensions and uses. IEEE Trans. VLSI Syst. 16(5), 554–564 (2008), doi:10.1109/TVLSI.2008.917538
2. Baboescu, F., Tullsen, D.M., Rosu, G., Singh, S.: A tree based router search engine architecture with single port memories. In: Proc. ISCA, pp. 123–133 (2005)
3. Basu, A., Narlikar, G.: Fast incremental updates for pipelined forwarding engines. IEEE/ACM Trans. Netw. 13(3), 690–703 (2005)
4. Bux, W., Denzel, W.E., Engbersen, T., Herkersdorf, A., Luijten, R.P.: Technologies and building blocks for fast packet forwarding. IEEE Communications 39(1), 70–77 (2001)
5. CACTI 5.3: `http://quid.hpl.hp.com:9081/cacti/`
6. Carli, L.D., Pan, Y., Kumar, A., Estan, C., Sankaralingam, K.: Flexible lookup modules for rapid deployment of new protocols in high-speed routers. In: Proc. SIGCOMM (2009)
7. Chabarek, J., Sommers, J., Barford, P., Estan, C., Tsiang, D., Wright, S.: Power awareness in network design and routing. In: Proc. INFOCOM, pp. 457–465 (2008)
8. Chen, B., Morris, R.: Flexible control of parallelism in a multiprocessor pc router. In: Proc. of the General Track: 2001 USENIX Annual Technical Conference, pp. 333–346 (2001)
9. Chen, G., He, W., Liu, J., Nath, S., Rigas, L., Xiao, L., Zhao, F.: Energy-aware server provisioning and load dispatching for connection-intensive internet services. In: Proc. NSDI, pp. 337–350 (2008)
10. Gupta, M., Singh, S.: Greening of the Internet. In: Proc. SIGCOMM, pp. 19–26 (2003)
11. Jiang, W., Prasanna, V.K.: Towards green routers: Depth-bounded multi-pipeline architecture for power-efficient IP lookup. In: Proc. IPCCC, pp. 185–192 (2008)
12. Jiang, W., Prasanna, V.K.: Reducing dynamic power dissipation in pipelined forwarding engines. In: Proc. ICCD (2009)
13. Jiang, W., Prasanna, V.K.: Architecture-aware data structure optimization for power-efficient ip lookup. In: Proc. HPSR (2010)

14. Jiang, W., Wang, Q., Prasanna, V.K.: Beyond TCAMs: An SRAM-based parallel multi-pipeline architecture for terabit IP lookup. In: Proc. INFOCOM, pp. 1786–1794 (2008)
15. Kaxiras, S., Keramidas, G.: IPStash: a set-associative memory approach for efficient IP-lookup. In: INFOCOM, pp. 992–1001 (2005)
16. Kennedy, A., Wang, X., Liu, Z., Liu, B.: Low power architecture for high speed packet classification. In: Proc. ANCS, pp. 131–140 (2008)
17. Kleinberg, J., Tardos, E.: Algorithm Design. Addison-Wesley Longman Publishing Co., Inc, Amsterdam (2005)
18. Kokku, R., Shevade, U.B., Shah, N.S., Dahlin, M., Vin, H.M.: Energy-Efficient Packet Processing (2004), `http://www.cs.utexas.edu/users/rkoku/RESEARCH/energy-tech.pdf`
19. Kumar, S., Becchi, M., Crowley, P., Turner, J.: CAMP: fast and efficient IP lookup architecture. In: Proc. ANCS, pp. 51–60 (2006)
20. LBNL/ICSI Enterprise Tracing Project: `http://www.icir.org/enterprise-tracing/download.html`
21. Le, K., Bianchini, R., Martonosi, M., Nguyen, T.D.: Cost- and Energy-Aware Load Distribution Across Data Centers. In: Proc. HotPower (2009)
22. Luo, Y., Yu, J., Yang, J., Bhuyan, L.N.: Conserving network processor power consumption by exploiting traffic variability. ACM Trans. Archit. Code Optim. 4(1), 4 (2007)
23. Lyons, A.M., Neilson, D.T., Salamon, T.R.: Energy efficient strategies for high density telecom applications. Princeton University, Supelec, Ecole Centrale Paris and Alcatel-Lucent Bell Labs Workshop on Information, Energy and Environment (2008)
24. Mahadevan, P., Sharma, P., Banerjee, S., Ranganathan, P.: A power benchmarking framework for network devices. In: Proc. Networking, pp. 795–808 (2009)
25. Mandviwalla, M., Tzeng, N.F.: Energy-efficient scheme for multiprocessor-based router linecards. In: Proc. SAINT (2006)
26. Nedevschi, S., Popa, L., Iannaccone, G., Ratnasamy, S., Wetherall, D.: Reducing network energy consumption via sleeping and rate-adaptation. In: Proc. NSDI, pp. 323–336 (2008)
27. RIS Raw Data: `http://data.ris.ripe.net`
28. Ruiz-Sanchez, M.A., Biersack, E.W., Dabbous, W.: Survey and taxonomy of IP address lookup algorithms. IEEE Network 15(2), 8–23 (2001)
29. Shi, W., MacGregor, M.H., Gburzynski, P.: Load balancing for parallel forwarding. IEEE/ACM Trans. Netw. 13(4), 790–801 (2005)
30. Taylor, D.E.: Survey and taxonomy of packet classification techniques. ACM Comput. Surv. 37(3), 238–275 (2005)
31. Valancius, V., Laoutaris, N., Massoulié, L., Diot, C., Rodriguez, P.: Greening the internet with nano data centers. In: Proc. CoNEXT, pp. 37–48 (2009)
32. Zane, F., Narlikar, G.J., Basu, A.: CoolCAMs: Power-efficient TCAMs for forwarding engines. In: Proc. INFOCOM (2003)

Chapter 9
Energy Consumption Analysis and Adaptive Energy Saving Solutions for Mobile Device Applications

Martin Kennedy, Hrishikesh Venkataraman*, and Gabriel-Miro Muntean

Abstract. Recent trends, motivated by user preferences towards carrying smaller and more complex devices, have focused on integrating different user-centric applications in a single general-purpose mobile hand-held device. Hence, Laptops, smart phones and PDAs are rapidly replacing computers as the most commonly-used Internet-access devices. This has resulted in much higher energy consumption and consequently, a reduced battery life of a wireless device. In fact, the biggest problem today in the mobile world is that they are battery driven and the battery technologies are not matching the required energy demand. This chapter focuses on different energy consuming components in the high-end wireless devices, with specific emphasis on adaptive energy efficient display and decoding mechanisms.

9.1 Introduction

The amount of computations and services in smart phones has increased exponentially over the last couple of years. Currently, a Moore's law-style growth is observed in the design of codecs, video compression techniques, efficient display screens, etc. and will continue to become better over time. However, the battery depletion problem still remains the biggest drawback of the electronic world in general; and smart phones/wireless devices in particular. There are quite a few default energy-saving techniques in iPhone and smart phones which allow the user to adapt certain application layer functionalities. For example, use of an on-device light sensor to monitor the ambient light and lower the display brightness. Another example is to manage CPU-intensive background applications. However, these techniques do not provide any step-wise change and real-time change in the energy consumption and is an inherent limitation of the system.

Martin Kennedy · Hrishikesh Venkataraman · Gabriel-Miro Muntean
Performance Engineering Laboratory, School of Electronic Engineering,
Dublin City University (DCU), Dublin 9, Ireland
e-mail: hrishikesh@eeng.dcu.ie

* Corresponding author.

9.1.1 Background

It is predicted that by 2013, mobile devices such as smart phones and PDAs will overtake PCs as the most popular devices used for accessing the Internet [1]. Most modern mobile phones are capable of playing video and audio, provide high-speed Internet access, enable photography and also support video capturing and video streaming. More advanced devices can interface with GPS systems and include additional sensors (for e.g. accelerometers). While these devices follow a functionality improvement rate similar to Moore's law, developments in battery life have lagged behind considerably. Panasonic, one of the world's leading battery manufacturers, estimates the annual improvement in the life of their batteries to be just 11% [2]. In 2011, a Deloitte study explained that progress in battery life for existing battery technologies is a slow process and that big improvements are observed only when a new battery technology or electrical storage technology is discovered [3]. A classic example of the gap between functionality and power-supply is the iPhone 4. When used continuously, for web browsing over 3G, the battery life lasts a mere 6 hours [4].

9.1.2 Current Solutions

Several research works have been carried in recent years to optimize the required power while simultaneously providing all the requisite functionalities. The different energy consuming operations are targeted without causing severe degradation to the user quality-of-experience (QoE), also defined as the user perception of the quality-of-service (QoS).

i. *Specialized Hardware*

In a smart phone/high end mobile device, there are several specific hardware configurations that can be exploited to achieve energy savings. Bahl *et al.* [5] investigated a technique using a low-power radio in conjunction with a regular 802.11 Wireless Network Interface Card (WNIC). One benefit of this mechanism is that it allows a mobile device and its radio to be powered off, while the low-power radio maintains a network presence that can be used to wake the device up, in the event of any data reception. One example where this approach would be effective is voice over IP (VoIP), where a user can maintain an online presence with minimal bandwidth and only requires a high speed connection while interacting with another user. Another example is the use of hardware acceleration in video decoding. Adobe Flash Player has utilized this functionality since version 10.1. The player uses both hardware H.264 decoding and hardware graphics. The main benefit is that by offloading all the processing from device's CPU to purpose designed hardware, the performance and energy efficiency increase dramatically.

ii. *Wireless Interface Sleep Mode*

The IEEE 802.11 standard outlines a built-in Power Save Mechanism (PSM) when operating in infrastructure mode. A simple energy saving technique is to put the WNIC of a device into sleep mode when it is not in use [6]. However, it is not always feasible and in fact, the savings depend on the application in use. Multimedia streaming and video-on-demand applications have different QoS requirements in comparison to traditional data transmission applications. Saidi *et al.* [7] have proposed a battery-aware localization mechanism for the wireless networks wherein the trade-off between saving energy in a wireless node by lengthening the sleep-cycle period is investigated while still allowing it to perform accurate localization calculations. The main issue with this scheme is that it inherently leads to a lower QoS unless the device knows exactly when it will receive data.

iii. *Energy Efficient Network Selection and Handover*

Most mobile devices are currently shipped with multiple heterogeneous wireless network interfaces, such as Wi-Fi, UMTS, GPRS and Bluetooth. Each of these networks has different energy consumption characteristics. For instance, in [8] measurements show that the energy consumption per unit time of communications over a UMTS and IEEE 802.11b/g are similar. However, the energy consumption as a function of the data transferred can be up to 300 times larger over the UMTS network interface. Trestian *et al.* [9] investigated new techniques in order to exploit these energy characteristics. In this work, a utility function (1) was been proposed for ordering the vertical/horizontal handover between different networks within range based on the predicted energy that will be consumed on each network.

$$U^i = (u_e^i)^{w_e} * (u_q^i)^{w_q} * (u_c^i)^{w_c} * (u_m^i)^{w_m} \tag{9.1}$$

Where: i - the candidate network, U - overall utility for network i and u_e, u_q, u_c, and u_m are the utility functions defined for energy, quality, monetary cost and user mobility for network i. w_e, w_q, w_c and w_m are the weights assigned for each of the four considered criteria.

Mahkoum *et al.* take another approach and propose a power management framework which enables a device to maintain a network presence across multiple heterogeneous networks while powering-off all but one of the network cards on a device [10]. This is similar to that discussed in [5] but would work with the heterogeneous network interfaces that come with most modern mobile devices and does not require specialized hardware. The efficiency is achieved by utilizing proxies on each of the heterogeneous networks to feign the connectivity of the device's network interfaces. If a connection is made through the proxy for any of the sleeping network interfaces, the proxy contacts the device's active interface, which in turn wakes up the interface required.

9.1.3 Limitations of Current Solutions

The above solutions focus on hardware (specialized hardware, network interface card) and network related processing in the device (network selection handover, etc). However, given the numerous functionalities in a handheld wireless device, there are several other avenues in the device where a significant energy is consumed. These include, audio, screen, signal processing, efficient information decoding, etc. However, given the nascent area, there is not enough material and not enough analysis on the energy consumed in the different wireless components. The next section provides a detailed and a comprehensive analysis of different energy-intensive components in a high-end mobile device/smart phone.

9.2 System Based Modeling and Testing

Battery depletion depends on both the hardware and software of a device. The exact amount of energy consumed by each of these components is dictated by the device characteristics and the nature of the applications running on the device. In order to obtain a comprehensive analysis of the energy consumption behavior in a high-end wireless device, a specific device, a HTC Nexus One phone, is considered. This device was selected because of its wide range of functionality and because it runs Android, which is Open Source. The test system comprises of the Nexus One phone (running Android 2.1) and a video streaming server in a wireless 802.11g network. The phone is connected to an external measurement setup that monitors and logs its power consumption during the execution of various tests. These tests, specifically designed for the analysis of the consumption, include receiving and playing video streams over the wireless network as well as applications to monitor the CPU usage and to automatically change device settings to put the phone into different states (e.g. changes in the screen brightness).

Fig. 9.1 shows a smart-phone and the potential major energy consuming components in a hand-held device. The power consumption is measured as shown in Fig. 9.2. A shunt resistor (1.24 Ω; ± 1%) is inserted between the phone and the battery to allow calculating the current by measuring the voltage drop across the resistor. The battery voltage is measured and multiplied by the current to obtain the power. All of these voltages are sampled by an Arduino micro-controller, which logs the instantaneous power-consumption of the device onto a computer. In order to break down the power consumption, experiments for each component are performed by changing the parameters of one component while leaving those for the other components constant. Additional information is provided by Android's battery usage statistics which give a rough indication of the percentage of battery usage attributed to each of the major consumers. Because of background processes the CPU usage tends to fluctuate. A background process is running to push the CPU usage to a constant 100%, which greatly stabilizes the power consumption in order to get an accurate current reading.

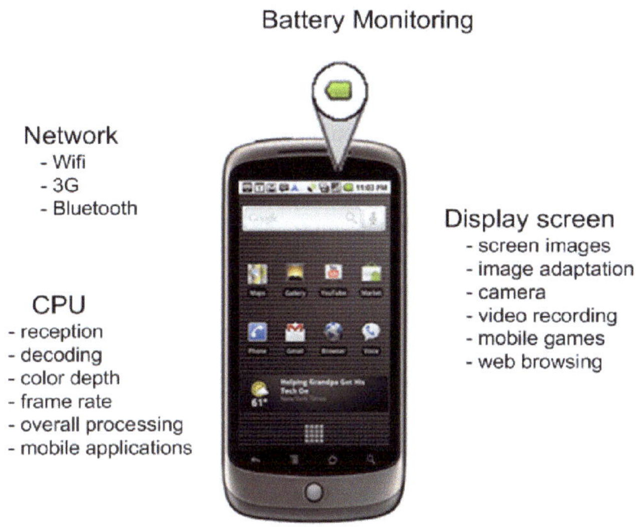

Fig. 9.1 Smart-phone and its different components

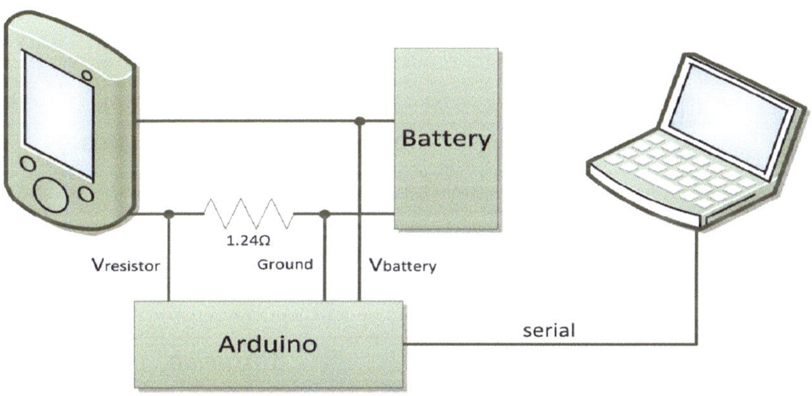

Fig. 9.2 Schematic of a Measurement Set up

1. The screen is tested by measuring different pairs of brightness levels and pixel colours.
2. The audio consumption is tested by playing the same song at different volume levels using the speakers and earphones.
3. The effects of the network interface and the video quality in video streaming are tested in a single experiment in which the same video is played using different quality settings once over a wireless network stream and also from a local file.
4. The CPU is tested by running a background service that gradually changes the CPU activity. Measured power values are then associated with the different CPU usage levels.

In addition to the data obtained from the experiments which focus on the dependency of energy consumption on device settings, further tests are performed in order to break down the overall consumption into the individual components. This is achieved by finding the minimum and maximum consumption values for each component. The screen energy draw is calculated by measuring the power of the same process with the screen enabled first and then disabled. By monitoring the CPU usage and subtracting the screen's consumption, the power drawn by CPU is obtained. Given these values, the other components' energy draw is easily acquired by calculating the power difference of the additional draw that is caused when another component is in use. Fig. 9.3 shows the measured powered distribution among the major energy consuming components. The meaning of the minimum and maximum values is explained in Table 9.1. The dependence of each component's consumption on device settings is discussed as follows.

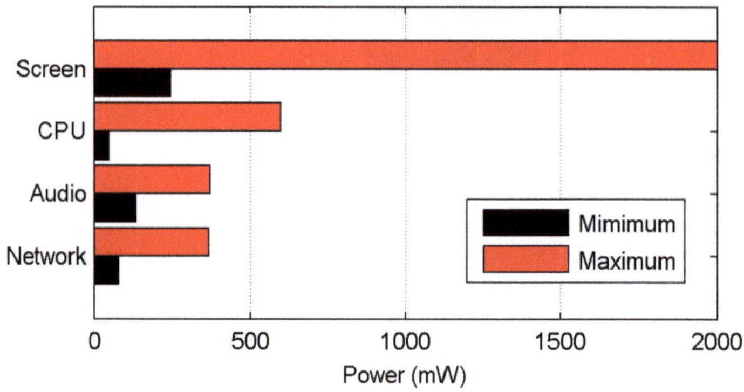

Fig. 9.3 Energy distribution of different components in the Nexus One mobile device

Table 9.1 Minimum and maximum power values in Nexus One mobile device

	Minimum	*Maximum*
CPU	CPU usage at 0%	CPU usage at 100%
Screen	Brightness at 1%, black pixel color	Brightness at 100%, white pixel color
Audio	Audio playback muted	Audio playback at highest volume using speakers
Network	Connected to a WiFi network, idle	Connected and receiving a high quality video stream

9.2.1 Energy Consumption in Screen

The screen's power consumption ranges from about 0.25W to 2W. For the tests, the red, green and blue pixel components were kept at identical levels; the energy consumption of different colours on the display was not measured but is discussed in section 9.2.3. The energy consumption was found to depend on both the brightness level of the screen as well as the brightness of the pixels' colours. From the measurements in Fig. 9.4, it can be observed that the consumption increases approximately in linear fashion with the screen's brightness and exponentially with the colour brightness. This is because of the energy characteristics of the particular OLED display. Hence, the higher the brightness level, the more important it is to know the average pixel brightness in order to accurately estimate the power consumption. Without taking the colour into account, the error of an estimate can be as high as 300% as the power difference between a black and a white screen reaches 1.77W at the highest brightness level.

9.2.2 Energy Consumption in CPU

The CPU's power consumption depends on its usage and ranges from about 50mW to 600mW. The results from four executions of the same test can be seen in Fig. 9.5. The power rises with the usage in a linear fashion but increases sharply at about 80% usage. The exact cause of this behaviour is unclear but could be a result of increased heat in the device. Thus, reducing the CPU usage from 70% and lower, yields relatively little power savings compared to a reduction that takes the usage from any value higher than 80% to any value lower than 70%. However, this assumes that the usage information provided by the Android API is always accurate. Nevertheless, the measurements give an indication about the energy saving potential of the CPU and put it into relation with the other components.

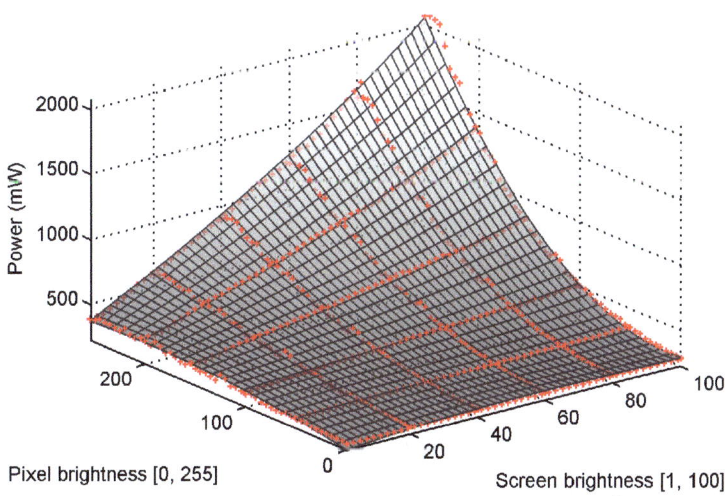

Fig. 9.4 Power measurements in the device screen

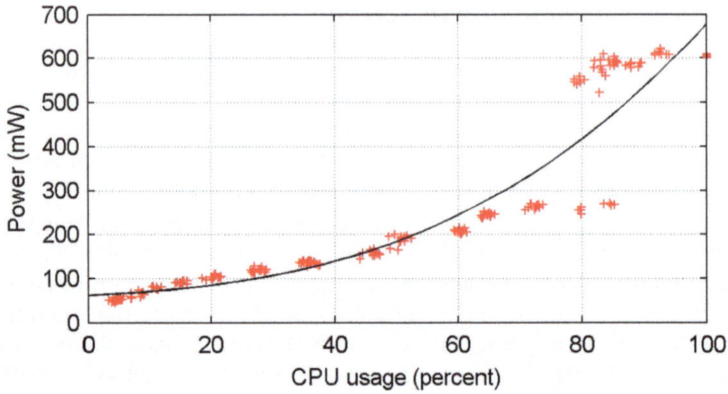

Fig. 9.5 Power consumption in CPU

9.2.3 Energy Consumption in Audio

When playing back audio the power consumption increases from roughly 135mW at the lowest to 375mW at the highest volume (around 40mW of these values are caused by increased CPU usage). Naturally the consumption increases with a higher volume but the difference in energy consumption when playing audio through earphones is rather marginal whereas the consumption seems to increase exponentially when using speakers as can be observed in Fig. 9.6. This implies that the audio interface offers nearly no savings when earphones are used and reducing the volume in this case to save energy is not advisable.

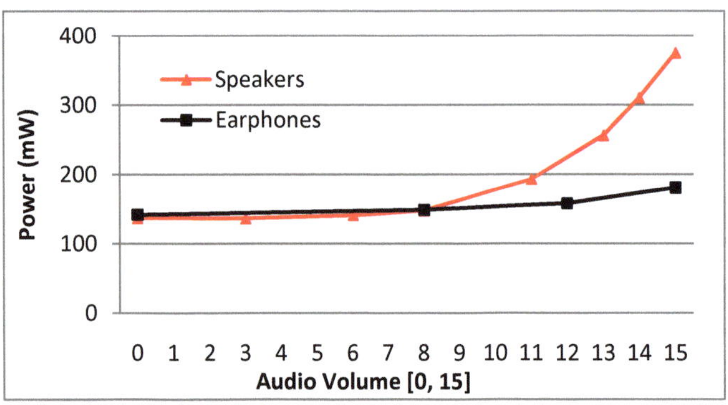

Fig. 9.6 Power consumption in audio playback

9.2.4 Network and the Effect of Video Quality during Streaming

In order to investigate video streaming, the same video clip was encoded using low and high quality settings as in Table 9.2. In order to deduce information about both the quality and the network effects on the energy consumption, each clip was played both locally and streamed. Fig. 9.7 puts the measurements for each clip into relation. The power difference between the low quality and high quality video is 255mW for the local playback and 325mW for the stream. The difference is greater for the stream as a higher quality not only increases the computational power but also the data rate the stream is received with. The power difference between the local playback and the stream is 305mW for the low quality video and 372mW for the high quality video. From this it can be seen that the network interface accounts for about 370mW while receiving the high quality stream and offers savings up to 70mW when the quality is reduced. Quality reduction also lowers the computational power which results in savings up to 255mW. No absolute power increase for video playback is given because the screen's consumption during playback is unknown. The average pixel color brightness would have to be calculated for each frame to estimate the screen's consumption.

Table 9.2 Video encodings used

	Low Quality(LQ)	*High Quality(HQ)*
Codec	MPEG-4	MPEG-4
Video bitrate	128 kbps	1536 kbps
FPS	10 frames/ second	23 frames/ second
Dimen- sions	200 x 120 px	800 x 480 px
Audio bitrate	32 kbps	128 kbps

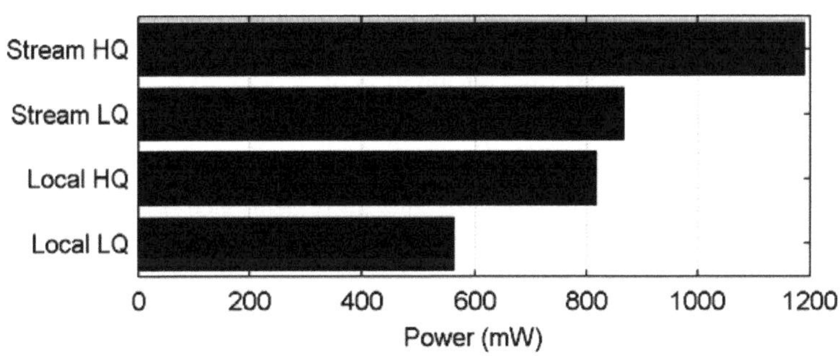

Fig. 9.7 Video playback comparison

9.2.5 Analysis of Solutions

As can be seen in Fig. 9.3, the components of a mobile device that consume the most energy are the screen, the network interfaces and the CPU [11]. From our measurements, it has been estimated that it takes the battery of Nexus One approximately 16 hours to deplete fully (with moderate usage[1]). This in itself is not a very long period of time but it implies that the device will function approximately one business day before requiring a recharge. In contrast though, if an energy intensive application is running non-stop on the device, such as a video streaming app, the battery will deplete in as little as four hours (using Wi-Fi). This would simply not be acceptable for a device that also acts as the owner's point of contact for the whole day. Over the recent years (2009 and 2010), several battery saving software solutions have been launched. These application layer based solutions are designed for specific operating systems (OSs) and provide some basic energy saving features in smart-phones:

1. "***Energy-Saver***" developed by Fedoroff Soft, USA [12] and focuses mainly on auto switch OFF and ON during night/unused time. It is available for the last 12 months at a price of 4.95 Euros.
2. "***Green-Phone***" application developed by Mobi-Monster [13]. *Green-Phone* offers few features like energy savings in the backlight display mechanism and automatic charger disconnection after charging. It supports Windows Mobile and is highly successful commercially. However, it does not provide any adaptive energy savings.
3. "***Power Manager***" is an application developed by X-Phone [14]. It adds basic dynamic power settings to the phone, like how long the screen is on during a call, if device stays on while the keyboard is open, etc. It automatically changes several settings of the device as a group. It is one of the most sought after Apps in Android but unfortunately, does not cater specifically to graphic-intensive battery drain in the device.
4. "***Power Control Plus***" currently available only in Android phones is a widget that lets you turn on/off more than 30 settings in the device [15]. This is a great mobile App but requires quite an amount of manual settings. It cannot and does not provide for any adaptive energy savings in the device middleware.
5. "***Green Mode 2.0***" is an app concept for iPhone [16] that explores the idea of switching off active functions and other apps when the user does not need them. It highlights the energy consumption of various Apps and functions and allows the user to manage their battery budget. It currently only supports ON/OFF behaviour (similar to '*Energy Saver*').

However, the biggest drawback of the aforementioned five energy saving applications is that none of the solutions go beyond the application layer. Importantly, the adaptive energy savings in decoding, display screen, etc. that can be bought in the device need to go beyond the application layer, into the protocol

[1] 30 minutes 3G talk-time, Wi-Fi and Synchronization enabled, 3 hours audio playback, 40 minutes browsing over Wi-Fi Interface, GPS and Bluetooth Disabled, 30 minutes game-play.

stack and the middleware design. Hence, there is a need for further research in these aspects. The next section targets three energy saving aspects in a high-end device. These are:

a. Adaptive streaming
b. Dynamic decoding
c. Dynamic screen control

9.3 Research Directions towards Adaptive Schemes

This section describes in detail the adaptive mechanism and energy savings in each of the three aforementioned approaches, along with the different challenges that need to be addressed in the future.

9.3.1 Adaptive Streaming

Adaptive streaming is the process whereby the quality of a multimedia stream is altered in real-time while it is being sent from server to client. This adaptation of quality is decided in decision modules on either the client or server. The adaptation may come as a result of weighing different network or device metrics, for example, with a decrease in network throughput, an adaptation to a lower quality of video may improve the QoS. Energy relevant metrics can also be considered in order to decide whether an adaptation would be beneficial or not. Energy savings are achieved on the WNIC because less data is being received over the interface. Additionally this opens up a larger window of free time that can be used to put the WNIC in sleep mode more frequently.

The functionality to serve adaptive video streams has become far more available in recent years, since Microsoft Silverlight, and subsequently Adobe Flash, began supporting it. Flash's HTTP dynamic streaming is of particular interest to the mobile device field now though since they began rolling out Flash player 10.1 for Android, BlackBerry, webOS, Windows Phone, LiMo, MeeGo and Symbian OS in 2010. Even though Flash has not been considered energy efficient in the past, version 10.1 was completely overhauled with energy conservation in mind [17]. Version 10.2 of the player has just been released for desktop computers and preliminary tests show that it is up to 34 times more efficient than version 10.1 [18]. As a result, Flash is quickly becoming a viable alternative for consuming video streams in an energy-aware way.

An emerging platform-independent video streaming technology is HTML5, which now comes with a <video> tag. While it is currently possible to use this technology to stream videos natively in HTML pages, the functionality of the <video> tag is still very limited. Internet Explorer does not support HTML5 yet. Not only that but different browsers are choosing to support different video codecs, which means multiple versions of streams are necessary for universal browser support. There is no support for dynamic streaming yet. Having said that, the development of HTML5 will be extremely interesting; it may eventually provide significant benefits over Flash as an energy-efficient streaming mechanism. In [19], Venkataraman, *et al.* proposed and developed a simple algorithm that analyses the remaining stream-duration and the remaining battery-life in order to decide whether or not to send an adaptation order to the dynamic streaming server. When the remaining stream duration exceeds the remaining battery life, the video quality is adapted. This is the simplest example of an energy-aware dynamic streaming mechanism for video. In order to improve upon this algorithm, additional metrics need to be considered for efficient dynamic streaming in real-time. This is very much an open problem and still needs considerable research work.

9.3.2 Dynamic Decoding

Dynamic decoding refers to operations on the client device that alter the default decoding process in real-time in order to maximize energy efficiency when required. This can involve simplifying the decoding process or parsing metadata to concentrate the decoding process on important regions in the screen, image or video.

Scalable Video Coding (SVC) is an extension to the H.264/MPEG-4 video standard. It details a mechanism for decoding an SVC video stream dynamically at one of multiple quality levels [20]. The dynamic scaling is achieved through any combination of three scaling mechanisms:

1. **Temporal Scalability:** Changing the frame-rate of the received video stream by dropping whole frames. In MPEG videos, B-Frames can be dropped without affecting any of the previous or following frames in a Group of Pictures.
2. **Spatial Scalability:** Changing the resolution of the video.
3. **Quality Scalability:** Changing the quantization parameter for each macroblock in the video decoder. This has been proven to yield a 42% decrease in energy consumption during video-decoding with a mere 13% quality degradation in the video [21].

A Region of Interest (RoI), in video terminology, is classified as an area of a video frame that attracts the most amount of attention from viewers. As an example, while watching a football match, the viewer may be most interested in looking at the area around the football (though this may not be the case for everybody). There are several methods for discovering the RoI in a video sequence, which include eye-tracking (with cameras), and video processing and analysis. However, these are not very practical on a battery-powered device as they would consume too much energy for mobile devices. An alternative to this would be to find the RoI on the streaming server itself, which could then transmit RoI metadata alongside the video. Ji *et al.* have shown how RoI processing can be combined with quality scalability to maximize energy-savings and QoE [22]; up to a 15% energy reduction can be achieved with minimal degradation in video quality. In fact, in this chapter, the RoI is assumed to be the centre of each video frame. While this assumption can be tolerated to initially explore whether energy savings may be obtained from RoI-aware decoding, a more advanced system would be required for determining the RoI in a real application [23]. Additionally, the proposed algorithm for implementing the decoding adaptations does not make use of temporal scalability or spatial scalability. It would be interesting to see if the adaptation of the video stream on the server could be performed using SVC and RoI mechanisms to maximize the QoE-to-bit-rate ratio. This is a significant research challenge that would have to be addressed in the coming years.

9.3.3 Dynamic Screen Control

As has been shown in our study in Section 9.2, the display screen is one of the largest energy consuming components on a mobile device. It also yields the largest range for energy saving possibilities. Different screen technologies have very different energy consumption characteristics. Unlike older LCD screens, OLED displays do not require a backlight as their pixels are light-emitting. In an LCD display, the backlight accounts for the majority of the device's power draw. For OLED devices, energy consumption depends on intensity and chromaticity of each pixel being displayed. As a result, OLED devices consume almost no power when displaying black pixels but consume far more than an LCD when displaying white pixels.

Dong *et al.* have done significant research in this domain. They were the first to begin research on manipulating the colours of pixels on OLED screens in order to conserve energy on mobile devices [24]. The first step was to create a model for an OLED device so that an optimal solution can be achieved. This involved devising a device independent mechanism for assessing, pixel-by-pixel, the energy required in order to display each of all the available colours on the OLED screen. This is quite a long and computationally intensive process to be performed in an iterative fashion. Hence, the authors created a shortened and simplified approximation algorithm. The new algorithm decreased the computational cost of the calculation by 1600 times while still achieving 90% accuracy.

Having devised the algorithm above, Dong *et al.* then investigated using colour transformations. In [25], the current colours of the different GUI themes are assessed in terms of their energy efficiency. They are then altered to different colours which maintain the overall contrasts of the different colours on the page while providing significant energy savings. Energy reductions of over 75% were achieved on the display while still showing a GUI to users that they felt comfortable using. In [26], the same authors succeeded in combining all of the work from their previous two papers in order to create a fully functional Android application. The application creates an energy-colour model of the device's OLED screen and then uses it to perform colour transformations to websites. The application, Chameleon, reduced the total system power consumption by over 41% during web-browsing. It would be interesting to perform mild colour-adaptations to videos which may have a significant effect on power consumption without compromising the user's QoE.

9.3.4 Additional Techniques

The list of techniques explained above is not exhaustive. Other approaches include dynamic background-process control, limiting CPU utilization, energy efficient routing and data caching, and energy efficient cooling techniques. For all of these energy-saving techniques future research will be concerned with the constructive combination of as many approaches as possible to optimize the energy consumption while maximizing the QoE. Design of a comprehensive energy-saving algorithm is discussed in more detail in the next section (9.4).

9.4 Context-Aware Algorithm Design

Context-aware energy-conscious computing is the process of recording inputs from various metrics and sensors available on a mobile device. These data in conjunction with past data are then fed into a utility-function which adapts and learns from past experiences in order to maximize the user's QoE while maximising the energy savings in a device. An important point about context-aware computing is that the aim is to find an optimal and complete solution by combining all available mechanisms. Vallina-Rodriguez *et al.* conducted a study where-by the day-to-day interaction of 20 subjects with their android devices was recorded for analytical purposes [26]. The goal was to investigate whether the users' interactions with mobile devices could be modelled easily in order to algorithmically create a unified energy conservation mechanism. Statistics about the CPU, memory, battery, network interfaces, the screen and other sensors were logged on each user's device. The authors concluded that as a result of widely varying user-device interaction behaviour, that static-algorithmic control of the device's resources is an insufficient approach.

There are three steps that must be taken in order to design a context-aware energy-conscious solution for mobile devices. Firstly, the loss-of-quality-of-experience-to-energy-savings ratio of each mechanism and combination of mechanisms outlined above must be evaluated in a range of different real-life situations. This entails designing and conducting significant user testing in order to gather enough information to model how each mechanism and combination of mechanisms perform in relation to each other. This model will later be used for selecting the combinations of mechanisms to be used in order to achieve the energy savings required on the mobile device. The lower the ratio is, the higher the mechanism's efficiency in terms of reducing the energy consumption while maximising the QoE. Testing similar to this as this can be seen in [27] where the authors describe the GUIs investigated in terms of user popularity and energy-reductions. The ratio calculation is an extension of this work. Secondly, a utility-function for assessing the residual power of the mobile device in the context of its current application must be created. This involves considering application-specific QoS requirements, the current rate of energy consumption, past experience, user preferences and predictions formed from the collection and analysis of past data. The utility function will give the system utility, U, which can then be assessed (using thresholds) in order to select the most effective of the mechanisms/combinations above. Finally, the algorithm must maintain a cache of device usage statistic and trends. This data is required for dynamic alteration of utility function weights and the threshold values in order to ensure the optimality of the solution. This step is where the algorithm becomes context-aware by conducting real-time learning and behaviour-monitoring.

9.5 Summary

This chapter provides a comprehensive insight into the energy consumption behaviour in a smart phone/high-end mobile device. In order to measure the energy consumption across the different functionalities and computations across the device, a particular model – an Android based HTC Nexus One phone has been used for extensive testing. Of the different components – the WNIC, display screen and the CPU are the most significant in terms of energy consumption, while the audio component consumes a negligible amount of energy. Furthermore, even though there are different energy saving mechanisms currently in the literature, there is a fundamental problem in the energy consumption space that is completely overlooked – i.e., there are no sophisticated adaptive processes in the device that prolong the battery life in real-time, depending on the nature of application and current battery level. Even though the energy saving mechanisms have improved considerably over the years, a dynamic adaptation and seamless increase in the battery life is still offered nowhere. Importantly, this chapter demonstrates that by investigating deeper into the protocol stack and by modifying the middleware design in the mobile terminal, significant and yet, context-aware energy savings can be obtained in the streaming, decoding and display components of the handheld device, by using appropriate mechanisms.

References

[1] Gartner Inc., Gartner's Top Predictions for IT Organizations and Users, 2010 and Beyond: A New Balance (2009)

[2] Kume, H.: Panasonic's New Li-Ion Batteries Use Si Anode for 30% Higher Capacity, http://techon.nikkeibp.co.jp/article/HONSHI/20100223/180545 (accessed: May 04, 2010)

[3] Deloitte: Squeezing the electrons in: batteries don't follow Moore's Law (2011)

[4] Apple Inc., Apple (Republic of Ireland) Technical specifications of iPhone 4, http://www.apple.com/ie/iphone/specs.html (accessed: April 08,2011)

[5] Bahl, P., Adya, A., Padhye, J., Walman, A.: Reconsidering wireless systems with multiple radios. ACM SIGCOMM Computer Communication Review 34(5) (2004)

[6] Kravets, R., Krishnan, P.: Application driven power management for mobile commu nication. ACM Wireless Networks 6(4), 263–277 (2000)

[7] Saidi, A., Zhang, C., Chiricescu, S., Rittle, L.J., Yu, Y.: "Battery-Aware Localization in Wireless Networks. In: Proceedings of IEEE Local Computer Networks, Zürich, Switzerland (October 2009)

[8] Petander, H.: Energy-aware network selection using traffic estimation. In: Proceedings of the 1st ACM Workshop on Mobile Internet through Cellular Networks, pp. 55–60 (2009)

[9] Trestian, R., Ormond, O., Muntean, G.M.: Power-friendly access network selection strategy for heterogeneous wireless multimedia networks. In: IEEE International Symposium on Broadband Multimedia Systems and Broadcasting (BMSB 2010), pp. 1–5 (2010)

[10] Mahkoum, H., Sarikaya, B., Hafid, A.: A framework for power management of handheld devices with multiple radios. In: Wireless Communications and Networking Conference, 2009, pp. 1–6. IEEE, Los Alamitos (2009)

[11] Theurer, D., Kennedy, M., Venkataraman, H., Muntean, G.-M.: Analysis of Individual Energy Consuming Components in a Wireless Handheld Device. In: China-Ireland International Conference on ICT, Wuhan, China (2010)

[12] FedoroffSoft. T-Reminder and Energy Saver official site, `http://www.fedoroffsoft.com/` (accessed: April 10, 2011)

[13] MobiMonster2.0: Business and Personal utility software for your Mobile, `http://www.mobimonster2.com/` (accessed: April 10, 2011)

[14] Power Manager, `http://www.xphonesoftware.com/pm.html` (accessed: March 09, 2011)

[15] `http://www.appbrain.com/app/power-control-plus-widget/com.siriusapplications.eclairwidgets` (accessed: April 10, 2011)

[16] `http://www.thegreenswitch.org/wp-content/uploads/2010/05/Green-Mode-App-V2.0.pdf`

[17] Microsoft Inc., Developing Energy Smart Software Case Study - Adobe Flash Player

[18] Nguyen, T.: Flash Player 10.2 is Here: Available Now for Windows, Mac, and Linux, `http://blogs.adobe.com/flashplayer/2011/02/flash-player-10-2-launch.html`

[19] Kennedy, M., Venkataraman, H., Muntean, G.-M.: Battery and Stream-Aware Adaptive Multimedia Delivery for Handheld Devices. In: IEEE Local Computer Networks, Denver, USA (October 2010)

[20] Alt, A.M., Simon, D.: Control strategies for H. 264 video decoding under resources constraints. In: Proceedings of the Fifth International Workshop on Feedback Control Implementation and Design in Computing Systems and Networks, pp. 13–18 (2010)

[21] Park, S., Lee, Y., Lee, J., Shin, H.: Quality-adaptive requantization for low-energy MPEG-4 video decoding in mobile devices. IEEE Transactions on Consumer Electronics, 51(3), 999–1005 (2005)

[22] Ji, W., Chen, M., Ge, X., Li, P., Chen, Y.: ESVD: Integrated energy scalable framework for video decoding. Journal on Wireless Communications and Networking (2010)

[23] Grois, D., Kaminsky, E., Hadar, O.: Dynamically adjustable and scalable ROI video coding. In: 2010 IEEE International Symposium on Broadband Multimedia Systems and Broadcasting (BMSB 2010), pp. 1–5 (2010)

[24] Dong, M., Choi, Y.S.K., Zhong, L.: Power modeling of graphical user interfaces on OLED displays. In: IEEE/ACM Design Automation Conference, pp. 652–657 (2009)
[25] Dong, M., Choi, Y.S.K., Zhong, L.: Power-saving color transformation of mobile graphical user interfaces on OLED-based displays. In: Proceedings of the 14th ACM/IEEE International Symposium on Low Power Electronics and Design, pp. 339–342 (2009)
[26] Vallina-Rodriguez, N., Hui, P., Crowcroft, J., Rice, A.: Exhausting Battery Statistics. In: Proceedings of ACM Mobiheld, New Delhi, India (August 30, 2010)
[27] Dong, M., Zhong, L.: Chameleon: A Color-Adaptive Web Browser for Mobile OLED Displays. Arxiv preprint arXiv:1101.1240 (2010)

[illegible]

Chapter 10
Energy Efficient GPS Emulation through Compasses and Accelerometers for Mobile Phones

Ionut Constandache, Romit Roy Choudhury, and Injong Rhee

Abstract. This paper identifies the possibility of using electronic compasses and accelerometers in mobile phones, as a simple and scalable method of localization. The idea is not fundamentally different from ship or air navigation systems, known for centuries. Nonetheless, directly applying the idea to human-scale environments is non-trivial. Noisy phone sensors and complicated human movements present practical research challenges. We cope with these challenges by recording a person's *walking patterns*, and matching it against possible *path signatures* generated from a local electronic map. Electronic maps enable greater coverage, while eliminating the reliance on WiFi infrastructure and expensive war-driving. Measurement on Nokia phones and evaluation with real users confirm the anticipated benefits. Results show a location accuracy of less than $11m$ in regions where today's localization services are unsatisfactory or unavailable.

10.1 Introduction

Recent years have witnessed the explosion of location based applications (LBAs). The iPhone App Store now features 3,000 LBAs. The Android community already lists 400 services, with the number growing rapidly every month [1]. Localization technology is projected to play a critical role in the future, ushering in applications such as location-based advertising, friend-tracking, micro-blogging, etc. At the

Ionut Constandache
Duke University
e-mail: ionut@cs.duke.edu

Romit Roy Choudhury
Duke University
e-mail: romit.rc@duke.edu

Injong Rhee
North Carolina State University
e-mail: rhee@ncsu.edu

outset of this explosion, GPS was primarily used for localization. However, Place Lab [2] and Skyhook [3] identified problems with GPS, including poor indoor operations, short battery life, and long acquisition time. Alternative solutions were proposed to exploit pre-existing infrastructure for localization. The basic idea is to *war-drive* an area to create a map of existing WiFi/GSM access points – this map is then made available to mobile devices. As a mobile device enters a mapped area, it computes its location by detecting WiFi/GSM access points, and searching for them in its stored radio map. Localization became feasible even in indoor environments while the location acquisition time reduced significantly. Overall, it has been a valuable enhancement to GPS based localization. However, the system leaves room for considerable improvement.

(1) *War-driving is an expensive calibration operation, but offers limited localization coverage.* Skyhook currently employs 500 drivers who continuously war-drive to create WiFi/GSM maps of new regions and update the existing ones[1][4]. Still, a large portion of space remains uncovered, including walking paths in university campuses, shopping plazas, apartment complexes, theme parks, etc. War-driving on these walking paths is impractical while war-walking is intolerably time consuming. Moreover, the recurring financial cost of war-driving is excessive, and its impact on environment is undesirable. Complementary solutions are necessary that are cheap, environment-friendly, but scale to regions where Skyhook cannot reach.

(2) *Independence from infrastructure.* WiFi based localization is a useful idea in urban regions covered with dense AP deployments. However, large portions of the world do not have WiFi coverage, especially rural regions in US and many developing regions. Cell tower based localization produces poor localization accuracy while on-phone GPS has serious energy ramifications discussed next. An infrastructure-independent solution would be ideal for global scalability.

(3) *Energy consumption with GPS and WiFi based localization.* Our prior research showed that GPS and WiFi pose a serious tradeoff between localization accuracy and energy [5, 6]. While GPS offers high accuracy (about 10m), it drains a (Nokia N95) phone's battery within around 10 hours. Skyhook's solution lasts for 16 hours although at the expense of a degraded accuracy of 30m (and a high variance). With continuous usage of location services, energy-efficiency is an important concern.

In this paper, we present a simple scheme, called *CompAcc*, to address the above deficiencies in today's localization systems. Our target is to enable energy-efficient localization over walking paths (outside the purview of Skyhook) without relying on war-driving or WiFi infrastructure. The main idea is to leverage the mobile phone's accelerometer and electronic compass to measure the walking speed and orientation of the mobile user. These readings can produce a *directional trail* that is matched against *walkable path segments* within a local area map. The local map is downloaded based on a rough location of the phone, easily available from the cell tower. The path segment with the best match yields the phone's approximate location.

[1] Updates are necessary because WiFi access points change over time as people shift in/out of apartments, homes and offices.

Using only infrequent Assisted-GPS (AGPS) readings, the phone can periodically recalibrate its location, and use it as a reference for subsequent position estimation. We have implemented CompAcc on Nokia phones, and have run live experiments in the Duke University campus. Evaluation results demonstrate that CompAcc achieves average localization accuracy of around $11m$, even in areas without WiFi. This is in contrast to Skyhook's accuracy of $70m$, computed on Duke campus with dense WiFi coverage. We also observed 23.4 hours of continuous operation with CompAcc, a 40% battery life improvement over Skyhook.

CompAcc is among the early attempts to jointly utilize the phone's accelerometer and compass for infrastructure-independent localization. Our system is not yet ready for wide-scale deployment. We are currently planning large scale testing for detailed parameter tuning, particularly for (unpredictable) phone orientations inside pockets, bags, holsters. Nevertheless, results in this paper are adequately promising to justify this large-scale experimentation effort. The promise is particularly pronounced because CompAcc is complementary to existing localization solutions. While Skyhook targets urban regions near roads and streets, CompAcc is focuses on areas not close to drivable roads or devoid of WiFi infrastructure. We believe that in conjunction with Skyhook, CompAcc may be an important step towards a complete localization technology.

10.2 System Design

Figure 10.1 presents the overall CompAcc architecture. We sketch an overview first, followed by description of the three main components: (i) generating path signatures, (ii) generating directional trails, and (iii) matching signatures with trails.

10.2.1 Architectural Overview

CompAcc initializes by obtaining the phone's location through either GPS or AGPS (we discuss the pros and cons later). The location is sent to a remote CompAcc server which then sends back a map of the small area around that location – we call this a *map tile*. A map tile is expected to include all walking paths in that region. Now, if the phone is detected to be stationary based on its accelerometer readings, the phone's location is naturally known from the initial GPS reading. If the phone begins to move, CompAcc acts on the accelerometer readings to estimate the user's displacement. The estimation algorithm exploits the rhythmic nature of human walking patterns, and computes the number of steps walked. The distance traversed can be derived by multiplying the step count with the user's step size possible to be approximated based on user's height and weight [7]. Alongside accelerometer readings, the phone's compass orientations are also recorded. Combining these time-indexed $< distance, orientation >$ tuples, CompAcc creates a *directional trail* of the user. Distance is expressed in *meters*, and orientation in *clockwise angular degree* with respect to magnetic north. With the starting point at the known AGPS location, this sequence of tuples presents a natural opportunity to estimate the user's position.

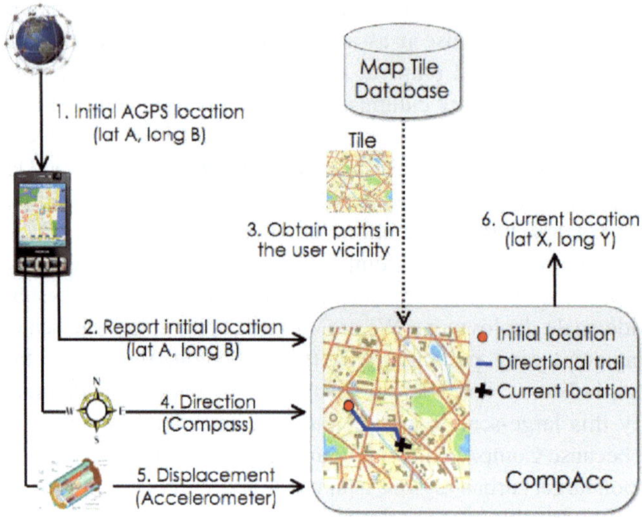

Fig. 10.1 Flow of operations in CompAcc.

With a perfect compass and an accelerometer, the user's location can be trivially computed over time. Unfortunately, electronic compasses and accelerometers are highly noisy [8] (unless extremely expensive). The user's movement variability and the phone often jiggling in pockets and backpacks further adds to this noise. Localization becomes erroneous when the device noise drowns the changes in motion and orientation, particularly along soft turns on curved paths.

CompAcc approaches this problem by "matching" the directional trail with possible walking paths around the phone's known location. The paths are extracted from Google Maps[9], and suitably formatted for matching. The best matching path is declared to be the path of the user. Matching continues, allowing CompAcc to identify when the user turns at an intersection or moves on a curved path. Even though the approximate step count may accumulate error over time, *direction changes reset the error*. Put differently, CompAcc recognizes the user making a direction change, and thereby learns the user's location with better accuracy. Of course, error can still arise due to walking up/down stairs, detours from walking paths, and user's leaps and jumps. CompAcc occasionally falls back on AGPS to reacquire its current location.

10.2.2 Generating Path Signatures and Map Tiles

Matching the user's movement against electronically generated paths eliminates the need for war-driving. The approach scales globally because the path generation is a faster process at the server's side. We describe how a map tile is constructed from path segments, and is sent to the mobile user.

Digital maps, including Google Maps, represent roads and paths as polylines. In computer graphics, a polyline is defined as a continuous line composed of one or more linear segments. The polylines are superimposed on the map to display directions. Figure 10.2 shows an example polyline, with segment ends marked with large blue icons. The latitude and longitude for each of the segment markers are available through the Google Maps API. The intermediate coordinates within each segment is computed based on the end points. The curvature of the earth disallows Cartesian formulas for distance computation. CompAcc employs the haversine formula [10] known to be accurate for small distances (equations omitted in the interest of space). We input the earth's radius as 6367 km. Figure 10.2 shows the intermediate points with small black icons. The separation between the small icons is $20m$ for visual clarity, but is actually $1m$ in our implementation.

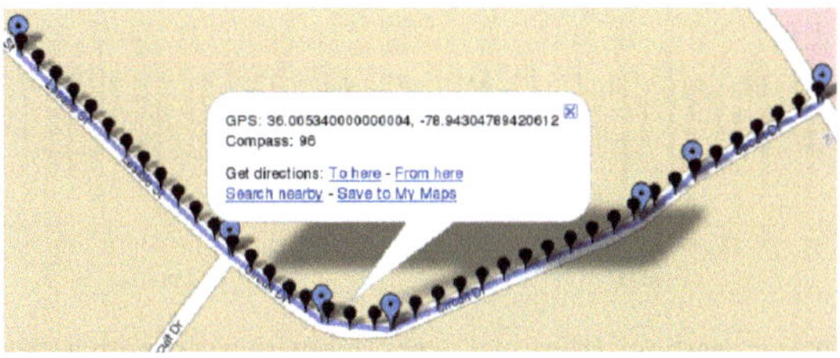

Fig. 10.2 Large blue icons mark the piecewise segments of the polyline, while black icons are computed intermediate points. Each location associated with the latitude, longitude and the angle from magnetic north.

Given the end coordinates of any path segment, the precise path orientation can be computed using the formulas in [11]. A path running east-west will result in an orientation of 90° or 270°. An ideal mobile phone compass is expected to reflect this value while the user is walking on this path segment.

When the mobile phone sends its AGPS location, the CompAcc server computes a *map tile* with that location as the center. The tile includes all the path segments in a data structure, each segment consisting of its end markers, intermediate coordinates, and orientations (a path signature). The tile is downloaded to the phone, and used to match against the user's directional trail. If Internet connectivity is a concern, a large tile may be downloaded in advance.

10.2.3 Generating Directional Trails

Modern phones come equipped with accelerometers, mostly used to rotate pictures between landscape and portrait mode. The accelerometer is always on and CompAcc exploits it for computing the user's displacement. The natural computation method is to double-integrate acceleration, $\int\int a(t)dtdt$, where the first integration computes speed, and the second computes displacement. Unfortunately, several factors cause fluctuations in acceleration, resulting in erroneous displacements [8].

Departing from this approach, CompAcc identifies a rhythmic acceleration-signature in human walking patterns. The rhythm is evident in the raw measurements in Figure 10.3; each spike in the negative direction, roughly corresponding to a step. Around -300 units of acceleration is due to gravity, g. In addition to g, we observe two spike patterns. The small ones (at around -500) correspond to the foot being lifted off the ground; the large ones (at around -700) are caused by the leg settling back on the ground. We observed this rhythm across 15 test users, only the height and periodicity of the spikes varied. CompAcc computes the number of steps using this signature, and multiplies it with the step size of the individual. The individual's step size can be approximately derived from the individual's height and weight [7].

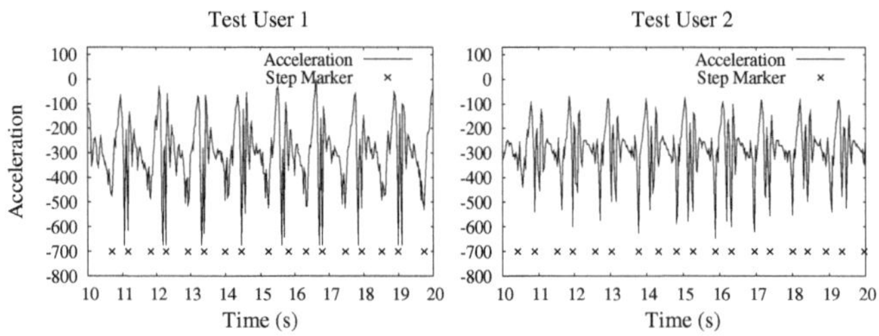

Fig. 10.3 Accelerometer readings from two users (steps marked with a cross)

The phone's compass records the user's orientation, expressed as an angle with respect to magnetic north. With Nokia 6210 Navigator models, this angle is the direction faced by the phone's display, when the phone is held vertically. Ideally, while the user is moving in a straight line (say towards East), the readings should be 90°. However, as with accelerometers, measurements yield noisy data with random fluctuations and slow responses to quick/sharp movements. CompAcc caches a window of previous N compass readings, utilized in constructing the user's directional trail.

Forming the Directional Trail

The directional trail is defined as a series of the last N compass readings and an associated set of displacements between them. The compass readings are collected with a time separation of 1 second. Denote the compass trail by a set $C_N = \{c_0, c_{-1}, c_{-2} ... c_{-N}\}$. Here, c_0 is the compass reading at the user's *estimated* current location, c_{-1} is the previous reading, and so on. Also, let the accelerometer based displacements between compass readings be denoted by a set $D = \{\delta_0, \delta_{-1}, ... \delta_{(-N+1)}\}$. Thus, the user walked a distance of δ_j between two compass measurements, c_{j-1} and c_j. The user covered a total displacement of $TD = \sum_{j=0}^{-N+1} \delta_j$ in the past N seconds.

Similarly, a sequence of compass readings is also obtained from the path signatures in the tile. Denote this sequence by $P_N = \{p_0, p_{-1}, p_{-2} ... p_{-N}\}$ where p_0 is the actual orientation (known from the tile) at the user's estimated current location.

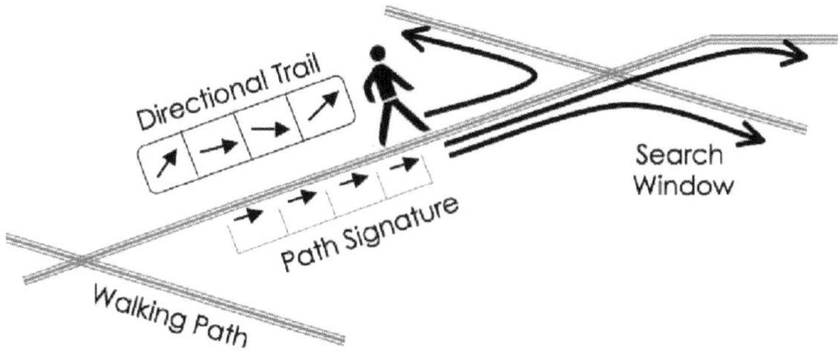

Fig. 10.4 CompAcc slides the path signature to a maximum of *Search-Window* steps, forward and backward. The sliding needs to turn into all possible directions within this window. For every step-wise slide, the "Dissimilarity" is computed between the given directional trail and the path signature. The path signature with minimum *Dissimilarity* yields the user's current location.

Similar to the compass trail, the values of p_k and p_{k-1} are also separated by distance δ_k. The goal of CompAcc is to slide the P_N window forward and backward, and find the maximum similarity between P_N and C_N. Figure 10.4 shows the trail and the (sliding) path signature. The details on matching are explained next.

10.2.4 Matching Path Signatures with Directional Trails

Matching is triggered only when CompAcc finds that the user's estimated location is within a threshold distance from an imminent direction change. Direction change is possible not only at path intersections, but also at the end of linear segments in a polyline. When near a segment end, CompAcc records the current compass trail C_N and the corresponding path signature P_N. The "Dissimilarity" ϕ is computed between the two sequences using the following simple equation:

$$\phi_0 = \sqrt{\frac{\Sigma_{i=0}^{N}(c_i - p_i)^2}{N+1}} \tag{10.1}$$

Of course, P_N is computed based on the user's estimated current location. This estimate can be erroneous (due to step count inaccuracies), and the user may either be ahead or behind the estimated location. If the user is actually ahead, then a *slide-forward* version of P_N is likely to better match C_N (and the vice versa). We use P_N^{+j} to represent a forward slide of j steps on P_N. Similarly, P_N^{-j} represents a backward slide of j steps on P_N. The maximum extent of the forward/backward slide is called the *Search Window* where W is the size of this window. CompAcc computes

$$k : \phi_k = min\{\phi_j\} \quad \forall j \in [-W, W] \tag{10.2}$$

Given the value of k, CompAcc computes P_N^k, and the corresponding value of p_0^k. The location corresponding to p_0^k is the new estimate of the user's location. The accelerometer based step count is reset, and restarted from this point. The next section discusses parameter choice in CompAcc. Evaluation results are reported thereafter.

10.2.5 Fallback Mechanisms

CompAcc continuously estimates the user's location. In certain scenarios, the difference between the estimated and the actual location (i.e., the localization error) can create confusion. Consider Figure 10.5(a) where CompAcc believes that the user is located between the 5th and 7th Street intersection, and walking northward. Now, assume that the actual user has taken a turn, hence, the compass values reflect it. Since CompAcc will search both forward and backward, the trail may match "a right turn" onto both 5th and 7th Streets. The actual user could be ahead or behind the estimated location, and hence, there is no way of telling the difference.

One way to resolve this is to ensure that CompAcc's Search Window covers at most one intersection at any given time. Let us denote the minimum distance between two intersections in a map tile as L_{min}. This value can be an attribute of the map tile, known to CompAcc. If CompAcc chooses a Search Window, $SW < \frac{L_{min}}{2}$, then the confusion in Figure 10.5(a) will not happen. As long as the location error is not too large, CompAcc will always search over the same intersection around which the user actually is located.

If the location error, however, grows large, the SW parameter choice will not be sufficient to prevent confusion. Consider Figure 10.5(b) where the user has actually

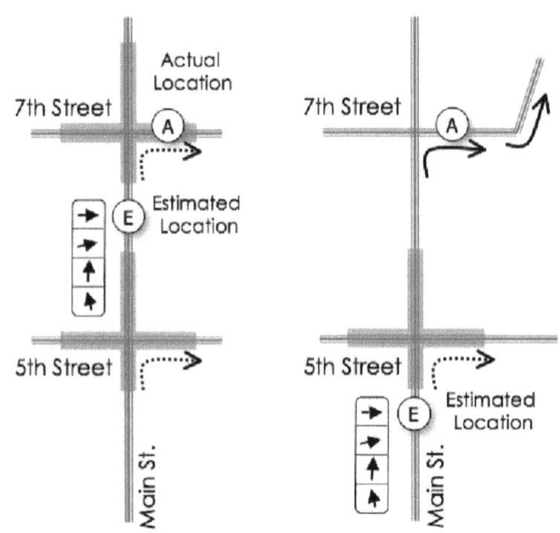

Fig. 10.5 "E" is the CompAcc-estimated user location, and "A" is the user's actual location. The scenarios lead to possible confusion (Search Window shown in gray).

turned onto 7th Street, but CompAcc's estimated location is before the 5th Street intersection. Observing that 5th Street is imminent, CompAcc will match its directional trail with path signatures, and is likely to find a good match with a "right turn on 5th Street". Clearly, the estimated location will be on a parallel path to the actual user, resulting in error.

We propose an AGPS based fallback mechanism to recover from such situations. Observe that the actual user's directional trails are unlikely to continuously match the path taken by the estimated user. The actual user may take a left turn while the estimated user has no left turn within its search window. The *minimum matching Dissimilarity* (ϕ) (between the directional trail and sliding path signatures) will be large in such situations. Figure 10.6 computes these distributions for the Duke University map tile. Essentially, the "Dissimilarity" is low when the actual directional trail is along the actual path signature, but much higher when the matching is against an incorrect path turn. Therefore, when CompAcc observes a large value of ϕ (45 in the case of the Duke University tile), it triggers an AGPS reading. The estimated user is relocated close to the actual location, step count is reset, and the operation continues. This ensures that the localization error does not grow (except perhaps in rare cases such as a Manhattan grid). Results on walking paths in the Duke University campus did not require any AGPS reading due to the inherent (directional) diversity in the walking paths.

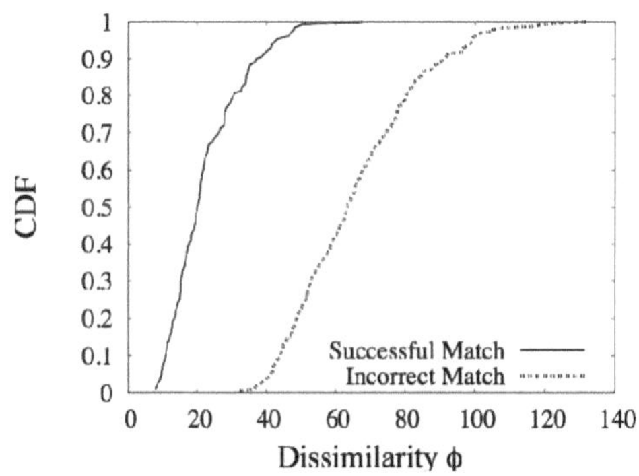

Fig. 10.6 *Dissimilarity* is high when the user's directional trail is matched against an incorrectly estimated path signature. This is a trigger for AGPS.

10.3 Experimentation and Evaluation

10.3.1 Experimental Methodology

We implement CompAcc on Nokia N95 and 6210 phone models using Python as the programming platform. Two phone models were necessary because N95 does not have a compass while 6210 does not have WiFi. Newer phones (like iPhone 3Gs and Android G1) have all the required sensors, but were not available when we started the project. Hence, we pretend that the two Nokia phones are the same device, gather data from each of them, and analyze the consolidated traces for localization.

We assume GPS to be the global truth, and obtain location readings through an high end Garmin 60CSx handheld GPS receiver. N95 and GPS were placed in the test-users' pockets, while the 6210 was carried vertically, either in hands or in trouser pockets. Map tiles of the Duke University west campus were generated from Google Maps. Google Maps already contains a few of the walking paths, so the rest were manually marked by clicking on start and end points of different polyline segments (Figure 10.7). Marking the paths was easy through Google's Satellite view, and took less than a minute to cover a $300m \times 300m$ tile. This projects to less than 5 hours to cover the Duke University campus (2.9 *sqkm*). We note that this is orders of magnitude less expensive (in time and money) compared to war-walking *all* paths within the campus.

To evaluate CompAcc, 5 students walked on random paths in the tile, carrying the equipment described earlier. We collected 25 traces in total covering all segments of the tile. We use two metrics for evaluation, namely, *Instantaneous Error* (IE) and *Average Localization Error* (ALE), defined as:

$$IE(t_i) = distance(CompAcc(t_i), GPS(t_i)) \quad (10.3)$$

$$ALE = \sum_{i=1}^{T} \left(\frac{IE(t_i)}{T} \right) \quad (10.4)$$

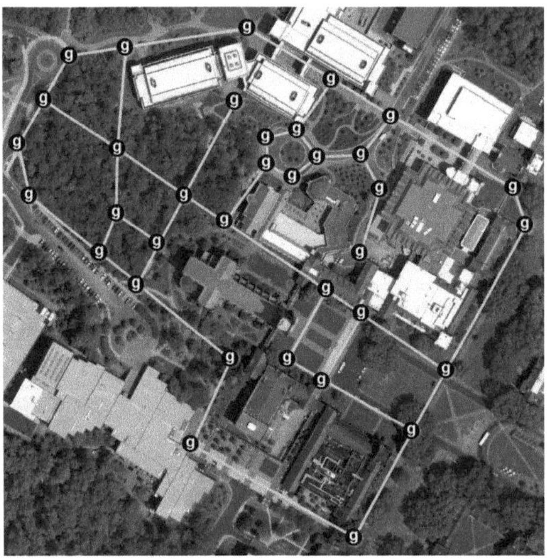

Fig. 10.7 A *map tile* from the Duke University west campus, used for evaluation. Light gray lines show some of the walking paths created manually.

Here $CompAcc(t_i)$ and $GPS(t_i)$ represent the user locations reported by CompAcc and GPS, respectively. A trace has T equally spaced discrete time-points, and $t_i - t_{i-1} = 5s$ (i.e., we sample the user's location every 5s).

10.3.2 Design Decisions and Parameter Choices

This subsection justifies the parameter choices in CompAcc.

Computing Initial Location through AGPS
CompAcc occasionally needs the phone's accurate location as a resetting mechanism. This ensures that undetected direction changes (or other unanticipated glitches) can be recovered. GPS is the natural solution, since it is most accurate. Unfortunately, GPS can experience a large delay for acquiring (adequate) satellite locks for localization. This behavior can be attributed to a variety of factors, including cheap GPS sensors, cloudy skies, or a "cold start". Assisted GPS (AGPS), has a faster acquisition time while retaining comparable accuracy. Table 10.1 compares the location acquisition time between GPS and AGPS for 10 experiments on 10 different days. Evidently, AGPS averages 13.1 seconds, in contrast to GPS's 98.6 seconds. All the experiments were obtained in summer with clear skies; our experience with GPS is worse in winter. This motivates the use of AGPS in CompAcc for infrequently calibrating the user's location, as well as for detecting deviations from walking paths (e.g., when the user walks into a building).

Table 10.1 Location acquisition delay: AGPS avg. is 13.1s, GPS is 98.6s.

Delay(s)	1	2	3	4	5	6	7	8	9	10
A-GPS	6	14	15	16	16	12	15	11	13	13
GPS	21	216	143	108	117	57	97	53	103	71

How to cope with different walking styles?
The accelerometer signature of human footsteps is characterized by rhythmic oscillations. We parameterized the oscillation by two parameters, h and Δt. Briefly, h is the minimum height of the oscillation, and Δt is the minimum periodicity. With comparable noise in the accelerometer and variations in walking styles, the values of h and Δt need to be chosen empirically. Error in the parameter value will result in step count error, however, CompAcc can tolerate part of this error through signature matching. Thus, an approximate user-displacement is acceptable.

We inspected 10 traces collected from 5 student test users (*different from those who will evaluate CompAcc later*). After compensating for gravity and simple noise reduction, we consistently found that step-induced deceleration drops below -50, i.e., $h \leq -50$. The step count was more sensitive to parameter Δt, and hence, we monitored its behavior for different values. Table 10.2 shows the results only for $\Delta t = 0.3s$, $0.4s$ and $0.5s$. $\Delta t = 0.4$ was found to exhibit less than 10% error and proved robust in tests with other arbitrary users. Thus, CompAcc increments the

Table 10.2 Calibrating Δt by measuring the time interval between spikes.

Test	1	2	3	4	5	6	7	8	9	10
count	10	50	56	64	70	71	77	89	113	645
$0.3s$	12	53	62	64	98	80	101	112	111	661
$0.4s$	11	53	61	61	77	72	77	88	111	618
$0.5s$	11	50	60	47	68	62	75	73	85	522

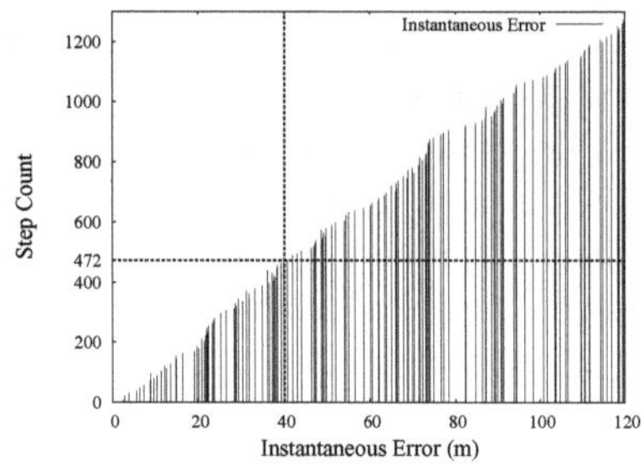

Fig. 10.8 The growing error from Step Count observed over a straight walk.

step count only when the oscillation has a magnitude less than -50 and a time separation greater than $0.4s$ (from the previous spike). Ongoing work is attempting to learn user-specific parameters based on step count and a few consecutive AGPS readings.

What if step count error keeps increasing?
As discussed earlier, the step count error is likely to get reset when users take turns. However, if the user walks on a long straight segment, the error continues to grow. An AGPS reading resolves this problem ensuring that the error is lowered before it exceeds a user-specified threshold. To meet this threshold, say δ, CompAcc must trigger the AGPS when the accumulated error is close to δ. We calibrate the accumulated error due to step count by requesting test users to walk along a long straight path. Figure 10.8 shows the variation. Evidently, for any value of δ (on the x-axis), we can find a corresponding value of step count, Y_δ, on the Y axis. CompAcc takes an AGPS reading if the user's compass orientation does not show a turn for Y_δ consecutive steps. In our evaluations, we used $\delta = 40m$, corresponding to around 472 steps (approximately $400m$ in physical distance). If users pause in between, CompAcc is aware of it, and does not increment the steps unnecessarily. For our tests in the Duke campus, the longest straight line path was around $300m$, so AGPS was never triggered. As we shall see later, this is desirable for energy conservation.

What happens if the phone shakes in pockets?
An ideal compass in ship or aircraft navigation is expected to track the device's orientation through turns and straight paths. However, with noisy compasses, swaying human movements, and the phone jiggling in pockets, the compass output is less

ideal. Figure 10.9(a,b) show an example recording from a trace with 7 turns. The compass readings are modulo 360 (i.e., −5 and 355 are along same northward directions). The ground truth is pre-computed from the map and shown as a solid black line. Other readings are from three placements of the phone, namely, handheld, in a trouser pocket, and in a running shorts (the phone display facing outward). Carrying the phone in shorts' pockets induces wide variations (segments *III*, *V* and *VI* show heavy fluctuations). For our evaluations in this paper, we assume that the phone is handheld or in a trouser pocket. Our future work will address the problems with carrying the phone inside shorts-pockets.

How large is the Directional Trail?

We select the value of $|C_N|$ based on error measurements for different window sizes (Figure 10.10). We include 4 out of 25 traces for visual clarity. ALE (Average Localization Error) is consistently large for small values of SW. This is because small windows are more susceptible to compass noise, resulting in degraded location quality. For window sizes 8 or larger, the compass shows a clearer (statistical) trend, and the ALE value drops sharply. Twenty one other traces show similar trends, guiding us to assign $|C_N|$ to 10.

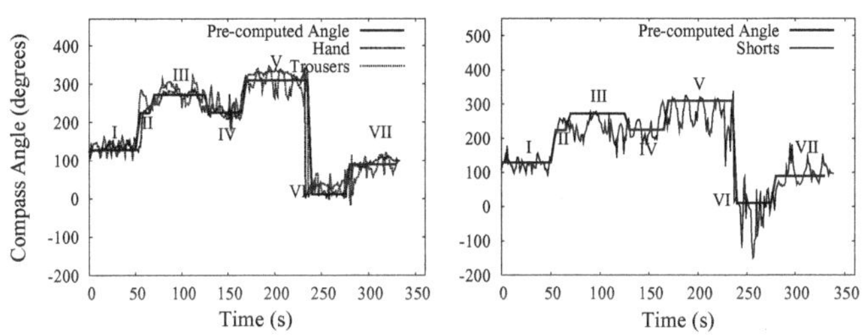

Fig. 10.9 Compass readings can be tolerated when phone is handheld or in trouser pockets. The phone jiggles excessively in shorts pockets.

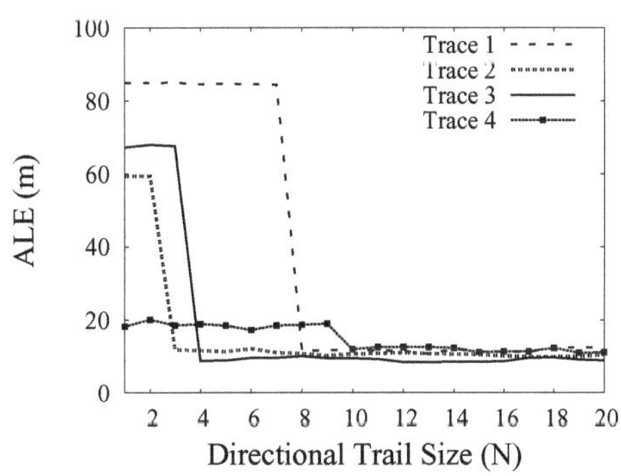

Fig. 10.10 Variation of localization error with the size of the directional trail.

10.3.3 Performance Evaluation

This section presents the performance of CompAcc in comparison to Skyhook and WiFi-War-Walk. We downloaded the *Skyhook* software from the Internet, and installed it on a Nokia N95 model (equipped with all the necessary sensors). Since Skyhook relies on war-driving to create the radio map, its performance within university campuses and other walking paths reflects today's achievable performance. *WiFi-War-Walk* is a customized scheme that captures the improvements if a Skyhook-like scheme was augmented by war-walking the footpaths. Although war-driving is expensive and time-consuming, this provides an optimistic estimate of Skyhook's performance. We war-walked the Duke University west campus, and implemented Place Lab's localization algorithm on the WiFi map (Skyhook's algorithm is proprietary, but similar to Place Lab). The metrics of interest are Average Localization Error (ALE), and energy consumption due to localization. We also qualitatively contrast localization coverage through Google Maps.

Figure 10.11 presents the per-trace ALE from all 3 schemes. The average across all traces for each scheme is included in the caption. CompAcc outperforms Skyhook by a factor of 6.4; the improvement over WiFi-War-Walk is 2.8. Further, we note that Duke University is densely covered with WiFi, and hence, similar comparisons in off-campus areas are expected to yield wider performance gaps. Figure 10.12 zooms into the CompAcc performance and shows the Instantaneous Error (IE) over time. Two arbitrary segments from distinct traces are shown – vertical lines denote instants when the user turned. The trend shows the error increasing as the user walks downs a path segment, but drops at turns. The error does not become zero because compass and accelerometer noise miscalculates the user's precise position. Nevertheless, it prevents the error from accumulating.

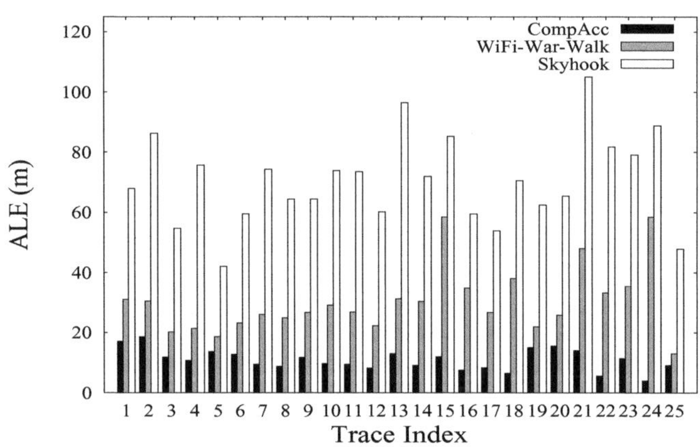

Fig. 10.11 ALE for CompAcc, WiFi-War-Walk, and Skyhook. Average localization errors across all traces and schemes are: CompAcc 10.95*m*, WiFi-War-Walk 30.24*m* and Skyhook 70.6*m*.

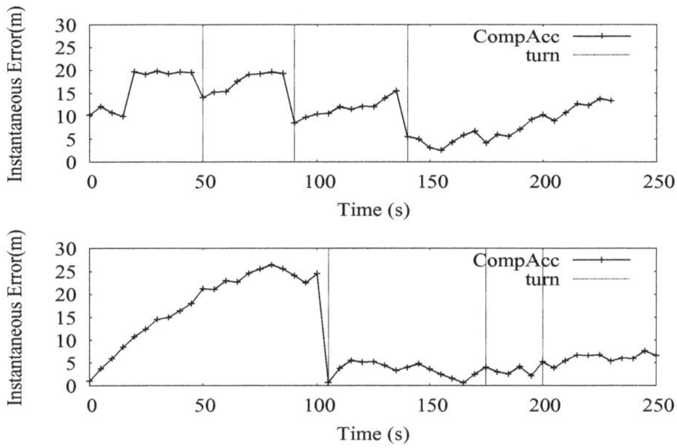

Fig. 10.12 CompAcc's Instantaneous Error (IE) drops at path turns.

Figure 10.13(a,b) visualizes the comparison between CompAcc and Skyhook on Google Satellite Maps. GPS locations are shown with a "g", Skyhook with "s", and CompAcc with "c". The light gray lines joining g–s or g–c denote the location errors. Figure 10.14 shows the variation of Instantaneous Error (IE) over time from 3 randomly selected traces. The differences between the schemes are visually apparent. Nevertheless, we make a few important observations:

(1) *Skyhook's localization is biased towards roads and streets (regions A, B and C).* Users walking within the campus were localized to the nearest street. Often, a street had a dead end within the university, and a large number of location readings were found to be clustered at these dead ends. This is an outcome of Skyhook war-driving only on the streets, resulting in poor localization on walking trails. CompAcc's independence to war-driving is evident in the consistent accuracy.

(2) *Skyhook relies heavily on WiFi availability (region E).* A popular campus walking path bordering a wooded region does not have WiFi connectivity. Skyhook was unable to localize the phone on this path. Although GSM localization should have been triggered, we suspect that the time to detect WiFi failure was too long. CompAcc, however, shows near-accurate localization.

(3) *Skyhook exhibited high error variations (regions B, C).* We are unsure why this occurs, however, open parking lots may be susceptible to signal strength variations. Labs may induce similar impacts due to electromagnetic radiations that interfere with WiFi. CompAcc remains unaffected by such environmental factors.

(4) *Skyhook performed well on a footpath beside a drivable path (region F).* CompAcc too achieved high accuracy in these regions.

Energy Consumption. We report energy consumption due to localization through the Nokia Energy Profiler (logging energy usage every 0.25 seconds). AGPS and Skyhook were made to sample the location every 5 seconds. For CompAcc, the accelerometer and compass were continuously probed. Figure 10.15 shows the power draw for AGPS, Skyhook, accelerometer, and compass. The 3G network was enabled for AGPS and Skyhook (GPS was turned off for SkyHook for fairness). The compass measurements were taken from the Nokia 6210 and superimposed on this graph. The heavy energy footprint of AGPS (baseline of $0.4W$) precludes it as a stand-alone localization method. Skyhook displays periodic spikes of around $1W$, separated by smaller spikes of $0.2W$. The accelerometer and compass sensors have similar power consumption signatures, with a baseline of around $0.1W$. Table 10.3 shows the individual battery life for each sensor/localization scheme, assuming only the profiler running in the background. The compass measurements were executed on 6210 which has a battery capacity of approximately 80% of a N95. We combined their individual recordings proportionately to derive the CompAcc battery life. Results show an improved battery life with CompAcc.

10.4 Limitations and Future Work

CompAcc is not yet ready for deployment. A number of issues still need to be resolved efficiently.

Map Generation. Currently, map tile generation includes a manual component of marking out the start/end points of footpaths. Although significantly less expensive than war-walking, an ideal system should automate this process. One option is to design Internet based labeling games as proposed by Von Ahn [12]. Another option could be to launch this into Mechanical Turk [13] where many Internet users could label paths for small financial incentives. A more complex approach could be to extract the paths through sophisticated image processing on Google Satellite view. We plan to investigate the viability of these and other approaches.

Improved Signature Matching. CompAcc resets a signature mismatch through AGPS. This is required because CompAcc keeps no record of possible directions the user may follow. Particle filters [14, 15] applied to localization systems operate on movement probabilities in multiple directions. Employing particle filters may provide further improvements.

Table 10.3 Battery lifetime: CompAcc (Acc, Comp), Skyhook, AGPS.

Sensor	Acc.	Comp.	CompAcc	Sky	AGPS
lifetime(h)	39.1	48.7	23.4	16.3	10.6

Multiplexing between Localization Methods. CompAcc needs to handoff localization to Skyhook/GPS when the user is in a vehicle or enters indoors. The control can be regained once the user is walking again. Switching between these localization modes need to be triggered automatically. The current system can activate handoff when the *matching Dissimilarity* is excessive (as used for fallback mechanisms). However, quicker activation is necessary.

Extending beyond Walking Paths. CompAcc targets regions that lie outside the coverage of war-driving based localization. However, the system can be extended to vehicular motions, or even towards micro-mobility within buildings, malls, stores[16]. Roads and streets' path signatures can be extracted directly from Google Maps without any manual intervention. One part of the ongoing work is focused towards these extensions. CompAcc (for vehicles) will be an effective solution for many countries in the world that do not have WiFi coverage, or remain to be war-driven.

10.5 Related Work

Previous research on localization makes different assumptions about infrastructure and calibration effort. In general, previous localization systems rely on deployed radios (e.g. APs) or require installation of specialized hardware in the environment (radio or bluetooth devices). A calibration effort maps overheard radio signals to location coordinates. Then, overheard signals are matched with the data recorded during the calibration phase and the user location is estimated. Cricket [17], Nokia Labs, VOR [18] and Pinpoint [19] rely on specially installed hardware for indoor localization. While effective in high-budget enterprises, these systems are expensive to deploy and maintain in outdoor environments.

Radar [20], Active Campus [21] and PlaceLab [2] rely on devices already present in the surroundings for localization. The Radar system [20] operates on WiFi fingerprints, and is capable of achieving high accuracy in indoor deployments. Nevertheless, Radar needs to calibrate WiFi signal strengths at many physical locations in a building. The calibration process is time-consuming and may not scale over larger areas. PlaceLab [2] uses WiFi and GSM signals for localization. A radio map is created by war-driving car-accessible roads and mapping APs/GSM towers to GPS coordinates. The radio map is distributed to mobile devices which localize themselves by comparing overheard APs/GSM towers with those recorded in the map. Active Campus [21] is similar to PlaceLab, but assumes that the locations of the WiFi access points are known *a priori*. Unlike the previous systems, CompAcc does not require war-driving, signal calibration, or deployed infrastructure for localization. Our approach is straightforward, relying on digital maps, compasses, and accelerometers, available in modern mobile phones.

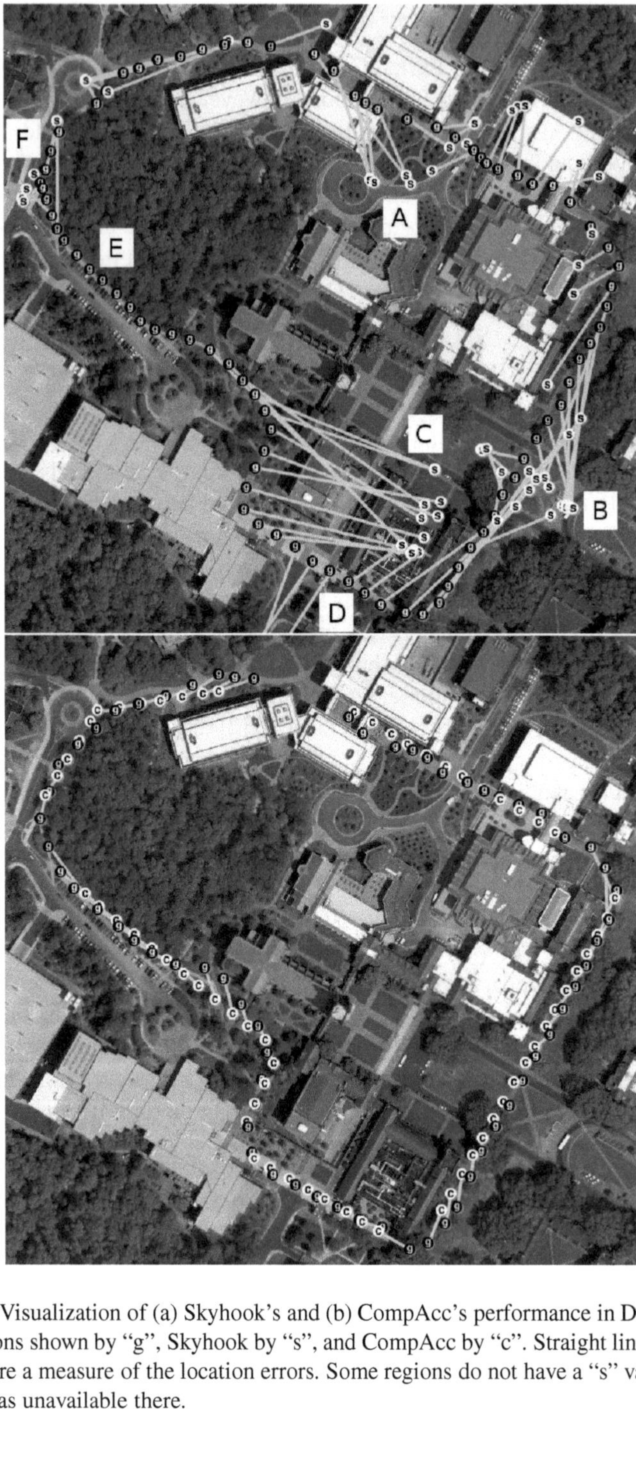

Fig. 10.13 Visualization of (a) Skyhook's and (b) CompAcc's performance in Duke campus. GPS locations shown by "g", Skyhook by "s", and CompAcc by "c". Straight lines joining s–g and c–g are a measure of the location errors. Some regions do not have a "s" value because Skyhook was unavailable there.

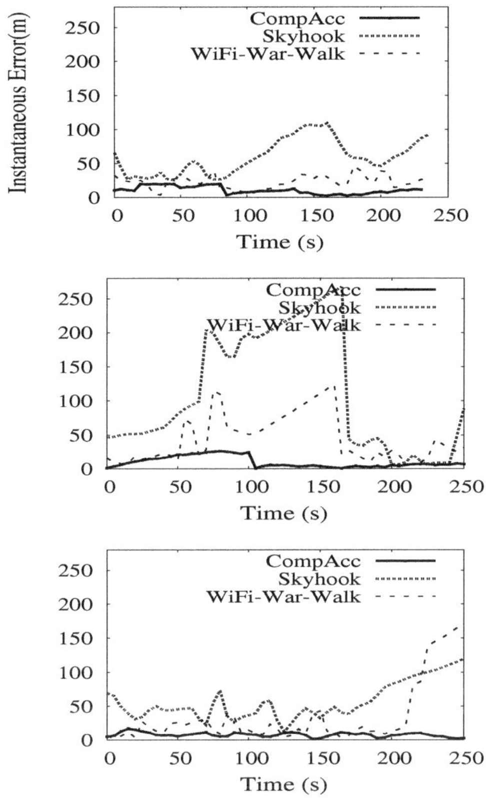

Fig. 10.14 Variation of localization accuracy with CompAcc, WiFi-War-Walk, and Skyhook for three test traces

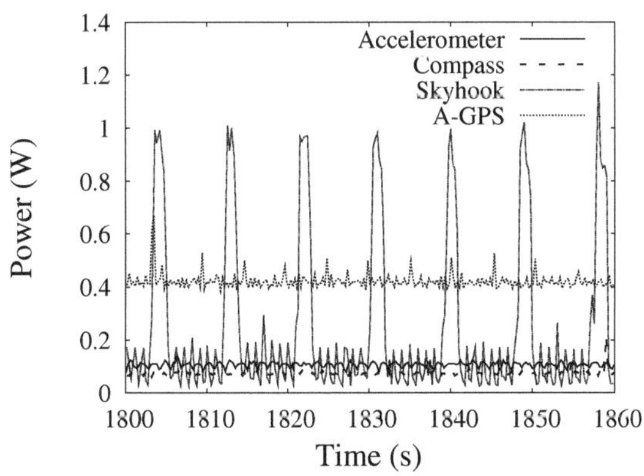

Fig. 10.15 Power consumed by Compass, Accelerometer, Skyhook, AGPS.

10.6 Conclusion

The growing popularity of location based services will call for improved quality of localization, including greater ubiquity, accuracy, and energy-efficiency. Current localization schemes, although effective in their target environments, may not scale to meet the evolving needs. This paper proposes *CompAcc*, a simple and practical method of localization using phone compasses and accelerometers. CompAcc's core idea has been known for centuries, yet, its adoption to human scale localization is not obvious. We believe that CompAcc could complement current localization technology, taking an important step towards a pervasive location service for the future.

References

1. LBA counts in apple store and android marketplace, `http://www.skyhookwireless.com/locationapps/`
2. Chen, Y., Chawathe, Y., LaMarca, A., Krumm, J.: Accuracy characterization for metropolitan-scale wi-fi localization. In: ACM MobiSys (2005)
3. Skyhook wireless, `http://www.skyhookwireless.com/`
4. Times, N.Y.: Cellphone locator system needs no satellite, `http://www.nytimes.com/2009/06/01/technology/start-ups/01locate.html`
5. Constandache, I., et al.: EnLoc: Energy-efficient localization for mobile phones. In: IEEE INFOCOM Mini Conference (2009)
6. Gaonlkar, S., et al.: Micro-blog: Sharing and quering content through mobile phones and social partiipation. In: ACM Mobisys (2008)
7. Step size in pedometers, `http://walking.about.com/cs/pedometers/a/pedometerset.htm`
8. Jekeli, C.: Inertial Navigation Systems with Geodetic Applications. Walter de Gruyter (2000)
9. Google maps API, `http://code.google.com/apis/maps/`
10. Haversine formula, `http://en.wikipedia.org/wiki/Haversine_formula`
11. Williams, E.: Aviation formulary, `http://williams.best.vwh.net/avform.htm`
12. Von Ahn, L., Dabbish, L.: Labeling images with a computer game. In: ACM CHI (2004)
13. Amazon Mechanical Turk, `http://aws.amazon.com/mturk/`
14. Hightower, J., Borriello, G.: Particle filters for location estimation in ubiquitous computing: A case study. In: UbiComp (2004)
15. Gustafsson, F., et al.: Particle filters for positioning, navigation, and tracking. IEEE Transactions on Signal Processing (2002)
16. Azizyan, M., et al.: Surroundsense: Localizing mobile phones via ambience fingerprinting. In: ACM MobiCom (2009)
17. Priyantha, N.B., et al.: The cricket location-support system. In: MobiCom (2000)
18. Niculescu, D., Nath, B.: VOR base stations for indoor 802.11 positioning. In: ACM MobiCom (2004)
19. Youssef, M., et al.: Pinpoint: An asynchronous time-based location determination system. In: ACM MobiSys (2006)
20. Bahl, P., Padmanabhan, V.N.: Radar: an in-building rf-based user location and tracking system. In: IEEE INFOCOM (2000)
21. Griswold, W.G., et al.: ActiveCampus: Experiments in community-oriented ubiquitous computing. IEEE Computer (2003)

Chapter 11
Energy-Efficiency Networking Games

Manzoor Ahmed Khan and Hamidou Tembine

Abstract. The concept of energy efficiency has moved in and out of favor with the public over the years, but recently has gained renewed broad-based support. The confluence of economic, environmental and fairness concerns around offering the same quality of service by reducing energy has moved efficiency to the fore. In the context of networking and communications, there are different energy-efficiency issues in terms of quality of service, quality of experience, energy consumption, pricing etc. This chapter will focus primarily on the game theoretical formulations of energy-efficiency metrics, with applications to networking problems. We will first present a broadly inclusive notions of energy-efficiency and then explore a variety of ways to analyze the strategic behaviors of the players depending on the information and the dynamics of the system. Applications to power management with stochastic battery state modelling and network selection problems with energy-efficiency criteria are presented.

11.1 Introduction

Recently, there has been extensive researches that employ distributed optimization, learning in games, and distributed control to model and analyze the performance that take into consideration the energy-consumption of various networks, such as communication networks, neural networks, computer networks, social networks, bio-inspired networks, cognitive radio networks, opportunistic mobile ad hoc networks etc. There are several successful examples where game theory provide deeper understanding of energy-efficiency based network dynamics and leads to better design

Manzoor Ahmed Khan
DAI-Labor/Technical University, Berlin Germany

Hamidou Tembine
Ecole Supérieure d'Electricité (Supelec), France
e-mail: tembine@ieee.org

of efficient, green, scalable, stable and robust networks. Still, there remain many interesting open research problems to be identified and explored, and many energy-efficiency issues to be addressed in both wireline and wireless networked games. In this chapter, we develop energy-efficiency game theoretic models for networking and communication problems. The chapter starts in designing suitable "green" performance metrics for interacting networking systems and present both static and dynamic game formulations as well as their solution concepts. Then, we focus on distributed strategic learning for energy-efficiency solutions. We apply these solution concepts to various energy-efficiency networking scenarios. The first example focuses on battery-state dependent energy-efficiency games. The scenario is generalized to backoff-state dependent energy-efficiency games which takes into consideration not only the energy level but also the number of retransmission until successful reception.

The second example considers user-centric network selection in wireless networks. In both cases, the energy-efficiency aspect plays important role(s) in the outcome of the interactions.

11.2 Energy Efficiency in Networking Games

In this chapter, the term energy efficiency refers to *"using less energy to provide the same service"*. The best way to understand this basic idea is through a simple example.

When you replace an office equipment such as a computer or a laptop with a more energy-efficient model, the new equipment provides the same service, but uses less energy. This saves you money on your energy bill, and reduces the amount of greenhouse gases going into the atmosphere. *This simple example captures the basic idea of energy-efficiency.* However energy-efficiency is different than *energy conservation*. Energy conservation is reducing or going without a service to save energy. Thus, the quality of service and the quality of experience are important aspects of *energy-efficiency* compared to energy-conservation. Turning off a light or fixing the maximum amount of power to something smaller is an energy conservation operation. Replacing an incandescent lamp with a compact fluorescent lamp which uses much less energy to produce the same amount of light is an energy efficiency operation. Both operations (energy-efficiency and energy-conservation) can reduce greenhouse gas emissions but may not give the same guarantee to the users, consumers, nodes, operators, service providers etc.

In multihop networks, intermediate nodes are often used as relays to reduce the transmission energy required to deliver a message to an intended destination. The selfishness of autonomous nodes, however, raises concerns about their willingness to expend energy to relay information for others. However it is clear how efficient will be the outcome of the distributed decision making compared to a centralized case. A coalitional structure of nodes have been proposed in order to offer at least the same service by reducing energy consumption of the system or a node. Then, the formation of coalitions, the cost of exchanging messages, the incentive to collude needs to be explained and designed.

11.2.1 How Efficient Are Energy-Efficiency Games?

The term energy-efficient game does not mean that the game is efficient. The outcome of the game may be inefficient not only in term of payoff but also in term of energy consumption. The reader may ask why it is called energy-efficiency payoffs?

The term *energy-efficiency payoff* is used here to capture the energy consumption aspect into standard payoff functions. Energy-efficiency function is then a generic function that takes into consideration a tradeoff between payoffs and energy consumption cost. Then, a game with energy-efficiency function as payoff function is an *energy-efficiency game*. As a consequence, this definition has nothing to do with the efficiency of the outcomes of the game.

11.2.2 Designing Energy-Efficiency Metrics

There are several ways to take into consideration the energy consumption issue in wireless networks. The first approach is to formulate the problem with cost constraints. The global metric depends on the payoffs and the costs. This can be obtained by taking the Lagrangian of the constrained problem. The second approach is to write directly an appropriate metric such as ratio between benefit and cost; probability of success per cost, capacity per cost, cost per capacity etc.

11.2.3 Energy-Efficiency One-Shot Games

We consider an energy-efficiency one-shot game described by

- A finite set of players $\mathcal{N} := \{1,2,\ldots,n\}$ where $n < +\infty$ is the cardinality of $\mathcal{N}$. For $n = 1$, the game is reduced to a single decision-maker problem. For $n \geq 2$ one has interaction between at least two players.
- Each player $j \in \mathcal{N}$ has an action set $\mathscr{A}_j$ which can be discrete or continuous. We assume that $\mathscr{A}_j$ is non-empty. An action profile is an element of $\prod_{j\in\mathcal{N}} \mathscr{A}_j$.
- Each player j has an *energy-efficiency payoff function* $u_j : \prod_{j\in\mathcal{N}} \mathscr{A}_j \longrightarrow \mathbb{R}$ to be optimized.

The collection $(\mathcal{N}, (\mathscr{A}_j)_{j\in\mathcal{N}}, (u_j)_{j\in\mathcal{N}})$ defines a game in normal form. If in addition, specific information structures, state spaces and order of play were present, these structures should be added in to collection as well as in the strategies of the players (the ways the players will make decisions).

11.2.4 Energy-Efficiency Payoff Functions

Pricing

We take the difference between reward and cost function. The payoff function has the form $\tilde{u}_j(a) - c_j(a)$ where $\tilde{u}_j$ is the benefit/satisfaction of player j and $c_j(.)$ is the cost function of player j. In individual pricing case, the pricing function has the form $c_j(a) = c_j(a_j)$. The cost functions are designed in order to get non-trivial outcomes that are close to a given target.

Satisfaction/Cost Ratio

In place of the difference between the converted satisfaction and cost units, we take the ratio between reward and cost function. Thus, the payoff function has the form $\frac{\tilde{u}_j(a)}{c_j(a)}$. The function $\tilde{u}_j$ is well-chosen in a way such that the ratio is well-defined. Examples of such consideration include the probability of success per unit cost, the bit-error-rate per unit consumption power, rescaled Shannon capacity per unit cost.

Cost per Unit Reward

Now, we take the ratio between cost and reward function. Thus, the payoff function has the form $\frac{c_j(a)}{\tilde{u}_j(a)}$. Examples of such consideration include the power consumption per probability of success, power consumption per bit-error-rate or power consumption per unit capacity.

11.3 Solution Concepts

In this section we present basic solution concepts: equilibrium, maximum solution, Bargaining solution and Pareto optimality.

An action profile $a^* = (a_1^*, \ldots, a_n^*)$ is a pure (Nash) equilibrium point if no player can improve its payoff by unilateral deviation:

$$\forall\, j \in \mathcal{N},\ a_j^* \in \arg\max_{a'_j \in \mathcal{A}_j} u_j(a'_j, a^*_{-j}),$$

where a^*_{-j} denotes the profile $(a^*_{j'})_{j' \neq j}$.

Most of the energy-efficiency games may not have a pure equilibrium. In that case, one can look at randomized actions. Let $\Delta(\mathcal{A}_j)$ be the set of probability distributions over the set $\mathcal{A}_j$ equipped with the canonical sigma-algebra. Then, the payoff function can be extended over the product space $\prod_{j \in \mathcal{N}} \Delta(\mathcal{A}_j)$. Denote by $U_j : \prod_{j \in \mathcal{N}} \Delta(\mathcal{A}_j) \longrightarrow \mathbb{R}$ the expected payoff of player j.

A (mixed) Nash equilibrium is a profile $x^* \in \prod_{j\in\mathcal{N}} \Delta(\mathcal{A}_j)$ satisfying

$$\forall\, j \in \mathcal{N},\; x_j^* \in \arg\max_{x_j\in\Delta(\mathcal{A}_j)} U_j(x_j,x_{-j}^*).$$

We say that x^* is a fixed-point of a (multi-valued) mapping Φ if $x^* \in \Phi(x^*)$. When Φ_j corresponds to be argmax defined above, one gets a characterization of equilibria as fixed-points. Then, important results can be obtained:

- Assume that the set of actions is non-empty and finite and the entries of the polymatrix are finite. Then, the game has at least one equilibrium in mixed strategies. One of classical proofs of this result uses a fixed point theorem (Brouwer, Schauder, Kakutani, Fan, Glicksberg, Debreu etc). Nash has constructed a mapping based on normalized excess payoff to use the Brouwer/Schauder fixed-point theorem. The mapping is also related to the Brown-von Neumann-Nash dynamics.
- The set of actions is non-empty, convex and compact and if u_j is quasi-concave relatively to a_j and continuous in the action profile then there exists at least one pure equilibrium (fixed-point theory).
- The set of actions is non-empty, convex and compact and u_j is continuous then there exists at least one equilibrium in mixed strategies (fixed-point theory).

A maximin solution is a profile that maximizes the minimum of the payoffs i.e

$$a^* \in \arg\max_{a\in\prod_j \mathcal{A}_j} \min\left(u_1(a_1,a_{-1}),\ldots,u_2(a_2,a_{-2}),\ldots,u_n(a_n,a_{-n})\right).$$

Maximin solutions are well-used for fairness issues.

Another notion that takes into consideration fairness issue is the bargaining solution. There are several Bargaining solutions (Nash, Kalai-Smorodinsky, Egalitarian etc). Below we define the Nash bargaining solutions.

A (Nash) "strategic" bargaining solution is a profile

$$a^* \in \arg\max_{a\in\prod_{j\in\mathcal{N}} \mathcal{A}_j} \prod_{j\in\mathcal{N}} \left(u_j(a) - \bar{u}_j\right)^{\omega_j},$$

where $\bar{u}_j$ is a reference payoff for player j when the negotiation process fails (disagreement point) and ω_j is a negotiation power associated to player j. Note that in this definition, the set of feasible payoffs may not be convex.

A Pareto optimal point is configuration where it is not possible to improve the payoff of one player without decreasing the payoff of another player. Example of Pareto optimal solutions include maxmin solutions and bargaining solutions.

11.3.1 Constrained Energy-Efficiency Games

Consider a classical one-shot game problem with satisfaction function given by $\mathbb{1}_{\{\tilde{u}_j \geq \gamma_j\}}$. The associated energy-efficiency game problem consists to ask if it is possible to " *use less energy consumption while providing the same service?*". We formulate this problem as an energy-constrained game. Each player optimizes its cost $c_j(a)$ subject to the constraints $\tilde{u}_j(a) \geq \gamma_j$ where γ_j is a satisfaction target of player j. The problem of player j is

$$\begin{cases} \inf_{a_j \in \mathscr{A}_j} c_j(a), \text{ subject to} \\ \qquad \tilde{u}_j(a) \geq \gamma_j \end{cases}$$

The system designer target could be to satisfy as large as possible players with a minimal energy consumption cost:

$$\begin{cases} \inf_{a \in \prod_j \mathscr{A}_j} \sum_{j \in \mathscr{N}} c_j(a), \text{ subject to} \\ \qquad \tilde{u}_j(a) \geq \gamma_j \\ \qquad j \in \mathscr{N} \end{cases}$$

Note that the solution(s) of the above problems can be different than the ones obtained by maximizing the functions $\tilde{u}_j$ with cost constraint $c_j(a) \leq \bar{c}_j$ where $\bar{c}_j$ is a fixed positive real number.

11.3.2 Energy-Efficiency Long-Term Games

We consider long-term interaction between the players. Time is discrete and time space is the set of natural numbers. At each time slot, the players interact. Each player chooses an action based on its previous experiences and observations. All the players get a payoff and the game moves to the next time slot. As it is well known without constraint and under observations structures, one can observe an emergence of cooperation between the players in the long-run (see the Folk Theorems in dynamic games). However, we need to take into consideration long-term constraints as it is usual in wireless networking. This is because the actions set may not be the same at each step. For example, if the players are battery-limited or if they are using renewable energy such as Solar-energy, they may not be able to transmit all the time slots. The constraint in that case will be the total amount of energy available or the weather condition in the case of solar systems.

11.3.3 Energy-Efficiency Issues in Large-Scale Systems

Many real-world problems modeled by stochastic games have huge state and/or action spaces, leading to the well-known curse of dimensionality. The complexity of the analysis of large-scale systems is dramatically reduced by exploiting mean field limit and dynamical system viewpoints. Under regularity assumptions and specific time-scaling techniques, the evolution of the mean field limit can be expressed in terms of deterministic or stochastic equation or inclusion (difference or differential). The energy-efficiency issue can be taken into consideration in the mean field stochastic game (MFSG) formulations. Considering long-term payoffs, one can characterize the mean field optimality equations by using mean field dynamic programming principle and Kolmogorov forward equations.

11.4 Learning in Energy-Efficiency Games

11.4.1 Learning under Perfect Monitoring

In this subsection we explain how the players can learn their optimal strategies based on observation about the past actions of the others. These classes of strategic learning are called partially distributed strategy-learning. Sophisticated learning algorithms in which each player adapts its strategy based on observations of the other players' actions. The origin of partially distributed learning goes back at least to the work by Cournot (1838), in the so-called Cournot tatonnement, which is known as *best response dynamics*. Partially distributed learning algorithms include classical fictitious play (FP), best response, classical logit learning, Adaptive play, Joint strategy FP, Stochastic FP, Regret matching, sequential asynchronous FP learning etc.

We briefly explain the basic idea of the most used version of fictitious play. Assume that at each time slot, each player observes its actions, its payoffs but also the last actions played by the other players. At each time slot, each player thus best responds to the empirical frequency of play (joint) of the other players (this presumes that he observes to last actions chosen by the other players). The empirical frequency is defined by

$$s_{-j} \in \prod_{j' \neq j} \mathscr{A}_{j'}, \ \bar{f}_{j,t}(s_{-j}) = \frac{1}{t} \sum_{t'=1}^{t} \mathbb{1}_{\{a_{-j,t'} = s_{-j}\}}$$

The sequence $\{\bar{f}_{j,t}\}_{t \geq 1}$ satisfies the recursive equation

$$\bar{f}_{j,t+1}(s_{-j}) = \bar{f}_{j,t}(s_{-j}) + \frac{1}{t+1} \left(\mathbb{1}_{\{a_{-j,t+1} = s_{-j}\}} - \bar{f}_{j,t}(s_{-j}) \right),$$

where $\mathbb{1}_{\{a_{-j,t+1} = s_{-j}\}}$ denotes the indicator function. It is equal to 1 if the joint actions chosen by the players than j at time t is s_{-j} and 0 otherwise.

Then the algorithm is defined by:

$$a_{j,t+1} \in \arg\max_{s_j \in \mathscr{A}_j} U_j(\mathbf{e}_{s_j}, \bar{f}_{j,t}) := \arg\max_{s_j \in \mathscr{A}_j} \sum_{s_{-j} \in \prod_{j' \neq j} \mathscr{A}_{j'}} \bar{f}_{j,t}(s_{-j}) u_j(s_j, s_{-j}) \quad (11.1)$$

The relation in (11.1) gives an Hannan set.

11.4.2 *Learning under Measurement via Numerical Value*

Now, we address, how learn the optimal strategy based only on numerical values of own-payoffs?

The main challenge here is limited information. Players may have limited knowledge about the status of other players, except perhaps for a small subset of neighboring players. The limitations in term of information induce robust stochastic optimization, bounded rationality and inconsistent beliefs. The basic learning schemes have the form of Robbins-Monro [2] or Kiefer-Wolfowitz [3].

11.4.2.1 Reinforcement Learning for Equilibria

The standard reinforcement learning (Bush-Mosteller, 1949-1955, [1]) consists in updating the probability distribution over the possible actions as follows: $\forall\ j \in \mathcal{N}, \forall s_j \in \mathcal{A}_j$,

$$x_{j,t+1}(s_j) = x_{j,t}(s_j) + \lambda_{j,t}\sigma_j(u_{j,t})\left(1\!\!1_{\{a_{j,t}=s_j\}} - x_{j,t}(s_j)\right) \tag{11.2}$$

where $\lambda_{j,t}$ is the learning rate of player j at time slot t. σ_j is a real-valued mapping. The parameter $\lambda_{j,t}$ and the payoff $u_{j,t}$ can be normalized such that $\sigma_j(u_{j,t}) > 0, 0 < \lambda_{j,t}\sigma_j(u_{j,t}) < 1$.

11.5 State-Dependent Energy-Efficiency Games

In this section, we assume that the payoff function has the form $u_j(w,a)$ where w describes a state. The set of states is denoted by $\mathcal{W}$. The state can represent a backoff state, SINR level, an energy level, a channel state, storage state etc.

11.5.1 *The Current Full State Is Known by All the Players*

Each player j knows the full state w as well as the mathematical structure of the payoff functions. Each player can then construct a state-feedback action i.e the choice of actions depends on the value of w. We denote this game by $G^1 = \{\{w\}, \mathcal{N}, \mathcal{A}_j, u_j^1(w,.)\}$. Equilibria of G^1 will be state-dependent equilibria. However, if the full state w is not available, the payoff $u_j^1(w,.)$ is difficult to optimize. In the next subsection, we formulate the case where each player j observes its component w_j of the state vector $w = (w_1, \ldots, w_n)$.

11.5.2 *A Partial Component of the Current State Is Known*

Here we assume that each player j knows its own state w_j and the distribution μ_{-j} over the states $w_{-j} = (w_{j'})_{j' \neq j}$ of the other players. Each player j is able to compute the payoff function given by

$$u_j^2(w_j, \mu_{-j}; a) = \mathbb{E}_{w_{-j} \sim \mu_{-j}} \left[u_j(w, a) \mid w_j, \mu_{-j} \right].$$

Let $\Delta(W_{-j})$ be the set of probability measures over W_{-j}, equipped with the canonical sigma-algebra. Then, $\mu_{-j} \in \Delta(W_{-j})$. Note that u_j^2 is defined over $W_j \times \Delta(W_{-j}) \times \left(\prod_j \mathscr{A}_j\right)$ where W_j denotes the state space observable of player j.

A pure strategy of player j in this game with partial state observation is a mapping from the given own-state w_j, and the distribution μ_{-j} to the action $\mathscr{A}_j$. We denote this game by

$$G^2 = (\mathscr{N}, \mathscr{A}_j, (w_j, \mu_{-j}), (u_j^2)_{j \in \mathscr{N}}).$$

A pure equilibrium for the game G^2 can be seen as a Bayesian-Nash equilibrium, and it is characterized by $\sigma_j : W_j \times \Delta(W_{-j}) \longrightarrow \mathscr{A}_j$,

$$\sigma_j(w_j, \mu_{-j}) \in \arg \max_{\tilde{a}_j \in \mathscr{A}_j} u_j^2(w_j, \tilde{a}_j, \sigma_{-j}), w_j \in W_j, \forall j \in \mathscr{N}.$$

Note that w_j is an own-state and does not necessarily "plays the role" of a type in the classical game theoretic formulations because the individual state is a realization of a certain random variable, not an endogenous element of a player. A second remark is that the consistency relationship between the types and the beliefs should be checked in games with incomplete information.

11.5.3 No State Component Is Known But a Distribution Is Available

Now, none of state components is known but a distribution μ over $\mathscr{W}$ is available. Each player j is able to compute the following payoff function

$$u_j^3(a, \mu) = \mathbb{E}_{w \sim \mu} u_j(w, a),$$

where $\mu \in \Delta(\mathscr{W})$. This leads to an expected robust game which we denote by

$$G^3 = (\mathscr{N}, \mathscr{A}_j, \mathscr{W}, \mu, (u_j^3)_{j \in \mathscr{N}}).$$

A pure strategy for player j in the game G^3 corresponds to a function of μ that gives an element of $\mathscr{A}_j$. We refer this game as *expected robust game*. This action spaces are finite and if the expectation is well defined for each entry (play) then the game G^3 has at least one equilibrium in mixed strategies.

11.5.4 Only the State Space Is Known

In this subsection we assume that even the distribution over the state is not available. Only state space $\mathscr{W}$ is known. Then, each player can adopt different behaviors depending on its way to see the state space. The well-known approaches in that case are the maximin robust and the maxmax approaches (pessimistic or optimistic approaches) and their variants. The payoff of player j in the maximin robust game is given by

$$u_j^4(a) = \inf_{w \in \mathscr{W}} u_j(w,a).$$

We denote the game by $G^4 = (\mathscr{N}, \mathscr{A}_j, \mathscr{W}, (u_j^4)_{j \in \mathscr{N}})$. The equilibria of G^4 are called maximin robust equilibria. Similarly, The payoff of player j in the maximax robust game is given by

$$\bar{u}_j^4(a) = \sup_{w \in \mathscr{W}} u_j(w,a)$$

11.5.5 Dynamic State-Dependent Games

In this subsection, we assume that none of the state, the distribution, the state space, the mathematical structure of the payoff function is available. Under such conditions, how to solve the problem?

Time is discrete. Time horizon is large. At each time, each player acts and observes a numerical measurement of its payoff. We look at the long-run game. To simplify the analysis, we assume that the state follows an independent and identically distributed random variable.

We distinguish three cases:

11.5.5.1 Stochastic Games with Perfect State Observation

The channel state vector $w(t)$ is observed by all the players and the instantaneous payoff functions are known. In this case, one can use the classical stochastic game framework to solve the long-run interaction problem. For example, Bellman's dynamic programming principle can be used.

11.5.5.2 Stochastic with Partial State Information

At each time each player observes a component: Its own state i.e $w_j(t)$ is observed by player j. Instantaneous payoff functions are also known. In this case, one can use *partially observable stochastic game* tools and dynamic programming over continuous state space to solve the long-term problem.

Each of the above formulations can be with public or private signals, perfect or imperfect monitoring for the actions chosen by the others.

11.5.5.3 Stochastic Games without State Information

The players do not observe any component of the state (no partial component is observed). The players do not know the state space. They do not know $\mathscr{N}$, they do not know the mathematical structure of the payoff functions. They do not even know if they are in a game.

However, they are not in a blind environment. We assume that they are able to discover progressively their own action space $\mathscr{A}_j$ and we assume that each of the players is able to observe an ACK/NACK as it is the case with the carrier sense multiple access (CSMA) with collision avoidance (CA) with ACK/NACK. Now, a private history up to T' is a collection of own-actions and own ACKs up to T'. We denote σ_j^5 a behavioral strategy of player j. We consider the time average payoff of the dynamic game i.e $\liminf_{T'} \mathbb{E}_{w(0),\sigma^5} F_{j,T'}^5$ where

$$F_{T'}^5 = \frac{1}{T'+1} \sum_{t=0}^{T'} u_{j,t}.$$

In this case, different variants of combined fully distributed payoff and strategy reinforcement learning (CODIPAS-RL) have been developed in [7]. The CODIPAS-RL is a joint and interdependent iterative scheme to learn both payoff function and the associated optimal strategies. The CODIPAS-RL is known (see [7] and the references therein) to converge in particular finite games:

- dominance solvable game,
- $2 \times 2-$ coordination games,
- 2×2 anti-coordination games,
- common interest games,
- stable games or games with monotone gradient payoffs,
- potential games
- potential population games,
- expected robust potential games,
- any symmetric two-player robust game with 3 actions which satisfies the Dulac's criterion,
- some classes of supermodular games,
- Lyapunov robust games: there exists a Lyapunov function associated the dynamics of the expected payoffs.

11.6 Power Management

We propose different energy-efficiency payoffs in state-dependent game where the state represent the channel state. The payoff function is now state-dependent and is denoted by $u_j(w,a)$.

11.6.1 Battery-State Dependent Energy-Efficiency Games

A stochastic modelling of battery-limited energy management have been proposed in [4]. In it the objective is to optimize not only the current satisfaction but also a long-term satisfaction based on a total budget of battery with finite lifetime. The more a user consumes power, the more its battery state transition decreases in probability. Considering a simple scenario of Aloha-based access network with power selection, the authors have demonstrated playing the maximum power may lead to suboptimal configuration and may not be an equilibrium. Detailed analysis of power allocation in Multiple-Input-Multiple-Output (MIMO) can be found in [4, 5, 6].

11.6.2 Backoff-State Dependent Energy-Efficiency Games

In addition to be the battery state criterion, we need to take into consideration the number of retransmissions for same until it succeeds. The backoff state account this number. Due to expiration delay, the total number of retransmission is bounded. If the transmission have successfully received then backoff state back to zero to restart with another packet. A packet is successfully transmitted if the received Signal-to-Interference-plus-Noise-Ratio (SINR) is greater than some acceptable threshold.

11.7 Energy-Efficiency and User-Centric Network Selection

Different definitions of energy-efficiency can be found in the literature e.g., *energy efficiency can be defined as using less energy to provide the same service.* However, we confine the definition of energy conservation to the field of wireless communication in general and to network selection in wireless networks in special. The motivation to confine ourselves to the mentioned paradigm is the envisioned ubiquitous communication environment in future, where a user equipment is connected to the internet anywhere and anytime through the heterogeneous interfaces. The general term heterogeneous here is self explanatory, however the heterogeneity in network selection provisions illustration. In network selection (or more appropriately, user-centric network selection) is commonly known as the decision mechanism hosted at the user terminal, which takes the decision over selecting the suitable (or best available) access network out of the available ones. Intuitively operators' communication footprint in a geographical coverage area mostly comprises of the access network coverage by technologies of different characteristics e.g, Wireless Local Area Networks (WLAN), Cellular technologies etc. (hence the term heterogeneous). Let us consider even a concrete scenario, where a user is faced with the decision of network selection in the coverage area covered by the WLAN and 3G networks. It should be noted that these technologies differ in their coverage areas i.e., WLAN covers comparatively smaller coverage area but provides higher data rates. Owing to the just mentioned fact, when a user is under coverage area of both WLAN and 3G cellular network, the preferred network is WLAN, however, it should be noted that in this case we assume that user is indifferent to service pricing. As this would lead to an increased performance. But at the same time we can not neglect that this operation (i.e., being associated to the WLAN) costs users more energy consumption.

We now illustrate on the energy consumption pattern of the different interfaces specifically when it comes to simultaneous use of the interfaces. This way, the end users benefit from better quality services (e.g., high bit rates) and network operators get more grained means for resource management as they might assign different flows of one user to different access technologies. In order to realize the parallel usage of interface while being mobile, mobile IPv6 protocol can be extended too.

On the other hand, the operation of more than one NIC at the same time increases the power consumption. As mobile devices are usually power driven, this is often used as an argument against the use of multiple interfaces. Although the increased power consumption is a matter of fact, a clear understanding of the order of magnitude is missing. In this chapter, we also take an opportunity to establish an initial knowledge about the effects of the use of multiple interfaces on power consumption.

From 802.11 protocols, each frame length is described in detail. It would help calculating power consumption of single frame or specific data length (e.g., 20MB) exactly working under specific working modes. This theoretical value of energy consumption can be calculated using vendor specification of a card working with 802.11 protocols. One can obtain and compute the energy value from data obtained from real devices. To analyze the energy consumption of multiple interfaces, a clear understanding of power consumption of a single wireless interface is necessary. In some regime, the per-packet energy consumption of network interfaces can be modeled in a linear equation

$$E = mz + b, \tag{11.3}$$

where E represents the energy, z represents the size, m and b are linear coefficients which can be determined by experimental results. b is a fixed component associated with device state change and channel acquisition overhead. The authors make use of popular DSSS Lucent IEEE 802.11 WaveLan PC "bronze" and "silver" cards are used as test cards on the IBM Thinkpad 560. The tcp-dump facilities are used to ensure that no other traffic is present on the channel. The energy consumption in this configuration is determined by the direct measurement of input voltage and current drawn at the network interface. By inserting small resistance in series with the device, the current drawn is determined, whereas the input current is determined by measuring the voltage across the test resistance. The results show that sending point-to-point and broadcast traffic have the same incremental cost, but point to point traffic has a higher fixed cost associated with the IEEE 802.11 control protocol. However, when it comes to receiving point-to-point then it differs from the broadcast traffic only in their fixed costs i.e., receiving point-to-point traffic has a high fixed cost, due to the costs of sending two control messages. Receiving broadcast traffic has the lowest fixed cost, representing the MAC header in the data massage and physical overhead.

We take into account the combination of parameters like RF transmission power, modulation schemes (transmission rate), packet size, and distance influence that affect the power consumption of a WLAN interface. Authors vary the packet size, transmission rate, and the RF power level in their experiments. The MAC packet size ranges from 64 to 2312 bytes. The transmission rate from 1 to 11 Mbps and the RF level accepts values of 1, 5, 20 and $50mW$. Furthermore, the work measured the average power consumption and the throughput during the Tx and Rx phase while the above mentioned parameters varied. This work provides two kinds of power dissipation results. The first is referred as instantaneous power consumption that describes the actual power consumption of the NIC for a particular working mode and for a particular set of parameters. As mentioned above, there are four different working modes and three parameters of variation. The second kind of power consumption is referred as average power consumption. It describes the power consumption for the Tx and Rx processes as a whole, including the transmission and reception of data and acknowledgement, respectively, as well as idle time. Power consumption is an unbiased measure since it can not deliver any insights on how long the battery will last. Therefore, the work derives the energy that is used to transmit one bit of goodput data. The authors realized that there is a strong dependence between the power consumption and the RF level used in the Tx mode. The higher the power level, the higher the power consumption. If the RF power level varies from 1 to $50mW$, the increase in power consumption comes to $500mW$. The results show that the power amplifier takes a major portion of the overall power consumption. Higher transmission rates cause a slight increase in power consumption which is probably caused by a slightly higher power consumption of the baseband processor. The RF does not have any influence in the reception mode as shown in the second figure. It can be seen that the Tx mode costs more power than the Rx, IDLE and SLEEP mode. There is only a small difference in power consumption between the Rx and the IDLE modes. The reason for this is that all of the reception hardware turn on within the IDLE mode to scan for valid RF signals. The difference is likely caused by the MAC processor, which is assumed to be idle during the IDLE mode of the NIC dissipating less power. This is further illustrated by the Table-11.1.

Table 11.1 Interface energy consumption in different access technologies

Technology	Transmission	Reception	Idle
CDMA NIC	2.8 W	495 mW	82 mW
IEEE 802.11b NIC	1.3 W	900 mW	740 mW

The first look at the table-11.1 dictates that when energy conservation is taken in to account for the network selection, the preferred strategy for a mobile user would be to stay connected to 3G networks. There are several definitions of energy-efficiency.

11.7.1 Energy Efficient Network Selection

The basic objective of the energy efficient network selection is to reduce the energy consumption in the mobile terminals without degrading the user perceived QoE. The decision of network selection in this case is derived by the fact, as to select the interface that consumes less amount of energy. Observing the power consumption of up(down)loading data through different network access technologies, one can in the first place suggest a *policy-based* like network selection solution, where the user defines a policy such that for bandwidth hungry applications, WLAN interface should be selected, whereas for smaller data or communication with greater idle periods the 3G network technologies are chosen.

Case Scenario

Consider the scenario depicted in Fig 11.1, where a user is under the coverage of access heterogeneous wireless network technologies of two different operators. As can be seen in the Fig 11.1 that various coverage areas are define. We assume that user is mobile and follows the given trajectory through different coverage areas, this dictates that user is faced with network selection decision when losing or gaining the coverage of any access network technology(ies).

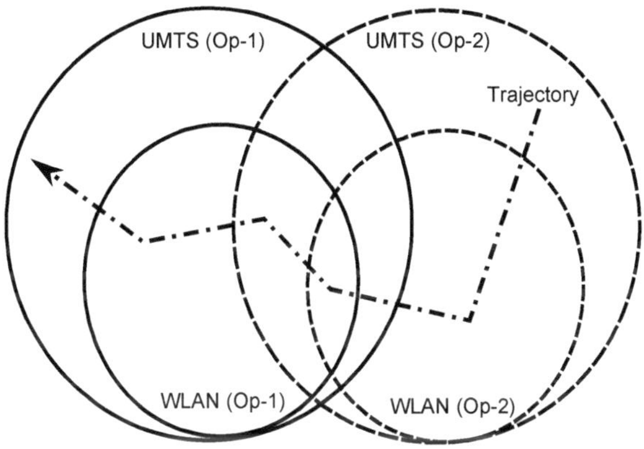

Fig. 11.1 Energy efficient network selection scenario

In this scenario, the user starts moving from geographical area A with a single access network technology i.e., UMTS (Op-1). A user when moves to the coverage area B under the coverage of UMTS and WLAN of operator-1, he is faced with the decision of interface selection, however the operator competition in this coverage does not come into picture in the decision making. Moving to areas C, D, and E introduce the operator competition in the network selection problem too, however the interface options available to the users differ in different areas. Geographical areas F and G are the replica of areas A and B. It should be noted that the network selection problem in this case is abstracted to emphasize the energy efficiency effect of the network selection problem. The two different operators are included in the scenario so that we can study the operators role in conservation of energy (by motivating users to select different energy efficient interfaces by offering different service prices through their interfaces).

11.8 Concluding Remarks

In this chapter, we have presented game theoretic and strategic learning formulations of energy-efficiency networking problems. Both deterministic and stochastic modelling can be included in the concept of offering the same service by consuming less energy. However, the game theoretical models need to be adapted to more realistic network scenarios under energy-efficiency criterion.

References

1. Bush, R., Mosteller, F.: Stochastic Models of Learning. John Wiliey & Son, New York (1955)
2. Robbins, H., Monro, S.: A Stochastic Approximation Method. Annals of Mathematical Statistics 22(3), 400–407 (1951)
3. Kiefer, J., Wolfowitz, J.: Stochastic Estimation of the Maximum of a Regression Function. Annals of Mathematical Statistics 23(3), 462–466 (1952)
4. Hayel, Y., Tembine, H., Altman, E., El-Azouzi, R.: Markov Decision Evolutionary Games with Individual Energy Management. In: Breton, M., Szajowski, K. (eds.) Annals of the International Society of Dynamic Games, Advances in Dynamic Games: Theory, Applications, and Numerical Methods, pp. 313–335 (2010)
5. Tembine, H.: Dynamic Robust Games in MIMO systems. IEEE Transactions on Systems, Man, and Cybernetics, Part B: Cybernetics (forthcoming 2011)
6. Tembine, H., Altman, E., ElAzouzi, R., Hayel, Y.: Evolutionary Games in Wireless Networks. IEEE Transactions on Systems, Man, and Cybernetics, Part B: Cybernetics 40(3), 634–646 (2010)
7. Tembine, H.: Distributed strategic learning for wireless engineers, notes, 250 pages, supelec (2010)

Part 2
Green Computing Technology

Chapter 12
Energy Efficiency of Data Centers

László Gyarmati and Tuan Anh Trinh

Abstract. As the power consumption has a significant and continuously increasing part of the operational expenses of data centers, energy efficient data center networking has received special attention from the academic and industrial research community recently. The complex design of data centers provides several directions toward more energy efficiency, including the consumption of servers and network equipments. We address the issue of power consumption of data centers from a higher-level point of view by analyzing the energy efficiency of data center architectures. We review the state-of-the-art data center architectures including BCube, DCell, fat-tree, and Scafida and present evaluate their energy-efficiency quantitatively. In addition, the trade-off between the power consumption and the performance of the data center is investigated. Next, other aspects of data centers' energy efficiency is revealed including thermal control techniques, energy management systems. Finally, a standard data center energy efficiency metric called PUE (Power Usage Effectiveness) is presented.

12.1 Introduction

The widely adoption of the novel Internet services of the last decade, e.g., web 2.0 services, cloud services, and cloud computing, modified the structure of the whole Internet ecosystem. Contrary to the earlier disperse structure, where each service had its own server to be operated on, the infrastructures of the current cloud services are highly centralized; numerous services are run by a single infrastructure. These facilities are commonly known as data centers.

The operators' profit-awareness causes the recent golden age of the data centers. Due to the economics of scale principle, the expenditures (both capital and operational) can be reduced with these highly concentrated architectures. The power consumption of the data centers is accounted for 15 percent of the total expenditures of a data center [8] while data centers have a non-negligible share of the total power consumption of the society. Based on the study of J. Kroomey

László Gyarmati · Tuan Anh Trinh
Network Economics Group
Department of Telecommunications and Media Informatics
Budapest University of Technology and Economics
2 Magyar tudósok körútja, Budapest, H-1117, Hungary

[14], the power consumed by data centers in 2005 was 6.4, 4.7, and 1.8 GW in US, Western Europe, and Japan, respectively. The power consumption of data centers was as high as 1.5% of the total power consumption in the US in 2006. Moreover, these ratios are increasing resulting from the recent data center deployments. As the price of electricity is continuously augmenting [24] and the environment aware operation of the companies is becoming more and more desirable by the customers, the energy efficient operation of data centers is required both from a financial and a social viewpoint. Therefore, the operators of the data centers are interested in more energy-efficient data center infrastructures.

The energy efficient operation of data centers is a complex task with several facets. First, the power consumption of servers, network equipments can be reduced; moreover, middleware can be applied. Second, the arrangement of the data centers' equipments, i.e. the architecture, has an impact on the power consumption. Third, cooling and heat control play a crucial role in the data center facilities: they are necessary to precede hardware failures; however, these equipment utilizes significant amount of energy. Finally, power management systems reduce the power consumption of the data centers by dynamically switching on and off parts of the topology based on the actual load of the system.

This chapter focuses on these aspects expect the first one, which was discussed in the previous chapters. We review the state-of-the-art data center architectures including BCube [9], DCell [10], fat-tree [3], NaDa [25] and Scafida [11] and present evaluate their energy-efficiency quantitatively. In addition, the trade-off between the power consumption and the performance of the data center is investigated. Next, the heat control proposals and techniques are discussed that aim to reduce the power consumed by the cooling facilities of the data centers. Afterwards, energy management systems like ElasticTree [13] are investigated and their power saving capabilities are revealed. Finally, widely adopted energy-efficiency metric called PUE (Power Usage Effectiveness) is presented that expresses the efficiency of data centers.

12.2 Power Consumption of Data Center Architectures

12.2.1 Symmetric Data Center Topologies

In this section, we investigate the energy efficiency of symmetric data center architectures based on simulations and numerical evaluations. In order to compare the properties of the architectures, we first shortly review these data center structures, namely BCube, DCell, and fat-tree topologies while balanced tree is covered as it is a widely used reference model After that, the power consumption of these systems is evaluated; in particular, the energy proportionality of these designs and the trade-off between the throughput capability and power consumption is analyzed.

Several data center architectures have been proposed recently; all of them are based on symmetric structures. One of the design principles of fat-tree topologies

[3,7,18] is to build a data center network using small, commodity switches. The fat-tree topology, also known as Clos topology, consists of three structural layers. The core level contains $n^2/4$ n-port switches at the top. The medium layer has n pods, each containing two layers of $n/2$ switches, i.e., one pod has n n-port switches. The higher level switches are connected to the core switches; each core switch has one port connected to each n pods; specifically, port i of every core switch is linked to the pod i. On the other hand, the lower switches of the medium layer have direct connection to $n/2$ servers. Accordingly, a fat-tree topology supports $n^3/4$ servers. The diameter of the fat-tree topology is 6 regardless the number of connections a switch can have. The fat-tree topology can be built using commodity Ethernet switches [3]; where the flows are leveraged using multiple paths. Portland [18] is a scalable, fault tolerant layer-2 data center built on multi-rooted tree topologies (including fat tree as well) that supports easy migration of virtual machines. VL2 [7] is an agile data center architecture with properties like performance isolation between services, load balancing, and flat addressing.

The BCube data center architecture [9] is designed to be applied in container based, modular data centers, which have a few thousands of servers. BCube is defined recursively, a $BCube_0$ is created from n servers connected to an n-port switch; a $BCube_k$ is structured from n $BCube_{k-1}$ and n^k n-port switches. The links are formed as follows. The $BCube_{k-1}$s are numbered from 0 to $n-1$ while the servers in each $BCube_{k-1}$ are denoted from 0 to n^k-1. Then, the level-k port of the ith server $(i \in [0,n^k-1])$ in the jth $BCube_{k-1}$ $(j \in [0,n-1])$ is connected to the jth port of the ith level-k switch. This redundant structure can have n^{k+1} servers. There are multiple edge-disjoint paths between any two servers; therefore, BCube can distribute the load efficiently. The diameter of the BCube structure is $k+1$, i.e., it is proportional to the number of switching levels. BCube has more links than a tree-based structure, which augments the wiring cost. MDCube architecture proposes a method to interconnect BCube containers to create a mega-data center [26].

The DCell data center architecture [10] is created out of commodity mini-switches to scale-out. The construction of DCell is recursive too. The smallest building block is called $DCell_0$ that consists of an n-port switch and n servers, which are connected to the switch. The servers of a $DCell_1$ has two links, the first is connected to its switch in the $DCell_0$ while the second is linked to a server of another $DCell_0$. Each $DCell_0$ is treated as a node in the $DCell_1$; these nodes form a complete graph. Accordingly, a $DCell_1$ can have $n+1$ $DCell_0$. The $DCell_k$ structures are constructed in the same way. Higher-level $DCell_k$ structures are made out of $g_k=t_{k-1}+1$ $DCell_{k-1}$s, while $DCell_k$ has $t_k=g_k\, t_{k-1}$ servers. These recursive expressions scale up rapidly; therefore, DCell can have enormous number of servers with small structural levels (k) and switch ports (n). The DCell structure contains switches only in the lowest hierarchy level; thus, the servers are actively involved in the routing process; each server has $k+1$ ports. The diameter of the DCell structure is less then $2log_kN-1$, where N denotes the number of servers within the structure.

As a reference, we incorporate the balanced tree structure into our analyses as well. A balanced tree distributes its leaves as evenly as possible between its branches. A balanced tree has a single switch in the core, which has n ports;

analogously, the switches in the intermediate levels have n ports too. The servers are at the leaves, connecting to one of the switches. If the balanced tree has k levels the architecture can have n^k servers. The diameter of the structure is $2k$.

We express analytically the number of servers, the number of switches, and the diameter of the state-of-the-art structures in Table 12.1. We note that the power consumption of the data center architectures is proportional to the number of servers and switches located in the topology. In all cases, the number of servers and switches in the topology highly depend on the parameters of the architecture, namely the number of ports that a server or switch can have, denoted by n, and the number of structural levels, denoted by k.

Table 12.1 Properties of symmetric state-of-the-art data center architectures (n ports, k levels)

Architecture	Number of servers	Number of switches	Diameter
Balanced tree	n^k	$\sum_{i=0}^{k-1} n^i$	2k
Fat-tree	$\frac{n^3}{4}$	$\left(\frac{n}{2}\right)^2 + n^2$	6
DCell	$\approx (n+1)^{2k}$	$\approx \frac{(n+1)^{2k}}{n}$	$2^{k+1} - 1$
BCube	n^{k+1}	$\sum_{i=1}^{k+1} n^i$	k+1

We illustrate the abovementioned data center structures in Figure 12.1, where the number of servers is 16 except DCell, which has 20 servers. Rectangles denote the servers, while switches are presented with circles. All of them have a precisely designed symmetric structure; this property has a drawback in terms of energy proportionality, as we will see shortly.

The energy efficiency of data centers can be assessed in several ways depending on the viewpoint of the investigation. Next, we introduce how the power consumption, measured in Watts, is computed for the data center structures in our analysis. Our goal is to investigate the energy efficiency of data center structures; our computation focuses on the network related power consumption in data centers. Therefore, the total power consumption incorporates the energy consumed by the parts of the data center fabric, i.e. the energy requirement of the switches, and the extra energy consumed at the ports of the servers as the servers have diverse number of ports at different architectures. However, the power consumption of the servers and the devices that are related to other operational devices; e.g., the cooling facility, are excluded from the analyses.

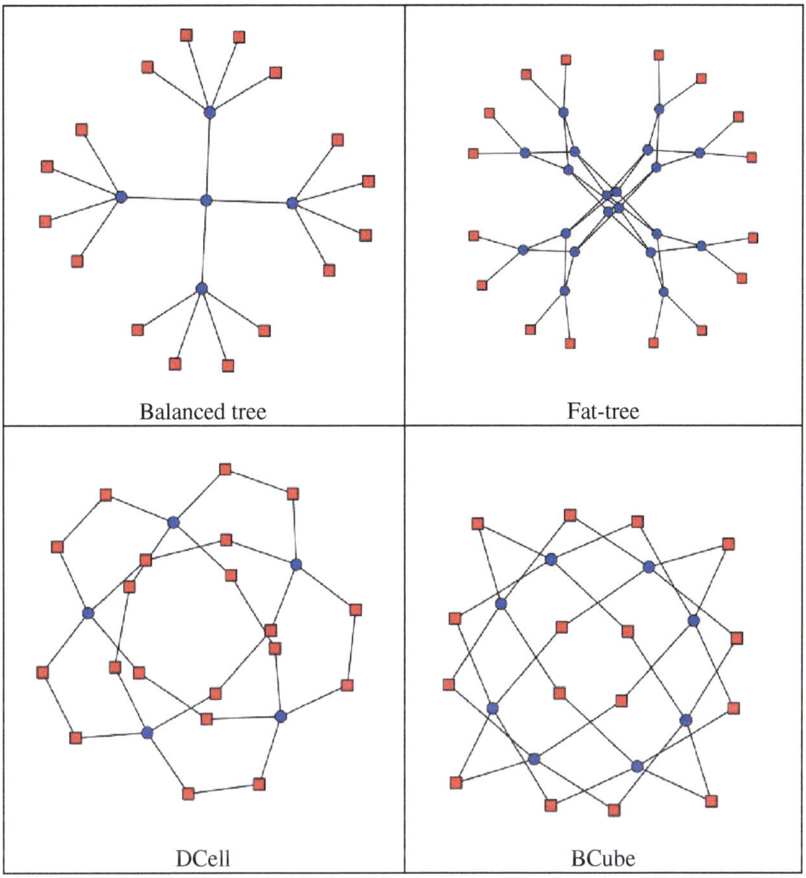

Fig. 12.1 Illustration of symmetric state-of-the-art data center topologies

We carried out simulations in order to investigate the trade-off between the power consumption and throughput capabilities of data center architectures; we present the results in the followings. As current trends show operators of data centers tend to apply commodity network equipments to reduce the capital expenditures; accordingly, our analysis are based on commodity equipments. Several switching fabric are used in the simulations; to be realistic, we apply power consumption values of switches currently available on the market. The products used in the simulations are the 8-port (2960-8TC-L), 24-port (2960-24TC-L), and 48-port (2960-48TC-L) Cisco switches and the 5-port (DGS-2205) D-Link switch; their power consumptions are 12W, 27W, 39W, and 5.12W, respectively. In addition, we assume that the energy consumption of a server increases with 1W per used ports, interpolated based on the power consumption of the 5-port switch. The power consumption values used in the simulations are presented in the specification of the products; however, the switches' power consumption depends also on the traffic that traverses them.

The power consumptions of the symmetric data center architectures are shown in Figure 12.2 as a function of the number of servers in the structure. The topologies are generated by scaling the parameters of the data center structures; thus, increasing the number of servers in the topology. As the topologies have only one or two parameters, there is only moderate possibility to adjust the properties of the architectures. Therefore, the structures of these data centers are rigid. This rigidity causes the large steps in the curves as in order to include for example one additional server to a data center, which is completely utilized considering the actual generation parameters, a much larger network structure has to be deployed that consumes more power. For example, in case of the BCube structure if we want to have a topology with 2305 servers instead of 2304, the power consumption almost triples from 11.3 kW to 32 kW. Similar significant jumps exist in case of DCell and fat-tree too. From an energy efficiency point of view, a preferred property would be if the power consumption of the data centers were a linear function of the number of servers that are located in the structures. This property can be called as energy proportionality. The simulations reveal that the investigated data centers are not energy proportional. The implication of this finding is that from an energy efficiency point of view these data center structures can be used efficiently when the demand of the data center can be predicted accurately, i.e. when it is not required to increase the size of the systems continuously. On the other, if the load of the data center is expected to be increasing, the applicability of these data center architectures is moderate. Several scalability extensions, which will allow affecting the size of the data centers on a finer grained scale, might improve the energy efficiency of these data centers in the future.

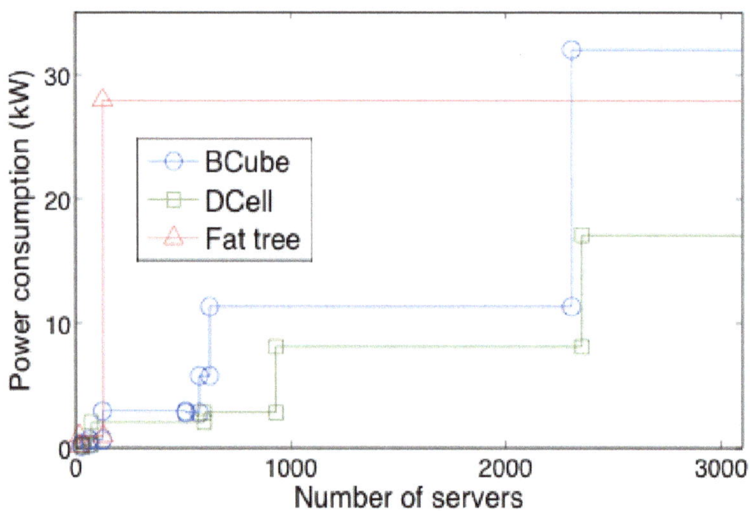

Fig. 12.2 Power consumption of symmetric state-of-the-art data centers (BCube, DCell, and fat-tree topologies); these designs are not energy proportional to the number of server that exist in the systems

Moreover, based on the simulation results presented Figure 12.2, the energy efficiency of the analyzed data center structures can be compared with each other. In case of small data centers that have less than 100 servers, the power consumptions of the three methods are similar. In case of larger architectures, until the topology has around 2000 servers, DCell and BCube architectures consume significantly less power than fat-tree. However, the power consumption of fat-tree might be preferable in case of data centers with around 3500 servers. This implies that the operators have to select the data center topology based on its required size in order to operate the service green.

The relation between the power consumption and the throughput capabilities is presented in Figure 12.3. We illustrate several different data center architectures with exactly the same number of servers (1000) to be able to compare their energy efficiency. The values next to the illustrating shapes refer to the structural parameters of the data center architectures; the first denotes the number of ports of the switches while the second shows the number of structural levels. The throughput capability is a crucial property of the data center networks. Some applications (e.g., MapReduce [6]) require intensive communication among the servers of the data center; bottlenecks in the topology would cause performance degradation. The throughput capability of a data center can be measured with bisection bandwidth that can be computed as follows. The servers are divided into two groups, in total 200 times in our simulations, afterwards the maximal flow between the two parts is computed. It is not surprising that data centers with larger bisection bandwidth values are better from the throughput point of view.

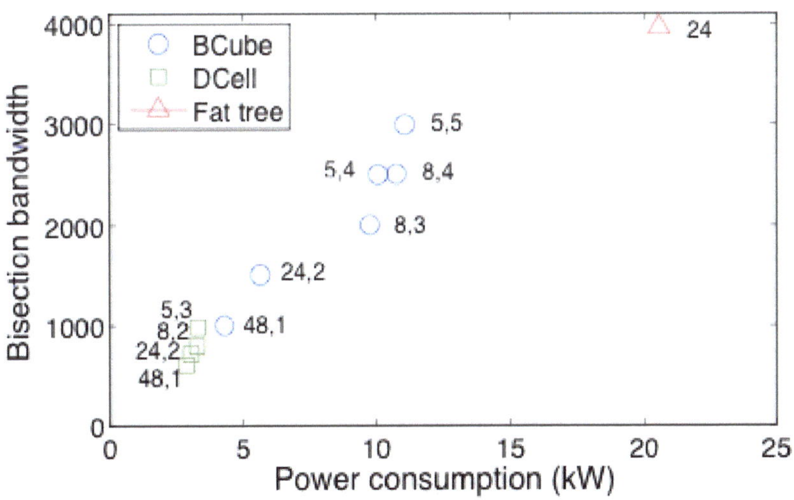

Fig. 12.3 Trade-off between bisection bandwidth and power consumption in case of different data center topologies with 1000 servers; the first value denotes the number of ports that the switches have while the second denotes the number of structural levels

Figure 12.3 shows different number of topologies for each data center architecture due to the limited number of generating parameters. Further data centers could be constructed with the required number of servers with increasing the values of the parameters. However, the power consumption of these topologies is significantly larger than that of the presented ones; therefore, these structures are not presented. DCell architectures consume the fewest power; however, the bisection bandwidth of these topologies is the fewest too. The type of the applied network equipments does not have a significant impact either on the power consumption or on the throughput capability of the DCell topologies. Contrary, BCube offers a possibility to the operators to select among data centers with diverse properties. If the inter-server traffic of the data center is moderate, the operator can establish a network out of 48-port switches, which consumes less energy. On the other hand, if it is necessary to communicate among the server of the data center intensely, it is better to deploy a system using 5-port commodity switches. Regardless of the data center architecture, there exists a trade-off between the power consumption and the throughput capability of the data center. In order to increase the bisection bandwidth of the network, additional switches have to be incorporated to the topology causing elevated power consumption. The capabilities of the architectures can be analyzed based on other metrics like load balancing and path lengths as well [12].

12.2.2 A Biology-Inspired Data Center Architecture: Scafida

In this section, we present a novel data center architecture, which is inspired by biological networks, and show why it is energy efficient due to its highly scalable and flexible design. Innumerable examples exist in the nature, where networks are formed out of components. From the protein of the cells throughout the cells and organs to the organisms and beyond networks are formed. The common sense of these networks is their development, as they all survived the natural selection of the evolution.

Surprisingly, some of the biological networks share a common structural characteristic: the degree distribution of the networks' vertices follows power-law distribution. These networks are called scale-free networks [2]. Barabási and Albert proposed a scale-free network generation algorithm [4] known also as preferential attachment. The algorithm generates the network structure iteratively, i.e. the nodes are added one by one; a new node is attached to an existing node probabilistically proportional to the node's degree; e.g., a new node attaches to a node that has ten links with a probability ten times larger than to a node that has only a single connection. Scale-free networks have two principle properties that may play an important role in the evolutional success of this structure: the diameter of scale-free networks is extremely small and scale-free networks are extremely resistant against random failures. These characteristics of scale-free networks are in accordance with the required properties of data center networks. On the one hand, short paths result fewer intra-network traffic, which enhances the throughput performance of the system. On the other hand, due to the large number of equipments, failures happen round the clock; accordingly, the high error tolerance of the proposed method could motivate its application in practice.

To create a scale-free network-inspired data center architecture, the original scale-free network generation algorithm of Barabási and Albert [4] is extended. The method–called Scafida [11]–artificially constrains the number of links that a node can have; i.e. the maximal degree of the nodes, in order to meet the port number of existing network routers and switches. The algorithm generates a data center topology, which has a determined number of servers and made out of the specified number of different switches. Surprisingly, although the maximal degree of the network's nodes is constrained, the preferable properties of scale-free networks can be sustained, namely short distances and high error tolerance. In particular, neither the average length of the shortest paths nor the load of the switches does increase significantly due to the degree constraining. For more details about the Scafida data center architecture, we refer to [11].

Contrary to the symmetric state-of-the-art data center architectures, Scafida has asymmetric structure with the properties of scale-free networks, where the nodes are heterogeneous. A heterogeneous data center would consume only as much power as required for the given data center size. Two Scafida data center architecture is presented in Figure 12.4 in order to illustrate the asymmetric topologies; the nodes of the scale-free network-inspired data center have at most 8 connections. The design patterns of state-of-the-art data center architectures stick to symmetrical structures, which is an artificial constraint on the data center design. Symmetrical network architectures can be implemented in small sized networks; however, if a network has tens of thousands of nodes the symmetry causes network overprovisioning as we have seen in the previous section. The biological networks, where energy efficiency is vital, are usually asymmetric, which may also warns us that symmetric architectures may not always be the best possible solutions.

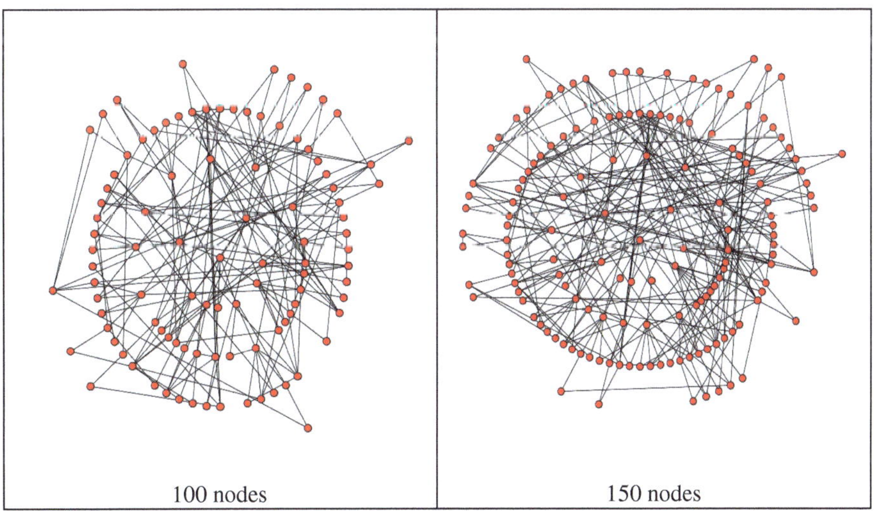

Fig. 12.4 Illustration of the Scafida data center architecture; the nodes have at most 8 ports

Due to the heterogeneous structure of Scafida, the network equipments do not need to be distinguished; a Scafida data center architecture has in total *2Nm* ports if the architecture has *N* servers while m denotes the parameter of the scale-free network inspired topology.

Next, we illustrate the high scalability and flexibility of the scale-free network-inspired data center design based on simulation results focusing on the power consumption of the architecture. On scalability we mean that the Scafida data center generation algorithm can be used to produce a system with few hundreds or hundreds of thousands servers; the desired properties of the structure are irrespective of the size of the data center. The method is flexible in the sense that a particular data center, which has for example 5000 servers, can be generated in enormous number of ways; i.e., using diverse sets of switches.

The proposed Scafida data center architecture is highly scalable; therefore, any size of data center can be generated. Similar to the state-of-the-art architectures, we present the power consumption of several Scafida topologies by scaling the number of servers within the structure (Figure 12.5). Topologies are generated with the above-introduced 5-, 8-, 24-, and 48-port switches; the servers are attached to the network with only one link; i.e., the value of parameter *m* is 1. The power consumption of Scafida data centers is proportional to the size of the system regardless of the type of the switches. Contrary to the symmetric state-of-the-art data centers, where the steps in the power consumption were caused by the structure of the generation methods, in case of Scafida the steps are only due to the scaling of the simulation parameters. Thus, if Scafida topologies would be generated for all the possible number of servers, the curves of Figure 12.5 would be

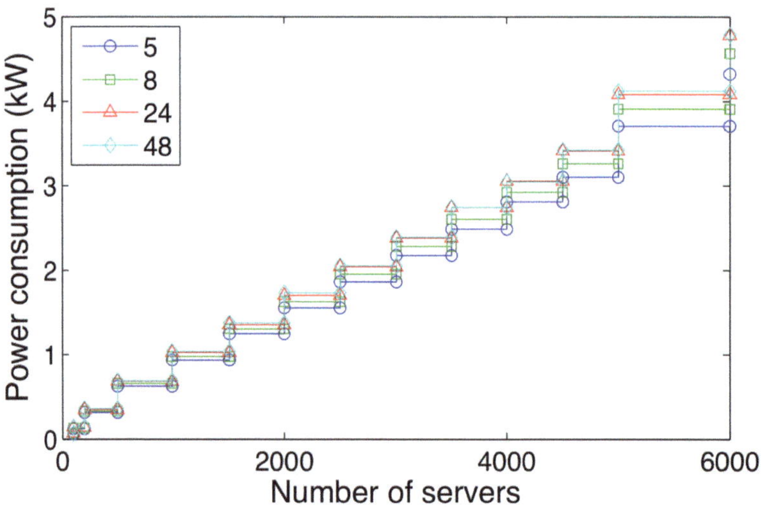

Fig. 12.5 Power consumptions of different scale-free network-inspired data centers consisting of diverse types of switches; Scafida has an energy proportional design

linear without any significant jumps. This implies that the Scafida data center structure is energy proportional; therefore, it can be applied in those cases too, where the size of the data center increases frequently. In case of the current parameters, the power consumption of Scafida is significantly less than that of symmetric state-of-the-art data centers; e.g., in case of a 3000-server data center Scafida consumes only 2 kW contrary to 32, 17, and 28 kW for BCube, DCell, and fat-tree, respectively.

In order to reveal the effect of parameter *m*, which denotes the number of links with a newly added node attaches to the network, the bisection bandwidth and the power consumption of several Scafida topologies is shown in Figure 12.6. The throughput capability of the data center is larger if the value of the parameter is increased, as the larger the *m* is the more links are in the topology. The implication of this result is that using the parameter, the bisection bandwidth of the Scafida method can be set by the operator. Due to the fact that the power consumption of a single port of a switch decreases as the number of ports of the switch increases, more energy efficient networks can be made out of switches with lots of ports. In case of larger *m* values, the usage of switches that have only few numbers of ports consumes much more power than a topology with for example 48 ports. The bisection bandwidth of such topologies does not compensate the increased power consumption.

The Scafida algorithm is capable to create data centers out of any set of network switches; accordingly, the operator of the system is able to specify the power in advance that can be consumed by the Scafida data center. Therefore, the Scafida design method can create energy efficient data center topologies that are in accordance with the traffic characteristic of the data center.

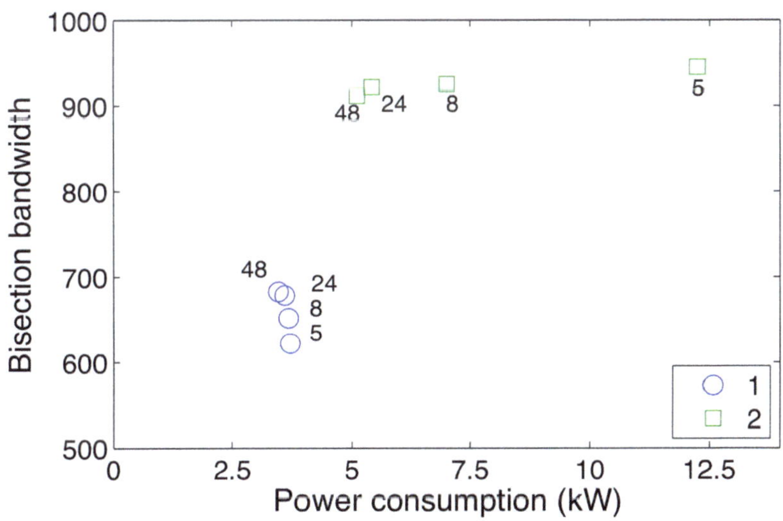

Fig. 12.6 The impact of the m parameter of Scafida on power consumption and throughput capabilities

12.2.3 Nano Data Centers

Contrary to ordinary data centers consisting of servers and switches, nano data centers offer a radically different approach to create data center architectures. Nano data centers can be made out of Internet Service Provider controlled home gateways to form a distributed, peer-to-peer data center structure [25]. The architecture shares storage and computational resources among the participants; the solution uses the underutilized resources and the already committed power consumption of the equipments. The first order goal of nano data centers is to form an energy-efficient content delivering data center. To exploit the advantage of the peer-to-peer structure, the users' requests are served from home gateways whenever it is possible; thus, the load of the content servers, located in the facilities of the operator, is decreased. The energy efficiency of the structure arises from two properties: as the gateways are located in the residence of the subscribers, the heat dissipation is solved without extra cooling facilities; the demand and the services are co-located that reduces the intra-network traffic. Valancius et al. claim that the power consumption can be decreased by at least 20 percent compared to traditional data center architectures.

12.3 Other Aspects of Data Centers' Energy Efficiency

In this chapter, we mainly focus on the structural power consumption of data centers; however, there exist several other parts of data centers where more energy efficient solutions can be reached. The energy efficiency of data centers can be enhanced with methods like reducing the power consumption and heat dissipation of servers and switches, energy-efficient replication strategies that may reduce the internal traffic of data centers, and resource allocation strategies that allow power-off currently underutilized parts of data center throughout virtualization techniques [23]. Next, we review some of these proposals; moreover, we present a standard metric of data centers' energy efficiency.

12.3.1 Heat Control and Cooling

Moore et al. [17] initiated the discussion about the heat management problem of data centers. Contrary to earlier proposals, a system level solution is proposed to control the heat generation of the data centers by placing the workloads temperature-aware. Based on a theoretical thermodynamic formulation run with steady-state thermal data, the authors developed a real-time scheduling method that is aware of the thermal consequences of workload placement. The application of the proposed solution may reduce significantly the power consumption of the cooling facilities of data centers.

A similar approach has been proposed by Sharma et al. [20] where the data center's load is distributed dynamically to reduce the power consumption of the cooling infrastructure. The presented method utilizes the real-time measurement data of sensors, located at different places of the data center. Accordingly, a heat minimizing load distribution is created with the proposed method.

Inspired by these works several thermal-aware load distribution methods have been proposed recently including [5,16].

12.3.2 Energy Management Solutions

In this section we review several energy management solutions that share a common goal, i.e., how to reduce the power consumption of a given data center by utilizing diverse algorithms. These proposals function as a middleware, i.e., they can be applied on top of the presented data center architectures.

ElasticTree [13] is a network-wide power manager; its goal is to reduce the power consumption of data centers by effectively switching of unnecessary servers and network elements. The algorithm maintains the throughput capability of the data center while reduces its power consumption. ElasticTree exploits the power usage characteristics of network equipments: more than half of the total power consumption of the devices is irrespective of the load. Therefore, powering off underutilized switches and servers can save significant amount of energy. Moreover, as the number of operating devices decreases due to the usage of ElasticTree, the power consumed by cooling facilities is reduced too. Based on a test-bed-based evaluation it has been shown that up to 60 percent of the data centers' power consumption can be saved with ElasticTree depending on the traffic matrices.

Abts et al. [1] proposes methods to reduce the power consumption of data center networks. The key of their solution is to dynamically adjust the data rate of the network equipments in order to reduce their power consumption. The presented heuristic method is evaluated based on real data traces; we authors claim that the power saving can be as high as 85 percent in an optimal scenario. Abts et al argue that the power consumed by the data center' network will be the first-order operating expenditure in future data centers. For example, if only 15 percent of a data center is utilized the network related power consumption can be significant–nearly 50 percent in total.

Shang et al. [19] introduces a routing mechanism with an analogous goal, i.e., reduce the power consumption of the data center network. To this extent, an energy-aware routing method is proposed that dynamically powers off some of the network equipments while the performance of the data center network does not decreases significantly. As an optimal solution of the problem would be unfeasible in real-time systems (i.e., it is shown to be an NP hard problem) the authors propose a heuristic method to augment the energy efficiency of the data centers.

VMFlow [15] considers both the topology of the data center and the network traffic to place and migrate the virtual machines to reduce the power consumption. The power aware virtual machine placement problem is NP complete; therefore, the authors propose a heuristic-based solution. It is shown based on simulations that power aware VM migration can even more decrease the power consumption of the data centers than the ElasticTree method, where only the traffic characteristic is exploited to create a more energy efficient data center.

12.3.3 Energy Efficiency Metrics

Along with the widespread utilization of data centers, several metrics have been proposed to describe the energy efficiency of data centers. Next we shortly review the most widely accepted metric called Power Usage Effectiveness then as an outlook we reveal the power consumption of supercomputer systems.

The Green Grid consortium [21] proposes a metric called Power Usage Effectiveness (PUE) to evaluate the power consumption of data centers. The PUE metric considers the power consumed by the whole data center infrastructure including servers, network switches, and cooling facilities. The PUE of a data center is computed as the aggregate power consumption of the whole facility divided by the power consumed by the IT equipments. Accordingly, the PUE ratio of a highly energy efficient data center would be 1. The typical PUE values ranges between 1.25 and 3. The Power Usage Effectiveness is used as a common, official metric by the US, Europe, and Japan to measure the energy efficiency of data centers since February 2010.

The Green500 list deals with the energy efficiency of supercomputers [22]; the 500 most efficient architectures are ranked. Although supercomputers and data centers are different, they share common properties as well, e.g., both facilities can be applied to compute the results of problems with huge resource needs. Therefore, the power consumption of the supercomputers can be used as a reference when the energy efficiency of data centers is investigated. The Green500 list presents both the aggregate (in KWs) and the relative (in MFlops/W) power consumption of the data centers. In terms of the relative metric, the system located at IBM Thomas J. Watson Research Center has an impressing 1684.20 MFlops/W value, based on the November 2010 rank. We note that the total power consumption of the listed 500 supercomputers is as high as 287,36MW.

12.4 Conclusion

In this chapter, we addressed the issue of energy efficiency in data centers. In terms of symmetric design, the state-of-the-art data center architectures were analyzed; not only the total power consumptions of the systems were compared but also the trade-off between the throughput capabilities and energy consumption were analyzed quantitatively. Contrary to these structures, a recently proposed data center design called Scafida was presented, which has asymmetric and heterogeneous structure. It has been shown that scale-free network-inspired data centers are highly scalable and flexible; therefore, their power consumption is configurable in fine-grained quantities. Thus, the power consumption of the Scafida data center is proportional to the number of its servers. Afterwards, other aspects of data centers' energy efficiency were addressed such as the power consumption of the cooling facilities, thermal control, dynamic energy management systems that are capable to power of parts of the data centers to reduce the power consumption. In addition, we introduced the Power Usage Effectiveness metric, which is a standard metric of data centers' energy efficiency. Although significant steps has been made towards energy efficient data center architectures, there still exist issues that have to be addressed including energy efficient replication and virtual machine allocation strategies, routing algorithms, and temporal server switch off methods.

References

[1] Abts, D., Marty, M.R., Wells, P.M., Klausler, P., Liu, H.: Energy proportional datacenter networks. In: Proceedings of the 37th Annual International Symposium on Computer Architecture, pp. 338–347 (2010)

[2] Albert, R., Jeong, H., Barabási, A.-L.: Internet: Diameter of the World-Wide Web. Nature 401(6749), 130–131 (1999)

[3] Al-Fares, M., Loukissas, A., Vahdat, A.: A Scalable, Commodity Data Center Network Architecture. In: ACM SIGCOMM 2008, pp. 63–74 (2008)

[4] Barabási, A.-L., Albert, R.: Emergence of Scaling in Random Networks. Science 286(5439), 509–512 (1999)

[5] Bash, C., Forman, G.: Cool job allocation: measuring the power savings of placing jobs at cooling-efficient locations in the data center. In: Proceedings of the USENIX Annual Technical Conference, pp. 29:1–29:6 (2007)

[6] Dean, J., Ghemawat, S.: MapReduce: Simplified data processing on large clusters. Communications of the ACM 51(1), 107–113 (2008)

[7] Greenberg, A., Hamilton, J.R., Jain, N., Kandula, S., Kim, C., Lahiri, P., Maltz, D.A., Patel, P., Sengupta, S.: Vl2: a scalable and flexible data center network. In: ACM SIGCOMM 2009, pp. 51–62 (2009)

[8] Greenberg, A., Hamilton, J., Maltz, D.A., Patel, P.: The Cost of a Cloud: Research Problems in Data Center Networks. ACM SIGCOMM Computer Communications Rev. 39(1), 68–73 (2009)

[9] Guo, C., Lu, G., Li, D., Wu, H., Zhang, X., Shi, Y., Tian, C., Zhang, Y., Lu, S.: BCube: a High Performance, Server-centric Network Architecture for Modular Data Centers. In: ACM SIGCOMM 2009, pp. 63–74 (2009)

[10] Guo, C., Wu, H., Tan, K., Shi, L., Zhang, Y., Lu, S.: DCell: a Scalable and Fault-tolerant Network Structure for Data Centers. In: ACM SIGCOMM 2008, pp. 75–86 (2008)

[11] Gyarmati, L., Trinh, T.A.: A Scale-Free Network Inspired Data Center Architecture. ACM SIGCOMM Computer Communications Review 40(5), 5–12 (2010)

[12] Gyarmati, L., Trinh, T.A.: How can architecture help to reduce energy consumption in data center networking? In: Proceedings of the 1st International Conference on Energy-Efficient Computing and Networking, pp. 183–186 (2010)

[13] Heller, B., Seetharaman, S., Mahadevan, P., Yiakoumis, Y., Sharma, P., Banerjee, S., McKeown, N.: ElasticTree: Saving Energy in Data Center Networks. In: USENIX NSDI 2010, pp. 1–16 (2010)

[14] Koomey, J.: Worldwide Electricity Used in Data Centers. Environmental Research Letters 3(3), 1–8 (2008)

[15] Mann, V., Dutta, P., Kalyanaraman, S., Kumar, A.: VMFlow: Leveraging VM Mobility to Reduce Network Power Costs in Data Centers. IBM Research Report RI10013 (2010)

[16] Moore, J., Chase, J.S., Ranganathan, P.: Weatherman: Automated, online and predictive thermal mapping and management for data centers. In: ICAC 2006: IEEE International Conference on Autonomic Computing, pp. 155–164 (2006)
[17] Moore, J., Chase, J., Ranganathan, P., Sharma, R.: Making scheduling cool: Temperature-aware workload placement in data centers. In: Proceedings of the Annual Conference on USENIX Annual Technical Conference (2005)
[18] Mysore, R.N., Pamboris, A., Farrington, N., Huang, N., Miri, P., Radhakrishnan, S., Subramanya, V., Vahdat, A.: Portland: a scalable fault-tolerant layer 2 data center network fabric. ACM SIGCOMM 2009, 39–50 (2009)
[19] Shang, Y., Li, D., Xu, M.: Energy-aware routing in data center network. In: Proceedings of the First ACM SIGCOMM Workshop on Green Networking, pp. 1–8 (2010)
[20] Sharma, R.K., Bash, C.E., Patel, C.D., Friedrich, R.J., Chase, J.S.: Balance of power: Dynamic thermal management for internet data centers. IEEE Internet Computing 9(1), 42–49 (2005)
[21] The Green Grid Consortium (2011), http://www.thegreengrid.org/
[22] The Green500 List (2010), http://www.green500.org
[23] Towards Green ICT. ERCIM News (2009)
[24] U. S. Energy Information Administration: Average Retail Price of Electricity to Ultimate Customers by End-Use Sector (2010), http://www.eia.doe.gov/cneaf/electricity/epa/epat7p4.html
[25] Valancius, V., Laoutaris, N., Massoulié, L., Diot, C., Rodriguez, P.: Greening the Internet with Nano Data Centers. In: ACM CoNEXT 2009, pp. 37–48 (2009)
[26] Wu, H., Lu, G., Li, D., Guo, C., Zhang, Y.: MDCube: a high performance network structure for modular data center interconnection. In: ACM CoNEXT 2009, pp. 25–36 (2009)

Chapter 13
Energy-Conservation in Large-Scale Data-Intensive Hadoop Compute Clusters

Rini T. Kaushik and Klara Nahrstedt

Abstract. Data-intensive computing is gaining rapid popularity given the rampancy and fast growth of Big Data. It's myriad use cases range from clickstream processing, mail-spam detection, credit-card fraud detection to meteorology, and genomics. Google's MapReduce is a programming model designed to greatly simplify Big Data processing. Hadoop, an implementation of MapReduce is increasingly becoming popular because of its open-source nature. The ever-increasing and large-scale deployments of Hadoop clusters bring in their wake huge energy costs, thereby making energy-conservation a priority. Scale-down is an attractive technique to conserve energy and it also allows energy proportionality with non-energy-proportional components such as the disks. Hadoop presents unique challenges to the popular scale-down based cluster-energy-management techniques. The same features that lend high performance to Hadoop, such as fine-grained load-balancing of data across servers in the cluster and data-locality of computations, also complicate scale-down. In this chapter, we go over the scale-down approaches that have been proposed specifically for the Hadoop clusters. Then, we present a case study of GreenHDFS, which uses a data-centric scale-down approach to save overall operating energy costs in a Hadoop compute cluster.

13.1 Introduction

IDC predicts that the digital data will surpass 35 zettabytes by 2020 [2]. Given the explosion in the Big Data [11, 12], computing today is increasingly becoming *data-intensive* in nature and revolves around performing computations over extremely large data sets. Data-intensive computing [24] has myriad use cases such as fraud detection in financial transactions, building indexes and search rankings for the internet-scale search engines, log processing, satellite imagery, mail anti-spam

Rini T. Kaushik · Klara Nahrstedt
University of Illinois, Urbana-Champaign, 201 N Goodwin Avenue, Urbana, IL
e-mail: {kaushik1, klara}@illinois.edu

detection, clickstream processing, machine learning and data mining to build predictive user interest models for advertising optimizations, improved security through better intelligence gathering, portfolio risk management, and classified usage in government undertakings.

Google's MapReduce is a programming model designed to greatly simplify Big Data processing [25]. It allows automatic parallelization and execution of tasks on a large-scale commodity cluster. An increasing number of companies and academic institutions have started to rely on Hadoop [7], which is an open-source version of MapReduce [25] framework for their data-intensive computing needs [63].

With the increase in the sheer volume of the data that needs to be processed [68], storage and server demands of computing workloads are on a rapid increase. For example, Yahoo!'s computing infrastructure already hosts 170 petabytes of data and deploys over 38000 servers [4]. Over the lifetime of IT equipment, the operating energy cost is comparable to and may even exceed the initial equipment acquisition cost [10] and constitutes a significant part of the total cost of ownership (TCO) of a data center [52, 17, 40]. Environmental Protection Agency [27] indicates that *'U.S. data centers will consume 100 Billion Kilowatt hours annually by 2011'*, resulting in an annual energy cost of $7.4 Billion.

Studies have shown that the computing power can range from 33% to 75% of the overall power [30, 31] and that cooling energy costs amount to almost half of the total energy costs. For every one watt of power consumed by the computing infrastructure, another one-half to one Watt is consumed in powering the cooling infrastructure [51]. Hence, reduction in *overall* (i.e., both server and cooling) operating energy costs of the extremely large-scale, server farms has become an urgent priority [18, 27].

Scale-down is an attractive approach to conserve energy as detailed in Section 13.3 and hence, scale-down based energy management of the data-intensive Hadoop compute clusters is the focus of this chapter. The remainder of the chapter is structured as follows. In Section 13.2, we present a brief background on Hadoop. In Section 13.3, we give a background on the scale-down approach, its mandates and the unique challenges presented by the Hadoop clusters to scale-down. In Section 13.4, we present the scale-down approaches that have been proposed specifically for data-intensive Hadoop clusters. In Section 13.5, we present a case study on GreenHDFS, an energy-conserving variant of the Hadoop Distributed File System (HDFS) that uses a data-centric energy-management approach to allow effective scale-down in Hadoop clusters. Finally, we conclude.

13.2 Hadoop Background

Hadoop is an open-source implementation of MapReduce [7]. It is logically separated into two subsystems: a highly resilient and fault-tolerant Hadoop Distributed File System (HDFS) [38] which is modeled after the Google File System [28], and a MapReduce task execution framework. Hadoop runs on clusters of low-cost commodity hardware and can scale-out significantly.

The centralization inherent in the client/server model provided by the traditional storage solutions such as Network File Systems (NFS) [53] has proven a significant obstacle to scalability. More recent distributed file systems have adopted architectures based on object-based storage [43, 67, 19]. Clients interact with a metadata server (MDS) to perform metadata operations (open, rename), while communicating directly with intelligent object storage devices (OSDs) to perform file I/O (reads and writes). The main motivation for decoupling the namespace from the data is the scalability of the system. Metadata operations are usually fast, whereas data transfers can last a long time.

HDFS is an object-based distributed file system and is designed for storing large files with streaming data access patterns [68]. It aims to provide high throughput of data accesses and hence, it is more suitable for batch-processing applications than for interactive applications which require low latency data access. HDFS cluster has two types of servers: NameNode (MDS) and a number of DataNodes (OSDs). HDFS supports a single namespace architecture which consists of a hierarchy of files and directories. Files and directories are represented on the NameNode by inodes, which contain permissions, modification and access times. The NameNode maintains the namespace tree and the mapping of the file blocks to the DataNodes.

Each file in HDFS is split into chunks of typically 128MB in size and each chunk is replicated n-way for resiliency. The default replication factor is 3 in HDFS. HDFS then distributes the data chunks and the replicas across the DataNodes for resiliency, high data access throughput, and fine-grained load-balancing.

In data-intensive computing with Big Data, it is no longer feasible or scalable to move data to the computations given the huge and ever-growing data set sizes. Now, computations need to move to the data and efficient data access from disks is an essential part of the computation. This is different from the traditional High Performance Computing (HPC), where the data typically lies in a shared Storage Area Network (SAN) and data has to be brought to the server where computation is happening.

Data locality of the computations is the most important feature of the MapReduce framework and is responsible for its high performance[1]. MapReduce tries to colocate the computation with the data to allow for fast local data access [68]. Thus, every time computation needs to happen on a data, HDFS determines the nearest DataNode containing a replica of the data and computation is then directed to that DataNode. In the next section, we discuss scale-down technique, its mandates and the challenges posed by Hadoop clusters to scale-down.

13.3 Scale-Down

There is significant amount of research on increasing energy-efficiency of the individual components of the server such as the processor [32, 64, 22], storage subsystem [66, 54], memory [45] and networking [6]. However, in a typical commodity

[1] A performance test performed at Google showed that MapReduce is capable of sorting 1TB on 1,000 computers in 68 seconds [29], thereby, illustrating its power.

server in a compute cluster, no single server component (i.e., CPU, DRAM or HDD) contributes significantly to the overall power consumption and hence, an energy-management scheme that encompasses the entire system is needed.

Energy-proportionality is increasingly becoming an important consideration [8]. In an energy-proportional system, almost no power is consumed when the system is idle and power consumption increases in proportion to the increase in the activity level.

In reality, instead of consuming negligible power, the server components consume almost 40%-60% [8] of the peak power during idle utilization. The CPU contribution to system power is nearly 50% when at peak but drops to less than 30% at low activity levels, making it the most energy-proportional of all main subsystems. By comparison, the dynamic range of memory systems, disk drives, and networking equipment is much lower.[2] This suggests that energy proportionality at the system level cannot be achieved through CPU optimizations such as Dynamic Voltage and Frequency Scaling (DVFS) alone, but instead requires improvements across all server components.

Some non energy-proportional components such as the disks will require greater innovation to be energy-proportional. Disk drives consume significant amount of power simply to keep the platters spinning, possibly as much as 70% of their total power for high RPM drives.[3] Creating additional energy-efficiency and energy-proportionality may require smaller rotational speeds, or smaller platters. Furthermore, a processor running at a lower voltage-frequency mode can still execute instructions without requiring a performance-impacting mode transition. Unfortunately, as of now, there are no other components in the server with *active* low-power modes. The only low-power modes currently available in mainstream DRAM and disks are inactive modes such as sleep/stand-by modes. These inactive modes involve paying a wake-up latency and energy penalty for an inactive-to-active mode transition.

Scale-down, that involves transitioning server components such as the CPU, DRAM, and disks to an inactive, low power-consuming sleep/standby state during periods of idle utilization, is an attractive technique to conserve energy. Given the non energy-proportional nature of some of the state-of-the-art server components, scale-down is one of the most *viable* options for yielding energy-proportionality during idle periods. A typical server consumes only 13.16W (i.e., 3% of peak power) in an inactive sleeping state vs. 132.46W (i.e., 30% of peak power) in active idle power mode.

---

[2] Approximately 2.0X for memory (DRAM DIMM consumes 3.5-5W at peak utilization and 1.8-2.5W at idle utilization), 1.2X for disks (Seagate Barracuda ES.2 1TB consumes 11.16W at peak utilization and 9.29W at idle utilization), and less than 1.2X for networking switches.

[3] In idle mode, a drive is active, which means that platters are spinning and any new request gets served immediately without any delay, the amount of power saved is as much as 50% of peak power. In the sleep mode, the drive heads are "parked away" from the drive platters (unloaded), and the platters are completely spun down, resulting in negligible power consumption. However, bringing drive back can take as long as 10 seconds.

13.3.1 Scale-Down Mandates

Scale-down cannot be done naively. Energy is expended and transition time penalty is incurred when the components are transitioned back to an active power mode. Scale-down can be done only if the idleness interval meets the criterion illustrated in the following equations.

Energy spent on computation is given by:

$$E = (k * V^2 * f + R_0) * t \tag{13.1}$$

Where, k is a constant, V is the voltage and f is the frequency. R_0 is the idle power and t is the execution time.

In any idle period t, there are two power state options for the servers: 1) scale-down, i.e., transition to low-power inactive mode, 2) remain in active power mode.

$$E_{sleep} = P_{sleep} * t + E_{wake} \tag{13.2}$$

$$E_{nosleep} = (k * V^2 * f + R_0) * t \tag{13.3}$$

Where, E_{sleep} is the energy consumed if the server is scaled-down for the entire t duration and is transitioned back to active power mode at the end of t interval. $E_{nosleep}$ is the energy consumed if the server just stays in the active power mode. E_{wake} is the energy expended upon a server wakeup. And, P_{sleep} is the power consumed during the inactive power mode.

To ensure energy savings, scale-down should be done during an idle period t, only if $E_{nosleep} - E_{sleep} >> 0$. This happens when t meets the length requirements illustrated below:

$$t > \frac{E_{wake}}{k * V^2 * f + R_0 - P_{sleep}} \tag{13.4}$$

An effective scale-down technique mandates the following:

- *Sufficient idleness* to ensure that energy savings are higher than the energy spent in the transition as shown in Equation 13.4.
- *Less number of power state transitions* as some components (e.g., disks) have limited number of start/stop cycles (e.g., 50,000) and too frequent transitions may adversely impact the lifetime of the disks.
- *No performance degradation.* Disks take significantly longer time to transition from active to inactive power mode (as high as 10 seconds). Frequent power state transitions will lead to significant performance degradation. Hence, steps need to be taken to reduce the power state transitions and also, to amortize the performance penalty of the unavoidable power state transitions that do need to occur.
- *No performance impact of load-unbalancing.* Steps need to be taken to ensure that load concentration on the remaining active state servers does not adversely impact overall performance (including data read and write access performance) of the system.

13.3.2 Scale-Down Challenges in Hadoop Clusters

Hadoop clusters present unique scale-down challenges. With data distributed across all nodes and with the data-locality feature of Hadoop, any node may be participating in the reading, writing, or computation of a data-block at any time. Such data placement makes it *hard* to generate significant periods of idleness in the Hadoop clusters even during low activity periods. A study of the average CPU utilization of 5,000 Google servers during a 6-month period showed absence of significant idle intervals despite the existence of periods of low activity [8].

One of the most popular approach for scaling-down servers is by manufacturing idleness through migrating workloads [20, 48, 57, 62, 21, 23, 9]. Unfortunately, migrating workloads and their corresponding state to fewer machines during periods of low activity does not work in context of Hadoop and other MapReduce clusters. Workload placement/migration can be relatively easy to accomplish when servers are state-less (i.e., serving data that resides on a shared NAS or SAN storage system). Servers in a MapReduce cluster (e.g., Hadoop cluster) are not state-less. Data-locality is an important feature in MapReduce that allows for high performance data processing. Computations are colocated with the data and hence, computation migration is limited to only the servers hosting a replica of the data that needs to be processed. Hence, it is necessary to explore new mechanisms for effective scale-down in MapReduce clusters keeping the unique scale-down challenges in mind.

In the next section, we present some of the existing scale-down based energy-management approaches that have been recently proposed specifically for the data-intensive Hadoop compute clusters.

13.4 Existing Hadoop Scale-Down Approaches

The existing scale-down based approaches to reduce energy costs in data-intensive Hadoop compute clusters can be broadly classified into three categories: 1) *Replica placement approaches*, 2) *Workload scheduling approaches*, and 3) *Data migration/placement based approaches*. In addition to the scale-down techniques, Vasudevan et. al. [65] and Hamilton [31] have proposed data-intensive clusters built with low power, lower performance processors (Wimpy Nodes) that aim to reduce the peak power consumption of the cluster. While promising, Lang et. al. [40] do point out that for more complex workloads, the low power, lower performance clusters result in a more expensive and lower performance solution.

13.4.1 Replica Placement

As mentioned in Section 13.2, HDFS maintains replicas (default replication factor is 3-way) of each data chunk in the system for resiliency, and reliability. Recent research on scale-down in HDFS managed clusters [5, 41] seeks to exploit the replication feature of these file systems and proposes energy-aware replica placement techniques. Leverich et. al. [41] propose maintaining a primary replica of the data on a "Covering Subset" (CS) of nodes that are guaranteed to be always on. The rest of the servers can be then scaled-down for energy savings. However, using just the CS servers for all the data accesses may result in degraded data access performance (response time will increase because of the increase in the queuing delays in the disks of the CS servers).

Amur et. al. [5] extend Leverich et. al.'s work by providing ideal power-proportionality in addition by using an "equal-work" data-layout policy, whereby replicas are stored on non-overlapping subsets of nodes. Their proposed file system Rabbit is capable of providing a range of power-performance options. The lowest power, lowest performance option in Rabbit is achieved by keeping just the servers with the primary replicas on. More servers are powered up as performance needs increase.

While promising, these solutions do suffer from degraded write-performance as they rely on write off-loading technique[4] [50] to avoid server wakeups at the time of file writes. Write-performance is an important consideration in Hadoop. Reduce phase of a MapReduce task writes intermediate computation results back to the Hadoop cluster and relies on high write performance for overall performance of a MapReduce task.

Furthermore, a study of a production[5] Hadoop cluster at Yahoo! observed that the majority of the data in the cluster has a news-server like access pattern [35]. Predominant number of computations happens on newly created data; thereby mandating good read and write performance of the newly created data. Given, the huge data set sizes, good write performance happens when the incoming data chunks are written in parallel across a large number of DataNodes in the cluster. Write off-loading is just not a performant and a scalable option for such writes. If these techniques do try to wakeup servers to absorb the new writes, they will still suffer performance degradation due to the power state transition penalty. Furthermore, given the significant number of file writes that happen in a day on a production cluster, waking up servers to absorb the writes will also adversely impact the lifetime of the components such as the disks.

13.4.2 Workload Scheduling

Lang and Patel propose an "All-In" strategy (AIS) for scale-down in MapReduce clusters [39]. AIS uses all nodes in the cluster to run a workload and then powers down the entire cluster. The main advantages of the AIS technique are: 1) It is a simple approach and does not need any code changes or over provisioning of the storage on a subset of servers, and 2) It offers same data access throughput as the baseline Hadoop and does not need any data layout changes.

[4] Write off-loading allows write requests on spun-down disks to be temporarily redirected to persistent storage elsewhere in the data center.

[5] Production clusters are the truly large-scale compute clusters [15] and are the ones that will benefit the most from the energy-conservation techniques.

This technique makes an underlying assumption that all the workloads happen simultaneously on the system. However, a typical production cluster has several workloads running on the system with varying start and stop times. Given, the globalization rampant today, significant number of clusters are in use 24/7 and hence, such a technique may not see enough idleness in the system to justify a scale-down. The authors also propose batching intermittently arriving jobs and then, submitting all the jobs in the batch simultaneously. Some of the jobs may have Service Level Agreements (SLA) considerations and it may not be acceptable to batch and execute such jobs at a later time. Resource contention may also arise if jobs are all invoked simultaneously on the cluster. For example, the cluster may not have enough map/reduce compute slots available to be able to service all the jobs simultaneously.

Sharma et. al. [56] have extended upon Meisner et. al.'s work [44] to allow energy savings via blinking (i.e., transitioning between high-power active state and low-power inactive state). However, both these approaches have assumed non-hard disk clusters. Disk-based clusters may suffer from significant performance degradation and impact on disk longetivity with frequent state transitions. Given, the low capacities and high costs of the SSD drives, clusters comprising entirely of SSD drives are not feasible at the moment, especially given the petascale storage demands of a single production compute cluster [15].

13.4.3 Data Migration/Placement

Kaushik et. al. [36, 35, 37] have taken a data-centric approach in GreenHDFS, an energy-conserving variant of HDFS. GreenHDFS focuses on *data-classification* techniques to extract energy savings by doing *energy-aware* placement of *data*. GreenHDFS trades cost, performance and *power* by separating cluster into logical *Hot* and *Cold* zones of servers according to the data frequency usage patterns. Each cluster zone has a different *temperature* characteristic where temperature is measured by the power consumption and the performance requirements of the zone. GreenHDFS relies on the inherent heterogeneity in the access patterns in the data stored in HDFS to differentiate the data and to come up with an energy-conserving data layout and data placement onto the zones. Since computations exhibit high data locality in the Hadoop framework, the computations flow naturally to the data in the *Hot* zone. As a result, the *Cold* zone servers receive minimal accesses and can be effectively scaled-down. A detailed case study, along with evaluation results using one-month long HDFS trace from a large-scale (2600 servers, 5 Petabytes, 34 million files) production Hadoop cluster at Yahoo! is presented in Section 13.5 to show how GreenHDFS is able to meet all the scale-down mandates. The large-scale evaluation in GreenHDFS is in contrast to the approaches presented earlier as they were all evaluated using a very small cluster and data set size.

Table 13.1 Per-Zone Policies in the GreenHDFS

	Hot Zone	Cold Zone
Storage Type	SATA	SATA
Processor	Intel Xeon	Intel Atom
Number of Disks	4,1TB	12,1TB
File Striping Policy	Performance-Driven[28]	Energy-Efficiency Driven,None
Server Power Policy	Always-on	Aggressive, Scale-down
Replication Policy	n-way	n-way
Data Classification	Frequently accessed	Rarely accessed
Power Transition Penalty	None	Medium
Energy Savings	None	High
Performance	High	Low
Thermally-Aware Location	Cooling-Efficient	Cooling-Inefficient

13.5 Case Study: GreenHDFS

GreenHDFS is an energy-conserving variant of HDFS [36]. GreenHDFS trades cost, performance and *power* by separating cluster servers into logical *Hot* and *Cold* zones as shown in Table 13.4.3. *Hot* zone contains hot, i.e., frequently accessed, and the *Cold* zone contains cold, i.e., dormant or infrequently accessed data.

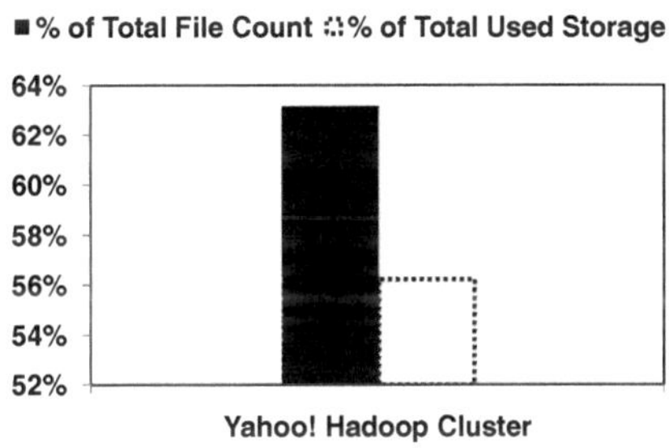

Fig. 13.1 56% of Used Capacity is Cold in the Yahoo! Production Hadoop Cluster.

As per a study of a production Hadoop cluster at Yahoo! [36], 56% of the storage in the cluster is cold (i.e., not accessed in the entire one-month long analysis duration) as shown in Figure 13.1. A majority of this cold data needs to *exist* for regulatory and historical trend analysis purposes and cannot be deleted[36]. In baseline HDFS, as shown in Figure 13.2, no data-differentiation-based data placement is performed. As a result, cold data is intermingled with hot data on the servers cluster-wide. GreenHDFS, on the other hand, differentiates between the data, and as shown in Figure 13.3, it separates out the hot and the cold data.

This separation of the data aids in energy-conservation. Since, computations exhibit high data locality in the Hadoop framework [68], energy-aware data placement translates into energy-aware computation placement. The computations flow naturally to the data in the *Hot* zone, resulting in maximal data accesses in the *Hot* zone

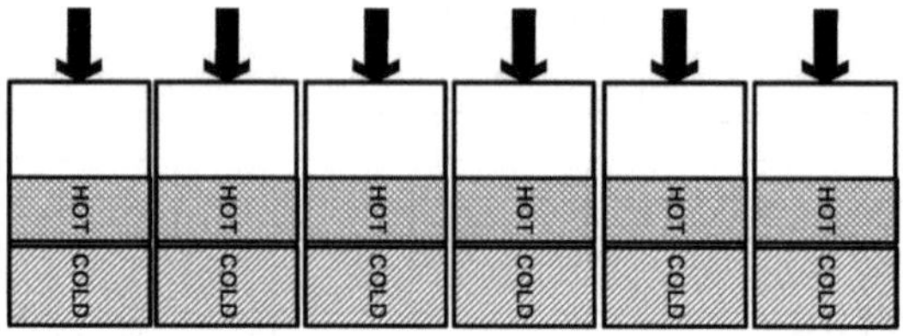

Fig. 13.2 There is no Data-Differentiation in the Baseline HDFS and Hot and Cold Data is Intermixed in the Cluster.

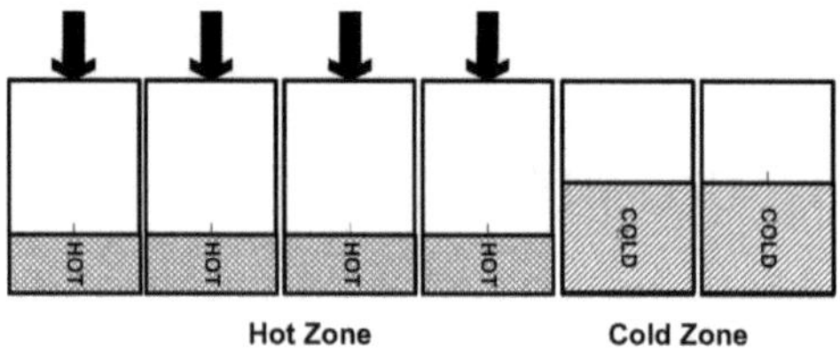

Fig. 13.3 In GreenHDFS, Data is Differentiated into Hot and Cold Data. Rarely Accessed Data is Migrated to the Cold Zone Servers Periodically.

and minimal accesses to the *Cold* zone. This results in significant idleness in the *Cold* zone servers which can then be effectively scaled-down.

In servers in the large-scale Internet services, the lowest energy-efficiency region corresponds to their most common operating mode, as shown by a six-month average CPU utilization study done at Google [8]. Per this study, servers operate most of the time between 10 and 50 percent of their maximum utilization levels. There is a significant opportunity to consolidate computations on the *Hot* zone and push the servers to operate closer to their upper energy-efficient range (closer to 70-80% of peak utilization). In the *Cold* zone, on the other hand, scaling-down servers aids in deriving energy-proportionality during idle utilization. Thus, GreenHDFS allows energy-proportionality with non-energy-proportional components.

Hot zone has strict SLA requirements and hence, performance is of the greatest importance. We trade-off power savings in interest of higher performance and no energy management is done in the servers in *Hot* zone. The data is chunked and replicated on the servers in the *Hot* zone similar to the baseline HDFS for high performance and resilience. Majority (70% +) of the servers in the cluster are assigned to the *Hot* zone upfront to ensure high performance of the hot data.

The *Cold* zone consists of rarely accessed, cold data. We trade-off performance for higher energy-conservation in this zone. Since, SLA requirements for *Cold* zone are not strict, and the performance in this zone is not critical, we employ aggressive power management policies in the *Cold* zone to scale-down servers. For optimal energy savings, it is important to increase the idle times of the servers and limit the wakeups of servers that have transitioned to the inactive-power-saving mode. Keeping this rationale in mind and recognizing the low performance needs and infrequency of data accesses to the *Cold* zone; this zone *does not chunk* the data. By not chunking the data, we ensure that future access to a particular data is limited to just one server hosting that data. However, an access to a file residing on the *Cold* Zone may suffer performance degradation in two ways: 1) if the file resides on a server that is not powered ON currently–this will incur a server wakeup time penalty, 2) transfer time degradation courtesy of no data chunking in the *Cold* zone.

To save on power, GreenHDFS deploys low cost, low performance and low power processors in the servers in the *Cold* zone such as Intel's Atom processors. The number of DRAMs is reduced from 8 to 2 in the *Cold* zone to further reduce the power consumption. The *Cold* zone servers can still be used for computations in situations where the *Hot* zone servers are not sufficient to absorb the entire workload. Low power processors have been proposed in the servers (Wimpy Nodes) of data-intensive clusters [60, 42, 47, 65].

Green Hadoop Cluster uses a hybrid storage model in which *Cold* zone servers are storage-heavy (12, 1TB) disks and *Hot* zone servers are storage-lite (4, 1TB disks). Storage-heavy servers significantly reduce the server footprint of the *Cold* zone servers. Otherwise, the sheer extent of the cold data (needs to exist for regulatory and historical trend analysis purposes and cannot be deleted) present in a production Hadoop cluster, would have required deployment of a significantly large number of storage-lite servers in the *Cold* zone. The hybrid storage model results in a difference in the cost and the power characteristics of the servers between the zones.

GreenHDFS relies on three policies to provide energy-conservation and hot space reclamation: 1) *Data Migration Policy*, 2) *Data Reversal Policy*, and 3) *Server Power Conservation Policy*. The Data Migration Policy runs in the *Hot* zone, and monitors the *Coldness* i.e., the *dormancy* of the files in the cluster on a daily basis during periods of low load. Dormant files that are chosen as the file migration candidates, are moved to the *Cold* Zone. The advantages of this policy are two-fold: 1) the policy leads to higher space efficiency as space is freed up in the *Hot* zone by moving rarely accessed files out of the servers in the *Hot* zone. This results in more space for files with higher Service Level Agreement (SLA) requirements and popularity, and 2) the policy allows significant energy-conservation.

Server Power Conservation Policy determines the servers which can be scaled-down in the *Cold* zone. We rely on Wake-on-LAN to wake the system upon arrival of a packet. Servers transition back to active power state upon receipt of data access, data migration or disk scrubbing events.

The *File Reversal Policy* runs in the *Cold* zone and ensures that the QoS, bandwidth and response time of files that become popular again after a period of dormancy is not impacted. We define the *hotness* of a file by the number of accesses that happen to the file. If the hotness of a file that is residing in the *Cold* zone becomes higher than a threshold $Threshold_{FRP}$, the file is moved back to the *Hot* zone. The file is chunked and placed unto the servers in the *Hot* zone in congruence with the performance-driven policies in the *Hot* zone. The threshold $Threshold_{FRP}$ helps in reducing unnecessary data oscillations between the zones.

13.5.1 Variants of General GreenHDFS

13.5.1.1 Reactive GreenHDFS

The Reactive GreenHDFS relies on a *reactive* Data Migration Policy to provide energy-conservation and hot space reclamation. The reactive Data Migration Policy runs in the *Hot* zone, and monitors the *Coldness* i.e., the *dormancy* of the files in the cluster as shown in Algorithm 1 on a daily basis during periods of low load. Files that have been dormant for more than threshold $Threshold_{FMP}$ days are deemed as cold files. Such cold files are then chosen as the file migration candidates and are moved to the *Cold* Zone. While the reactive Data Migration Policy leads to huge energy savings[36], it cannot really foretell the future and may result in inaccurately classifying files with intermittent periods of dormancy as cold files and may prematurely migrate them to the *Cold* zone. The premature movement of the files will result in an increase in the accesses to the *Cold* zone servers resulting in server wakeups and hence, lowered energy savings. The Predictive GreenHDFS, covered in next Section, overcomes this weakness of the Reactive GreenHDFS courtesy of its ability to predict the future from the past.

Algorithm 1. Reactive File Migration Policy

```
//For every file i in Hot Zone
for i = 1 to n do
    Coldness_i ⇐ current_time − last_access_time_i
    if Coldness_i ≥ Threshold_FMP then
        {Cold Zone} ⇐ {Cold Zone} ∪ {f_i}//file migrates to the Cold zone
        {Hot Zone} ⇐ {Hot Zone} / {f_i}//file is removed from the Hot zone
    end if
end for
```

13.5.1.2 Predictive GreenHDFS

The daily incoming log data (such as the clickstream web logs) and aggregated results of the data analysis performed on these logs are placed in very well-defined and time-differentiated directory hierarchies in a production Hadoop cluster at Yahoo!. It is observed that the file attributes (file size, file life span and file heat) are very strongly, and statistically associated with the hierarchical directory structure in which the files are organized in that cluster. Thus, the hierarchical directory structure provides strong, *implicit* hints about the file attributes to the underlying storage system which can be utilized in data management decisions. Similar well-defined, hierarchical directory layouts are in place at Facebook[16], at Europe's largest ad targeting platform Nugg.ad [33], and in other log processing production environments.

Predictive GreenHDFS uses a supervised machine learning technique to learn the correlation between the directory hierarchy and file attributes using historical data from the production cluster [69]. The predictor then utilizes the learning acquired during the training phase to predict various attributes of a new file as it is created, such as the new file's life span evolution. The life span attribute of interest is $Lifespan_{CLR}$ (in days), which is the life span between the file creation and last read.

At every new file creation, the file's $Lifespan_{CLR}$ is predicted and anticipated migration time of the file is calculated as a day more than the predicted $Lifespan_{CLR}$ value and is inserted into a cluster-wide *MigrateInfo* datastructure. These per-file predictions are then used to guide migration in a fine-grained, scalable and self-adaptive manner. The predictive file migration policy, shown in Algorithm 2, is invoked everyday to migrate files to the *Cold* zone.

Algorithm 2. Predictive File Migration Policy

//Lookup Policy Run Date's entry in the cluster-wide *MigrateInfo* Data Structure. *MigrateInfo* is populated at every new file creation.
$\{Migration\ List\} = Lookup(MigrateInfo, Policy_{RunDate})$
//Every file i in Migration List has $Migration_{Time} == Policy_{RunDate}$
//For every file i in Migration List
for $i = 1$ to n **do**
    $\{Cold\ Zone\} \Leftarrow \{Cold\ Zone\} \cup \{f_i\}$//file migrated to the Cold zone
    $\{Hot\ Zone\} \Leftarrow \{Hot\ Zone\}\ /\ \{f_i\}$//file is removed from the Hot zone
end for

13.5.1.3 Thermal-Aware GreenHDFS

A typical data center is laid out with a hot-aisle/cold-aisle arrangement by installing the racks and perforated floor tiles in the raised floor [58]. Computer Room Air Conditioner (CRAC) delivers cold air under the elevated floor. The cool air enters the racks from their front side, picks up heat while flowing through these racks, and exits from the rear of the racks. The heated exit air forms hot aisles behind the racks, and is extracted back to the air conditioner intakes, which, in most cases, are positioned above the hot aisles. Each rack consists of several chassis, and each chassis accommodates several computational devices (servers or networking equipment).

The goal of a cooling subsystem is to maintain the inlet temperatures within a safe operating range below the redline temperature. However, due to the complex nature of airflow inside data centers, some of the hot air from the outlets of the servers recirculates into the inlets of other servers. The recirculated hot air mixes with the supplied cold air and causes inlets of some of the servers in the data center to experience a rise in inlet temperature. This reduces the cooling-efficiency of these servers and results in an increase in their outlet exhaust temperature which ultimately results in an overall higher temperature in the data center. This makes the cooling subsystem work harder to reduce the temperature in the data center, resulting in higher cooling costs. Furthermore, the CRAC units vary in their ability to cool different places in a data center (e.g., a corner of the room, farthest from the CRAC), and further aid in thermal hot spot creation [9].

Such an uneven thermal profile and thermal hotspots are a ubiquitous issue in data centers. To offset the recirculation problem, in majority of the real-world datacenters, the temperature of the air supplied by the cooling system is set much lower than the redline temperature, low enough to bring all the inlet temperatures well below the redline threshold. Unfortunately, operating the air conditioning units (CRACs) at lower temperature reduces their coefficient of efficiency (COP)[6] and results in higher cooling costs.

Thermally-aware GreenHDFS is cognizant of the significant differences in the thermal-efficiency (i.e., cooling-efficiency) of the servers; i.e., some locations in the data center can dissipate heat better than the others [49].

Since, servers in the *Hot* zone are the target of majority of the computations in the compute cluster, consume significant energy and generate substantial heat. It becomes imperative to efficiently cool such servers to ensure that the servers don't exceed the redline temperature as exceeding the redline temperature runs the risk of damage to the server components and results in decrease in reliability.

Thermal-Aware GreenHDFS assigns thermally-efficient (i.e., cooling-efficient) servers to the *Hot* zone to ensure that computations happen on thermally-efficient servers which are capable of maximal heat reduction. Thermal-Aware GreenHDFS reduces the maximum exhaust temperature in the cluster, which in turn results in reducing the overall cooling costs of the cluster substantially. GreenHDFS assigns thermally-inefficient (i.e., cooling-inefficient) servers to the *Cold* zone. *Cold* zone servers in GreenHDFS contain unpopular data with infrequent data accesses and due to the data-locality consideration, do not have computations targeted to them. Thus, these servers experience significant idleness and as a result, consume significantly less power (power consumed at idle utilization is typically 40% of the peak power) and generate much less heat. Hence, using thermally-inefficient servers' inefficient cooling efficiency does not impact the *Cold* zone servers exhaust temperature.

To assist thermally-aware GreenHDFS in its cooling-aware server allocation, the cluster administrator needs to run an initial calibration phase. In the calibration run, the thermal-profile of the cluster is created by measuring the inlet temperature of each and every server in the cluster. Inlet temperature gives the best indication of the thermal-efficiency (i.e., cooling-efficiency) of a server. Higher the inlet temperature, lower is the thermal-efficiency and hence, lower is the cooling-efficiency of the server.

13.5.2 Evaluation

GreenHDFS is evaluated using a trace-driven simulator and floVENT [1], a computational fluid dynamics simulator. The trace-driven simulator is driven by real-world HDFS traces generated by a production Hadoop cluster (2600 servers with 34 million files in the namespace and a used capacity of 5 Petabytes) at Yahoo!. The GreenHDFS simulator is implemented in Java and MySQL distribution 5.1.41 and executes using Java 2 SDK, version 1.6.0-17. The servers in the Hadoop cluster are assumed to have Quad-Core Intel Xeon X5400 Processor [34], Intel Atom

[6] COP is the coefficient of performance of the cooling device. COP characterizes the efficiency of an air conditioner system; it is defined as the ratio of the amount of heat removed by the cooling device to the energy consumed by the cooling device. Thus, work required to remove heat is inversely proportional to the COP.

Z560 Processor [70] [34], Seagate Barracuda ES.2 which is a 1TB SATA hard drive (HDD) [3], and DRAM DIMM [46] [36] and both, performance and energy statistics in the simulator are calculated based on the information extracted from their datasheets. A typical Hot zone server (2 Quad-Core Intel Xeon X5400 Processor, 8 DRAM DIMM, 4 1TB HDD) consumes 445.34W peak power, 132.46W in idle mode and 13.16W in sleep mode. A Cold zone server (1 Intel Atom Z560 Processor, 2 DRAM DIMM, 12 1TB HDD) consumes 207.1W peak power, 151.79W idle power and 13.09W power in the sleep mode.

13.5.2.1 Evaluation Dataset

Log processing is one of the most popular use cases of data-intensive computing [13]. In log processing, a log of events, such as clickstream, log of phone call records, email logs [14] or even a sequence of transactions, is continuously collected and stored in the cluster. MapReduce is then used to compute various statistics and derive business insights from the data. Internet services companies such as Facebook, Google, Twitter, LinkedIn, and Yahoo! [13] rely on clickstream processing [59, 26, 13], an example of log processing, to calculate the web-based advertising revenues (source of majority of their income), and to derive user interest models and predictions. For this, daily huge web logs are analyzed in the production environments. For example, every day 60-90TB uncompressed raw log data is loaded into Hadoop at Facebook [61], 100+GB log data is loaded at Orbitz [55], and at Europe's largest ad targeting company [33].

In the production Hadoop cluster at Yahoo!, the clickstream data amounts to 60% of the total used storage capacity of 5 petabytes. The GreenHDFS analysis is done using the clickstream dataset given its monetizaton value and sheer size. To make the cluster size proportional to the dataset size, 60% of the total cluster nodes (i.e., 1560 nodes) are used in the simulator. 70% (i.e., 1170) servers are assigned to the *Hot* zone and 30% (i.e., 390) servers are assigned to the *Cold* zone.

13.5.2.2 Evaluation Results

Energy Savings: GreenHDFS results in a 26% reduction in energy consumption of a 1560 server cluster with 80% capacity utilization. The cost of electricity is assumed to be $0.063/KWh. Extrapolating, upwards of 2 million can be saved in the energy costs if GreenHDFS technique is applied to all the Hadoop clusters at Yahoo! (upwards of 38000 servers). Energy savings from off-power servers will be further compounded in the cooling system of a real datacenter. For every Watt of power consumed by the compute infrastructure, a modern data center expends another one-half to one Watt to power the cooling infrastructure [51].

Performance Impact: In Reactive GreenHDFS, 97.8% of the total read requests are not impacted by the power management. Impact is seen only by 2.1% of the reads. The predictive GreenHDFS, on the other hand, is able to completely cut down the accesses to the *Cold* zone and thus, incurs no performance penalty.

Power State Transitions: The maximum number of power state transitions incurred by a server in a one-month simulation run is just 11 times and only 1 server out of the 390 servers provisioned in the *Cold* zone exhibited this behavior. Most of the disks are designed for a maximum service life time of 5 years and can tolerate 50,000 start/stop cycles. Given the very small number of transitions incurred by a server in the *Cold* zone in a year, GreenHDFS has no risk of exceeding the start/stop cycles during the service life time of the disks.

Cooling Energy Cost Savings: The cooling costs reduce 38% with thermal-aware GreenHDFS by scaling down 90% of the *Cold* zone servers. The Hadoop cluster shows an overall temperature cool-down ranging from 2.58^{o}C to 6°C. This will allow the CRACs to be potentially run with an increased supply temperature. Operating the CRAC at 5°C higher supply temperature of 20°C would result in increasing CRAC's coefficient of efficiency (COP) from 5 to 6 [37]. This increase in the COP results in an additional 13% cooling costs savings and increases the cooling cost savings from 38% to 51%.

Data Migrations: Every day, on average 6.38TB worth of data and 28.9 thousand files are migrated to the *Cold* zone. Since, we have assumed storage-heavy servers in the *Cold* zone where each server has 12, 1TB disks, assuming 80MB/sec of usable disk bandwidth, 6.38TB data can be absorbed in less than 2hrs by one server. The migration policy can be run during off-peak hours to minimize any performance impact. The migration of cold data frees up valuable storage capacity on the *Hot* zone which can be used for better data placement of the hot data or for aggressive replication of really hot data.

Total Cost of Ownership (TCO) Reduction: The TCO analysis assumes a hybrid Hadoop cluster. Two server types are considered in the TCO calculations: 1) *Hot* zone Server with two Quad core Intel Xeon X5400 processors, 4 1TB SATA drives, and 8 DRAM DIMM and 3) *Cold* Zone Server with one Atom processor, 12 1TB SATA drives, and 2 DRAM DIMM.

GreenHDFS results in significant amount of TCO savings. When 90% of the servers in the *Cold* zone are scaled-down (the most common occurrence based on the evaluation), GreenHDFS is projected to save $59.4 million if its energy-conserving techniques are applied across all the Hadoop clusters at Yahoo! (amounting to 38000 servers). The 3-yr TCO of the 38000 servers is $308,615,249.

13.5.2.3 Discussion on GreenHDFS

GreenHDFS is able to meet all effective scale-down mandates (i.e., generates significant idleness, results in few power state transitions, and does not degrade read data access performance) despite the significant challenges posed by a Hadoop compute cluster to scale-down. All the newly created data is written on the *Hot* zone. Since, *Hot* zone has the largest share of servers in the cluster and is performance-driven, newly created data is written at high speed. Scale-down allows GreenHDFS to save 26% of the server operating energy costs.

Thermally-aware variant of GreenHDFS further aids in reducing the cooling energy costs significantly by 1) Reducing the amount of the heat generated in the system. One way is to reduce the server power consumption so that the CRAC needs to extract and cool down less amount of heat, 2) Increasing the temperature of the air supplied by the CRAC (of course, while ensuring no server exceeds the redline temperature. This increases the COP, and allows CRAC to operate with higher efficiency, 3) Lowering the maximum exhaust temperature. This reduces the temperature at the inlet of the CRAC. Since, cooling costs are proportional to the difference in the temperature between the inlet of the CRAC and the temperature of the air supplied by the CRAC, lower temperature results in reduced cooling costs, and 4) Reducing hot air recirculation and by maintaining more uniform thermal profile in the data center. Thus, GreenHDFS is able to reduce the *overall* (i.e., both server and cooling) energy costs in the Hadoop clusters.

13.6 Conclusion

Data-intensive computing is growing by leaps-and-bound given the explosion in Big Data and has myriad use cases. MapReduce offers an extremely scalable, shared-nothing, high performance and simplified Big data processing framework on large-scale, commodity server clusters. Scale-down is an attractive technique to conserve energy and also allows energy proportionality with non-energy-proportional components such as the disks. Unfortunately, the very features such as data locality, fine-grained load-balancing, complicate scale-down in such large-scale compute clusters. This chapter covered the energy-management techniques that have proposed especially for data-intensive MapReduce based Hadoop compute clusters. It also presented a case-study of GreenHDFS that relies on a data-classification driven energy-aware data placement and a hybrid multi-zoned cluster layout to conserve energy in Hadoop clusters.

References

1. floVENT: A Computational Fluid Dynamics Simulator, `http://www.mentor.com/products/mechanical/products/flovent`
2. IDC, `http://www.idc.com`
3. Seagate, ES.2 (2008), `http://www.seagate.com/staticfiles/support/disc/manuals/NL35Series&BCESSeries/BarracudaES.2Series/100468393e.pdf`
4. Eric Baldeschwieler, Hadoop Summit (2010)
5. Amur, H., Cipar, J., Gupta, V., Ganger, G.R., Kozuch, M.A., Schwan, K.: Robust and Flexible Power-Proportional Storage. In: Proceedings of the 1st ACM Symposium on Cloud Computing, SoCC 2010, pp. 217–228. ACM, New York (2010)
6. Ananthanarayanan, G., Katz, R.H.: Greening the Switch. Technical Report UCB/EECS-2008-114, EECS Department, University of California, Berkeley (September 2008)
7. Hadoop, `http://hadoop.apache.org/`

8. Barroso, L.A., Hölzle, U.: The Case for Energy-Proportional Computing. Computer 40(12) (2007)
9. Bash, C., Forman, G.: Cool Job Allocation: Measuring the Power Savings of Placing Jobs at Cooling-Efficient Locations in the Data Center. In: Xiao, B., Yang, L.T., Ma, J., Muller-Schloer, C., Hua, Y. (eds.) ATC 2007. LNCS, vol. 4610, pp. 363–368. Springer, Heidelberg (2007)
10. Belady, C.: In the Data Center, Power and Cooling Costs more than the IT Equipment it Supports (February 2010)
11. Bell, G., Hey, T., Szalay, A.: Beyond the data deluge
12. Bigdata (2010), http://en.wikipedia.org/wiki/Big_data
13. Blanas, S., Patel, J.M., Ercegovac, V., Rao, J., Shekita, E.J., Tian, Y.: A Comparison of Join Algorithms for Log Processing in MapReduce. In: Proceedings of the 2010 International Conference on Management of Data, SIGMOD 2010, pp. 975–986. ACM, New York (2010)
14. Boebel, B.: RACKSPACE's MapReduce Usage (2008), http://highscalability.com/how-rackspace-now-uses-mapreduce-and-hadoop-query-terabytes-data
15. Borthakur, D.: Largest Hadoop Cluster at Facebook Personal Communication, http://hadoopblog.blogspot.com/2010/05/facebook-has-worlds-largest-hadoop.html
16. Borthakur, D.:Hierarchical Directory Organization at Facebook. Personal Communication (2011)
17. Brill, K.G.: Data Center Energy Efficiency and Productivity (2007)
18. Brill, K.G.: The Invisible Crisis in the Data Center: The Economic Meltdown of Moore Law (2007)
19. Carns, P.H., Ligon III, W.B., Ross, R.B., Thakur, R.: PVFS: A Parallel File System for Linux Clusters. In: ALS 2000: Proceedings of the 4th Annual Linux Showcase & Conference, pp. 28–28. USENIX Association, Berkeley, CA, USA (2000)
20. Chase, J.S., Anderson, D.C., Thakar, P.N., Vahdat, A.M., Doyle, R.P.: Managing Energy and Server Resources in Hosting Centers. In: Proceedings of the Eighteenth ACM Symposium on Operating Systems Principles, SOSP 2001, pp. 103–116. ACM, New York (2001)
21. Chen, G., He, W., Liu, J., Nath, S., Rigas, L., Xiao, L., Zhao, F.: Energy-Aware Server Provisioning and Load Dispatching for Connection-Intensive Internet Services. In: Proceedings of the 5th USENIX Symposium on Networked Systems Design and Implementation, NSDI 2008, pp. 337–350. USENIX Association, Berkeley, CA, USA (2008)
22. Chen, S., Joshi, K.R., Hiltunen, M.A., Schlichting, R.D., Sanders, W.H.: CPU Gradients: Performance-Aware Energy Conservation in Multitier Systems. In: International Conference on Green Computing, pp. 15–29 (2010)
23. Chen, Y., Das, A., Qin, W., Sivasubramaniam, A., Wang, Q., Gautam, N.: Managing Server Energy and Operational Costs in Hosting Centers. In: Proceedings of the 2005 ACM SIGMETRICS International Conference on Measurement and Modeling of Computer Systems, SIGMETRICS 2005, pp. 303–314. ACM, New York (2005)
24. Data-intensive computing, http://research.yahoo.com/files/BryantDISC.pdf
25. Dean, J., Ghemawat, S., Inc, G.: MapReduce: Simplified Data Processing on Large Clusters. In: Proceedings of the 6th Conference on Symposium on Operating Systems Design and Implementation, OSDI 2004, USENIX Association (2004)
26. Department, P.C., Chatterjee, P., Hoffman, D.L., Novak, T.P.: Modeling the Clickstream: Implications for Web-Based Advertising Efforts. Marketing Science 22, 520–541 (2000)

27. EPA. EPA Report on Server and Data Center Energy Efficiency. Technical report, U.S. Environmental Protection Agency (2007)
28. Ghemawat, S., Gobioff, H., Leung, S.-T.: The Google File System. In: Proceedings of the Nineteenth ACM Symposium on Operating Systems Principles, SOSP 2003, pp. 29–43. ACM, New York (2003)
29. Google. Official Google Blog, http://googleblog.blogspot.com/2008/11/sorting-1pb-with-mapreduce.html
30. Greenberg, S., Mills, E., Tschudi, B., Rumsey, P., Myatt, B.: Best Practices for Data Centers: Results from Benchmarking 22 Data Centers. In: Proceedings of the ACEEE Summer Study on Energy Efficiency in Buildings (2006)
31. Hamilton, J.: Cooperative Expendable Micro-Slice Servers (CEMS): Low Cost, Low Power Servers for Internet-Scale Services (2009)
32. Horvath, T., Abdelzaher, T., Skadron, K., Member, S., Liu, X.: Dynamic Voltage Scaling in Multitier Web Servers with End-to-End Delay Control. IEEE Transactions on Computers (2007)
33. Hutton, R.: Hadoop in Europe (2010), http://www.cloudera.com/blog/2010/03/why-europes-largest-ad-targeting-platform-uses-hadoop
34. Intel: Quad-Core Intel Xeon Processor 5400 Series (2008), http://www.intel.com/Assets/en_US/PDF/prodbrief/xeon-5400-animated.pdf
35. Kaushik, R.T., Bhandarkar, M.: GreenHDFS: Towards an Energy-Conserving, Storage-Efficient, Hybrid Hadoop Compute Cluster. In: Proceedings of the 2010 International Conference on Power-Aware Computing and Systems, HotPower 2010, pp. 1–9. USENIX Association, Berkrley (2010)
36. Kaushik, R.T., Bhandarkar, M., Nahrstedt, K.: Evaluation and Analysis of GreenHDFS: A Self-Adaptive, Energy-Conserving Variant of the Hadoop Distributed File System. In: Proceedings of the 2nd IEEE International Conference on Cloud Computing Technology and Science (2010)
37. Kaushik, R.T., Nahrstedt, K.: Thermal-Aware GreenHDFS. Technical Report, University of Illinois, Urbana-Champaign (2011)
38. Konstantin, S., Kuang, H., Radia, S., Chansler, R.: The Hadoop Distributed File System. In: MSST (2010)
39. Lang, W., Patel, J.M.: Energy Management for MapReduce Clusters. In: Proceedings VLDB Endow., vol. 3, pp. 129–139 (2010)
40. Lang, W., Patel, J.M., Shankar, S.: Wimpy Node Clusters: What about Non-Wimpy Workloads? In: Proceedings of the Sixth International Workshop on Data Management on New Hardware, DaMoN 2010, pp. 47–55. ACM, New York (September 2010)
41. Leverich, J., Kozyrakis, C.: On the Energy (In)efficiency of Hadoop Clusters. SIGOPS Operating Systems Review 44, 61–65 (2010)
42. Lim, K., Ranganathan, P., Chang, J., Patel, C., Mudge, T., Reinhardt, S.: Understanding and Designing New Server Architectures for Emerging Warehouse-Computing Environments. In: Proceedings of the 35th Annual International Symposium on Computer Architecture, ISCA 2008, pp. 315–326. IEEE Computer Society, Washington, DC, USA (2008)
43. Lustre (2010), http://www.lustre.org
44. Meisner, D., Gold, B.T., Wenisch, T.F.: PowerNap: Eliminating Server Idle Power. In: Proceeding of the 14th International Conference on Architectural Support for Programming Languages and Operating Systems, ASPLOS 2009, pp. 205–216. ACM, New York (2009)

45. Merkel, A., Bellosa, F.: Memory-Aware Scheduling for Energy Efficiency on Multicore Pocessors. In: Proceedings of the 2008 Conference on Power Aware Computing and Systems, HotPower 2008, pp. 1–1. USENIX Association, Berkeley, CA, USA (2008)
46. Micron. DDR2 SDRAM SODIMM (2004)
47. Microsoft. Marlowe (2009), http://bits.blogs.nytimes.com/2009/02/24/microsoft-studies-the-big-sleep/
48. Moore, J., Chase, J., Ranganathan, P., Sharma, R.: Making Scheduling Cool: Temperature-Aware Workload Placement in Data Centers. In: Proceedings of the Annual Conference on USENIX Annual Technical Conference, ATEC 2005, pp. 5–5. USENIX Association, Berkeley, CA, USA (2005)
49. Moore, J., Sharma, R., Shih, R., Chase, J., Patel, R., Ranganathan, P., Labs, H.P.: Going Beyond CPUs: The Potential of Temperature-Aware Solutions for the Data Center. In: Proceedings of the Workshop of Temperature-Aware Computer Systems (TACS-1) held at ISCA (2004)
50. Narayanan, D., Donnelly, A., Rowstron, A.: Write Off-Loading: Practical Power Management for Enterprise Storage. ACM Transaction of Storage 4(3), 1–23 (2008)
51. Patel, C., Bash, E., Sharma, R., Beitelmal, M.: Smart Cooling of Data Centers. In: Proceedings of PacificRim/ASME International Electronics Packaging Technical Conference and Exhibition, IPACK 2003 (2003)
52. Patel, C.D., Shah, A.J.: Cost Model for Planning, Development and Operation of a Data Center. Technical report, HP Labs (2005)
53. Pawlowski, B., Juszczak, C., Staubach, P., Smith, C., Lebel, D., Hitz, D.: NFS version 3: Design and Implementation. In: Proceedings of the Summer 1994 USENIX Technical Conference, pp. 137–151 (1994)
54. Riska, A., Mi, N., Casale, G., Smirni, E.: Feasibility Regions: Exploiting Tradeoffs between Power and Performance in Disk Drives. In: Second Workshop on Hot Topics in Measurement and Modeling of Computer Systems (HotMetrics 2009), Seattle, WA (2009)
55. Seidman, J.: Hadoop at Orbitz (2010), http://www.cloudera.com/resource/hw10_hadoop_and_hive_at_orbitz
56. Sharma, N., Barker, S., Irwin, D., Shenoy, P.: Blink: Managing Server Clusters on Intermittent Power. In: Proceedings of the Sixteenth International Conference on Architectural Support for Programming Languages and Operating Systems, ASPLOS 2011, pp. 185–198. ACM, New York (2011)
57. Sharma, R.K., Bash, C.E., Patel, C.D., Friedrich, R.J., Chase, J.S.: Balance of Power: Dynamic Thermal Management for Internet Data Centers. IEEE Internet Computing 9, 42–49 (2005)
58. Sullivan, R.: Alternating Cold and Hot Aisles Provides more Reliable Cooling for Server Farms (2000)
59. Sweiger, M., Madsen, M.R., Langston, J., Lombard, H.: Clickstream Data Warehousing. John Wiley and Sons, Inc, Chichester (2002)
60. Szalay, A.S., Bell, G.C., Huang, H.H., Terzis, A., White, A.: Low-Power Amdahl-Balanced Blades for Data Intensive Computing. SIGOPS Operating Systems Review 44, 71–75 (2010)
61. Thusoo, A., Shao, Z., Anthony, S., Borthakur, D., Jain, N., Sen Sarma, J., Murthy, R., Liu, H.: Data Warehousing and Analytics Infrastructure at Facebook. In: Proceedings of the 2010 International Conference on Management of Data, SIGMOD 2010, pp. 1013–1020. ACM, New York (2010)

62. Tolia, N., Wang, Z., Marwah, M., Bash, C., Ranganathan, P., Zhu, X.: Delivering Energy Proportionality with Non Energy-Proportional Systems: Optimizing the Ensemble. In: Proceedings of the 2008 Conference on Power Aware Computing and Systems, HotPower 2008, Berkeley, CA, USA, pp. 2–2. USENIX Association, Berkeley, CA, USA (2008)
63. List of Companies Using Hadoop, `http://wiki.apache.org/hadoop/PoweredBy`
64. Varma, A., Ganesh, B., Sen, M., Choudhury, S.R., Srinivasan, L., Jacob, B.: A Control-Theoretic Approach to Dynamic Voltage Scheduling. In: Proceedings of the 2003 International Conference on Compilers, Architecture and Synthesis for Embedded Systems, CASES 2003, pp. 255–266. ACM, New York (2003)
65. Vasudevan, V., Franklin, J., Andersen, D., Phanishayee, A., Tan, L., Kaminsky, M., Moraru, I.: Fawndamentally Power-Efficient Clusters. In: Proceedings of the 12th Conference on Hot Topics in Operating Systems, HotOS 2009, pp. 22–22. USENIX Association, Berkeley, CA, USA (2009)
66. Weddle, C., Oldham, M., Qian, J., Wang, A.I.A.: PARAID: The Gear Shifting Power-Aware RAID. In: Proceedings of USENIX Conference on File and Storage Technologies (FAST 2007). USENIX Association (2007)
67. Weil, S.A., Brandt, S.A., Miller, E.L., Long, D.D.E., Maltzahn, C.: Ceph: A Scalable, High-Performance Distributed File System. In: Proceedings of the 7th Conference on Symposium on Operating Systems Design and Implementation, OSDI 2006(2006)
68. White, T.: Hadoop: The Definitive Guide. O'Reilly Media, Sebastopol (2009)
69. Kaushik, R.T., Abdelzaher, T., Nahrstedt, K.: Predictive Data and Energy Management in GreenHDFS. In: Proceedings of International Conference on Green Computing (IGCC). IEEE, Los Alamitos (2011)
70. Intel. Intel Atom Processor Z560, `http://ark.intel.com/Product.aspx?id=49669&processor=Z560&spec-codes=SLH63`

Chapter 14
Energy-Efficient Computing Using Agent-Based Multi-objective Dynamic Optimization

Alexandru-Adrian Tantar, Grégoire Danoy, Pascal Bouvry, and Samee U. Khan

Abstract. Nowadays distributed systems face a new challenge, almost nonexistent a decade ago: energy-efficient computing. Due to the rising environmental and economical concerns and with trends driving operational costs beyond the acquisition ones, *green computing* is of more actuality than never before. The aspects to deal with, e.g. dynamic systems, stochastic models or time-dependent factors, call nonetheless for paradigms combining the expertise of multiple research areas. An agent-based dynamic multi-objective evolutionary algorithm relying on simulation and anticipation mechanisms is presented in this chapter. A first aim consists in addressing several difficult energy-efficiency optimization issues, in a second phase, different open questions being outlined for future research.

14.1 Introduction

High Performance Computing (HPC) evolved over the past three decades into in creasingly complex distributed models. After attaining Gigaflops and Teraflops performance with Cray2 in 1986, respectively with Intel ASCI Red in 1997, the Petaflops barrier was crossed in 2008 with the IBM Roadrunner system [31]. And trends indicate that we should reach Exaflops in the next ten to fifteen years [15]. The shift towards decentralized paradigms raised nonetheless scalability, resilience and, last but not least, energy-efficiency issues [8]. Disregarded or seen as an extraneous factor in the HPC's beginnings, the carbon emissions footprint of data centers escalated to levels comparable to those of highly-developed countries [21]. Estimates place the energy consumption of an Exaflops scale system in the range

Alexandru-Adrian Tantar · Grégoire Danoy · Pascal Bouvry
Faculty of Sciences, Technology and Communication, University of Luxembourg
e-mail: {alexandru.tantar,gregoire.danoy,pascal.bouvry}@uni.lu

Samee U. Khan
North Dakota State University
e-mail: samee.khan@ndsu.edu

of hundreds of megawatts. Thus, not counting the raising environmental concerns, the trend inflicts important economical consequences. The scope of this chapter is therefore focused on this last aspect, i.e. minimizing energy consumption while delivering a high performance level, and addresses several difficult issues in the *energy-efficient dynamic* and *autonomous* management of distributed computing resources. The intricate interplay of factors shaping the problem calls nevertheless for solutions at the crossing of, among others, distributed computing, scheduling and dynamic optimization [26, 16]. Answering part of these aspects, we propose an agent-based dynamic evolutionary multi-objective approach dealing with the time-dependent dynamic and stochastic factors defining a distributed computing environment.

Without entering into details, a few key characteristics of large scale systems, of interest for the aims of this chapter, are introduced hereafter. We first focus on HPC core aspects, gradually extending the discussion to grid and cloud systems. Existing implementations rely on expensive institution-centered high-end computers, clusters or grids, with a high degree of complexity [4, 9]. Moreover, these resources may potentially be shared across administrative domains or multiple geographically-distributed centers. Therefore, as main points, one has to deal with complexity for performance, energy consumption and execution deadlines. The DOE/NNSA/LANL (Los Alamos National Lab) 100 million US$ BladeCenter Cluster, ranked first in Top500 in 2009, was wired through more than 10000 Infiniband and Gigabit Ethernet connections extending over almost 90 kms of fiber optic cable [18]. The system was estimated to deliver, according to IBM, 444.94 Megaflops/watt, ranking seventh in Green500[1]. Over the same period, the first Green500 system, a BladeCenter QS22 Cluster, delivered 536.24 Megaflops/watt. We have therefore two ends, opposing energy and performance, where an improvement in one of them leads to a degradation of the other. And the gap forcing energy requirements and performance apart does not cease to extend with time. At one year distance, in June 2010, the Oak Ridge National Laboratory's Cray XT5-HE was ranked first in Top500 with 2.331 Petaflops of computational power and only 56th in Green500 with 253.07 Megaflops/watt. At the same time, an ascending Flops per Watt tendency can be identified. Under development at IBM and to start running at the Lawrence Livermore National Laboratory (LLNL) in 2012, the Sequoia system, designed to include 1.6 millions of power-processors providing a peak performance of more than 20 Petaflops, will supposedly sustain 3000 Megaflops/watt for an input power of 6 Megawatts [19].

Cloud computing, while still dealing with complexity and energy consumption issues, brings into focus business and additional privacy constraints [1, 28]. As an outline, cloud computing can be described in terms of computational demand and offer where entities (individuals, enterprises, etc.) negotiate and pay for access to resources administered by a different entity that acts as provider. Here, demanding parties may have diverging requirements or preferences, given in contractual terms that stipulate data security, privacy or quality of service levels. Performance

[1] http://www.green500.org

metrics include in this case occupation of resources, users' satisfaction or service level agreement violations [30, 29]. What is more, dynamic and risk-aware pricing policies may apply where predictive models are used either in place or through intermediary brokers to assess the financial and computational impact of decisions taken at different moments in time. Furthermore, legal enforcements may restrict access to resources or data flow, e.g. data crossing borders or transferred to a different resource provider. As common examples, one can refer to Amazon Web Service or Google Apps Cloud Services.

Summarizing, various scenarios have to be dealt with in order to construct an energy-efficient optimization paradigm for distributed computing systems that offers scalability and resilience capabilities in an autonomous, transparent manner. An ideal approach would have to cope with time evolving performance and cost constraints while anticipating the long term effects of the decisions taken at specific moments in time. National security, surveillance and defence applications for which reaction time is critical impose performance as single and only criterion. Sharing computational power among high-priority applications running inside such a system may nonetheless require to make use of strategies which take into account not only the current contextual state but also future possible evolutions. At the opposite end, for the academic and public domains, reducing energy consumption may stand as a main factor [6].

We propose a decentralized, agent-based, dynamic multi-objective anticipative EA. The chapter identifies the main components one deals with in energy-efficient autonomic computing and, through abstraction and conceptualization, advances the idea of a generic application framework. Evolutionary Algorithms (EAs) represent a natural option to consider given their capability of dealing with highly multi-modal functions in both mono and multi-objective cases as well as their notorious success for various applications [13, 7]. Furthermore, the adoption of an agent oriented paradigm is clearly adapted to address the autonomous and decentralized management of energy in distributed systems. As mentioned in [25], multi-agent systems can be used as an approach to the construction of robust, flexible and extensible systems and as a modeling approach. Also, autonomic computing and, therefore, distributed systems management, is a "killer app" for multi-agent systems [11].

The remainder of this chapter is organized as follows. Related work is discussed in Section 14.2, followed by a brief introduction to basic optimization notions in Section 14.3 and a description of the model and concepts later used for experimentation in Section 14.4. Dynamic energy-efficient optimization aspects are discussed in Section 14.5, followed by results in Section 14.6 and conclusions.

14.2 Related Work

In the literature, energy-efficient approaches for large-scale distributed systems are typically divided into two classes: centralized and distributed. Centralized approaches [33] are historically the first ones, which allow close to optimum results but imply scalability and fault-tolerance problems as soon as the number of nodes increases. To answer those limitations, distributed approaches were introduced, which in turn brought new challenges such as ensuring maintainability and global performance.

In [23] Khargharia presented a holistic theoretical framework for autonomic power and performance management exploiting different components (CPU, memory, network and disks) and their interactions. Their autonomic resource management framework optimises the power/performance ratio at each level of the hierarchy using a mathematical approach for which simulation results on static and dynamic scenarios are provided. Similarly, Bennani *et al.* in [27] proposed a hierarchical approach to the data center resource allocation problem using global and local controllers. It uses a prediction model based on mutliclass queuing network models combined with a combinatorial search technique (i.e. the Beam search algorithm). Berral *et al.* in [5] introduced another predictive model for power consumption and performance for task scheduling based on a machine learning approach (i.e. linear regression algorithm). In this approach, the learning algorithm permits to model data for a given management policy whereas in the previous approaches it is used to learn management policies.

Bio-inspired algorithms have also been investigated by Barbagallo *et al.* in [3] who optimized energy in data centers using an autonomic architecture. However this approach is limited to single-objective optimization. Another natural approach to model autonomic power management is the multi-agent paradigm, as demonstrated by Das *et al.* in [10], in which power and performance in a real data center setting is managed by agents autonomously turning on/off servers. This approach also uses reinforcement learning to adapt the agents' management policies.

To summarize, it appears that various distributed approaches dedicated to energy optimization in large-scale distributed systems have already been investigated. However these only considered standalone approaches, such as static mathematical models, single-objective bio-inspired algorithm or prediction models. The contribution proposed in this chapter intends to combine and extend these through an anticipative dynamic multi-objective evolutionary algorithm for agent-based energy-efficient computing.

14.3 Optimization Introductory Notions

For an energy function F, defined over a decision space X and taking values in an objective space Y, $F : X \to Y$, one may consider the $\min_{x \in X} F(x)$ minimization problem. With no restriction for the topics of this chapter, we can additionally assume that F is nonlinear and that $Y \subseteq \mathbb{R}$. If F is continuous and double differentiable, x^+ is considered to be a local optimum point if the following relations stand:

$$\frac{\partial F}{\partial x^+} = 0, \; \frac{\partial^2 F}{\partial^2 x^+} > 0.$$

No straightforward formulation can be given for dynamic functions. Different classes can be designated here, with functions including time-evolving factors or enclosing random variables, functions depending on past states, etc. For the general case and assuming a minimization context, the goal is to identify a sequence of solutions $\mathbf{x}(t)$, i.e. solution $\mathbf{x}$ to be applied at the t time moment, with $t \in [t_0, t^{end}]$ leading to the following:

$$\min_{\mathbf{x}(t)} \int_{t_0}^{t^{\text{end}}} F(\mathbf{x}(t), t)\,\mathrm{dt}$$

In a more explicit form, this implies providing a sequence of solutions which, for the given time interval, minimize the cumulative objective function obtained by integration (for discrete time intervals the problem can be formulated by summing over all time moments). Assuming $\mathbf{x}(t)$ to be the sequence that minimizes the above relation, for a different sequence $\mathbf{x}^{\triangle}(t)$, with $t \in [t_0, t^{end}]$, the following stands:

$$\min_{\mathbf{x}(t)} \int_{t_0}^{t^{\text{end}}} F(\mathbf{x}(t), t)\,\mathrm{dt} \leq \min_{\mathbf{x}^{\triangle}(t)} \int_{t_0}^{t^{\text{end}}} F(\mathbf{x}^{\triangle}(t), t)\,\mathrm{dt}$$

For the multi-objective case, F is extended to define a vector of objective functions $F: X \to \mathbb{R}^k, \quad F(x) = [f_1(x), \ldots, f_k(x)]$. The set $X \subset \mathbb{R}^d$ of all the *feasible* solutions defines the *decision space*, while the function F maps *feasible solutions* into an *objective space*. In addition, the Pareto optimality concept is used, based on partial order relations defined as follows.

Definition 1. Let $v, w \in \mathbb{R}^k$. We say that the vector v is *less than* w ($v <_p w$), if $v_i < w_i$ for all $i \in \{1, \ldots, k\}$. The relation $\leq_p$ is defined analogously.

Definition 2 (Dominance). A point $y \in X$ is *dominated* by $x \in X$ ($x \prec y$) if $F(x) \leq_p F(y)$ and if $\exists i \in \{1, ..., k\}$ such that $f_i(x) < f_i(y)$. Otherwise y is called non-dominated by x.

Definition 3. A point $x \in X$ is called a *Pareto* point if there is no $y \in X$ which dominates x. The set of all Pareto solutions forms the *Pareto Set*.

In the multi-objective case, by extension, a scalarization function, given in explicit or implicit form, e.g. weighted sum of the objective functions, preference or expert based decisions, has to be used at each time step for selecting solutions out of the approximate Pareto front. Note that due to the dynamic nature of the problem *one solution and only one has to be selected and applied* per time step – this does not represent an option. The modeled problem can occupy a single state at a given time moment and no reset to a previous state can be done. In this case minimization is considered over a vector of cumulative objective function realizations and can be defined to obey Pareto optimality laws. In addition, the semantics of the minimization operator may be subject to context, e.g. deviation from a specified target. As a general model, the problem can be therefore stated as follows:

$$\min_{\mathbf{x}(t)} \left\{ \int_{t_0}^{t^{\text{end}}} f_i(\mathbf{x}(t),t)\,\text{dt} \right\}_{1\leq i\leq k} = \min_{\mathbf{x}(t)} \left\{ \int_{t_0}^{t^{\text{end}}} f_1(\mathbf{x}(t),t)\,\text{dt},\ \ldots, \int_{t_0}^{t^{\text{end}}} f_k(\mathbf{x}(t),t)\,\text{dt} \right\}$$

Last but not least, if solutions are dependent, e.g. with $x(t) = H(x(t-c))$, where $H(\cdot)$ is an arbitrary function and $c < t$, anticipation has to be considered. If only the current *context* is regarded, a solution minimizing the function F at the moment t is optimal, but this may not stand for the complete set of following solutions, i.e. a suboptimal overall fitness is attained. The function F must be consequently optimized by taking simultaneously into account the complete set of solutions (over time). A more detailed discussion is given in Section 14.5 also introducing algorithm and simulation related notions like strategy or scenario [6].

14.4 An Anticipative Decentralized Oriented Dynamic Multi-objective Evolutionary Approach

A framework for an anticipative decentralized dynamic multi-objective evolutionary optimization of energy has been designed for simulation purposes. Modeled using the agent paradigm, an example of the instantiation of the framework is illustrated in Fig. 14.1, with one agent assigned per computing node.

The *General Scheduler* assigns tasks to the different available nodes, i.e. in this case through the corresponding *Local Schedulers*. In addition, each node is provided to enclose decision mechanisms capable of autonomously managing local energy-efficiency with respect to the overall or partially observed states of the system. Nodes are independently controlled through dynamic frequency and voltage scaling, directly driven by local parameters such as local system load (SL), node overhead (NO), idle time (IT) or current node state (CS), and by global indicators which describe system load (SL), operation cost (OC) and energy vs performance ratio (EvP). The scaling is also indirectly driven by the type and rate of loaded and offloaded tasks exchanged among nodes. Note that although exclusive use of simulation is made in this work, the passage to a real-world environment would still require having simulation mechanisms in order to be able to carry anticipation over the future states of the system.

The next two subsections describe the computing environment in terms of tasks attributes, e.g. life-time and states, followed by computing nodes properties. The local scheduler agent is detailed in the last subsection.

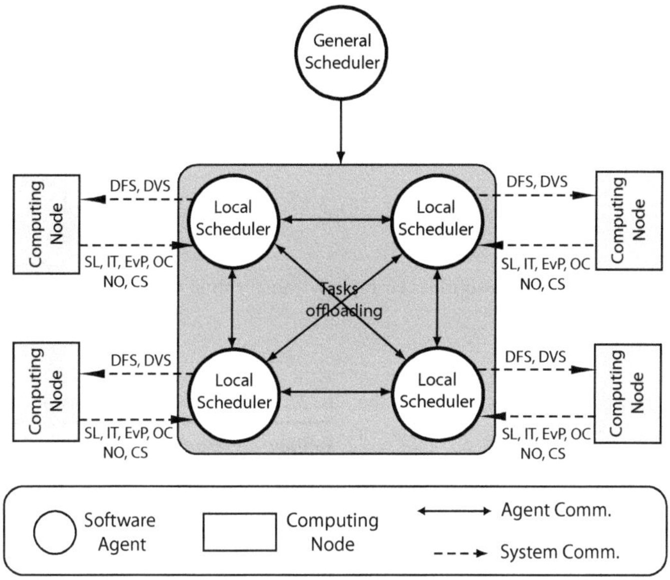

Fig. 14.1 Example of a *per node* system instance.

14.4.1 Computing vs. Communication Intensive Tasks

For a realistic modelling of the execution environment, the system is analyzed by considering computing and communication intensive tasks. We mainly focus on studying the behaviour of the simulated system under intense computational load and in the presence of communication or memory oriented tasks. Different scenarios can be thus envisaged where distinct policies apply, e.g. computing intensive tasks being assigned to independent machines, or where clusters of jobs are dynamically created in order to maximize occupancy and resource utilisation. As a side note, a heterogeneous environment has been preferred to a system running tasks with both computing and communication traits exclusively in order to study extreme situations. Finally, energy consumption is assessed by relying on load indicators with no direct account for communication.

The specifications defining a task are given in either invariant form, fixed for the entire lifetime of the task, or dynamic, subject to random variation at execution time. The first category includes due time, i.e. the latest point in time by which the task has to be completed, and an offloading flag. Migrating idle virtual machines to a secondary server may be coherent with load-balancing policies while offloading an application which depends on real-time processing or that requires access to security critical databases is not desirable. The offloading flag therefore serves for marking tasks which can not be offloaded once assigned to a specific resource. Estimations of the required execution time, computational load and average network transfer time fall in the second category. In addition, task completion and, if the case, transfer

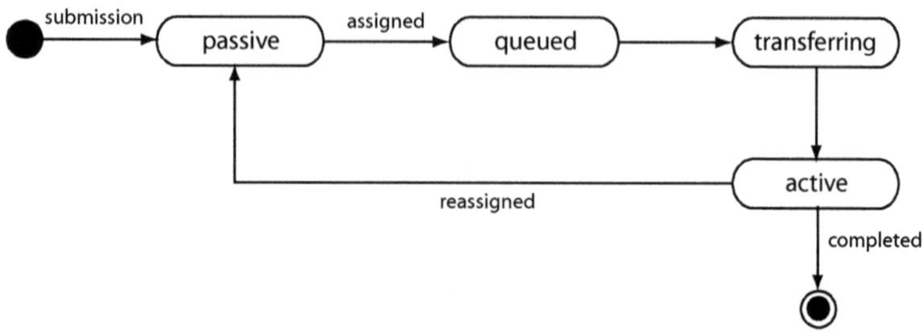

Fig. 14.2 Task life cycle – reassignments imply a transfer time overhead leading to potential delay penalties.

Table 14.1 Power states (p-states) for the Intel®Pentium®M processor (A) and a stock 3.2 GHz, 130W processor (B) with afferent frequencies, voltage and approximate power requirement.

A: Intel®Pentium®M

P-State	Frequency	Voltage	Power
P0	1.6 GHz	1.484 V	24.5 Watts
P1	1.4 GHz	1.420 V	17 Watts
P2	1.2 GHz	1.276 V	13 Watts
P3	1.0 GHz	1.164 V	10 Watts
P4	800 MHz	1.036 V	8 Watts
P5	600 MHz	0.956 V	6 Watts

B: 3.2 GHz, 130W stock processor

P-State	Frequency	Voltage	Power
P0	3.2 GHz	1.45 V	130 Watts
P1	3.06 GHz	1.425 V	122 Watts
P2	3.0 GHz	1.4 V	116 Watts
P3	2.8 GHz	1.375 V	107 Watts
P4	2.66 GHz	1.35 V	100 Watts
P5	2.6 GHz	1.325 V	94 Watts
P6	2.4 GHz	1.3 V	85 Watts
P7	2.2 GHz	1.275 V	77 Watts
P8	2.0 GHz	1.25 V	68 Watts
P9	1.8 GHz	1.175 V	55 Watts
P10	1.7 GHz	1.15 V	50 Watts
P11	1.5 GHz	1.125 V	42 Watts
P12	1.4 GHz	1.1 V	38 Watts

percentage indicators can be observed at run-time. Please note that none of these dynamic factors are seen as an *a priori* information, e.g. as for use with scheduling algorithms, being available for the purpose of simulation only.

The complete life cycle of a task can be described by several different states. A schematic representation is given in Fig. 14.2. After submission, the task is placed in a passive state, waiting to be assigned. Active state is reached after all afferent data is transferred to the target resource and processing starts. Subsequently, depending on the nature of the task and the employed optimization strategy, multiple reassignments may occur. Before migrating to a different resource, a task is placed back into a passive state, subsequently moving to queued and transferring states. Note that all transitions requiring to move a task from a prior passive state to an active one, imply a network transfer time overhead. As a last aspect, the explicit failure of a node leads to all assigned tasks to be reinitialised and queued back into the system, i.e. any derived results or application states are considered erroneous.

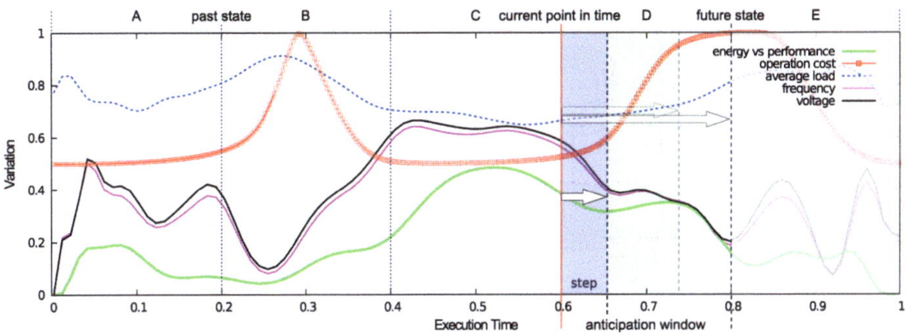

Fig. 14.3 Example of voltage and frequency scaling for a single node – normalized average values. A specified performance target (traced in the lower part of the figure) and operation cost direct the dynamics of the node. Other factors like overall system load or network traffic (not drawn here) may intervene.

14.4.2 Environment and System Nodes

The computational environment is defined over a collection of abstract nodes which can scale to approximate the behavior of a single processing core or that of hierarchical resource, referred to as "Computing Node" in Fig. 14.1. As detailed hereafter, we focus on the interaction connecting external factors, provided in a descriptive, informative form, with state indexes, internal to nodes and imperative in nature. No information is known at system level with respect to the functioning of the nodes, assumed nonetheless autonomous and capable of transiting different performance states. An example is given in Table 14.1, where, for the Intel®Pentium®M 24.5W processor, the frequency and voltage associated to each state is given along with the resulting power consumption [20]. For simulation and experimentation purposes we also rely on a 3.2 GHz, 130W stock processor, not discussed in detail here. All nodes are in addition subject to a system-wide synchronization, ensuring execution context coherency over time-dependent factors.

For the purpose of this study we consider that nodes define interconnected single processing units. Different techniques may apply at core level, including Dynamic Frequency Scaling (DFS) and Dynamic Voltage Scaling (DVS) [14, 22]. Extrapolating to clusters, Dynamic Power Management (DPM) schemes can be implemented or machines can be dynamically assigned to different states or roles, e.g. communication front-ends, dispatching or storage machines. A graphical representation capturing the evolution of a single node over a one-day span, normalized as an execution time varying from 0.0 to 1.0, is given in Fig. 14.3 (discussed in more detail later in this section). Energy vs performance and operation cost identify the balance to ensure between consumption and computational performance, respectively the cost associated with the operation of the node. Low energy vs performance values imply a request for minimizing consumption whereas values close to 1.0 demand performance. Definite semantics can be formulated upon the incurred delay, emergency level or load percentage, where high values lead to an increased

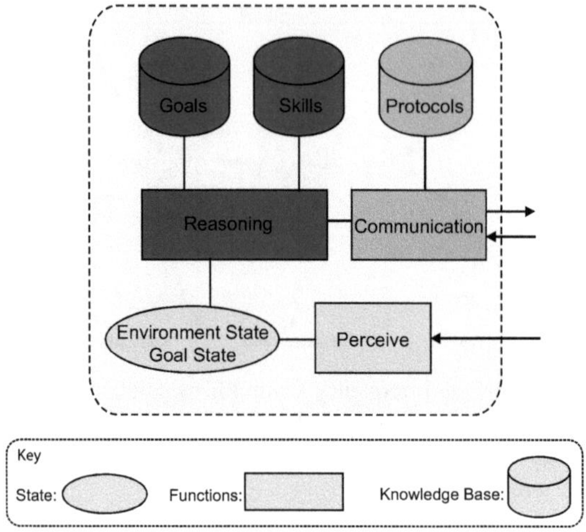

Fig. 14.4 A conceptual basic view of a node's architecture.

performance demand. Analogously, cost can be associated to direct financial indicators, e.g. variation expressed as a function of operational expenses. Alternative definitions may relate to carbon dioxide emissions footprint, energy supply levels (low battery translating into high costs), or environmental temperature where, for example, thermal sensors placed at different locations in a center room may limit execution from entering into a hazardous functioning condition. For the herein case we consider that both energy vs performance and operation cost terms are specified by an external independent source with arbitrary associated (unknown) semantics. No explicit dependency is assumed, i.e. energy vs performance and operation cost may vary independently.

The fundamentals defining a node at an internal level, for this study, relate to the different associated performance states, operating frequency, voltage and power specifications. Energy consumption, for a given activity factor, input power and effective capacitance, denoted respectively by α, C, f and V, is considered to scale linearly with frequency and to be quadratically proportional to voltage. An approximation can be obtained by integrating over $P \approx \alpha C f V^2$ (power dissipation) [24].

14.4.3 Local Scheduler

After the detailed description of the computing nodes provided in the previous subsection, we now describe the architecture of the local scheduler. A conceptual view of its architecture is given in Fig. 14.4, based on a cognitive agent model. The *goals* of a local scheduler are to minimize through local decisions the system's energy consumption and the overall delay.

To achieve these goals, the agent has a set of *skills* and a *reasoning* function. Skills include dynamic voltage and frequency scaling (DVS, DFS) and tasks offloading capabilities. A fuzzy inference system implements the reasoning function. A multi-objective evolutionary algorithm is used to evolve the weights of the fuzzy system while relying on an anticipation mechanism (detailes provided in the following). As an additional skill, tightly linked with the communication function, nodes have the possibility to offload tasks to other local schedulers. To communicate with other local scheduler agents, an agent relies on a set of *communication protocols* (assumed transparent here and not further detailed). Through the *perceive* function, the agent is aware of the environment state (i.e. the global system's performance) and of its local goals' status. The following provides a detailed description of the instantiation of the aforementioned local scheduler architecture.

As previously mentioned, a first effective energy-management skill consists in dynamically scaling frequency and voltage [14,17]. A linear decrease in energy consumption is attained by downscaling frequency, respectively quadratic by reducing voltage. Nonetheless, while a load-dependent scaling would stand as a straightforward approach, no direct formulation is possible here due to the additional energy vs performance and operation cost specifications. High load should result in a frequency and voltage increase but this has also to obey performance and operation constraints. Additional factors taken into account here define local and average system load, node overhead, idle time and current state (active or sleeping).

A second skill consists in enforcing local and system wide loading and offloading policies along with task assignment and distribution mechanisms [2]. The dynamics driving the loading and offloading of tasks, in this case, have a direct impact on a node's load and on the balance of the system while indirectly affecting the context from which scaling decision criteria are drawn. Energy-aware load-balancing techniques have to be consequently defined, controlling not only the distribution of tasks across resources but also the type of tasks assigned to or offloaded from nodes, e.g. regrouping one computational intensive task with multiple communication oriented ones on the same node. As an example, an Intel®Pentium®M processor operating at full load in High Frequency Mode (HFM), P0 state, is expected to drive energy consumption at a higher level than three identical processors operating in Low Frequency Mode (LFM), state P5, under an equivalent load.

Summarizing, we opted for a design where frequency and voltage scaling is driven (1) in direct manner by taking into account the average load over a short term period, idle time, energy vs performance and operation cost, and (2) indirectly by controlling the type and rate of loaded or offloaded tasks. Furthermore, given that a centralized approach would not stand the scaling requirements of a large distributed system, a *by* node driven paradigm has been preferred. The effective control of the systems is conducted upon the emergent result of all nodes' decisions where each node incorporates an independent functioning logic. For the herein case all nodes use a zero degree Takagi-Sugeno [32] fuzzy inference system composed of 41 disjunctive rules (13 rules for deciding when and how to increase voltage and frequency, 16 rules for downscaling control and 12 rules for task loading and offloading conditions). An excerpt of these rules is given hereafter with the title of example.

if *energyVsPerfTarget* **is** *energy* $\wedge$ *operationCost* **is** *high* **then**
decrease voltage and frequency by a large percentage

if *energyVsPerfTarget* **is** *perf.* $\wedge$ *consumption* **is** *low* $\wedge$ *load* **is** *high* **then**
increase voltage and frequency by a large percentage

if *energyVsPerfTarget* **is** *energy* $\wedge$ *consumption* **is** *high* **then**
offload a high number of tasks

if *energyVsPerfTarget* **is** *perf.* $\wedge$ *load* **is** *low* $\wedge$ *sysLoad* **is** *high* **then**
accept a large number of tasks to be loaded

if *energyVsPerfTarget* **is** *balance* $\wedge$ *consumption* **is** *low* **then**
accept a moderate number of tasks to be loaded

A brief functioning insight is offered hereafter by referring to the different sections of Fig. 14.3. The simulation starts from a LFM mode with a moderately high load (section A of the graph). Despite the slight operation cost increase, it is straightforward to observe the raise in voltage and frequency, in correlation with energy vs performance – note that scaling also depends on load. Analogously, for approximately the same energy vs performance level (transition between section A and section B) but for an increased operation cost and in spite of the load's escalation, voltage and frequency drop, given the steep ascension of the operation price. In the following steps, with the sustained energy vs performance increase, scaling reacts to raise voltage and frequency, also reflected by the load decline due to the completion of a larger number of tasks.

14.5 Dynamic Energy-Efficient Optimization

This section provides a detailed description of the last skill of the local scheduler agent, i.e. the dynamic multi-objective anticipative evolutionary algorithm. As described in the previous sections, the simulation model is designed to capture the dynamic nature of a real-life distributed environment. Tasks arrive at different moments in time, not known in advance, performance and cost parameters vary independently, load overhead may affect all performance factors, e.g. due to external users accessing the system, etc. Therefore, an energy-efficient approach, in this case, has to cope not only with instantaneous, static factors, but also needs to adapt over time to a continuously changing environment and, what is more, to perturbations induced by multiple stochastic sources. An additional aspect to consider is that different evolution paths are possible from any given point in time. Moreover, as load overhead, operation cost and failure are all modeled in stochastic form, the system can be defined as a random variable where, given an initial configuration, all possible realisations at the next step or over multiple steps have to be assessed. Having taken into account all possible outcomes, a decision can be formulated with respect to changes to perform in order to maintain the system in an optimal state (maximal performance at lowest possible energy consumption). Nonetheless, as a complete assessment of all possible realisations of the system is not feasible from a computational point of view, decisions have to be taken only upon a reduced number of samples. Furthermore, as a multi-objective approach is adopted, with energy consumption and delay as objectives to be minimized, solution selection and decision problems have to be addressed. We therefore rely on concepts like *scenario*, *strategy* and *anticipation window* [6], discussed in more detail in the following. Different open questions arise and solutions to the afferent problems are subject to further research – an outline of the main concepts and implications is given hereafter.

14.5.1 Strategies and Scenarios

The evolution of the distributed system as a whole is an emergent effect of all nodes' transitions. As we desire to minimize energy consumption and delay, the control parameters of each node have to be dynamically modified at each step. We therefore speak of decisions acting on frequency and voltage scaling, e.g. putting the node in a higher state than the current one, or local energy vs performance variation. Decisions can be formulated with respect to the factors defining the context of a node at a given moment in time and the optimality of these decisions can be heuristically assessed over their effective application life-time. The main difficulty is that a decision taken at a specific moment t, although optimal with respect to the outcomes induced over its effective processing time window, i.e. $[t, t+p]$, may not be optimal at long term. For a graphical illustration refer to Fig. 14.3. Otherwise stated, for a given configuration γ of the system to simulate at a given time t, and given two decision d_t^1, d_t^2, e.g. performing transitions towards a lower, respectively upper power state, the short term overall outcome of the first decision may be suboptimal with respect to the second decision, i.e. $\int_t^{t+st} F_{d_t^1}(\gamma)\mathrm{dt} \leq \int_t^{t+st} F_{d_t^2}(\gamma)\mathrm{dt}$, whereas long term overall effects may prove to be exactly the opposite with $\int_t^{t+lt} F_{d_t^2}(\gamma)\mathrm{dt} \leq \int_t^{t+lt} F_{d_t^1}(\gamma)\mathrm{dt}$, where $st < lt$ represent short and long term corresponding time window lengths. Therefore, instead of relying on punctual decisions, long term strategies are used as basis for constructing a dynamic approach capable of anticipating future consequences of current decisions. A first problem to be addressed, once the strategy concept developed, relates to the effective evaluation of these objects. For a system following a predictable evolution, which can be simulated in exact manner, i.e. with no stochastic factors, it is sufficient to carry the simulation over the specified anticipation time interval.

At the opposite end, when stochastic sources are part of the model, for any given number of processing steps, we can only approximate the most probable states or the average state of the system. Therefore, for a set of different candidate strategies and for a given configuration of the system at a specified moment in time, samples have to be drawn, termed hereafter as scenarios, and evaluated independently. A realization of the system is thus obtained for each scenario, with respect to the constraints and dynamics dictated by each strategy. For coherence and consistency, strategies are evaluated over the same set of scenarios, the final outcome of each strategy being expressed as the average energy consumption and delay.

Note that several questions arise, subject to further research, with respect to processing and anticipation time windows, strategy evaluation procedures, etc. First, we face two opposing directions. Long simulation times lead to an escalation of the anticipation error – simulating stochastic factors over a long period of time potentially leads to configurations which no longer follow or represent reality. One would therefore want to limit the extent over which anticipation is carried. At the opposite end, short anticipation windows may not capture correctly the effects of a current decision over the future state(s) of the system. What is more, having a strategy evaluated as the average result of different scenarios may not always be consistent, e.g. when clustered or sparse states are obtained as a result.

14.6 Experimentation and Results

In the remainder of this chapter we present several results, first by analyzing the behaviour of the fuzzy system at a node level, subsequently moving to a system-wide focused discussion.

A comparative view of energy consumption and overall delay is given in Fig. 14.5. For fixed energy vs performance and operation cost values, constant over the entire one-day simulation time, the corresponding energy consumption and delay are respectively illustrated in Fig. 14.5a and Fig 14.5b. Note that, for illustration purposes, axes are mirrored between the two figures. Each point, e.g. energy vs performance and operation cost set at 0.3, respectively 0.5, represents the average value of 30 independent simulations. Different plateaus can be identified for energy consumption and, correspondingly, for delay. A substantial energy minimization is obtained for an operation cost above 0.6, i.e. the equivalent of stating that energy is an important asset, and performance levels below 0.3. Recalling the performance states of Intel®Pentium®M, this is the equivalent of operating the processor in the P4, P5 power states. Moderate consumption is attained for performance levels between 0.3 and 0.8 (equivalent of P1, P2 and P3 states) when operation cost is superior to 0.6. For all other cases, energy consumption raises to reflect either a high performance demand (energy vs performance superior to 0.8, i.e. equivalent of P0 for Intel®Pentium®M) or a low operation cost which, in turn, allows a high voltage and frequency to be maintained. The behaviour of the system with respect to extreme values is also of interest, the following cases being possible:

- lowest possible energy consumption at the highest operation cost (energy vs performance and cost set at 0.0 respectively 1.0). The outcome of drastic frequency and voltage downscaling, e.g. processor constantly operated in lower performance states, close to or in LFM, is obvious and straightforward to anticipate: energy consumption is at its lowest while delay is driven to a maximum.
- energy vs performance and operation cost both set to 0.0, i.e. knowing that the price to pay for energy, for example, is highly affordable, minimize energy consumption. A first remark to be made here is that operation cost constrains the variation amount allowed for scaling. Nonetheless, as a second aspect, scaling

does also consider the load of a node. Therefore, as the system is executed at an affordable cost or at no costs at all, we only have to cope with load. As a consequence, this leads to a separation in two subclasses: 1) execution under high load – the node is put in a high execution state due to the affordable cost but this in turn drives energy consumption over a positive slope (case exposed in Fig. 14.5); 2) low load – scaling only copes with load; minimal energy consumption is attained for minimal load.

- maximal performance, knowing that operation cost is at its highest (energy vs performance and cost both set to 1.0). As specifications demand performance, with no regard for energy consumption, and given the high load, the node is operated in HFM, state P0. This reflects symmetrically in consumption and delay graphics (Fig. 14.5a and Fig. 14.5b).
- maximal performance with lowest operation cost (values set to 1.0, respectively 0.0). Scaling is performed by following energy vs performance and load indicators. The system is hence driven to respond to load stress, where from the high energy consumption illustrated in Fig. 14.5.

A less explicit aspect of the presented graphs is the influence of task scheduling and load-balancing. The order and priority associated to tasks (at node level) have a direct impact on delay whereas load-balancing (system level) can result in perturbations of both energy consumption and delay. The assignment of highly computational intensive tasks to low performance nodes may be a first source of delay. Similarly, the exclusive use of a high power consumption node to run communication intensive tasks may not be optimal. As the purpose of this study is to analyse the use of a dynamic optimization algorithm for distributed systems in the presence of different stochastic sources, we consider a minimal definition for scheduling and balancing policies. Furthermore, as a centralized approach does not fulfill the requirements of a large scale distributed environment, e.g. where administration policies may have only a limited acting power across domains, we preferred enforcing node-based balancing. Each node is thus responsible for deciding upon acceptance rates, e.g. for loading tasks, bias between computational and communication intensive tasks, as well as on offloading parameters. The process is controlled by a subset of rules within the fuzzy inference system, taking into consideration local and system load, energy vs performance and operation cost specifications. Subsequent to loading, all tasks receive equal priorities, the amount of dispatched active processing time varying in concordance with the specified computational load footprint.

The complete simulation environment, as described in the previous sections, relies on a collection of autonomous nodes. These nodes although capable of adjusting their state as dictated by the internal base of rules may not optimally scale to all possible scenarios. Or, otherwise put, the weights associated to the rules may not be optimal. A possible solution, as mentioned earlier in this chapter, consists in using an evolutionary algorithm to modify the firing weights of the rules dynamically. Thus, for each node, besides the enclosed fuzzy inference system, a multi-objective parallel NSGA-II [12] evolutionary algorithm is deployed. For each algorithm, individuals code the strategy to use for modifying rules' weights as an array of real

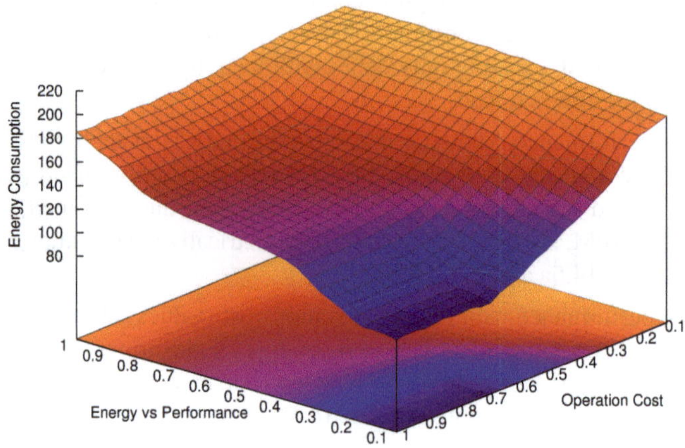

(a) Energy Consumption

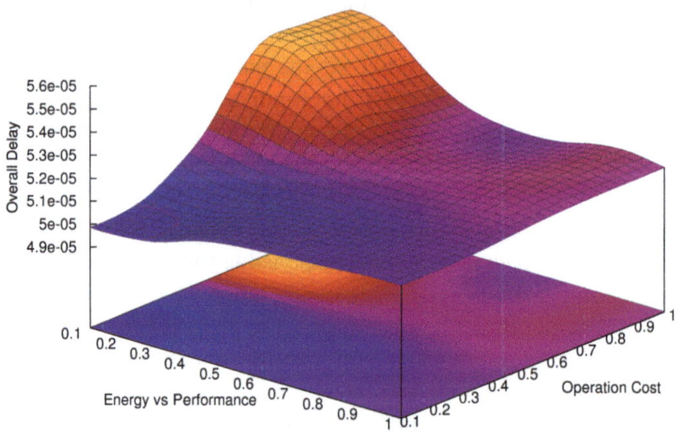

(b) Delay

Fig. 14.5 Energy consumption and delay for various energy vs performance and operation cost levels. A high energy consumption reflects increased voltage and frequency values which in turn lead to a reduced delay – pairwise points, axes mirrored between 14.5a and 14.5b. Symmetrically, a reduced energy consumption translates to increased delay levels.

transition values (initialized in uniform random manner in $[0,1]$). The fitness function is expressed as the energy and delay average values obtained over three different independent scenarios through simulation and anticipation in a specified time frame. A scenario is obtained as a simulation carried over a given length of time where random and stochastic factors are seeded by a given initial value. This allows for different strategies to be analyzed against an identical set of scenarios. Simulations and anticipation are performed in parallel on remote nodes to speedup the process. Please refer to Fig. 14.3 for an intuitive illustration of simulation and anticipation related notions.

Strategies are iteratively evolved by the algorithm having as performance measure the same anticipation mechanism, at the end of each iteration, weights being exchanged between the nodes. At following steps the simulation carried inside each node uses the updated weights hence maintaining context coherence. At every 15 iterations, for each node, out of the locally obtained front of solutions, a set of weights, best approximating the currently specified energy vs performance and operation cost constraints, is selected and made active inside the node. Having weights set for all nodes a processing step is performed, the algorithm being restarted from the new execution point. The duration of a processing step is set to approximately 30 minutes (the system being run under the direction of the inference system with the determined weights) with an anticipation time of 6 hours.

Besides the difficulty of static multi-objective optimization, the resulting online dynamic case demands, as previously described, selecting a solution out of the Pareto set at fixed discrete time moments. The selected solution is used to advance the system to a new state (processing step illustrated in Fig. 14.3). As the intervention of a decision maker does not represent a feasible nor a practical solution, we propose the use of an approach inspired from the interactive EMO context. The classic way of handling the preferences of an user in the interactive evolutionary multi-objective context is by means of an *achievement scalarizing function*, initially proposed by Wierzbicki [34] and defined as follows:

$$\sigma(z,z^0,\lambda,\rho) = \max_{j=1,\ldots,d}\left\{\lambda_j(z_j - z_j^0)\right\} + \rho\sum_{j=1}^{d}\lambda_j(z_j - z_j^0)$$

where,

$\sigma(\cdot,\cdot,\cdot,\cdot)$ is an application of Z into $\mathbb{R}$;
$z = (z_1, z_2, \ldots, z_j, \ldots, z_d)$ is an objective function vector;
$z^0 = (z_1^0, z_2^0, \ldots, z_j^0, \ldots, z_d^0)$ is a reference point vector;
$\lambda = (\lambda_1, \lambda_2, \ldots, \lambda_j, \ldots, \lambda_d)$ is a weighted vector;
ρ, is an arbitrary small positive number ($0 < \rho \ll 1$).

These functions have as basis Chebyshev definitions and have the role of projecting a given reference point $z^0 \in \mathbb{R}^d$ (feasible or infeasible solution) into the optimal Pareto set. Note that through the achievement scalarizing functions the multi-objective problem formulation together with the reference point coordinates are incorporated in a mono-objective optimization model. The reference point for

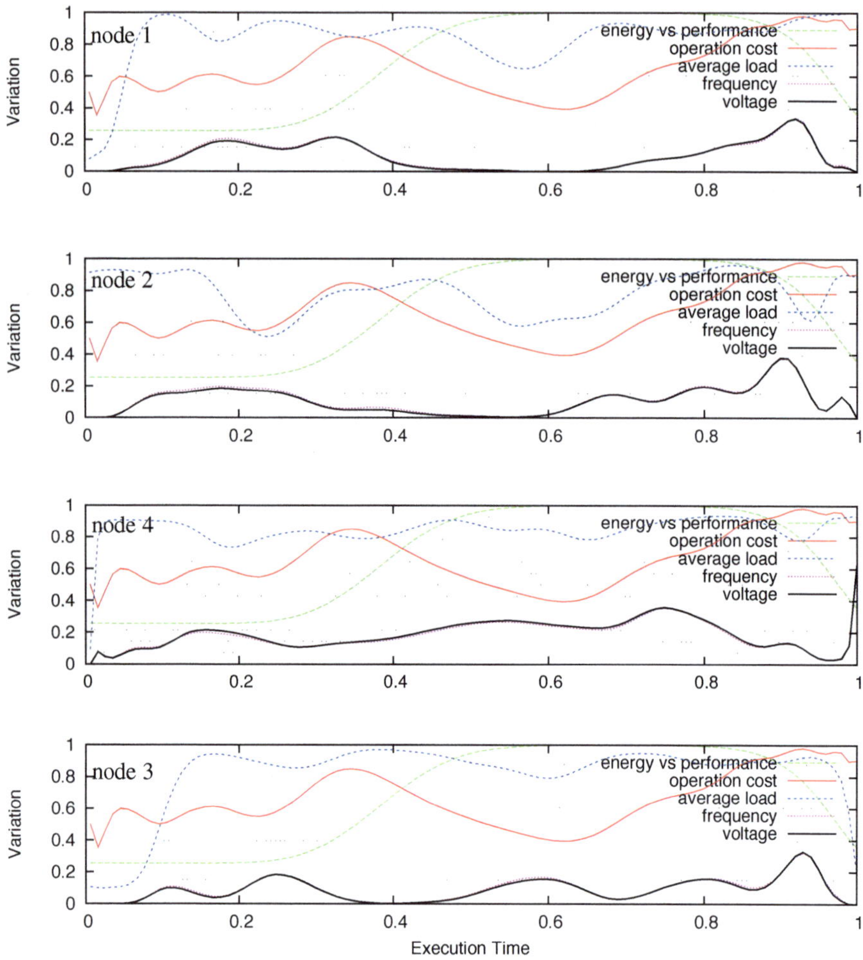

Fig. 14.6 Dynamic evolution of different fuzzy-enabled nodes (subset of the setup). For each node and for each processing step, weights are optimized by a dedicated, per-node multi-objective dynamic EA. Depending on the specific context, e.g. load, operation cost and energy vs performance levels, each node adapts its frequency, voltage and the acceptance or the offloading rates in order to assure an energy-efficient execution (outcome effects of varying exchange rates visible near 0.6). Please note that scenarios are determined by the operation cost, energy vs performance and the task arrival patterns, being generated and analyzed at execution time.

the herein case is given by the coordinates of the ideal point and provides the localization of the focus region on the Pareto-optimal front, while the weights λ_i, $i \in \{1,...,d\}$ used in the scalarization process supply the localization of the Pareto solutions of interest in the aforementioned region.

A graphical illustration of an example execution is given in Fig. 14.6 (partial view of the nodes). All simulations have been carried inside Grid'5000[2] [9] using a parallel decentralized parallel NSGA-II algorithm over an average number of 300 cores. Note that energy vs performance and operation cost are identical for all nodes as an expression of system-wide constraints. Nodes start from an idle state with no tasks assigned, the first steps showing a massive offload towards the third and fourth depicted nodes. Once these nodes reach a high load level, tasks are accepted only at low rates, resulting a first equilibrium point (time line from 0.1 to 0.15). As frequency and voltage are simultaneously increased, in concordance with the demanded energy vs performance level, a load decrease is recorded (close to 0.2). Next, the ascending trend of the operation cost with a peak at 0.35, leads to lower performance states for most of the nodes, where from the escalation of node load indicators. The remaining time frame follows similar dynamics.

14.7 Conclusions

A general agent-based paradigm relying on fuzzy inference enabled nodes and a multi-objective evolutionary algorithm has been presented in this chapter, outlining several key points and open questions. The results provided by the algorithm show that it is possible to cope with a highly stochastic and dynamic environment. Nodes adapt not only to load specifications but also follow arbitrary energy vs performance and operation cost constraints. A large number of extensions are possible where concrete cases are analyzed independently, e.g. as for thermal-aware systems or as with market rules driven environments. Additional insight has to be gained with regard to how to exploit anticipation in an efficient, effective manner, as well as on how to select a strategy out of a given set of points, e.g. potentially using decision-making derived techniques.

References

1. Armbrust: M., Fox, A., Griffith, R., Joseph, A.D., Katz, R.H., Konwinski, A., Lee, G., Patterson, D.A., Rabkin, A., Stoica, I., Zaharia, M.: Above the clouds: A berkeley view of cloud computing. Tech. Rep. UCB/EECS-2009-28, EECS Department, University of California, Berkeley (2009)
2. Aydi, H., Mejía-Alvarez, P., Mossé, D., Melhem, R.: Dynamic and aggressive scheduling techniques for power-aware real-time systems. In: Proceedings of the 22nd IEEE Real-Time Systems Symposium, RTSS 2001, p. 95. IEEE Computer Society Press, Washington, DC, USA (2001)

[2] https://www.grid5000.fr

3. Barbagallo, D., Di Nitto, E., Dubois, D.J., Mirandola, R.: A bio-inspired algorithm for energy optimization in a self-organizing data center. In: Weyns, D., Malek, S., de Lemos, R., Andersson, J. (eds.) SOAR 2009. LNCS, vol. 6090, pp. 127–151. Springer, Heidelberg (2010)
4. Basney, J., Welch, V., Wilkins-Diehr, N.: Teragrid science gateway AAAA model: implementation and lessons learned. In: Proceedings of the 2010 TeraGrid Conference, TG 2010, pp. 2:1–2:6 ACM Press, New York (2010)
5. Berral, J.L., Goiri, I., Nou, R., Julià, F., Guitart, J., Gavaldà, R., Torres, J.: Towards energy-aware scheduling in data centers using machine learning. In: Proceedings of the 1st International Conference on Energy-Efficient Computing and Networking, e-Energy 2010, pp. 215–224. ACM Press, New York (2010)
6. Bosman, P.A.N., Poutré, H.L.: Learning and anticipation in online dynamic optimization with evolutionary algorithms: the stochastic case. In: GECCO, pp. 1165–1172 (2007)
7. Branke, J.: Evolutionary Optimization in Dynamic Environments. Kluwer Academic Publishers, Norwell (2001)
8. Brown, D.J., Reams, C.: Toward energy-efficient computing. Commun. ACM 53, 50–58 (2010)
9. Cappello, F., Caron, E., Dayde, M., Desprez, F., Jegou, Y., Primet, P., Jeannot, E., Lanteri, S., Leduc, J., Melab, N., Mornet, G., Namyst, R., Quetier, B., Richard, O.: Grid'5000: A Large Scale and Highly Reconfigurable Grid Experimental Testbed. In: Proceedings of IEEE/ACM International Workshop on Grid Computing (GRID 2005), pp. 99–106. IEEE Computer Society, Washington, DC, USA (2005)
10. Das, R., Kephart, J.O., Lefurgy, C., Tesauro, G., Levine, D.W., Chan, H.: Autonomic multi-agent management of power and performance in data centers. In: Proceedings of the 7th international joint conference on Autonomous agents and multiagent systems: industrial track, AAMAS 2008, International Foundation for Autonomous Agents and Multiagent Systems, Richland, SC, pp. 107–114 (2008)
11. Das, R., Whalley, I., Kephart, J.O.: Utility-based collaboration among autonomous agents for resource allocation in data centers. In: Proceedings of the Fifth International Joint Conference on Autonomous Agents and Multiagent Systems, AAMAS 2006, pp. 1572–1579. ACM, New York (2006)
12. Deb, K., Pratap, A., Agarwal, S., Meyarivan, T.: A fast and elitist multiobjective genetic algorithm: Nsga-ii. IEEE Transactions on Evolutionary Computation 6, 182–197 (2002)
13. Eiben, A., Schut, M.: New ways to calibrate evolutionary algorithms. In: Advances in Metaheuristics for Hard Optimization. Natural Computing Series, ch. 8, pp. 153–177. Springer, Heidelberg (2008)
14. Flautner, K., Reinhardt, S., Mudge, T.: Automatic performance setting for dynamic voltage scaling. In: Proceedings of the 7th Annual International Conference on Mobile Computing and Networking, MobiCom 2001, pp. 260–271. ACM Press, New York (2001)
15. Gropp, W.: MPI at exascale: Challenges for data structures and algorithms. In: Ropo, M., Westerholm, J., Dongarra, J. (eds.) PVM/MPI. LNCS, vol. 5759, pp. 3–3. Springer, Heidelberg (2009)
16. Hatzakis, I., Wallace, D.: Dynamic multi-objective optimization with evolutionary algorithms: a forward-looking approach. In: Proceedings of the 8th Annual Conference on Genetic and Evolutionary Computation, GECCO 2006, pp. 1201–1208. ACM, New York (2006)
17. Herbert, S., Marculescu, D.: Analysis of dynamic voltage/frequency scaling in chip-multiprocessors. In: Proceedings of the 2007 International Symposium on Low Power Electronics and Design, ISLPED 2007, pp. 38–43. ACM, New York (2007)
18. IBM: Fact sheet & background: Roadrunner smashes the petaflop barrier (2008)

19. IBM: 20 petaflop sequoia supercomputer (2009)
20. Intel: Enhanced Intel Speedstep Technology for the Intel Pentium M Processor, Intel White Paper (2004)
21. Kaplan, J., Forrest, W., Kindler, N.: Revolutionizing data center energy efficiency (2009)
22. Kaufmann, R.K.: The mechanisms for autonomous energy efficiency increases: A cointegration analysis of the us energy/gdp ratio. The Energy Journal 25(1), 63–86 (2004)
23. Khargharia, B., Hariri, S., Yousif, M.S.: Autonomic power and performance management for computing systems. Cluster Computing 11, 167–181 (2008)
24. Le Sueur, E., Heiser, G.: Dynamic voltage and frequency scaling: The laws of diminishing returns. In: Proceedings of the 2010 Workshop on Power Aware Computing and Systems (HotPower 2010), Vancouver, Canada (2010)
25. Mcarthur, S.D.J., Davidson, E.M., Catterson, V.M., Dimeas, A.L., Hatziargyriou, N.D., Ponci, F., Funabashi, T.: Multi-Agent Systems for Power Engineering Applications - Part I: Concepts, Approaches, and Technical Challenges. IEEE Transactions on Power Systems 22(4), 1743–1752 (2007)
26. Mehnen, J., Wagner, T., Rudolph, G.: Evolutionary optimization of dynamic multiobjective functions. Tech. rep. (2006)
27. Bennani, N., Menasce, M.A.: Resource allocation for autonomic data centers using analytic performance models. In: Proceedings of the Second International Conference on Automatic Computing, pp. 229–240. IEEE Computer Society, Washington, DC, USA (2005)
28. Parkhill, D.F.: The challenge of the computer Utility. Addison-Wesley Pub. Co, Reading (1966)
29. Stantchev, V.: Performance evaluation of cloud computing offerings. In: Proceedings of the 2009 Third International Conference on Advanced Engineering Computing and Applications in Sciences, ADVCOMP 2009, pp. 187–192. IEEE Computer Society, Washington, DC, USA (2009)
30. Stantchev, V., Schröpfer, C.: Negotiating and enforcing qoS and sLAs in grid and cloud computing. In: Abdennadher, N., Petcu, D. (eds.) GPC 2009. LNCS, vol. 5529, pp. 25–35. Springer, Heidelberg (2009)
31. Sterling, T., Messina, P., Smith, P.H.: Enabling technologies for petaflops computing. MIT Press, Cambridge (1995)
32. Takagi, T., Sugeno, M.: Fuzzy Identification of Systems and Its Applications to Modeling and Control. IEEE Transactions on Systems, Man, and Cybernetics 15(1), 116–132 (1985)
33. White, R., Abels, T.: Energy resource management in the virtual data center. In: Proceedings of the International Symposium on Electronics and the Environment, pp. 112–116. IEEE Computer Society, Washington, DC, USA (2004)
34. Wierzbicki, A.: The use of reference objectives in multiobjective optimization. Lecture Notes in Economics and Mathematical Systems 177, 468–486 (1980)

Chapter 15
Green Storage Technologies: Issues and Strategies for Enhancing Energy Efficiency

Hyung Gyu Lee, Mamadou Diao, and Jongman Kim

Abstract. Energy cost in the data center is rising at a steep rate, and a significant portion of that cost comes from the storage system. Storage (disks and memory systems) accounts for more than 30% of the total power consumption. This number is expected to rise to 50% of the overall IT budget in the future. Enhancing the power efficiency of storage systems is of utmost importance, and, many techniques have been proposed and are being explored in both component and system levels. This chapter starts by providing a brief introduction on the basic power characteristics of storage components from a device-level perspective. Then several architectural and high-level power management techniques are covered.

15.1 Introduction

Along with the growth of Information Technology (IT), the amount of available information is increasing dramatically day by day. This makes high-bandwidth and high-capacity storage systems crucial to the IT industry. As a result, storage components (including memory and last level storage) account for a large portion of the power consumed in computing systems. Although the estimates vary by applications and system configurations, they range from 30% to 50% [2], and this portion is expected to grow rapidly. As a consequence, improving power efficiency has become a major challenge even in storage systems. There have been many approaches in building smart and energy efficient storage systems, from component-level techniques to system-level power management techniques.

One of the most distinct changes in storage devices was the use of flash memory based storage devices (systems) as an alternative solution for achieving high performance and low power consumption at the same time, from embedded systems to

Hyung Gyu Lee · Mamadou Diao · Jongman Kim
School of Electrical and Computer Engineering,
Georgia Institute of Technology, Atlanta GA
e-mail: {hyunggyu,diao,jkim}@gatech.edu

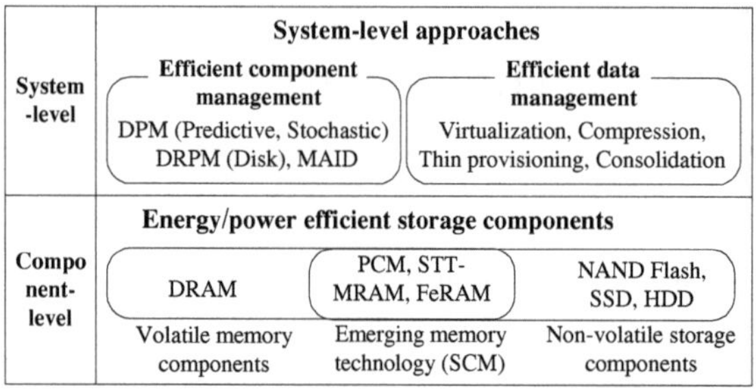

Fig. 15.1 Energy/power efficient storage techniques.

high-end enterprise machines. In addition, the new concepts of non-volatile memory technology such as PCM (Phase Change Memory) and STT-MRAM (Spin-Torque Transfer Magneto-resistive RAM) are now emerging as a storage class memory that can be used as a main memory device as well as a storage device. We expect that efficient use of storage class memory will dramatically increase the system performance while reducing the energy consumption both in main memory systems and in storage systems.

The efforts to reduce the energy consumption in storage systems can be broadly divided in component-level and system-level approaches as shown in Fig. 15.1. At the component-level, efforts evolve around improving energy efficiency of storage components (from traditional volatile memory devices including DRAM to non-volatile storage devices such as Hard Disk Drive (HDD)) through the exploration of new storage technologies and better energy-efficient designs. On the other hand, system-level approaches are often orthogonal and encompass a variety of techniques that operate at a higher abstraction level. In our context, a system is defined as a collection of storage components that can be controlled and managed. System-level techniques control storage components, explore new component layouts and storage hierarchies, and look at data layout, data movement and data processing operations in order to reduce the energy consumption of storage systems.

Based on this classification, this chapter surveys state of art techniques in developing energy-efficient storage systems. We start by presenting component-level techniques to achieve efficient storage systems. We believe this will be helpful in understanding the basic power consumption behaviors of storage components, and will further provide many helpful hints to devise system-level storage management techniques for achieving green storage technology.

The remainder of this chapter is organized as follows: Section 15.2 covers different storage technologies from the physical level to the device level with a particular focus on their power consumption and the tradeoffs between power and performance. Section 15.3 describes various methods at the system level used to improve the power consumption and energy cost of storage systems. Section 15.4 takes a big picture look and discusses current challenges and opportunities of green storage technologies.

15.2 Component-Level Approach for Energy Efficient Storage

To reduce the fuel consumption of a car, an easy way would be to use a high-gas-mileage vehicle. This naturally reduces the fuel consumption independently of how efficient our driving is. To further reduce our fuel consumption, we can drive the car within energy-efficient speed ranges by minimizing steep accelerations and sudden breaking. The same principle can be applied to low energy storage systems. Similar to our car analogy, a first way to reduce the energy consumption of a storage system is to use low-power storage components. Another way to further reduce the energy consumption is to control the components in an energy efficient way. The component manufacturer may have more responsibility in the former case, while system designers may have more responsibility in the later case. One thing is sure, proper knowledge of the power consumption behaviors in storage components will be helpful to system designers and administrators in their quest for greener storage systems.

This section mainly focuses on component level characteristics of several types of storage components, particularly their power consumption behaviors. Then micro architectural efforts for reducing the energy consumption will be introduced. Depending on the applications and system requirements, various types of memory devices have been used to store data, because there is no single memory device at the mass-production stage satisfying all the requirements such as read/write performance, density and non-volatility. In a broad sense, storage components include volatile main memory devices as well as permanent storage devices. Based on these two categories, we first explore DRAM devices which have been commonly used as volatile main memory devices, and then we explore HDDs and the NAND flash memory devices as permanent storage devices. Finally an introduction about some emerging memory technologies will be given with their impacts on storage systems in terms of energy consumption.

15.2.1 Volatile Main Memory Component - DRAM (Dynamic Random Access Memory)

Traditionally, DRAM-based main memory systems have been commonly used for the last several decades because the DRAM device supports fast read and write, and has relatively large capacities with reasonable cost, compared to other memory devices. But due to its internal structure, it consumes more power than other memory devices.

15.2.1.1 Basics of DRAM Devices

To know the power consumption behavior of the DRAM, it is useful to understand the basic structure and functionality of the device as shown in Fig. 15.2. DRAM cells are basically made up of a capacitor and a transistor. Data is stored in the cell by setting the bit line to a high or low, when the select line is activated. So when a cell is read, the stored charges in the cell are removed and restoring

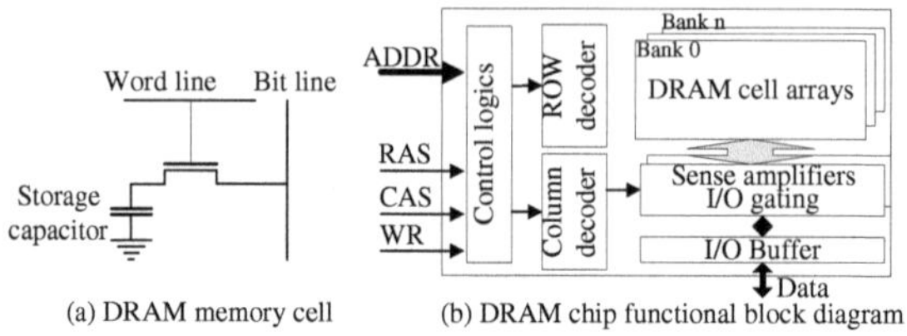

Fig. 15.2 DRAM cell and chip structure.

after each read must be performed in order not to lose the data. In addition, there is steady leakage in the capacitor-based cell. The charge must be restored periodically (refreshing). These operations require non-negligible power consumption during active states as well as idle states.

Generally, a DRAM operation sequence consists of two steps: row & bank active and column read/write. During row & bank active operation, the control logic selects a bank and row address using the row decoder, and transfers the selected row's cell data, which is stored in the array, to the sense amplifier. The data stays in the sense amplifiers until a precharge command to the same bank restores the data to the cell in the array. After row & bank active command, column reads and writes can be placed. A read command decodes a selected column along the row data that is stored in the sense amplifiers, and then the data from this column is driven to the internal I/O buffer through the I/O gating and the multiplexer logics. The process of a write is the opposite of the read process.

Fig. 15.3 shows a detailed operation sequence and power modes of DRAM devices. Basically, two standby modes are supported in most conventional DRAMs. One is called Precharge Standby (IDD2N) mode where none of the banks and rows are activated, and thus the power consumption is relatively low. The other is called Active Standby (IDD3N) mode where more than one bank and row are activated, and thus the power consumption is higher than that of precharge standby mode.

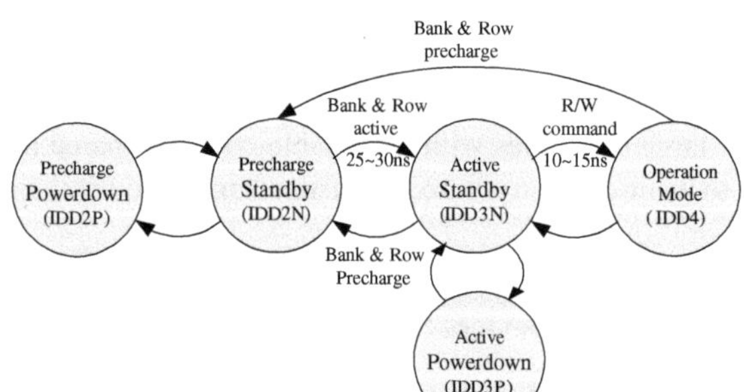

Fig. 15.3 DRAM operation sequence and power modes.

Table 15.1 Power consumption in DRAM (128M x 8 bit DDR2, mW/chip).

	IDD2P	IDD2N	IDD3N	IDD3P	IDD4	
					Read	Write
Samsung (K4T1G084QQ)	8	35	55	35	135	115
Micron (MT47H128M8)	7	50	70	40	160	160

To reduce the standby mode power consumption, most chip manufacturers provide several power-down modes (IDD2P and IDD3P) from each standby mode (IDD2N and IDD3N), which disables most components in a chip, and it is mainly controlled by CKE (ClocK Enable) signals. In these modes, the power consumption can be decreased, but in order to exit from the power-down modes, extra delay is indispensable, which causes performance degradations.

Table 15.1 shows the power consumption for each power mode. Depending on the chip vendor, power consumption is slightly different, but power down modes from each standby mode can reduce the power consumption significantly.

In addition to the steady power consumption in standby modes, the DRAM also consumes a significant amount of power during the active operation, called dynamic power consumption, normally right after the command issuing. At each command (bank & row active, column read/write or precharge), the device changes internal states and this consumes a significant amount of power. So minimizing the number of state changes is also one of useful ways to reduce the power consumption.

15.2.1.2 Dynamic Power Mode Management in DRAMs

As described previously, most DRAMs support 4 standby modes including power-down modes. Waiting on the lowest power mode (IDD2P) incurs the highest delay while waiting on the high power mode (IDD3N) incurs the lowest delay. For example, if the memory controller stands by on active standby-mode and the consequent row address is the same as the previous one, the data can be read from the sense amplifier without issuing another row & bank active command. But in this case, if the interval between the consecutive accesses is long, the device consumes more power than waiting on the precharge standby mode. This surely helps to increase the performance, but this may increase the power consumption unless the inter-arrival time of consecutive accesses is short enough, which means a trade-off between the performance and the power consumption. Initial low power techniques have exploited power mode changes depending on the performance and power requirements. The memory controller dynamically changes the standby mode considering the statistics of memory access request [26, 9].

However, it is difficult to find a right time to change the power mode because the memory access characteristics are very dependent on the applications. In order to have further energy reductions, several studies change the data placement, and reschedule the access sequence, which maximizes the number of individual DRAM component and the time staying in the power down mode [15, 6, 19].

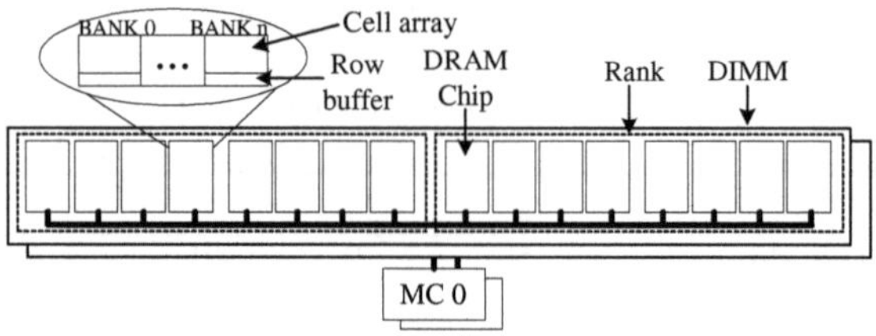

Fig. 15.4 An example of DRAM architecture. One DIMM consists of 2 ranks, and one ranks consists of 8 chips

The widespread use of multi-core architectures and the rapidly increasing data size have dramatically increased the requirement on memory bandwidth and capacity. In order to accommodate these demands, a lot of memory chips must be installed, and operated in parallel. Modern systems generally have multiple memory controllers, and each controller has multiple DIMMs (Dual In-line Memory Modules).

Fig. 15.4 briefly shows a conventional DIMM architecture. Each DIMM consists of several ranks, and each rank consists of several chips. Finally, each chip consists of multiple banks which are the minimum unit of interleaving operation. So there are many ways to map the CPU's logical address into the physical memory chip address. One possible mapping is to allocate the consecutive memory address to the different memory controller. Under each controller, memory address interleaving can similarly occur across the multiple physical memory banks. Striping consecutive access across multiple devices and controller can achieve higher bandwidth, but it may consume more power, which means a power and performance tradeoff. It has also been proved that row hit rate depends on the mapping strategy [26], which again affects the power consumption.

Unlike conventional DRAM ranks, one study proposed to break the rank into multiple smaller mini-ranks so as to reduce the number of devices involved in a single memory access [34]. As a result, they can reduce the memory power consumption by 44% with 7.4% performance penalty. In order to prevent the overfetch feature in DRAM, where a single request activates thousands of bitlines in many DRAM chip, [31] redesign a conventional DRAM chip by activating only necessary data line using a SBA (Selective Bitline Activation) and a SSA (Single Subarray Access) technique.

15.2.1.3 Use Leading-Edge Memory for Low Power Consumption

DRAM technology has been significantly enhanced during the last decade. Its technology scales down to 30 nm, and its operating clock frequency is enhanced up to 800 MHz. It is known that leading-edge DRAM technology offers low energy consumption per Gigabyte [2]. It reports that the use of leading-edge DRAM technology of a 30 nm class can reduce the total power consumption of server systems in a data center by almost 23% when compared to 50 nm class technology.

15.2.2 *Permanent Storage Component - HDDs (Hard Disk Drives)*

15.2.2.1 Basics of HDDs

HDD is the most common permanent storage device, from laptop computers to high-end servers, because it provides the most capacity with the lowest cost.

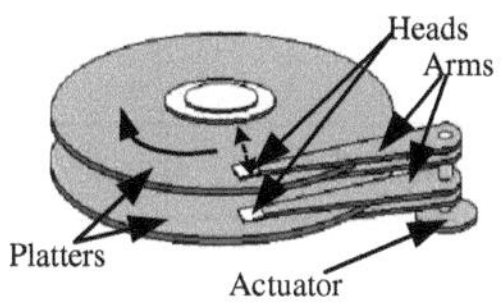

(a) HDD structure

Latency (seek time)	4-10ms
Throughput	140MB/s
Min. access unit	512B (Sector)
Max. density	2TB
Power consumption	
ACTIVE	8~12W
IDLE	4~8W
POWER DOWN	0.53~1.26W

(b) Specifications

Fig. 15.5 HDD basic structure and characteristics.

Fig. 15.5 briefly shows the internal structure and basic characteristics of conventional HDD. Unlike other semiconductor memory devices, it mainly consists of several mechanical parts including a spindle motor and an actuator for the head, as well as electrical components. The two mechanical parts consume a large portion of the power and they account for a big portion of the latencies, normally several *ms*, as shown in Fig. 15.5. Studies of power measurements on various disks demonstrated that the proportion of the power consumption due to spindle motor is almost 50% of the overall power consumption for a two-platter disk, and can increase to 81% for a ten-platter disk [12].

Fig. 15.6 shows that the power consumption of a disk is proportional to the capacity and rotational speed. Generally, high-capacity disks have more platters in a disk which requires more power on a spindle motor. In addition, the power consumption values in idle state are ranging from 59% to 71% of the power consumption values in operating state. These values roughly mean that the proportions of the power consumption on spindle motor is similar to the analysis in [12].

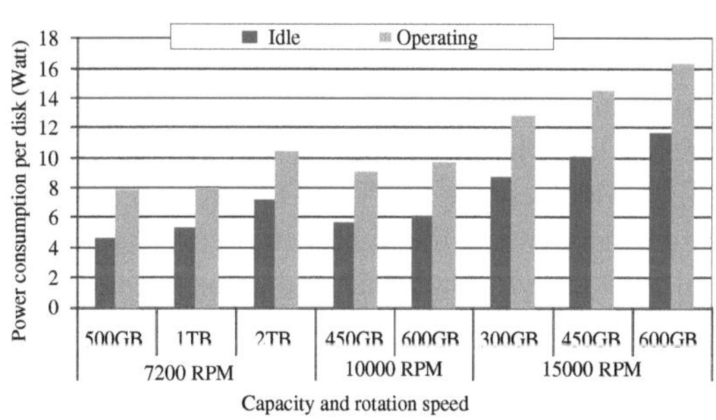

Fig. 15.6 Power consumption of HDD depending on capacity and RPM. Data is extracted from Seagate server and enterprise storage series.

15.2.2.2 Low Power Techniques for HDDs

Most component-level techniques for low power HDDs have concentrated on reducing the power consumption of the spindle motor. The initial research has focused on simple dynamic power management by shutting down the spindle motor during the idle time in the context of single disk systems. The existing work related to this will be presented in Section 15.3.

Table 15.2 Low power idle modes in state-of-the-art HDDs. PowerChioice^{TM} Technology from Seagate.

	Idle	Idle-A	Idle-B	Idle-C	Standby-Z
Servo	On	Off	Off	Off	Off
Head	Loaded	Loaded	Unloaded	Unload	Unload
Spinning	Full speed	Full speed	Full speed	Low speed	spin down
Recovery time (s)	N/A	0	0.5	1	8
Power (W)	2.82	2.82	2.18	1.82	1.29

Similar to DRAM, most current disks offer different power modes of operation such as operating where the disk is serving a request, idle where the disk is not serving, but spinning, and power down where the disk is not spinning.

Table 15.2 shows the low power idle modes supported in Seagate PowerChioic^{TM} technology. Depending on the number of component offs and spinning speed, power consumption varies. The lower power mode generally requires higher delay for recovering, hence the power and performance trade-off.

15.2.2.3 Hybrid HDD Technology

Over the last several years, one of the most distinct changes in permanent storage devices was the use of flash memory technology as an alternative solution for achieving high performance and low power consumption at the same time. Now, most embedded systems get equipped with this flash memory as a storage device, and started to be adapted to the high-end server systems. However, due to its cost and density, it cannot completely replace the conventional HDDs. However, several research results showed that hybrid use of flash-based storage with HDDs can increase the system performance as well as energy efficiency. There have been roughly two approaches: component level hybrid solutions and system-level hybrid solutions. The hybrid HDDs [21] which equips the NAND flash components in the HDDs belongs to the former category, while the hybrid storage system consisting of NAND flash memory-based SSD (Solid State Disk) and conventional HDDs belongs to the latter category. Since the latter category will be visited at the next section, this section will detail the former category.

One of the most critical drawbacks of the previously described dynamic power management in HDDs is that the delay of recovering from the power down state is too damaging if the idle time estimation is not accurate. So several studies proposed the hybrid use of flash memory component with HDDs. Initial approaches integrated the NAND flash memory component into the HDDs, called Hybrid HDD as shown in Fig. 15.7 (a), and it is now commercially available. The embedded NAND flash memory acts like as non-volatile cache. So the system can instantaneously read and write the data even when the spindle has stopped, as a result the total spindle down time is dramatically increased.

Another approach is to integrate the flash memory storage into the host system rather than into the HDDs as shown in Fig. 15.7 (b) [30, 14]. Their basic motivation is similar to the hybrid HDD technique, but support less latency than Hybrid HDDs.

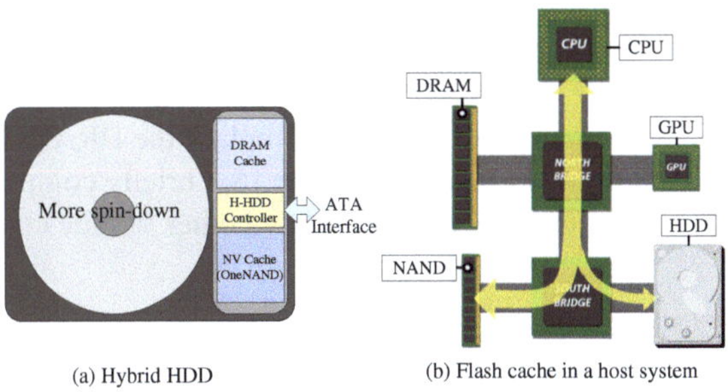

Fig. 15.7 Hybrid use of flash memory and HDDs.

15.2.3 Emerging Memory Technologies

Although there have been many studies to reduce the energy consumption of the main memory system mainly consisting of DRAMs, and the storage system mainly consisting of HDDs and partially NAND flash memory, any fundamental solutions for increasing the memory capacity have not been provided yet.

15.2.3.1 Storage Class Memory (SCM)

Recently, several types of new memory technologies have been introduced. Among them, PCM (Phase Change Memory) which is based on the material's phase changes between the amorphous and crystallize, is emerging as one of the most attractive solution for future memory systems because its scalability is better than DRAM. In addition, PCM consumes less power than DRAM, especially for read operations and during idle states [10]. There are other types of memory devices being developed but we will mainly cover PCM.

IBM defines PCM and flash memory as SCM (Storage Class Memory) which is a new class of data storage and memory devices [10]. Its ideal features are non volatile, fast access time like a DRAM, low cost per bit more like disk, solid state which means no moving parts. Fig. 15.8 (a) represents the characteristics of the SCM devices in the view of performance and capacity. Since the SCM can be efficiently used in-between the memory device and the storage device, the SCM is

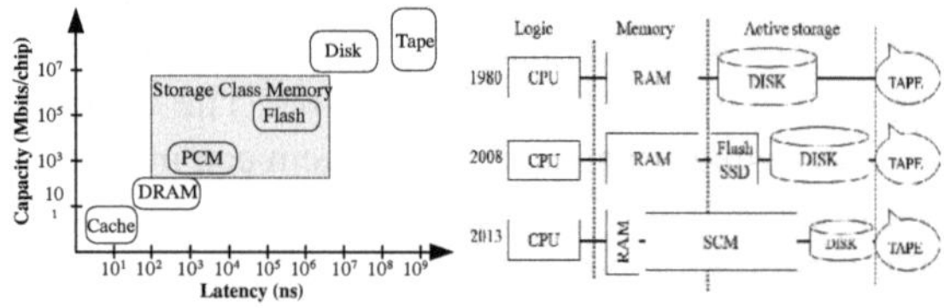

(a) Storage class memory (b) Future memory and storage system architecture

Fig. 15.8 Storage class memory.

expected to replace a large portion of memory and storage as shown in Fig. 15.8 (b). In this case, SCM can also greatly contribute to reduce the overall system power consumption because the SCM is a naturally low-power device, and does not require any significant power during idle state.

But there exist few limitations to completely replace the DRAM with PCM; low write performance and limited endurance. Table 15.3 briefly compares the characteristics of the PCM with conventional DRAM and other storage components.

Table 15.3 Memory and storage devices.

		DRAM	PCM	NAND Flash
Minimum. access unit		Word	Word	R/W: 2KB ∼ 8KB Erase: 128KB ∼ 256KB
Performance (Delay / Throughput)		R/W: 25 ∼ 30ns /∼300MB/s	R:50ns / 300MB/s W: ∼150ns/∼100MB/s	R: 25us / 40MB/s W: 200us / 10MB/s E: 1.5ms / 250MB/s
Power (/chip)	Active	130∼175mA	10∼70mA	15mA
	Idle	35mA	∼1mA	∼1mA
	P-down	6mA	∼0mA	∼0mA
Density		4Gb	1Gb	256 Gb
Scaldown limit		∼ 2x nm	∼ 5nm	∼ 1x nm
Endurance		10^{15}	$10^{6\sim 8}$	10^{5}

As shown in the table, PCM's write performance is relatively low, and this even may result in higher write energy consumption, compared to the DRAM. Nevertheless, its other characteristics are still attractive for enhancing storage performance and power consumption, and the weakness can be compensated by the hybrid use of existing memory technologies.

15.2.3.2 PCM-Based Main Memory Architecture

Although it has been just a few years since the initial PCM technology was announced, there has been lots of work to exploit its various benefits. [16] and [35] introduced PCM architectures as a DRAM alternative. They replaced the conventional DRAM main memory system with PCMs, but added more features such as buffer reorganization, partial write, redundant bit-writes, row shifting and segment swapping in order to compensate the low write performance and limited endurance. Their experiments demonstrated that PCM-based main memory systems can reduce the energy consumption of the memory system by almost 60% on average.

[24, 7] designed PCM-based main memory architectures with a use of small DRAM as a cache and buffer. The large size of PCM memory is dedicated to compensate the performance gap between the disk and main memory, and thus reducing the number of page faults, while the small size DRAM memory is dedicated to compensate the low write performance and endurance of PCM, using cache technology. Compared to similar cost of DRAM-based main memory systems, they can reduce the energy consumption in memory systems by around 30% to 50%.

Although the PCM has not matured as a technology yet and very limited versions of commercial products (NOR flash interface compatible version) are available in the market, DDR (Double Data Rate) and DDR2 interface supported versions are expected to be announced in the near future. We expect that efficient use of this PCM will contribute to reduce the energy consumption of the memory system as well as the storage system significantly.

15.3 System-Level Approach for Energy Efficient Storage

This section is concerned with system-level approaches in green storage technologies. The previous section covered, at the component-level, power and energy considerations of different storage technologies. Beyond component-level issues, there are a number of techniques that can be used at the system level to optimize the power consumption of storage in IT systems. A system is a collection of components that can be controlled and managed and that can operate under different states or conditions (Fig. 15.9). We are interested in storage systems, including memory and last level storage (disks).

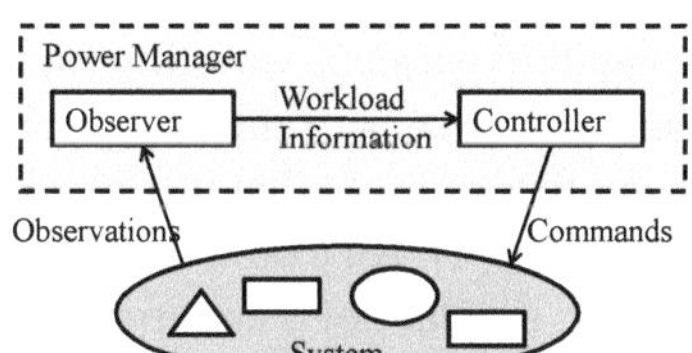

Fig. 15.9 System level power manager.

Clark and Yoder [4] have outlined a list of techniques for improving energy efficiency in data centers. The following summarizes some of those best practices relevant to storage energy efficiency:

- Use appropriate RAID level
- Leverage storage virtualization
- Use data compression, data deduplication and file deduplication techniques

- Use thin provisioning
- Use tiered storage (hierarchies of storage devices)
- Leverage solid state storage

At the system level, solutions for storage power management can be divided into the following categories [11]:

- Hardware based solutions: they explore new hierarchies and combination of storage components to reduce power consumption.
- Disk management solutions: these techniques control disk configuration and data layout. [23, 27, 36].
- Caching and data movement solutions: they explores caching mechanisms that allow a portion of the components of a storage system to remain idle for long periods of time.

The remainder of this section details some system-level approaches to reduce energy in storage systems and look at the effect on power performance of important issues in reliability and performance of storage systems.

15.3.1 Dynamic Power Management

Dynamic Power Management (DPM) approaches are techniques that can dynamically reconfigure a system to minimize the power consumption and stay within acceptable performance (response time, I/O bandwidth, etc). Many modern storage components can operate at different power modes. Disks and memory chips can operate at different power modes, spin at different speeds; processors can operate at different frequencies, etc. In the example of disks, they can operate under modes such as: *active* (disk is servicing requests), *idle*(disk platters are spinning but not servicing requests), *standby* (disk platters are not spinning) and possibly other operating modes. These modes may vary depending on the manufacturer and the type of application (embedded, servers, etc).

Systems and their components experience variations in their workload at runtime. This non uniform load is one of the basis of the applicability of DPM. In addition, DPM techniques repose on another premise that it is possible to predict, to a certain degree, the variations in workloads [1]. An example of a simple DPM technique can be to turn off a component (disk, display) of a laptop computer after a certain period of inactivity.

DPM can be implemented in both hardware and/or software. Software offers more flexibility. DPM is better handled when implemented in software at the OS level because the OS is responsible for managing resources (memory, disks, IO operations, etc). Benini et al. [1] note that the implementation of DPM is a hardware/software co-design problem because of the need to interface the OS-based power manager and the hardware resources.

We take the example of a single disk system power management system. Many of the principles will translate to other storage components. The role of the dynamic power manager is to timely transition the disk into power saving operating modes without adversely affecting performance. The DPM problem can be modeled as state transition diagram as illustrated in Fig. 15.10. It is important to note that transitioning between modes comes at a cost in performance. When a disk is set to sleep and a request arrives, the disk has to be spun up and then service the request, which causes a loss of performance in terms of I/O latency. A key decision the power manager has to make is to decide when to move the disk to low power modes. This decision is based on the best guess of the power manager of when the next requests will happen. Ideally, if we knew in advance all the requests to the disk, the power manager can devise an optimal strategy that minimizes the power consumption under certain tolerable loss (or none at all) of performance. In real systems, the DPM has to guess based on the previous I/O request. This prediction is called idle time predictions. Some DPM techniques use a fixed threshold and decide to spun down the disk when no I/O request has occurred for longer than a time threshold. The fixed time threshold method does not make use of the I/O request history. To make use of the I/O request history, the threshold can be adapted and adjusted according to the program behavior. In general, idle time prediction mechanisms can be categorized in two approaches [1]: predictive techniques and stochastic control techniques.

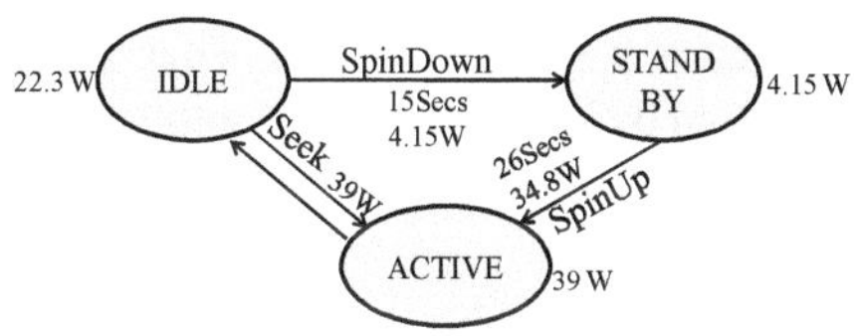

Fig. 15.10 Power modes and state transition diagram.

15.3.1.1 Predictive Techniques

Predictive techniques attempt to predict the idle period by exploiting the correlation between the history of the workload and its future. The simplest predictive technique is the fixed threshold method previously described also known as the *fixed timeout*. Timeout methods always waste power when waiting for the time to expire and pay a performance hit every time the disk wakes up. To address these issues, predictive shutdown and predictive wakeup policies have been proposed.

Predictive shutdown policies take power management decisions as soon as the idle period starts based on previously observed idle periods. In [1] the idle time is predicted from previous active and idle time durations in the following way:

$$T_{pred} = \Phi(T^{n}_{active}, T^{n-1}_{idle}, \ldots, T^{n-k}_{active}, T^{n-k-1}_{idle}) \quad (15.1)$$

Where $\Phi()$ is a non linear function. The system transitions to a low power mode as soon as it become idle if T_{pred} is greater than a threshold.

15.3.1.2 Stochastic Control Techniques

Predictive models have many limitations. They are based on a two-state system model, they assume deterministic response and transition times, they require parameter tuning. To address these issues, stochastic control approaches have been proposed. The stochastic control approach formulates policy optimization as an optimization problem under uncertainty. One class of stochastic methods uses Markov processes [1]. The system and the workload are modeled as Markov chains. Within this framework, it is possible to model the uncertainty in system power consumption and response times, to model many power states, to compute power management policies that are globally optimum and to explore tradeoffs between power and performance.

15.3.1.3 DRPM (Dynamic RPM)

Many disk power management use few power modes and often the choice is between stopping the disk rotation completely and keeping it rotating at a constant speed. Such methods pose problems when the idle times between disk requests is not long enough or less predictable or when we cannot afford the latency overhead (as can be the case in servers). To address this problem, Gurumurthi et al. [12] have proposed to adjust the rotational speeds of disks. They proposed a method called DRPM to modulate disk speed dynamically, and as a consequence adjust their power consumptions.

15.3.1.4 Main Memory Energy Reduction

While many of the techniques previously described also apply to dynamic power management of memory devices, there are specific power management techniques to the main memory. Dinis et al. [8] study dynamic power management methods for main memories of a stand-alone computer. They proposed and evaluated four DPM methods (Knapsack, LRU-Greedy, LRU-Smooth and LRU-Ordered) that dynamically adjust the power states of the memory devices depending on the load. They cast the DPM with a maximum power budget as a Multi-Choice Knapsack problem (MCKP) where the power budget represents the capacity, each memory device and power state represent an object, the power consumption represents the weights of objects and the overhead of transitioning from a power state to another represents the objects' cost. The goal is to pick a power state for each memory device to minimize potential performance degradation while staying under the power budget. MCKP is a NP-hard problem, but for small number of memory devices, a brute force solution is feasible. However, when the number of memory devices increases, approximate solutions are necessary. Diniz et al. propose thee such solutions (LRU-Greedy, LRU-Smooth and LRU-Ordered). See [8] for more details.

In data servers, memory accesses generated by network and disk DMAs (Direct Memory Access) are dominant [22]. Building from the specificity of DMA memory accesses, Pandey et al. [22] propose two performance-directed DMA-aware memory energy management techniques: One that temporally aligns DMA operations from different I/O buses to the same memory device to sequence these operations in lock-step and a second technique that lays pages in memory based on their logarithmic popularity curve in order to align DMA transfers.

15.3.2 Disk Arrays and Large Scale Storage Systems

Large scale server systems have particular characteristics in terms of their storage requirements, their workloads and their need for performance as noted by Ganesh et al. [11]. Server systems have short idle periods. This makes dynamic power management techniques that rely on long idle periods less effective unlike in devices like laptops. Gurumurthi et al. [12] have also shown that idle periods are harder to predict because of the low correlation between consecutive idle period durations. In addition server systems have higher performance constraints in general because they must abide by service level agreements to guarantee a certain quality of service.

To increase I/O performance, disk arrays have been proposed and are generally used in high performance storage systems. The data is stripped across many disks and the high throughput is achieved through parallel I/O access to many disks. The use of disk arrays is not a power conserving method. Disk arrays are a solution to the I/O bottleneck caused by the large and increasing gap between processor speed and disk bandwidth. To address the reliability problem caused by using multiple disks, many large scale storage systems rely on RAID (Redundant Array of Independent Disks). All those methods to increase I/O throughput and reliability come at an additional power consumption cost.

Several studies have been done in addressing the power consumption of large-scale storage systems [5]. The proposed methods revolved around improving energy efficiency of RAID systems, power efficient data movement and placement, and caching mechanisms. In the following sections, we will detail some of those approaches.

15.3.2.1 Caching and Efficient Data Movement

Massive Arrays of Idle Disks (MAID) [5] was used by Colarelli et al. as an alternative to tape backup drives. The main idea is to use a small number of cache disks next to the MAID disks. The small portion of the data that will be accessed at a time in the archives is copied in the cache disks and the other MAID disks can operate at lower power modes or shut down because they will be less solicited. The shutdown disks will be powered up only when a cache miss occurs. LRU can be used as a replacement policy for the caches. MAID is particularly suitable for *Write Once, Read Occasionally* (WORO) applications.

Along the same lines as MAID, Pinheiro and Bianchini [23] proposed an energy conservation technique for disk-array network servers termed Popular Data Concentration (PDC). PDC dynamically moves the most frequently accessed disk data to a subset of the disks. The goal is to skew the load toward few disks so that the remaining ones can be set to low power consumption states. PDC is built for Web server workload that have the tendency to solicit a small number of files.

Son et al. [27] proposed a new data layout mechanism with the purpose of increasing disk idle times significantly in order to enable more effective energy management. Their method analyzes array access patterns through trace profiling and determines the number of disks to use, the start disk for stripping the data, the strip unit and how the data is stripped.

An important issue in cache storage is the replacement policy. Zhu et al [37] looked at the effect of cache storage management on energy consumption and proposed several cache replacement algorithms using dynamic programming to minimize the disk energy consumption.

On a different work, Zhu et al. [36] proposed *Hybernator*, a disk energy management solution combining dynamic power management (that leverages multispeed disk drives and adapts the speed based on the workload), a data migration scheme called randomized shuffling and a multi-tier data layout where each tier corresponds to a set of disks operating at the same speed.

Data Prefetching has also been used as a form of data movement method to improve storage energy consumption. Son and Kandemir [28] proposed a compiler-directed energy-aware data prefetching schemes for scientific applications that process data residing in disks. Their proposed scheme sets the data stripping parameters, the prefetch distance and the disk speeds in a unified setting.

15.3.2.2 Enhancing the Power Efficiency of RAID

Many high performance storage systems rely on RAID for reliability and performance. However RAID systems have a high power consumption because conventional RAID balances the distribution of data across disks and causes all the disks to be active independently of the load. Gurumurthi et al. [13] performed an analysis of power performance implications of RAID on transaction processing workloads. One of their main conclusions is the fact that idle periods are generally low and difficult to exploit for power management. This observation has influenced many approaches that attempt to reduce the power consumption of RAID either by caching [20, 17] or efficient data stripping [32].

Weddle et al. [32] introduced power-aware RAID (PARAID), a skewed stripping pattern that allows RAID disks to use enough disks to meet the load. The number of powered disks can be varied based on the load to meet performance constraints.

Lee et al. [17] used a caching mechanism. They use an SSD as a large cache to alleviate the RAID disks and increase their idle periods. When a read to the RAID happens, the data is copied to the SSD and subsequent reads are serviced by the SSD. Writes are buffered in the SSD and copied to the RAID disks at longer time intervals.

Otoo et al. [20] proposed a dynamic block exchange algorithm for reducing the energy consumption of disk arrays configured as RAID-5. They use a queuing model and the observed workload to re-distribute the load (exchanging data block) among RAID groups so that the most frequently accessed data will reside in few RAID groups. A RAID group being a collection of disks organized in RAID fashion.

15.3.3 Virtualization

Virtualization has grown popular in many computing platforms. Virtualization provides increased efficiency and manageability, security isolation, better accessibility, etc. However, virtualization poses new challenges to power management in general and to I/O power management in particular. This is because of the additional layer (the Virtual Machine Monitor) between the operating system and the hardware. Several works have been done that attempt to bridge this gap and propose mechanisms to interface the workload behaviors of applications running on the virtual machines (VM) and the host operating systems (OS) power management layer [3, 29, 33].

Tian and Dong [29] analyzed how I/O virtualization challenges impact energy efficiency. They propose a power-aware I/O virtualization architecture called PAIOV to reunite VM workloads to the host power management framework. They validate their approach on a portable device and show reduced virtualization overhead and extended battery life.

Ye et al. [33] explored the disk I/O activities between Virtual Machine Monitor (VMM) and VMs to understand the I/O behavior of the VM system. They propose mechanisms to address the isolation between VMM and VMs. Their methods increase the burstness of hard disk accesses for better energy management of hard disks.

Chen et al. [3] proposed *ClientVisor*, a solution that leverages Commercial-Off-The-Shelf (COTS) OS functionalities for power management in virtualized desktop environment.

15.3.4 Compression, Deduplication

A number of techniques to reducing the energy consumption of storage systems attempt to do so by reducing the size footprint of the data to be stored. Such techniques include data compression and deduplication [4]. Data compression techniques compress the data at the bit level to minimize the amount of data to be stored. Data deduplication on the other hand, works at the data block level, and tries to avoid duplication. It is common to have duplicate identical blocks when copies of identical files or minor alterations of a file copy are present. In data deduplication, redundant data blocks are identified and referenced to a single data block. Both compression and deduplication introduce a layer of additional processing. At every write, the data should be compressed and at every read, the data should be decompressed. File deduplication operates with the same principle as data deduplication but at the file system level.

15.4 Challenges and Opportunities

The ever-growing storage requirements of large-scale scientific and multimedia applications have led industry and research communities to make substantial investments in procuring energy-conserving high-performance computing platforms based on high end storage farms and embedded systems. Digital data is growing at an astounding rate of 30% per year, rendering it almost impossible for a single storage platform to keep up. This evolution of storage hierarchy results in increased management costs and complexity, and thus data storage management is a growing concern from a green computing perspective. Rich Internet and Media (RIM) applications including Augmented Reality (AR), immersive communications, and personalization services lead to tremendous unstructured files that require new, more scalable and energy-effective solutions.

Even with the increasingly powerful technology, system resources are failing to keep up with what these modern applications (Big Data) demand, driving the need for larger data retention capacity and processing cost, which in turn drives energy consumption and cooling demands. Industry aggressively moves forward innovations on energy-aware fast compression/decompression, storage virtualization and high capacity green storage for enterprise market, beyond expandability, cooperative access control and availability. It may require forward thinking architectural solution.

HDD have been the primary storage media for large-scale file system for the last few decades. With advancements in the semiconductor technology, NAND flash memory based SSD has been widely used and evolving into SCM. SSD offers several advantages over HDD such as lower access latency, higher resilient to external shock and vibration and lack of noise, lower power consumption which results in lower operating temperatures. Further, recent reductions in cost (in terms of dollar per GB) accelerated the adoption of SSDs in a wide range of application areas from mobile and embedded systems to enterprise-scale storage systems. Flash technology is one promising storage media for various platforms, and hybrid approaches are widely explored in combining flash for persistency, high capacity, and low cost with DDR/RAM for performance. Further, introducing emerging memory devices such as PCM and STT-RAM has paradigm-shifting implications for new memory-hierarchies from both business and technological viewpoints. Technically speaking, PCM brings in several desirable properties [18], [25]. For example, first, its read performance will match that of SDRAM while achieving significantly low operational power. Second, its non-volatility will enable on-board storage for fast sleep and resume actions. Furthermore, PCM devices (banks) that store currently unused memory pages can be opportunistically powered off for low energy consumption. Third, PCM is projected to scale nicely in future technology generations, achieving high cost-effectiveness. Lastly, PCM devices are more robust than DRAM against soft errors. To use PCM, however, there are several challenges, including relatively slow performance (especially write) and finite write endurance. To manage the hybrid storage organization, we should address various HW/SW solutions, considering storage configurations, the quantity and speed of tiered-access I/O connectivity, the types and speeds of the processors, or even the amount of cache memory that determines performance.

Power consumption is a major concern in the design of modern systems and data centers. As with green storage systems, most of the recent research has focused on dynamic and intelligent power management system as a smart way to optimize and manage power. In the intelligent power management, main memory and storage have become a significant issue, contributing to as much as 40-50% of total consumption on modern server systems. Scalable and flexible clustered file server and storage systems require energy-efficient hybrid SCM, leveraging the inherent processing capabilities of constantly improving underlying hardware platforms.

References

1. Benini, L., Bogliolo, A., De Micheli, G.: A survey of design techniques for system-level dynamic power management. IEEE Trans. Very Large Scale Integr. Syst. 8, 299–316 (2000)
2. Blumstengel: P., Jordan, S., Koch, F., Korp, M., Gottschling, D., Block, H., Wurthmann, G.: Leveraging memory technology to cut data center power consumption. Whitepaper
3. Chen, H., Jin, H., Shao, Z., Yu, K., Tian, K.: ClientVisor: leverage COTS OS functionalities for power management in virtualized desktop environment. In: Proceedings of the 2009 ACM SIGPLAN/SIGOPS international conference on Virtual execution environments, VEE 2009, pp. 131–140 (2009)
4. Clark, T., Yoder, A.: Best practices for energy efficient storage operations. SNIA (2008)
5. Colarelli, D., Grunwald, D.: Massive arrays of idle disks for storage archives. In: Proceedings of the 2002 ACM/IEEE Conference on Supercomputing, Supercomputing 2002, pp. 1–11 (2002)
6. Delaluz, V., Kandemir, M., Vijaykrishnan, N., Sivasubramaniam, A., Irwin, M.J.: Hardware and software techniques for controlling DRAM power modes. IEEE Transactions On Computers 50, 1154–1173 (2001)
7. Dhiman, G., Ayoub, R., Rosing, T.: PDRAM: A hybrid PRAM and DRAM main memory system. In: Design Automatin Conference (2009)
8. Diniz, B., Guedes, D., Meira Jr., W., Bianchini, R.: Limiting the power consumption of main memory. In: Proceedings of the 34th annual international symposium on Computer architecture, ISCA 2007, pp. 290–301 (2007)
9. Fan, X., Ellis, C.S., Lebeck, A.R.: Memory controller policies for DRAM power management. In: Proceedings of the International Symposium on Low Power Electronics and Design (ISLPED), pp. 129–134 (2001)
10. Freitas, R.F., Wilcke, W.W.: Storage-class memory: The next storage system technology. IBM Journal of Research and Development 52(4/5), 439–447 (2008)
11. Ganesh, L., Weatherspoon, H., Balakrishnan, M., Birman, K.: Optimizing power consumption in large scale storage systems. In: Proceedings of the 11th USENIX workshop on Hot topics in operating systems, pp. 9:1–9:6 (2007)
12. Gurumurthi, S., Sivasubramaniam, A., Kandemir, M., Franke, H.: DRPM: dynamic speed control for power management in server class disks. In: Proceedings of the 30th Annual International Symposium on Computer Architecture, ISCA 2003, pp. 169–181 (2003)
13. Gurumurthi, S., Zhang, J., Sivasubramaniam, A., Kandemir, M., Franke, H., Vijaykrishnan, N., Irwin, M.J.: Interplay of energy and performance for disk arrays running transaction processing workloads. In: Proceedings of the 2003 IEEE International Symposium on Performance Analysis of Systems and Software, pp. 123–132 (2003)
14. Kgil, T., Roberts, D., Mudge, T.: Improving NAND Flash based disk caches. In: Proceedings of the International Symposium on Computer Architecture, pp. 327–338 (2008)
15. Lebeck, A.R., Fan, X., Zeng, H., Ellis, C.: Power aware page allocation. In: Architectural Support for Programming Languages and Operating Systems, pp. 105–116 (2000)

16. Lee, B.C., Ipek, E., Mutlu, O., Burger, D.: Architecting phase change memory as a scalable DRAM alternative. In: International Symposium on Computer Architecture (2009)
17. Lee, H.J., Lee, K.H., Noh, S.H.: Augmenting RAID with an SSD for energy relief. In: Proceedings of the 2008 conference on Power aware computing and systems, HotPower 2008, pp. 12–12 (2008)
18. Lee, K.J., Cho, B.H., Cho, W.Y., Kang, S., Choi, B.G., Oh, H.R., Lee, C.S., Kim, H.J., Park, J.M., Wang, Q., Park, M.H., Ro, Y.H., Choi, J.Y., Kim, K.S., Kim, Y.R., Shin, I.C., Lim, K.W., Cho, H.K., Choi, C.H., Chung, W.R., Kim, D.E., Yu, K.S., Jeong, G.T., Jeong, H.S., Kwak, C.K., Kim, C.H., Kim, K.: A 90nm 1.8V 512Mb diode-switch PRAM with 266MB/s read throughput. In: Solid-State Circuits Conference, ISSCC 2007, Digest of Technical Papers, IEEE International, pp. 472–616 (2007), doi:10.1109/ISSCC.2007.373499
19. Liu, S., Zhang, Y., Memik, S.O., Memik, G.: An approach for adaptive DRAM temperature and power management. IEEE Transaction on Very Large Scale Integration 18, 668–684 (2010)
20. Otoo, E., Rotem, D., Tsao, S.C.: Dynamic data reorganization for energy savings in disk storage systems. In: Proceedings of the 22nd International conference on Scientific and Statistical Database Management, SSDBM 2010, pp. 322–341 (2010)
21. Panabaker, R.: Hybrid hard disk and readydrive^{T}M technology: Improving performance and power for windows vista mobile PCs. Microsoft WinHec (2006)
22. Pandey, V., Jiang, W., Zhou, Y., Bianchini, R.: DMA-aware memory energy management. In: The Twelfth International Symposium on High-Performance Computer Architecture (2006)
23. Pinheiro, E., Bianchini, R.: Energy conservation techniques for disk array-based servers. In: Proceedings of the 18th Annual International Conference on Supercomputing, ICS 2004, pp. 68–78 (2004)
24. Qureshi, M.K., Srinivasan, V., Rivers, J.A.: Scalable high performance main memory system using phase-change memory technology. In: International Symposium on Computer Architecture (2009)
25. Raoux, S., Burr, G.W., Breitwisch, M.J., Rettner, C.T., Chen, Y.C., Shelby, R.M., Salinga, M., Krebs, D., Chen, S.H., Lung, H.L., Lam, C.H.: Phase-change random access memory: A scalable technology. IBM Journal of Research and Development 52(4.5), 465–479 (2008), doi:10.1147/rd.524.0465
26. Shim, H., Joo, Y., Choi, Y., Lee, H.G., Chang, N.: Low-energy off-chip SDRAM memory systems for embedded applications. ACM Transactions on Embedded Computing Systems 2, 98–130 (2003)
27. Son, S.W., Chen, G., Kandemir, M.: Disk layout optimization for reducing energy consumption. In: Proceedings of the 19th Annual International Conference on Supercomputing, ICS 2005, pp. 274–283 (2005)
28. Son, S.W., Kandemir, M.: Energy-aware data prefetching for multi-speed disks. In: Proceedings of the 3rd Conference on Computing Frontiers, CF 2006, pp. 105–114 (2006)
29. Tian, K., Dong, Y.: Power-aware I/O virtualization. In: Proceedings of the 2nd conference on I/O virtualization, WIOV 2010, pp. 8–8 (2010)
30. Trainor, M.: Overcomming disk drive access bottlenecks with Intel Robson technology. Intel (2006)
31. Udipi, A.N., Muralimanohar, N., Chatterjee, N.: Rethinking DRAM design and organization for energy-constrained multi-cores. In: Proceedings of the International Symposium on Computer Architecture Microarchitecture (2010)
32. Weddle, C., Oldham, M., Qian, J., Wang, A.I.A., Reiher, P., Kuenning, G.: PARAID: A gear-shifting power-aware RAID. Trans. Storage 3 (2007)

33. Ye, L., Lu, G., Kumar, S., Gniady, C., Hartman, J.H.: Energy-efficient storage in virtual machine environments. SIGPLAN Not. 45, 75–84 (2010)
34. Zheng, H., Lin, J., Zhang, Z., Gorbatov, E., David, H., Zhu, Z.: Mini-rank: Adaptive DRAM architecture for improving memory power efficiency. In: Proceedings of the International Symposium on Microarchitecture, pp. 210–221 (2008)
35. Zhou, P., Zhao, B., Yang, J., Zhang, Y.: A durable and energy efficient main memory using phase change memory technology. In: International Symposium on Computer Architecture (2009)
36. Zhu, Q., Chen, Z., Tan, L., Zhou, Y., Keeton, K., Wilkes, J.: Hibernator: helping disk arrays sleep through the winter. In: Proceedings of the twentieth ACM symposium on Operating systems principles, SOSP 2005, pp. 177–190 (2005)
37. Zhu, Q., Zhou, Y.: Power-aware storage cache management. IEEE Trans. Comput. 54, 587–602 (2005)

Chapter 16
Sustainable Science in the Green Cloud via Environmentally Opportunistic Computing

In-Saeng Suh and Paul R. Brenner

Abstract. The energy consumed by data centers is growing every year. Significant energy and cost savings are possible through modest gains in efficiency. These leverage solutions for both economical gains and improvement of their environmental footprint. One solution is Environmentally Opportunistic Computing (EOC), which is a sustainable computing concept that capitalizes on the physical and temporal mobility of modern computer processes and enables distributed computing hardware to be integrated into a facility to optimize the consumption of computational waste heat. In this work, we will review the EOC methodologies as applied in the Green Cloud (GC) framework, and describe application of the EOC concepts to the local and global grid infrastructures.

16.1 Introduction

Data centers are vital components in nearly every sector of the economy and deliver critical information technology services, including data storage, communications, and internet accessibility. The 2007 United States Environmental Protection Agency reported that data centers use a significant amount of energy, accounting for 1.5% of total U.S. electricity consumption at a cost of $4.5 billion annually [1]. The energy consumed by data centers is growing every year and expected to almost double over the next five years. Significant energy and cost savings are possible through modest gains in efficiency. These leverage solutions for both economical gains and improvement of their environmental footprint. One solution of this problem is Environmentally Opportunistic Computing (EOC) [2,3,4], which is a sustainable computing concept that capitalizes on the physical and temporal mobility of modern computer processes and enables distributed computing hardware to be integrated into facility infrastructures to optimize the consumption of computational waste heat.

In-Saeng Suh · Paul R. Brenner
Center for Research Computing, University of Notre Dame,
Notre Dame, Indiana 46556, USA
e-mail: isuh@nd.edu, pbrennel@nd.edu

The first implementation of EOC prototype at Notre Dame is the Green Cloud [2,3,4], a bridge between traditional data center and containerized data center, located at the City of South Bend Botanical Conservatory and Greenhouse. Much like a ground-source geothermal system, EOC performs as a "system-source" thermal system, with the capability to create heat where it is locally required, to utilize energy when and where it is least expensive, and to minimize a building's overall energy consumption. Instead of expanding active mechanical systems to contend with thermal demands, the EOC concept utilizes existing high performance computing and information communications technology coupled with system controls to enable energy hungry, heat producing data systems to become service providers to a building while concurrently utilizing aspects of a building's HVAC infrastructure to cool the machines.

Relating with the "system-source" thermal system, one of the important factors is a consistent computing job supply. In order to realize this in a production level, we propose to integrate the Green Cloud running as part of the Notre Dame Condor pool with the Open Science Grid (OSG), specifically, the Notre Dame CMS Tier 3 site which is connected with the Worldwide LHC Computing Grid (WLCG) through the Northwest Indiana Computational grid (NWICG).

In this work, we will first introduce the EOC methodologies implemented in the Green Cloud framework. In the EOC section we provide operational details, computing job control system through Condor, and thermal control concept. In the Integration with the Open Science Grid we introduce the OSG and NWICG and then describe the integration of the Green Cloud with the OSG and WLCG. Finally we conclude with a summary of the practical utility of our prototype and related work with the Green Cloud linked to the national cyberinfrastructure.

16.2 Environmentally Opportunistic Computing

Environmentally Opportunistic Computing recognizes that increased efficiency in computational systems must be realized beyond systems-side advancement, and that the aggressive growth of users – and the demand capability of those users – must necessarily be met with new, integrated design paradigms that look beyond optimization of the traditional, single facility data center. EOC integrates distributed computing hardware with existing facilities to create heat where it is already needed, to exploit cooling where it is already available, to utilize energy when and where it is least expensive, and to minimize the overall energy consumption of an organization. The EOC research focuses on the developing models, methods of delivery, and building/system design integrations that reach beyond current waste heat utilization applications and minimum energy standards to optimize the consumption of computational waste heat in the built environment. What must happen in order to push existing computation waste heat reclamation forward to be transformative is the development of a systematic method for assessing, balancing, and effectively integrating various interrelated "market" forces (see reference [3]) related to the generation and efficient consumption of computer waste heat.

The efficient consumption of computer waste heat must be closely coordinated with building HVAC systems. A sensor-control relationship must be established between the GC systems. The controls network must mediate not only the dynamic relationship between source and target but also the variation in source and target interaction due to governing outside factors such as seasonal variations. In the colder winter months the computational heat source can provide necessary thermal energy whereas the relationship inverts during the hot summer months when the facility can provide reasonably cool exhaust/make-up to the computational components.

16.2.1 EOC Prototype

As the first field application of EOC, the University of Notre Dame Center for Research Computing (CRC), the City of South Bend, Indiana, and the South Bend Botanical Society have collaborated on a prototype building-integrated distributed data center at the South Bend Botanical Garden and Greenhouse (BGG) called the Green Cloud (GC) Project [2], which evolved from the Grid Heating framework [5,6]. The Green Cloud prototype is a container that houses CRC high performance computing (HPC) servers and is located immediately adjacent to the BGG facility, where it is ducted into one of the BGG public conservatories. The GC hardware facilities are fully integrated into the Notre Dame Condor pool and are currently able to run typical campus-level research computing loads. The heat generated from the HPC hardware is exhausted into the BGG public conservatory, with the goal to offset wintertime heating requirements and reduce BGG annual expenditures on heating. In 2006, the BGG spent nearly $45,000 on heating during the months of from January to March, alone. With implementation of the EOC GC prototype, there was about 5% energy saving effect and this should be able to reach to about 15% with improved design [3].

As shown in Figure 16.1, the EOC prototype is a container that houses 100 servers. The container was custom manufactured and each entry way is heavily secured for the security of the HPC equipment. During moderate-temperature months, external air (~50 °F/10 °C) is introduced into the container through a single louver, heated by the HPC hardware, and expelled into the BGG conservatory. Conversely, during cold-temperature months, when external air is too cold (< 50 °F/10 °C) to appreciably heat for benefit to the conservatory, a return vent has been ducted to the conservatory to draw air directly from the conservatory into the container, heat it from the HPC hardware, and then return it directly into the conservatory. Air is driven by a set of three axial fans through two ducts into the BGG. For operation during summer months, when the conservatory does not require additional heating, the ductwork is disconnected and the container uses free air cooling for the HPC hardware (see reference [3] for details).

Fig. 16.1 (a) Layout of prototype EOC GC container integrated into BGG facility (left) and (b) Photograph of EOC GC prototype at the BGG conservatory.

16.2.2 Intelligent EOC

One of the big challenges in the development of the GC prototype is heat management. To address this problem, a suite of temperature management scripts (the thermal manager) was developed and designed to both efficiently run jobs on the servers as well as regulate the overall temperature of the hardware and the EOC container itself. The thermal measurement systems monitor various local temperatures throughout the GC container as well as the server temperature and server loads. The thermal manager also measures the overall temperatures and updates configurations of the machines in the GC pool. The technical details of thermal sensor-control management framework are described in [2,3,4]. In order to manage the computing job workload, all machines are additionally managed by an environmentally-aware GC Manager (GCM) control system [3] which is written in Python and provides rule-based control of the servers running in the Green Cloud. The GCM maintains each machine within its safe operating temperature and therefore maximizing temperature of the hot-isle air that is used for the greenhouse heating.

The GCM interfaces both Condor and xCAT (extreme Cloud Administration Toolkit) [7]. The Condor component manages all of the scientific workload management: deploying jobs on the servers, evicting jobs from machines, and monitoring the work state in the GC pool. The xCAT component handles the hardware's built-in service processors: power state control and measurements intake, memory, CPU temperature, fan speed, and voltages. Therefore, once compute-intensive jobs flow through a workflow mechanism into the EOC Green Cloud pool, the GCM dynamically manages the computing jobs and controls temperature of the Green Cloud.

An important issue is to ensure consistent job supply into the GC pool to provide the system-source thermal resource in a productive computation level during the cold season (from November to March). Currently, the Notre Dame Condor pool has been integrated to supply campus-scale research jobs. In order to improve reliability of the GC system, integration with larger-scale computing infrastructures is necessary. Therefore, in order to resolve this problem, we propose to expand GC to the local (NWICG) and global (OSG/WLCG) grid infrastructure.

16.3 Integration with OSG and NWICG

The Open Science Grid [8] is a distributed computing infrastructure for large-scale scientific research, built and operated by a consortium of universities, national laboratories, scientific collaborations, and software developers. The OSG Consortium's unique community alliance brings petascale computing and storage resources into a uniform grid computing environment. Researchers from many fields, including astrophysics, bioinformatics, computer science, medical imaging, nanotechnology and physics, use the OSG infrastructure to advance their research [9]. The OSG capabilities are also being driven by the needs of large scientific collaborations, such as ATLAS and CMS particle physics experiments at LHC [10] and LIGO [11], a facility dedicated to the detection of cosmic gravitational waves.

The NWICG [12] builds a cyberinfrastructure that supports the solution of breakthrough level problems, and enables continuing world-class advances in the underlying technologies of high performance computing. As a regional grid, NWICG is architected around the OSG. Recently the NWICG at Notre Dame has tied with the CMS WLCG through building the Notre Dame CMS Tier 3 site, becoming a science gateway for national infrastructures such as TeraGrid and OSG.

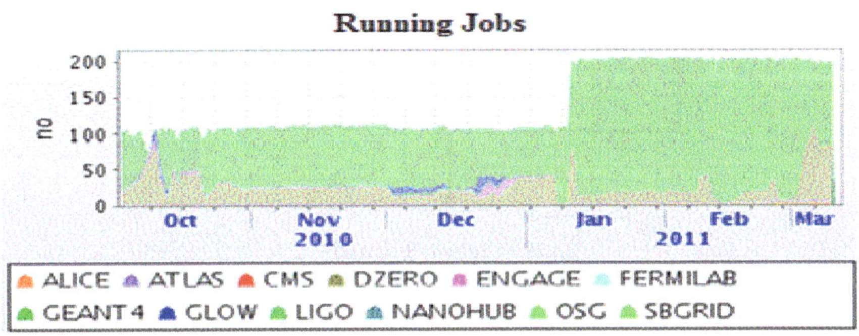

Fig. 16.2 Total number of running jobs at NWICG_NotreDame from various OSG VOs during 2010-2011 winter season.

As an example, Figure 16.2 shows the total number of running jobs at the NWICG_NotreDame site from various virtual organizations in the OSG during a period from Oct. 2010 to Mar. 2011.

As Figure 16.2 shows, consistent numbers of jobs are submitted from OSG VOs and running on the NWICG resources. Therefore, in order to keep supplying productive research computing jobs to the GC, we can integrate the GC with the local (NWICG) and global (OSG/WLCG) grid infrastructures as shown in Figure 16.3. As a testbed model, we propose to link a portion of the NWICG hardware into the Notre Dame Condor pool. This is denoted with a solid arrow line from NWICG to ND Condor. The dashed line denotes a redundant direct workflow model from NWICG to EOC Green Cloud.

To realize the constant stream of grid computing jobs between the grid infrastructure and the EOC Green Cloud, we need to develop and deploy campus grid technology. One model is the glidein-based Workflow Management System (glideinWMS) which is a general purpose workload management system (WMS) developed by US CMS and works on top of Condor. Details of the glideinWMS architecture and workflow mechanism are described in [13]. Through the glideinWMS technology, the local campus Condor pool can be integrated seamlessly with the global grid infrastructure. Therefore, grid jobs submitted from the global grid infrastructure are dynamically collected and submitted into the EOC Green Cloud through the glideinWMS Frontend. Once the jobs enter through the glideinWMS into the Green Cloud, the GCM in the EOC Green Cloud controls the job scheduling as described in Section 16.2.2.

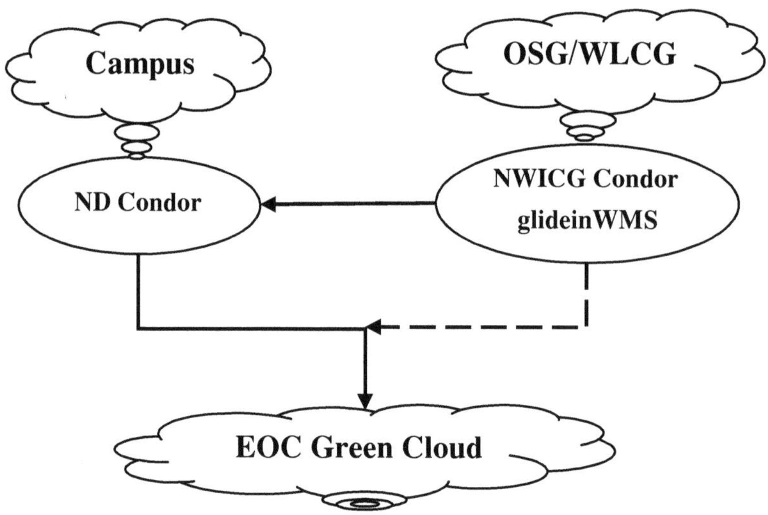

Fig. 16.3 A diagram depicting the EOC Green Cloud, campus, and global grid infrastructure integration.

We now are actively working on integration challenges, such as software stack locality, credentials, large-scale data locality, and network bandwidth, etc. of integrating the Notre Dame Tier 3 infrastructure with the campus condor grid.

16.4 Summary

In this work we have reviewed the basic structure of Environmentally Opportunistic Computing and the Green Cloud which is integrated with a City of South Bend greenhouse. Alongside growing utilization of cloud computing and virtualization, EOC could realize environmental and economic benefits to improve the sustainability of IT infrastructure. If computational waste heat can be recaptured, recirculated, and free cooling encouraged by explicit regulation, we may see an accelerated shift in the way that data centers are designed, driven by both economic and environmental conservation factors. The Green Cloud is a sustainable computing technology that works in conjunction with existing efficiency of IT resources at the application, operating system and hardware levels. In addition, we have introduced NWICG and OSG and provided an integration model of the Green Cloud with the grid infrastructure. In order to improve the reliability of the GC system, integration with larger-scale computing infrastructures is beneficial.

Acknowledgement

PB would like to recognize the generous donation of over 100 servers by the eBay Corporation for use in the Green Cloud prototype. The City of South Bend and South Bend Botanical Society have been outstanding partners in this first EOC prototype deployment. Partial support for this work was provided by Notre Dame Center for Research Computing and a Department of Energy grant for the Northwest Indiana Computational Grid. We would also like to thank collaborators Prof. Aimee Buccellato, David Go, and Douglas Thain.

References

1. US EPA Report to congress on server and data center energy efficiency public law 109-431. United State Environmental Protection Agency, Technical Report (August 2007)
2. Brenner, P., Jansen, R., Go, D.B., Thain, D.: Environmentally Opportunistic Computing: Transforming the Data Center for Economic and Environmental Sustainability. In: 2010 International Green Computing Conference, Chicago, IL, pp. 383–388 (2010)
3. Buccellato, A.P.C., Brenner, P., Go, D.B., Jansen, R., Ward, E.M.: Environmentally Opportunistic Computing: Computation as Catalyst for Sustainable Design. In: ASHRAE Winter Conference, ASHRAE2011-86119, Las Vegas, NV (2011)
4. Witkowski, M., Brenner, P., Jansen, R., Go, D.B., Ward, E.M.: Enabling Sustainable Clouds via Environmentally Opportunistic Computing. In: Cloud Computing, HCI, & Design: Sustainability and Social Impacts Workshop, 2nd IEEE International Conference on Cloud Computing Technology and Science (CloudCom 2010), Indianapolis, IN (2010)
5. Brenner, P., Thain, D., Lattimer, D.: Grid Heating: Transforming cooling constraints into thermal benefits. Technical Report 2008- 2009, University of Notre Dame, Computer Science and Engineering Department (April 2008)

6. Brenner, P., Thain, D., Lattimer, D.: Grid Heating Clusters: Transforming cooling constraints into thermal benefits. In: Uptime Institute – IT Lean, Clean, and Green Symposium, pp. 1–7 (2009)
7. Ford, E.: xCAT: extreme Cloud Administration Toolkit, IBM internal software development. Open Sourced 2007 (1999)
8. Pordes, R., Weichel, J.: Analysis of the Current Use, Benefit, and Value of the Open Science Grid. J. of Physics: Conf. Series 219(6), 062024:1–8 (2010)
9. Pordes, R., et al.: New Science on the Open Science Grid. J. of Physics: Conf. Series 125, 012070:1-6 (2070)
10. Collins, G.P.: Large Hadron Collider: the Discovery Machine. Scientific American 39–45 (January 2008)
11. Shawhan, P.S.: LIGO Data Analysis. Nuclear Instruments & Methods in Physics Research A502, 396–401 (2003)
12. Suh, I.-S., Bensman, E., Nabrzyski, J.: The Open Science Grid and Northwest Indiana Computational Grid: Challenges to Extreme Computing, UKC2009 Creative Minds for Global Sustainability, FPHY, KSEA, pp. 1620-1625 (2009)
13. Sfiligoi, I : glideinWMS - A general pilot-based Workload Management System. J. of Physics: Conf. Series 119, 062044:1–9 (2008)

Part 3
Smart Grid and Applications

Chapter 17
Smart Grid

George W. Arnold

Abstract. The basic architecture of the electric grid has not changed much since its development over 100 years ago: it is designed to move power from controllable centrally-generated sources through a transmission and distribution network to end users, supplying power needed to satisfy demand. About 68% of the world's power is generated from combustion of fossil fuels, and is a major source of CO_2 emissions. Additionally, the need to provide enough generation capacity to meet relatively short intervals of peak demand results in a very inefficient system. The "Smart Grid" refers to a modernized electric grid that is capable of supporting a high proportion of "uncontrollable" variable renewable carbon-free sources such as wind and solar, achieving greater system efficiency through simultaneous management of demand as well as generation, and greater reliability through extensive use of sensors and automation. The Smart Grid also promises further environmental benefits by supporting growing use of electric vehicles that can be charged using idle capacity in the grid. The application of information and communications technology is critical to the operation of the Smart Grid, which will operate much more dynamically than the present electrical system. The chapter will discuss the motivation, goals and benefits of the Smart Grid; its conceptual architecture and key enabling technologies; the critical role of standards to ensure interoperability of devices and systems that make up the grid; and discuss the status of its development in different parts of the world.

17.1 Introduction

Modern life would not be possible without reliable, affordable and universally available electric power. Studies have shown a strong correlation between electricity consumption and socio-economic indicators such as life expectancy, education, and GDP per capita[1]. It is not surprising, therefore, that in citing the greatest engineering achievements of the 20th century, the National Academy of

George W. Arnold
National Institute of Standards and Technology, U.S. Department of Commerce
100 Bureau Drive MS 8100 Gaithersburg, MD 20899-8100 U.S.A.
email: george.arnold@nist.gov

Engineering ranked the electric grid as the number one achievement, ahead of other marvels such as the automobile, airplane, telephony, and the internet. The quality of electric service has become increasingly important as we become more and more dependent on the pervasive application of microprocessors, electronics and software.

The electric grid dates back to 1882. In that year, Thomas Edison established the first investor-owned utility, the Edison Illuminating Company, and switched on the first 110V DC electric distribution station serving 59 customers on Pearl Street in New York City. In the ensuing century electricity became ubiquitous and affordable, powering the U.S., indeed the world's economic growth. As remarkable as it is, it is even more remarkable that the basic architecture of the modern electric grid is in many respects the same as it was a century ago. The grid has had many innovations and improvements over the years, but the fundamental architecture has not changed. The grid is designed to move electricity generated at large, primarily fossil-fueled centralized generating stations through a one-way transmission and distribution network to consuming devices that have no information about the cost of electricity or whether the grid is overloaded. This is in contrast to other infrastructures, such as telecommunications networks, that have undergone radical transformation in the last twenty years. Modernization of the aging electrical infrastructure has become a priority for many countries around the world.

17.2 Characteristics of the Present Electric Grid and the Need for Modernization

17.2.1 Greenhouse Gas Emissions

Generation of electricity is a significant cause of greenhouse gas emissions that contribute to global warming. Globally, 68% of electricity is generated from fossil fuels, mainly coal and natural gas[2] (for the United States, this proportion is 72%[3]). In the United States, electric power generation accounts for about 40 percent of human-caused emissions of carbon dioxide, the primary greenhouse gas[4]. The need to reduce carbon emissions has become an urgent global priority to mitigate climate change. If the existing power grid were just 5 percent more efficient, the resultant energy savings would be equivalent to permanently eliminating the fuel consumption and greenhouse gas emissions from 53 million cars[5]. Reducing the production of carbon in electricity generation will require a variety of clean or cleaner energy sources, including coal with carbon capture and sequestration, natural gas, nuclear, hydro, geothermal, biomass, and increased use of wind and solar. While still a small proportion of the overall portfolio, the penetration of wind and solar generation is increasing rapidly. Unlike conventional sources of energy, wind and solar are intermittent and cannot be controlled. As the proportion of these sources increases, new technologies and more dynamic controls need to be introduced into the grid to maintain reliable operation.

17.2.2 Aging Infrastructure

The electric grid in developed countries is also aging and equipment and facilities are nearing the end of their useful lives. In the U.S. about half of electricity is generated from coal [3], and half of these generating plants are more than 40 years old. They will need to be retired and replaced or modernized. The Brattle Group has estimated that the investment required in replacement or new generation is about $560 billion by 2030[6]. The overall investment in grid modernization and expansion, including transmission, distribution and generation is expected to be $1.5 to $2 trillion by 2030 in the U.S. The International Energy Agency estimates that globally, $13 trillion will be invested in the world's electric grids by 2030. Finding ways to make the grid more efficient could reduce the amount of investment that will be needed. For example, if greater efficiency in the transmission and distribution system and in end use could reduce the need for generation capacity by even a few percent, tens of billions of dollars could be saved.

17.2.3 Inefficiency

There are several sources of inefficiency in today's grid. Because the demand for electricity varies considerably according to time of day and season, sufficient generating capacity must be provided to handle peak periods that occur infrequently, for example on hot summer afternoons in areas where air conditioning represents a significant load. Less than half of the generation capacity in the U.S. comes from power plants designed to run all the time to meet demand. The rest – which are much more costly to operate - are in reserve to supply electricity during periods of high demand. It is estimated that capacity to meet demand during the top 100 hours in the year account for 10 to 20 percent of electricity costs [7]. If some of that demand could be shifted in time to non-peak periods, significant cost savings could be achieved.

Another source of inefficiency is waste in electricity consumption. Customers receive limited information about their own energy use that is helpful in monitoring and reducing their energy consumption. In most cases that information is limited to monthly usage readings. Electricity is consumed by three categories of users: residential, commercial (e.g. office buildings), and industrial (e.g. factories). Each category accounts for roughly one-third of overall consumption [3]. Automation and information technology have been employed for many years to increase efficiency or curtail load when necessary in commercial and industrial environments. However, there has been little attention placed on residential environments. There exist demand response programs that allow utilities, by agreement with customers, to control devices such as thermostats, pool pumps or hot water heaters during periods of peak demand, however these programs are not widely used and there are many additional opportunities to increase efficiency that have not yet been exploited.

A third source of inefficiency is a result of the limited situational awareness in today's grid. For example, most distribution grids lack sensors that provide utilities with real-time measurement of the actual voltages delivered to customers, so systems must be operated based on estimates of losses along the line. As a result utilities often supply higher voltage than necessary and energy is wasted. Deployment of automated sensors and controls that permit dynamic optimization of voltage levels and reactive power may permit reduction of voltage levels by a few percent and reduced power consumption [8].

17.2.4 Substituting Electricity for Oil in Transportation

While the preceding discussion focused on opportunities to reduce electricity generation and consumption, electrification of transportation presents an opportunity to achieve environmental benefits through increased use of electricity. Many nations that rely heavily on imported oil are concerned about the security of their energy supply. While oil represents less than 2% of the fuel used to generate electricity in the U.S. [3], transportation is heavily dependent on oil. Substituting "green" electricity for oil to provide heating and to power transportation has a double benefit by reducing carbon emissions while also increasing the security of energy supply.

17.2.5 Reliability

Improving the reliability of the electric system is an important priority. In the U.S. for example, distribution outages exceed two hours per year per customer [9]. These outages are estimated to cost the U.S. economy about $80 billion annually [10]. Significant improvement is possible – for example in Japan, power outages at the distribution level average only about 16 minutes per year per customer [11]. Lack of wide-area situational awareness in the transmission grid has resulted, on occasion, in large-scale cascading failures. Advanced sensors called synchrophasors that can provide remote real-time measurement of the condition of transmission facilities are just starting to be widely deployed.

17.3 Vision of the Future Smart Grid – Goals and Benefits

The development of the smart grid is intended to accomplish the following goals:

- Increase the efficiency and cost effectiveness of producing and delivering electricity

- Provide consumers with electronically-available information and automated tools to help them make more informed decisions about their energy consumption and control their costs
- Help reduce production of greenhouse gas emissions in generating electricity by permitting greater use of renewable sources
- Improve the reliability of service
- Prepare the grid to support a growing fleet of electric vehicles in order to reduce dependence on oil.

As discussed earlier, if usage during the peak hours could be shifted to non-peak periods or otherwise curtailed (a capability referred to as demand response), some of the infrequently used generation capacity could be saved. The U.S. Federal Energy Regulatory Commission estimates the potential for peak electricity demand reductions to be equivalent to up to 20 percent of national peak demand—enough to eliminate the need to operate hundreds of back-up power plants [12]. Demand response can be implemented either through direct load control by the utility, or indirectly through market forces with dynamic or time-of-use pricing of electricity.

The concept is very similar to peak/off-peak pricing plans and flow control protocols that have historically been used to smooth demand in telecommunications networks. Dynamic pricing can take many different forms, ranging from simple schemes such as scheduled time-of-use pricing, to schemes that set higher prices only during critical peak periods, to real-time interval pricing based on wholesale market rates. Evaluating which schemes are most effective is an important area of experimentation and research. Some pilot programs have demonstrated that well-designed schemes can achieve reductions in peak energy use and customer savings across all income levels and are well-received by customers [13].

Presently customers have little information available to understand or manage their energy use. An advanced metering infrastructure (AMI), a key element of the smart grid, will allow near real-time measurement of customer energy use. The smart grid will also provide customers with information management capabilities that permit smart appliances and energy management systems to minimize energy use and shift demand to less costly non-peak periods, saving money.

The smart grid will introduce new technologies and operating principles needed to support increasing use of intermittent renewable energy as part of a clean energy generation portfolio. In the U.S., most states have set goals for the fraction of electricity to be generated from renewable sources. California, for example, has a goal to achieve 33 percent by 2020 [14]. Solar and wind represent abundant sources of clean, renewable energy. However, unlike traditional energy sources, solar and wind are variable and intermittent. Integrating solar and wind into the grid presents a new challenge as the penetration increases because of the need to continually balance load with a varying and less predictable supply. Energy storage technologies that can buffer varying supply will become an important element of the smart grid. While some solar and wind generation will be deployed in centralized, large-scale "farms", a growing proportion of renewable generation will be distributed locally, for example solar rooftop panels. Some communities will function as "micro- grids" capable of generating, at times, enough power to satisfy their own needs, or selling excess power back into the grid at other times, or buying power from the grid at other times. The smart grid will have to support much more distributed power generation and two-way flow of electricity.

The smart grid will increase reliability through the deployment of sensors that provide real-time situational awareness and controls that can reconfigure parts of the grid when failures occur. Phasor Measurement Units (PMUs), now being widely deployed in the U.S. transmission grid, will allow earlier detection of and response to anomalies and make widespread cascading failures less likely. Sensors in the grid could also, for example, detect when a transformer is deteriorating and allow replacement before a failure occurs. Remotely operated reclosers and sectionalizers can speed service restoration following a failure.

Over the long term, the electrification of the transportation system has the potential to yield huge energy savings and other important benefits. Estimates of associated potential benefits include:

- Displacement of about half of U.S. net oil imports
- Reduction in U.S. carbon dioxide emissions by about 25 percent
- Reductions in emissions of urban air pollutants of 40 percent to 90 percent.

A U.S. Department of Energy study found that the idle capacity of today's electric power grid could supply 70 percent of the energy needs of today's cars and light trucks without adding to generation or transmission capacity—if the vehicles charged during off- peak times [15]. The smart grid will provide capabilities to monitor and manage the charging of electric vehicles to avoid overloading the grid and minimize cost for consumers.

17.4 Conceptual Architecture and Key Enabling Technologies

While definitions and terminology vary somewhat, all notions of an advanced power grid for the 21st century hinge on the integration of advanced information technology, digital communications, sensing, measurement and control into the power system. The smart grid evolves the architecture of the legacy grid which can be characterized as providing a one-way flow of centrally-generated power to end users into a more distributed, dynamic system characterized by two-way flow of power (centralized and distributed) and information. Bi-directional flows of information will enable an array of new functionalities and applications that go well beyond "smart" meters for homes and businesses.

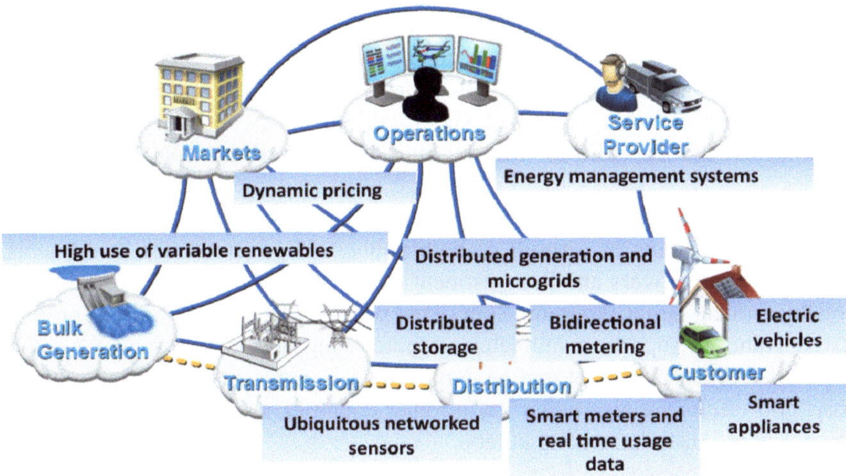

Fig. 17.1 Conceptual model of the smart grid

The National Institute of Standards and Technology in the U.S. has described a conceptual model of the smart grid (see Figure 17.1) that is comprised of seven interconnected domains. Four domains – bulk generation, transmission, distribution, and the customer – are linked together to provide a bidirectional flow of both electricity and information from the point of generation to end use. Three domains – markets, operations, and service provider – provide operations and information that deal with the buying and selling of electricity, operation of the system, and provision of retail and value-added services to customers.

Following are key technologies and capabilities that enable the operation of the smart grid:

- High penetration of renewable energy sources: 20% – 35% or more in some areas
- Distributed generation and microgrids
- Bidirectional metering – enabling locally-generated power to be fed into the grid
- Distributed storage to buffer variability and intermittency of renewable generation
- Smart meters that provide near-real time usage data
- Time of use and dynamic pricing to encourage smoother demand on the system
- Ubiquitous smart appliances communicating with the grid
- Energy management systems in homes as well as commercial and industrial facilities linked to the grid
- Growing use of plug-in electric vehicles
- Networked sensors and automated controls throughout the grid.

Maintaining cost effective, reliable electric service while modernizing the grid is a paramount concern. The new technologies and operating principles of the smart grid will not simply appear all at once. The legacy grid will undergo a transformation into the envisioned smart grid in an evolutionary way. These technologies and capabilities will be introduced over the course of several decades.

In the United States, the transition to the smart grid already is under way, and it is gaining momentum as a result of both public and private sector investments. The American Recovery and Reinvestment Act of 2009 (ARRA) included a Smart Grid Investment Grant Program (SGIG)[16] which provides $3.4 billion for cost-shared grants to support manufacturing, purchasing and installation of existing smart grid technologies that can be deployed on a commercial scale. Significant smart grid efforts are also underway in other countries, including China, the EU, Japan, Korea, Australia, among others.

Category	$ Million
Integrated/Crosscutting	2,150
AMI	818
Distribution	254
Transmission	148
Customer Systems	32
Manufacturing	26
Total	3,429

18 million smart meters
1.2 million in-home display units
206,000 smart transformers
177,000 load control devices
170,000 smart thermostats
877 networked phasor measurement units
671 automated substations
100 PEV charging stations

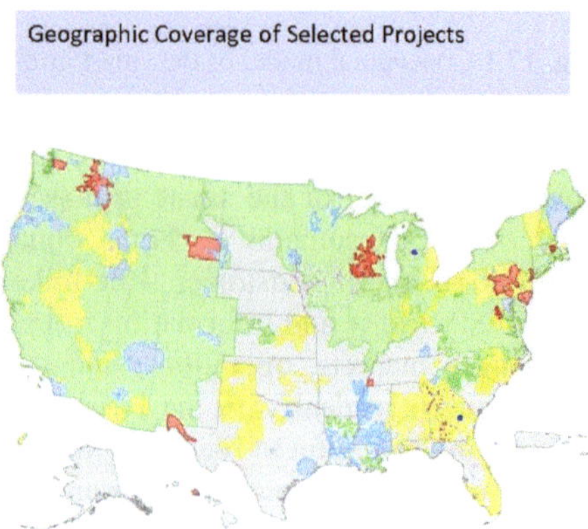

Fig. 17.2 Department of Energy Smart Grid Investment Grants 2009 [16]

17.5 Technical Challenges

The Smart Grid involves many new technologies and new operating paradigms that need to be integrated into the existing grid as it evolves while maintaining safe and reliable operation. This poses many significant technical challenges, a few of which are highlighted below.

17.5.1 Wide Area Measurement and Control Systems

Phasor Measurement Units that are now being widely deployed will provide time-stamped real time voltage and current phasor measurements at points throughout the grid. The widespread use of these measurements will significantly improve reliability, power quality, grid robustness, and resilience. Significant technical challenges need to be addressed to create a data measurement infrastructure to enable the effective use of this technology. The North American Synchrophasor Initiative (NASPI), a collaborative effort between the U.S. Department of Energy, the North American Electric Reliability Corporation, and North American electric utilities, vendors, consultants, federal and private researchers and academics, is working to address these challenges. Ensuring the accuracy of measurements from different vendors' synchrophasors and precise time synchronization is critical to ensuring accurate system-wide measurement. NIST's measurement research and synchrometrology testbed is supporting the development of new synchrophasor measurement and communication methods, and testing methods for new synchrophasor standards.

17.5.2 Achieving High Penetration of Renewable and Distributed Generation

Achieving a high penetration of renewable and distributed generation resources in the grid - such as wind and solar, either utility-scale or distributed at customer facilities – presents a different set of challenges. Unlike conventional generation, wind and solar resources have variable output that can change quickly and cannot be controlled. New technologies for cost-effective electrical storage and demand management will be needed to accommodate growing use of renewable generation if we wish to avoid underutilized back up conventional generation for periods when the wind does not blow or clouds obscure the sun.

Electricity storage could be used in place of back up conventional generation to buffer short-term imbalances between generation and demand. Development of improved grid-scale storage technologies is an active area of research and is being supported by he U.S. DoE's Advanced Research Projects Agency-Energy (ARPA-E). The ability to achieve high penetration levels of renewable resources and incorporate electricity storage technologies into the grid while ensuring stable system operation also requires enhanced interfaces to renewable generators and storage devices. These enhanced interfaces must be capable of providing dispatchable power levels, voltage and reactive power regulation, controllable ramp rates, grid disturbance ride-through, micro-grid capability, etc. NIST research efforts are developing measurement and testing methods to support development of next-generation power electronics and power conditioning systems to provide such functionality.

17.5.3 Dynamic Operation

In many respects, today's grid operates as a deterministic system. Generation capacity is planned to meet peak load and is dispatched as needed to meet real-time demand. Demand can be forecasted with a high degree of accuracy based on historical patterns and weather conditions. However, demand has not been treated as a controllable resource in the way generation is. In the future the grid will require more stochastic control, in which variable generation output can be balanced by dynamically managing demand using smart grid technologies. For example, smart appliances will be able to adjust energy usage dynamically in response to price or congestion signals. Building energy management systems can dynamically shift usage patterns by, for example pre-cooling the building or generating hot water when electricity demand is low. Maintaining reliable operation in such a dynamic system requires the development of new models to characterize performance at component and system levels and control strategies to maintain balance between generation and demand.

17.5.4 Electric Vehicles

Eventual large-scale deployment of electric vehicles represents both a challenge and an opportunity for the grid. A challenge that must be addressed is managing the additional load that vehicle charging will place on the grid. Generation and transmission resources do not present a problem because their idle capacity can be utilized provided that vehicle-to-grid communication and control protocols ensure that most charging occurs outside of peak periods. However, local distribution grids could become overloaded – an electric vehicle draws as much power as a typical house while it is charging. Utilities will have to monitor patterns in electric vehicle deployment and plan for distribution system upgrades where needed. The batteries in electric vehicles represent an opportunity to provide distributed storage resources for the grid, since vehicles generally sit idle most of the time. Electric vehicle batteries could be helpful in providing regulation service for the grid without requiring deep cycling that shortens battery life. However, the automotive industry is reluctant to support use of electric vehicle batteries for this purpose because little if any data is available to characterize the impact of such use on battery life. Research on measurement and characterization of battery performance in grid storage applications is needed to determine the potential role of electric vehicles as a support to the grid.

17.5.5 Cybersecurity

A great deal of attention is being placed on vulnerabilities and threats to the grid resulting from the application of information and communications technology to control system operation. A large array of cybersecurity tools, standards, and practices have been developed and applied to other critical commercial infrastructures, such as financial systems. NIST has played a significant role in the development of cybersecurity technology and standards for government and commercial applications. These tools are being adapted and applied to the electric grid. However research on cybersecurity threats and mitigation techniques for the electric grid will be an ongoing critical priority since threats and vulnerabilities will continue to evolve.

17.6 The Role of Standards and How They Are Being Developed

17.6.1 Importance of Standards

Transitioning the existing infrastructure to the smart grid requires an underlying foundation of standards and protocols that will allow this complex "system of systems" to interoperate seamlessly and securely.

Many suppliers from industry sectors that have not historically had to work together, such as electric equipment manufacturers, information and communication technology and service suppliers, and automotive manufacturers provide the equipment and systems that comprise the smart grid. Control systems operated by different electric utilities whose networks interconnect will need to exchange information. Customer-owned smart appliances, energy management systems, and electric vehicles will need to communicate with the smart grid. Standards defining the meaning, representation, and protocols for transport of data are essential for this complex "system of systems" to interoperate seamlessly and securely. Establishing the needed standards is a large and complex challenge.

17.6.2 Approach to Standardization in the United States

In the United States, Congress recognized the important role of standards and assigned the responsibility for coordinating the development of interoperability standards for the U.S. smart grid to the National Institute of Standards and Technology (NIST) through the Energy Independence and Security Act of 2007[17]. NIST, a non-regulatory science agency within the U.S. Department of Commerce, has a long history of working collaboratively with industry, other government agencies, and national and international standards bodies in creating technical standards underpinning industry and commerce.

There is an urgent need to establish standards. Some smart grid devices, such as smart meters, are moving beyond the pilot stage into large-scale deployment. The DoE Smart Grid Investment Grants will accelerate deployment. In the absence of standards, there is a risk that these investments will become prematurely obsolete or, worse, be implemented without adequate security measures. Lack of standards may also impede the realization of promising applications, such as smart appliances that are responsive to price and demand response signals. In early 2009, recognizing the urgency, NIST intensified and expedited efforts to accelerate progress in identifying and actively coordinating the development of the underpinning interoperability standards.

NIST developed a three-phase plan[18] to accelerate the identification of standards while establishing a robust framework for the longer-term evolution of the standards and establishment of testing and certification procedures. In May 2009, U.S. Secretary of Commerce Gary Locke and U.S. Secretary of Energy Steven Chu chaired a meeting of nearly 70 executives from the power, information technology, and other industries at which they expressed their commitment to support NIST's plan.

17.6.3 International Cooperation

Development of standards for the smart grid also requires efforts at regional and international levels. While electric utilities typically operate within national boundaries, there are interconnections across borders, such as between the United States, Canada, and Mexico in North America, and among member states of the European Union. In addition, many of the suppliers of equipment and systems used in the smart grid are global companies that seek to address markets around the world. Unnecessary variations in equipment and systems to meet differing national standards add cost, which eventually gets passed on to consumers. International standards promote supplier competition and expand the range of options available to utilities, resulting ultimately in lower costs for consumers.

17.6.4 Need for Coordination

Technical standards for the smart grid are under development by many standards development organizations (SDOs), most of them international in scope, including IEC, IEEE, ISO, ITU-T, IETF, among others. Since the standards need to work together to support an overall system, coordination of the standards work by the SDOs is critically important. During 2009, NIST engaged over 1500 stakeholders representing hundreds of organizations in a series of public workshops over a nine month period to create a high-level architectural model for the smart grid (see Fig. 17.3), analyze use cases, identify applicable standards, gaps in currently available standards, and priorities for new standardization activities. The result of this work, "Release 1.0 NIST Framework and Roadmap for Smart Grid Interoperability" was published in January 2010 [19]. To evolve the initial framework, late in 2009 NIST established a new public/private partnership, the Smart Grid Interoperability Panel (SGIP), which has international participation. Over 640 companies and organizations are participating in the SGIP. This body is also guiding the establishment of a testing and certification framework for the smart grid [20].

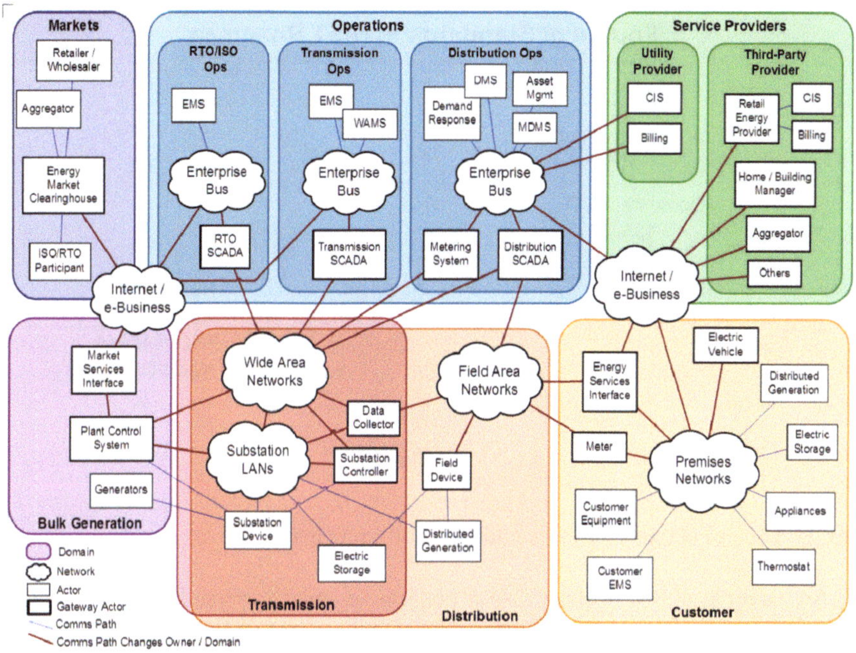

Fig. 17.3 NIST smart grid reference model

17.6.5 The NIST Framework

The NIST Release 1 framework identifies 75 standards or families of standards that are applicable or likely to be applicable to support smart grid development. The standards address a range of functions, such as basic communication protocols (e.g. IPv6), meter standards (ANSI C12), interconnection of distributed energy sources (IEEE 1547), information models (IEC 61850), cyber security (e.g. the NERC CIP standards) and others. The standards identified are produced by 27 different standards development organizations at the national and international level, such as the International Electrotechnical Commission (IEC), International Organization for Standardization (ISO), IEEE, SAE International, Internet Engineering Task Force (IETF), National Electric Manufacturers Association (NEMA), North American Energy Standards Board (NAESB), and many others. Seventy-seven percent of the identified standards are international standards (see Figure 17.4).

In the course of reviewing the standards during the NIST workshops, 70 gaps and issues were identified pointing to existing standards that need to be revised or new standards that need to be created. NIST has worked with the standards development community to initiate 18 priority action plans to address the most urgent of the 70 gaps.

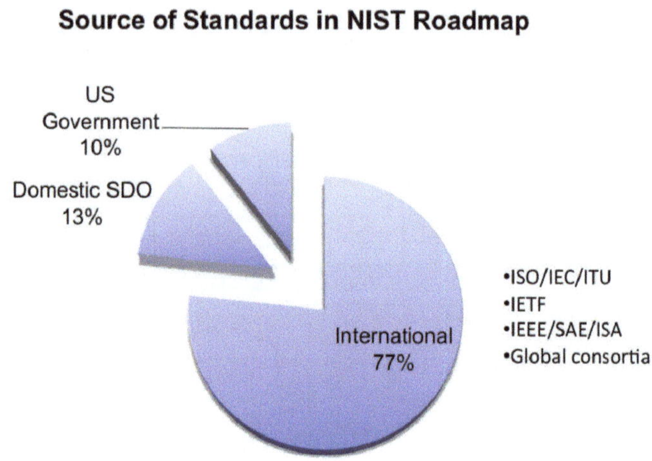

Fig. 17.4 Source of standards in the NIST Release 1.0 Framework

An example of one of these issues pertains to smart meters. The ANSI C12.19 standard, which defines smart meter data tables, is one of the most fundamental standards needed to realize the smart grid. Unless the data captured by smart meters is defined unambiguously, it will be impossible to create smart grid applications that depend on smart meter data. The existing ANSI C12.19 standard defines over 200 data tables but does not indicate which are mandatory. Different manufacturers have implemented various subsets of the standard, presenting a barrier to interoperability. In addition, the standard permits manufacturer-defined data tables with proprietary functionality that is not interoperable with other systems. To address this problem, one of the 18 priority action plans defined in the NIST roadmap was established to update the ANSI C12.19 standard to define common data tables that all manufacturers must support to ensure interoperability.

Manufacturers require lead-time to implement the revised standard. In the meantime, smart meters are in the process of being deployed and public utility commissions are concerned that they may become obsolete. To address the issue, NIST requested the National Electrical Manufacturers Association to lead a fast-track effort to develop a meter upgradeability standard. Developed and approved in just 90 days, the NEMA Smart Grid Standards Publication SG-AMI 1-2009, "Requirements for Smart Meter Upgradeability," is intended to provide reasonable assurance that meters conforming to the standard will be securely field-upgradeable to comply with anticipated revisions to ANSI C12.19.

Other priority action plans that are underway to accelerate and coordinate the work of standards bodies include:

- Standard protocols for communicating pricing information, demand response signals, and scheduling information across the smart grid
- Standard for access to customer energy usage information
- Guidelines for electric storage interconnection

- Common object models for electric transportation
- Guidelines for application of internet protocols to the smart grid
- Guidelines for application of wireless communication protocols to the smart grid
- Standards for time synchronization
- Common information model for distribution grid management
- Transmission and distribution systems model mapping
- IEC 61850 objects/DNP3 mapping
- Harmonize power line carrier standards for appliance communications in the home
- Standards for wind plant communication

17.6.6 Smart Grid Interoperability Panel

NIST established a public-private partnership, the Smart Grid Interoperability Panel (SGIP) to provide an organizational structure to support the ongoing evolution of the smart grid standards framework and priority action plans. The SGIP provides an open process for stakeholders to participate in the ongoing coordination, acceleration and harmonization of standards development for the smart grid. The SGIP does not write standards, but serves as a forum to coordinate the development of standards and specifications by many standards development organizations. The SGIP reviews use cases, identifies requirements, coordinates and accelerates smart grid testing and certification, and proposes action plans for achieving these goals.

The structure of the SGIP is illustrated in Figure 17.5. The SGIP has several permanent committees and working groups. One committee is responsible for maintaining and refining the architectural reference model, including lists of the standards and profiles necessary to implement the vision of the smart grid. The other permanent committee is responsible for creating and maintaining the necessary documentation and organizational framework for testing interoperability and conformance with these smart grid standards and specifications. A Cyber Security Working Group deals with the standards needed to secure the grid.

The SGIP is managed and guided by a Governing Board that approves and prioritizes work and arranges for the resources necessary to carry out action plans. The Governing Board's responsibilities include facilitating a dialogue with standards development organizations to ensure that the action plans can be implemented.

The SGIP and its governing board are an open organization dedicated to balancing the needs of a variety of smart grid related organizations. Any organization may become a member of the SGIP. Members are required to declare an affiliation with an identified Stakeholder Category (twenty-two have thus far been identified by NIST and are listed in Figure 17.6). Members may contribute multiple Member Representatives, but only one voting Member Representative. Members must participate regularly in order to vote on the work products of the panel. Membership in the panel is open to organizations from around the world, and includes participation from China, Japan, Korea, Canada, Mexico, Brazil, Europe, and other countries.

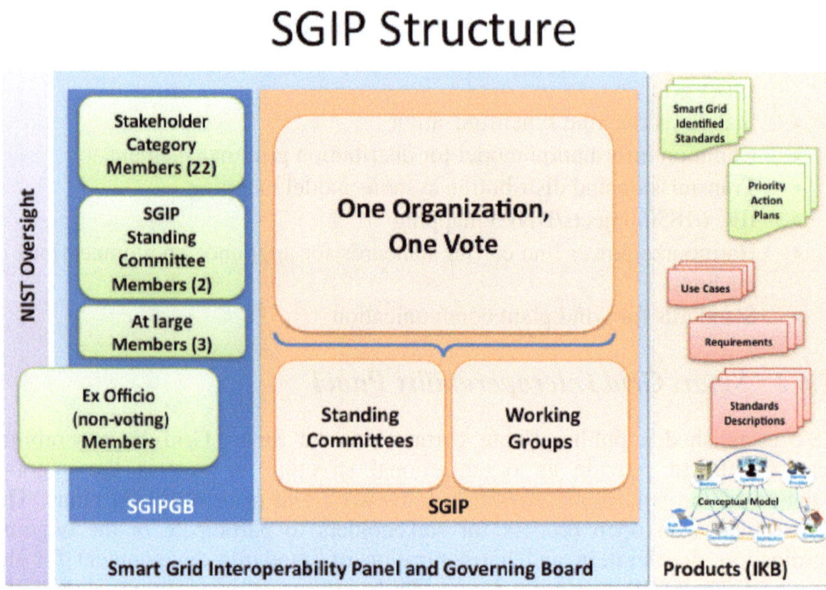

Fig. 17.5 Smart Grid Interoperability Panel structure

1	Appliance and consumer electronics providers	12	Power equipment manufacturers and vendors
2	Commercial and industrial equipment manufacturers and automation vendors	13	Professional societies, users groups, and industry consortia
3	Consumers – Residential, commercial, and industrial	14	R&D organizations and academia
4	Electric transportation industry Stakeholders	15	Relevant Federal Government Agencies
5	Electric utility companies – Investor Owned Utilities (IOU)	16	Renewable Power Producers
6	Electric utility companies - Municipal (MUNI)	17	Retail Service Providers
7	Electric utility companies - Rural Electric Association (REA)	18	Standard and specification development organizations (SDOs)
8	Electricity and financial market traders (includes aggregators)	19	State and local regulators
9	Independent power producers	20	Testing and Certification Vendors
10	Information and communication technologies (ICT) Infrastructure and Service Providers	21	Transmission Operators and Independent System Operators
11	Information technology (IT) application developers and integrators	22	Venture Capital

Fig. 17.6 Smart grid stakeholder categories

17.7 Cyber Security Considerations

Ensuring cyber security of the Smart Grid is a critical priority. Security must be designed in at the architectural level, not added on later. Cyber security must address not only deliberate attacks, such as from disgruntled employees, industrial espionage, and terrorists, but also inadvertent compromises of the information infrastructure due to user errors, equipment failures, and natural disasters. Vulnerabilities might allow an attacker to penetrate a network, gain access to control software, and alter load conditions to destabilize the grid in unpredictable ways. Additional risks to the grid include:

- Increasing the complexity of the grid could introduce vulnerabilities and increase exposure to potential attackers and unintentional errors
- Interconnected networks can introduce common vulnerabilities
- Increasing vulnerabilities to communication disruptions and introduction of malicious software could result in denial of service or compromise the integrity of software and systems
- Increased number of entry points and paths for potential adversaries to exploit
- Potential for compromise of data confidentiality, including the breach of customer privacy.

A NIST-led Cyber Security Working Group (CSWG) under the SGIP, consisting of more than 500 participants from the private and public sectors, has been leading the development of cyber security guidelines for the Smart Grid. Activities of the working group include identifying use cases with cyber security considerations; performing a risk assessment including assessing vulnerabilities, threats and impacts; developing a security architecture linked to the smart grid conceptual reference model; and documenting and tailoring security requirements to provide adequate protection. Release 1.0 of the Guidelines for Smart Grid Cyber Security (NIST IR 7628) was published in August 2010[21]. Ongoing research is needed to help ensure security of the grid. New threats will continually surface, and ongoing development of new technologies and methodologies to detect and mitigate threats and vulnerabilities will continue to be a top priority for smart grid efforts.

17.8 Status of Development Around the World

Many countries have begun or are planning to modernize their electric grids. NIST and the SGIP have established cooperative relationships with corresponding smart grid standardization initiatives that are underway in other parts of the world. In Europe, a European Joint Working Group for Standardization of Smart Grids has been established in which CEN, CENELEC, ETSI and the European Commission are participating. Japan has developed an initial standards roadmap for the smart grid and has also formed a Smart Community Alliance, which has extended the concept of the smart grid beyond the electric system to encompass energy efficiency and intelligent management of other resources such as water, gas and transportation. The government of South Korea has announced a plan to build a national smart grid network and is beginning work on a standards roadmap. In China, the State Grid Corporation has developed a draft Framework and Roadmap for Strong and Smart Grid Standards. International collaboration among these efforts is underway through a recently-established International Smart Grid Action Network (ISGAN) under the auspices of the Clean Energy Ministerial[22].

Bilateral and multilateral cooperation facilitates information sharing about national and regional requirements which are providing input to the technical specifications that are being developed by international organizations such as the IEC, IEEE, IETF, ISO, ITU-T, SAE and others, thus promoting international harmonization.

17.9 Conclusion

Creation of the smart grid is a historic engineering challenge requiring strong international collaboration that will span decades. Transforming the 20th century electric grid into the smart grid is essential to provide clean, reliable and affordable power to meet the world's needs for power in the 21st century.

References

[1] Leung, C.S., Meisen, P.: How electricity consumption affects social and economic development by comparing low, medium and high human development countries. Global Energy Network Institute, (July 2005), `http://www.geni.org/globalenergy/issues/global/qualityoflife/index.shtml`

[2] International Energy Agency. Key World Energy Statistics (2009) `http://www.iea.org/textbase/nppdf/free/2009/key_stats_2009.pdf` (accessed November 26, 2009)

[3] Energy Information Administration, Electric Power Annual 2009. U.S. Department of Energy (November 2010), `http://www.eia.gov/FTPROOT/electricity/034809.pdf`

[4] Energy Information Administration (2009). U.S. Carbon Dioxide Emissions from Energy Sources, 2008 Flash Estimate (2009), `http://www.eia.doe.gov/oiaf/1605/flash/pdf/flash.pdf` (accessed December 1, 2009)

[5] U.S. Department of Energy, The Smart Grid: an Introduction (2008), `http://www.oe.energy.gov/SmartGridIntroduction.htm`

[6] Fox-Penner, P.S., Chupka, M.W., Earle, R.L.: Transforming America's Power Industry: The Investment Challenge Preliminary Findings. The Brattle Group, Presented at the Edison Foundation Conference (April 21, 2008), `http://www.brattle.com/_documents/UploadLibrary/Upload678.pdf`

[7] Government Accountability Office, Electricity Markets - Consumers Could Benefit From Demand Programs, But Challenges Remain. Washington, D.C (2004)
[8] Volt-VAR Control Task Force. Volt-VAR Control Survey in United States and Canada, Presented at IEEE Smart Distirbution Working Group, 15.06.08 (2010), http://grouper.ieee.org/groups/td/dist/da/doc/Herve%20Delmas%20-%20Survey%20on%20VVC%20V3.pdf
[9] Distribution Reliability Working Group, IEEE Benchmarking 2009 Results, Presented at IEEE Working Group on Distribution Reliability (2010), http://grouper.ieee.org/groups/td/dist/sd/doc/2010-07-Benchmarking-Results-2009.pdf
[10] LaCommare, K.: Eto, J.:Understanding the Cost of Power Interruptions to U.S. Electricity Customers. Berkeley, CA: Lawrence Berkeley National Laboratory LBNL-55718 (2004)
[11] Nakanishi, Hironori.: Japan's Approaches to Smart Community. Presented at First IEEE Conference on Smart Grid Communications, (October 4-6, 2010), http://www.ieeesmartgridcomm.org/2010/downloads/Keynotes/nist.pdf
[12] The Brattle Group. A National Assessment Of Demand Response Potential (2009), http://www.brattle.com/documents/UploadLibrary/Upload775.pdf (accessed December 1, 2009)
[13] Smart Meter Pilot Program, Inc., PowerCentsDC Program Final Report. eMeter Corporation, Washington, DC (2010)
[14] Governor of the State of California. Executive Order S-21-09 (2009), http://gov.ca.gov/executiveorder/13269/ (accessed December 1, 2009)
[15] Kintner-Meyer, M., Pratt, K., Schneider, R.: Impacts Assessment of Plug-in Hybrid Vehicles on Electric Utilities and Regional U.S. Power Grids Part 1: Technical Analysis. Pacific Northwest National Laboratory, U.S. Department of Energy, Richland, WA (2006)
[16] U.S. Department of Energy. Press release (October 27, 2009), http://www.energy.gov/news2009/8216.htm (accessed January 31, 2009)
[17] U.S. Congress, Energy Independence and Security Act of 2007 (Public Law No: 110-140). U.S. Government Printing Office (2007), http://frwebgate.access.gpo.gov/cgibin/getdoc.cgi?dbname=110_cong_public_laws&docid=f:publ140.110.pdf
[18] National Institute of Standards and Technology. NIST Announces Three-Phase Plan for Smart Grid Standards, Paving Way for More Efficient, Reliable Electricity. Press release (2009), http://www.nist.gov/public_affairs/smartgrid_041309.html (accessed Decembeer 1, 2009)
[19] National Institute of Standards and Technology, NIST Framework and Roadmap for Smart Grid Interoperability Standards, Release 1.0 (Special Publication 1108). U.S. National Institute of Standards and Technology, Gaithersburg, MD (2010)

[20] Smart Grid Interoperability Panel. Smart Grid Interoperability Panel (November 2009), http://collaborate.nist.gov/twiki-sggrid/bin/view/SmartGrid/SGIP

[21] National Institute of Standards and Technology. Introduction to NISTIR 7628 Guidelines for Smart Grid Cyber Security. U.S. National Institute of Standards and Technology, Gaithersburg, MD (2010)

[22] Clean Energy Ministerial, Fact Sheet: International Smart Grid Action Network. U.S. Department of Energy (July 23, 2010), http://www.energy.gov/news/documents/ISGAN-Fact-Sheet.pdf

Chapter 18
Green Middleware

Hans-Arno Jacobsen and Vinod Muthusamy

18.1 Introduction

To foster innovation, address energy independence, react to global warming, and increase emergency resiliency, governments around the world are promoting efforts to modernize electricity networks by instigating smart grid, e-energy and e-mobility initiatives [5, 6, 3, 4]. A *smart grid* refers to efforts that enhance and extend today's electric power grids with information processing technology to more efficiently use the world's scarcest resource, *energy* [6].

A smart grid is envisioned to deliver electricity from suppliers to consumers guided by information communication technology (ICT) with two-way communications to monitor and control appliances and devices at consumers' homes, save energy and reduce emissions, increase reliability and transparency, and use distributed energy storage to level electricity supply and demand during peak and off-peak periods [6]. Smart grids also envision the integration of renewable and distributed energy resources, such as photovoltaic, wind, hydro, and tidal sources, in pursuit of the above outlined goals [2].

A smart grid overlays the electrical power grid with a massively distributed *information systems infrastructure* for large-scale monitoring and fine-grained control of all elements involved in the energy production and consumption pipeline. These elements can include energy generation, storage, transmission, distribution, and consumption. A smart grid ensures the secure, reliable and cost-effective flow of energy from distributed energy sources to energy consumers.

Hans-Arno Jacobsen
Middleware Systems Research Group (msrg.org), University of Toronto
e-mail: `jacobsen@eecg.utoronto.ca`

Vinod Muthusamy
Middleware Systems Research Group (msrg.org), University of Toronto
e-mail: `vinod@eecg.utoronto.ca`

It is the distributed information systems infrastructure that is the subject of this chapter. In particular, this chapter aims to highlight some of the research challenges and open problems faced by developers involved in building various information systems elements as well as applications enabled by smart grids. Furthermore, a potential approach to tackle some of the involved issues is presented.

Emerging smart grids will be ultra-large scale systems (ULS) architected as decentralized systems of systems. According to CMU's Software Engineering Institute [16], ULS are systems of unprecedented scale with respect to lines of code, number of interdependencies, number of interacting hardware, software and human elements, and amount of data stored and processed. ULS are characterized by operational and administrative independence, evolutionary development, emergent behaviour, geographic distribution, and heterogeneity.

The design, development and operation of smart grids require information systems abstractions that among others exhibit the following key properties and features:

1. *Decouple* involved systems and components to support the independent evolution and substitutability of sub-systems and components to reduce the impact on overall operational goals and metrics.
2. Similarly, system interactions should be *loosely coupled* to allow for continuous system evolution and support adaptive system behaviour.
3. Support *event processing* to detect and predict emergent behaviour.
4. Allow for *many-to-many* interactions to connect disparate systems.
5. Guarantee *interoperability* to mediate differences.
6. Finally, support *asynchronous operation* to enable performance and scale.

Traditionally, information system abstractions that facilitate the design, the development, the deployment, the operation and the integration of distributed applications in heterogeneous networked environments are referred to as *middleware*. We therefore refer to abstractions used in the development of the information communication technology elements of smart grids, in particular as pertaining to the distributed systems and application development aspects, as *green middleware*[1]. For example, *publish/subscribe* middleware offer many of the above postulated decoupling, loose coupling, many-to-many interaction, and asynchronous operation requirements and enable event processing, such as event detection and dissemination [26, 14, 17, 28, 9, 21, 22, 33, 23, 20].

However, the overall conception of an information systems infrastructure as information technology basis for smart grids is still an open problem. Furthermore, how to best address the above requirements is not know. While this chapter cannot answer these questions, we aim to raise a number of fundamental questions imposed by applications enabled by future e-energy initiatives for future middleware abstractions. From these questions, we will derive interesting research challenges.

[1] We acknowledge that the potential scope of *green middleware* could be far larger, also including making the middleware abstractions themselves more energy-aware and -efficient, for example.

This chapter is organized as follows. Section 18.2 reviews application scenarios that illustrate some of the potentials of smart grids and underline the features required from green middleware systems. Section 18.3 lays out research questions and discusses open issues whose resolution would contribute to the successful, reliable and secure development and operation of smart grids. Section 18.4 presents the PADRES system, positioning it as one viable solution to address many of the requirements demanded of green middleware. While PADRES has not yet been evaluated in the context of smart grids, we believe it exhibits many characteristics that are of fundamental importance in the support of an information systems infrastructure and of applications for smart grids. Section 18.5 concludes this chapter with a few recommendations for next steps.

18.2 Motivating Application Scenarios

In this section, we first review projections for the evolution of energy consumption worldwide to underline the severity of the energy problem to be addressed by smart grids enabled through green middleware. Then, through forward-looking application scenarios, we further motivate the need for green middleware enabling these scenarios in future smart grids as well as illustrate the research challenges that need to be overcome in order to develop green middleware.

In terms of the energy consumption projections, the International Energy Agency (IEA) reports that in many western homes, energy use from Information and Communication Technology (ICT) exceeds that of traditional appliances [7]. This load-shift is due to the aggregate electrical power draw from devices such as flat screen TVs, digital VCRs, computers, home routers, modems, chargers etc. ICT now accounts for almost 42% of energy consumption in western homes and, given a forecasted annual growth rate in consumption of 15%, could consume 40% of the world's electricity by 2030 [32]. Even with the less optimistic growth rate projection of 6% per annum, power consumption would double every decade [7]. Similar figures are reported for commercial data centers that power today's information economy. Data centers are projected to consume 12% of all electricity in the U.S. by 2020 and data center electricity costs tallied up to about $5 billion dollars in 2006 according to findings from Lawrence Berkeley National Labs [31].

Energy consumption policies

The above projections are alarming at best. Of fundamental concern are the development of approaches to help reduce power consumption, to exploit and integrate renewable energy resources into power grids, to incentivise consumers to save energy, and to develop innovative solutions to save energy without much consumer input.

For instance, imagine a thermostat that allows its owner to opt for saving 15% on her electricity bill, either by letting the owner select the margin, or by having the thermostat offer discounts based on analysis of past usage patterns, and controlling the home in appropriate ways. Owner consumption violations could be taxed at higher prices to account for the extra cost of provisioning surplus energy resources. A similar scenario applies at a larger scale, where a corporation decides to shave off a few percent of its monthly electricity spending by providing high-level policies to its power hungry equipment, office environment or HVAC operations management.

Across entire regions, utilities could make use of these signals to plan power generation schedules, to reduce operating cost of expensive stand-by generation resources, to lower emissions, and to comply with emission standards. Additionally, it is not inconceivable that consumers may wish to express resource preferences, possibly driven by regulatory policies, such as preferences over type or location of energy resources, degree of reliability, price and environment consciousness.

Appliance monitoring

With a smart grid infrastructure in place, applications can be developed to the customer experience beyond simply saving on energy costs.

For example, one often leaves the house, wondering after some time, if the oven had been shut off, or a person suffering from Alzheimer's may inadvertently forget to turn off the oven or other critical appliances. This is both wasteful and dangerous. Critical conditions of this kind could be detected either by monitoring for unusually long durations of oven use, or for certain correlated activities, such as the oven being on but the kitchen lights off (i.e., no one is in the kitchen) or the front door opening (i.e., someone is leaving the house), or no movement in the house for prolonged periods of time (i.e., someone either left the house or is resting.) Possible actions include automatically shutting off the appliance, alerting the person with a recorded warning, or calling a friend or relative.

Energy use pattern detection

A monitoring system could detect suspicious energy use. For example, lights coming on while the household is on vacation, or off at work and school, could indicate a break-in. Conversely, the absence of certain routine activities (such as an elderly person who watches TV or has a shower at the same time every day), may indicate a health issue that deserves attention. Again, the system can report this to the authorities or caregiver. Detecting these cases requires policies be written and/or the system learns about usage patterns and compares current usage with historical trends.

The system could recommend when appliances need maintenance or replacement. For example, based on the current temperature and thermostat settings, the air conditioner should operate within a range of duty cycles or energy use. Anything outside this range could indicate the air conditioner is not operating efficiently and needs to be replaced or repaired. Similarly, sensors placed around the house can detect irregular temperature readings. For example, large temperature differences among rooms could indicate clogged or leaky vents; and on a more fine-grained level, temperature gradients within a room could suggest worn out window seals.

Aggregate pattern detection

A smart grid offers opportunities to mine the energy use patterns of many consumers. For example, health care professionals could detect disease outbreaks by monitoring deviations in aggregated energy use such as people going to bed earlier (all appliances and lights off), staying home (appliances used during the day), or using more hot water (longer showers).

Similarly, marketers can infer household lifestyles based on changes in energy use. For example, a sudden increase in the number of times lights are turned on for short periods late at night may indicate a new baby in the house. Similarly, certain devices may exhibit unique power fingerprints; for example, a certain brand of personal video recorder may turn itself on at 3 am every morning to update itself. Such knowledge can be used to deliver better targeted advertisements.

We are fully aware that these examples tread on very sensitive privacy issues that are out of the scope of this chapter. As we point out later, however, these concerns can potentially be addressed if the green middleware offered a way to detect such aggregate patterns without revealing personally identifiable information that may violate users' privacy expectations.

Middleware requirements

The examples above illustrate many of the green middleware properties listed in Section 18.1. For example, loosely coupled interactions among components are required for there to be any reasonable chance of managing such a large ecosystem of power generators, transmission lines, storage nodes, and appliances, all of which are owned, operated, administered, and adapted independently. Many of the above scenarios also rely on real-time monitoring of the status of individual components, a task that calls for some sort of event processing capabilities. Moreover, the above applications can largely be implemented in an asynchronous manner. For example, monitoring appliances for unusual usage only requires that these appliances report their status, perhaps by emitting events; the component that analyzes these event streams can consume and process these events at its own pace. Notably, there is no synchronous, step-by-step conversation among the components, as this would quickly become infeasible as the number of components increases.

These scenarios also reveal certain challenges for a green middleware. For example, it is not clear how to architect a middleware that can support the ultra large scale of a smart grid. Also there are seemingly conflicting requirements, such as the need for both efficiency and resilience in the middleware itself as well as in the generation and consumption of energy; the very redundancy in the system that improves resilience can also be seen as an inefficient use of resources. Furthermore, managing the components in a smart grid requires detailed knowledge of their behavior, but this knowledge may violate users' privacy regulations. The following section discusses these research challenges in more detail.

18.3 Research Challenges and Opportunities

The long-term goals of a green middleware roadmap are the development of advancements inspired from the demands of future smart grid information systems infrastructures. These advancements will lead to new standards that enable an open ecosystem for applications offered by third parties that exploit the capabilities offered by smart grids.

In this sense, the objectives are to contribute to the design of an information system infrastructure that supports a distributed energy grid enabling energy consumers and suppliers to negotiate supply and demand in real-time, to perform this negotiation at the device level driven by system-wide constraints, to reduce energy consumption and save cost, and to adhere to regulatory constraints.

The fundamental questions that must be answered are as follows:

1. *What is the architecture of the information systems infrastructure that supports the requirements of future smart grids, especially the ultra-large scale nature of these systems?*
   The emerging integration of distributed energy resources, the envisioned two-way flow of monitoring and control information, and the novel application scenarios present a paradigm shift away from the existing centrally managed and controlled, proprietary and closed, and one-way communication-based energy infrastructures of today.
2. *How to balance efficiency and resiliency to accommodate disturbances in ultra-large scale systems?*
   The lack of resiliency, for the benefit of efficiency and cost reduction, is often detrimental, especially when manifest as cascading failures such as blackouts.
3. *How to determine the aggregate system behaviour without needing to reveal individual behaviour?*
   For example, in order to prevent cascading failures, it is critically important to be able to detect and predict emergent behaviour and to react accordingly. In the context of smart grids, where transparency is paramount, this question becomes one of knowing how consumers act collectively without needing to know how they are acting individually. In other words, how can the system detect emergent global behaviour while preserving consumer privacy?

These questions lead to a number of interesting research challenges:

- *Designing the information systems infrastructure for the expected scale of operation.*
  This scale amounts to millions of households, similar amount of electric cars as potential energy storage buffers, dozens of devices per household, and hundreds of thousand of distributed energy resources. The expected load amounts to millions of events per second, and the expected communication patterns constitute asymmetric event flows with lots of monitoring and a disproportionally smaller amount of control data.

- *Enabling real-time visibility into an ultra-large scale system architected as a system-of-systems.*
  In existing infrastructures, the lack of visibility is often used as an argument to fend of off customer complaints about unexpectedly large charges. However, visibility and transparency in smart grids are essential to their success.
- *Ensuring fine-grained security constraints, including privacy and access control.*
  In ultra-large scale systems, different entities will inevitably need different views of information. In the context of smart grids, for instance, the billing department only needs to know how much the aggregate energy consumption per household over a large time window, but should and need not be concerned who and with how the energy was used. On the other hand, energy providers do not need to know who uses the energy but need to have require fine-grained knowledge of usage patterns. Each consumer, however, should have immediate access to all information including which per-device energy usage, ideally in real-time.
- *Designing an information system infrastructure that can accommodate growth and evolution over decades.*
  The scale of operation mandates provisions for gradual roll-out over time. Similarly, the use of open standards is needed to enable competition and choice.
- *Unifying information routing and storage, which are traditionally handled separately.*
  The content-based routing paradigm supports the routing and propagation of short-lived information, while content-centric networking concerns addresses the efficient storing and querying of long-lived data in large-scale distributed environments. Both models are content-based, as opposed to the predominant addressed-based model of today's networks. The Besides the benefits of increased privacy, logical addressing is a good abstraction for mobile entities. In both cases, there is a need to filter and aggregate information for privacy and efficiency.

18.4 Towards Meeting the Research Challenges

Many of the challenges of a green middleware for smart grids arise from the need to support complex interactions among the many components in the system, including the enormous number of power generating sources, energy storage sites, transmission lines, consumer appliances, and plug-in electric vehicles.

The publish/subscribe interaction model offers the loose coupling and flexible many-to-many communication requirements of a green middleware. Moreover, implementations of the publish/subscribe model have been designed to provide scalable complex event processing that are also necessary for smart grids. This section provides an overview of the PADRES system, a open-source distributed publish/subscribe platform.

18.4.1 Publish/Subscribe

The publish/subscribe (pub/sub) paradigm provides a simple and effective method for disseminating data while maintaining a clean decoupling of data sources and sinks [14, 22]. This decoupling can enable the design of large, distributed, and loosely coupled systems that interoperate through simple publish and subscribe invocations. A large variety of emerging applications benefit from the expressiveness, the filtering, the distributed event correlation, and the complex event processing capabilities of content-based publish/subscribe.

These applications include RSS feed filtering [27, 29], system and network management, monitoring, and discovery [8, 15, 34], business process management and execution [30, 25], business activity monitoring [15], workflow management [13, 25], and automatic service composition [19, 18].

Typically, content-based publish/subscribe systems are built as application-level overlays of content-based publish/subscribe brokers, with publishing data sources and subscribing data sinks connecting to the broker overlay as clients. In content-based publish/subscribe, message routing decisions are based on evaluating subscriptions over the content of a message, and are not based on IP-address information. Events relevant to applications are conveyed as publications to the publish/subscribe system and routed based on their content to interested subscribers.

18.4.2 The PADRES *Publish/Subscribe System*

PADRES is a distributed content-based publish/subscribe system, developed by the Middleware Systems Research Group at the University of Toronto [1].

Figure 18.1 depicts the major layers in the PADRES system as they relate to a smart grid. At the bottom physical layer, around the core compute resources such as routers and compute servers, are entities such as households, consumer appliances, the electricity transmission infrastructure, and power generating and storage devices. The compute resources are used to build a distributed overlay network of publish/subscribe brokers, and the remaining entities participate as clients that interact with one another through the publish/subscribe broker network. Finally, the top layer includes the application logic that processes the event flows in the smart grid. The details of these layers will become clearer by the end of this section.

PADRES provides a number of novel features including composite subscriptions, composite event detection, historic query capability, load balancing, fault detection and repair, and monitoring support.

The historic data access module allows clients to subscribe to both future and historic publications [24]. This may be used to detect if the future energy consumption habits of a user deviate from their past. The load balancing and client relocation modules handle the scenarios in which a broker is overloaded by a large number of publishers or subscribers [11, 12, 19]. Such a feature would alleviate the need for a system administrator to manually reallocate resources as new energy producers or consumers enter the system. The fault tolerance module detects failures in the publish/subscribe layer and initiates failure recovery [20]. In the smart grid can

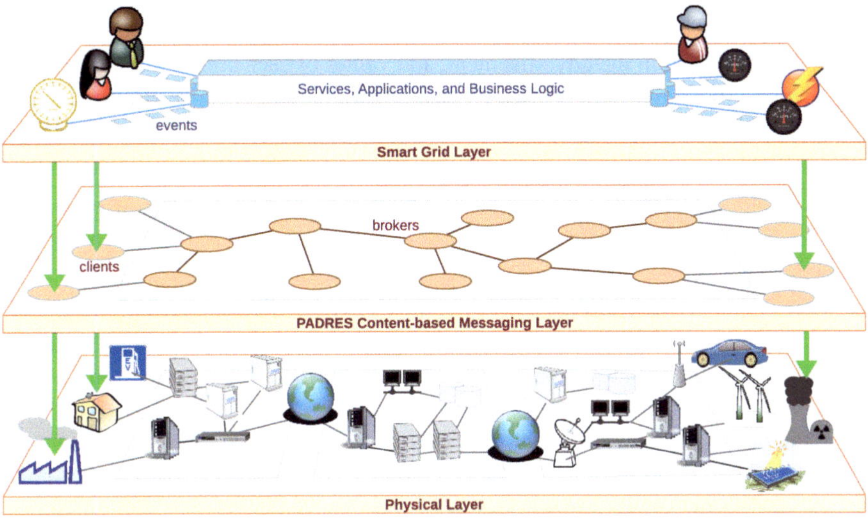

Fig. 18.1 PADRES as information systems infrastructure for smart grids

continue to operate despite isolated faults in the system. A monitoring module, which is an administrative client in PADRES, can be used to visually display the broker network topology, trace messages, and measure the performance of the network. These facilities can be used to inspect, at a fine-grained level, the operation of the green middleware itself or the devices in the smart grid,

18.4.3 The PADRES *Language and Data Model*

PADRES provides a SQL-like syntax PADRES SQL (PSQL) [24], which allows users to uniformly access data produced in the past and the future. The PSQL language includes the specification of notification semantics, and it can filter, aggregate, correlate and project any combination of historic and future data as described below. In PADRES, a message has a message header and a message body. The header includes the message identifier, which is unique throughout the system, the last hop and next hop information, which indicates where the message comes from and where it will be forwarded to, and the time stamp when the message is generated. There could be four types of objects in the message body: publications, advertisements, subscriptions, and notifications.

Advertisements: Before each publisher issues its publications, it specifies a template that describes the publications it will publish. This is done by issuing an *advertisement* message. The advertisement is an indication of the data that the publisher is going to publish in the future. In this sense, an advertisement is like a database schema or a programming language type. An advertisement is said to *induce* publications. That means the attribute set of an *induced* publication is a subset of attributes defined in the advertisement, and values of each attribute in an induced publication must satisfy the predicate constraint defined in the advertisement. Note that only publications induced from an advertisement of a publisher are allowed to be published by the publisher. We adapt the equivalent SQL table creation statement to express advertisements.

```
CREATE TABLE (attr op val[, attr op val]*)
```

PSQL's `CREATE TABLE` differs slightly from the same statement in SQL. Tables are unnamed since they need not be referred to by subscriptions or publications. Also, the range of values of each attribute (or column) can be specified. Moreover, regardless of the attribute value constraint, each attribute can implicitly be a null value.

In the rest of this section, we use an energy consumption management example consisting of a thermostat that periodically emit events indicating the duty cycle of the air conditioners under its control, and an external weather service that reports on current weather conditions.

```
CREATE TABLE (class = thermostat, id = *, time = *
               ac_model = *, duty_cycle = *)
CREATE TABLE (class = weather, id = *,
               time = *, location = *,
               degrees = *, humidity = *)
```

In the above example, * is a wildcard that indicates that the corresponding attribute may have any value.

Publications: A publication is expressed using a construct similar to SQL's `INSERT` statement.

```
INSERT (attr[, attr]*) VALUES (val[, val]*)
```

The following publication is compatible with the advertisement schema defined above.

```
INSERT (class, time, duty_cycle)
  VALUES (thermostat, 9:00, 20%)
```

Notice that only a subset of attributes defined in the schema need to be specified. For example, the above publication does not include personally the identifiable attributes such as the user id and air conditioner model information.

A publication may also contain a *payload*, which is an optional data value delivered to subscribers. The payload cannot be referenced by a subscription's constraints.

In many applications, publications represent events. Often, both terms are used synonymously in the publish/subscribe literature.

Subscriptions: Subscribers express interests in receiving publications by issuing *subscriptions*. Subscriptions set constraints on matching publications. PADRES not only allows subscribers to subscribe to individual publications, but also allows correlations or joins across publications. In that sense, subscriptions can be classified into *atomic subscriptions* and *composite subscriptions*.

Subscribers issue SELECT statements to query both historic and future publications. With SELECT, a subscriber can specify a set of attributes or functions that she wants to receive once the subscription is matched. The WHERE clause indicates the predicate constraints applied to matching publications. The FROM and HAVING clauses are optional and are used to express joins and aggregations.

```
      SELECT [ attr | function ], ...
       [FROM src, ...]
       WHERE attr op val, ...
     [HAVING function, ...]
   [GROUP BY attr, ...]
```

A traditional publish/subscribe subscription for future publications would look as follows in PSQL.

```
SELECT *
 WHERE class = thermostat, duty_cycle > 50%
```

Note that the above statement does not query a single table, so the results may have any number of attributes. The only guarantee is that all notifications will have the *class* and the *duty_cycle* attributes with values constrained as specified.

Reserved attributes *start_time* and *end_time* specify time constraints, and are used to query for publications from the past, the future, or both. For example, after replacing one of her air conditioners, a landlord may want monitor how it is performing. The following subscription queries data in a time window that begins two days before the time the query is issued and extends into the future.

```
SELECT *
 WHERE class = thermostat, ac_model = ACME3000,
       start_time = NOW - 2 days,
       end_time = NOW + 1 week
```

The system internally splits the above subscription: one purely historic subscription that is evaluated once, and one ongoing future subscription. A subscription for both historic and future data is a *hybrid* subscription.

Publish/Subscribe composite subscriptions [22] can be expressed with simple join conditions. The event correlation is supported using the FROM clause, where the event pattern can be specified using Boolean expressions. For instance, the subscription below monitors how air conditioners that have historically exhibited inefficient duty cycles perform during hot days in the future. In this case, the subscription is both a hybrid and composite subscription.

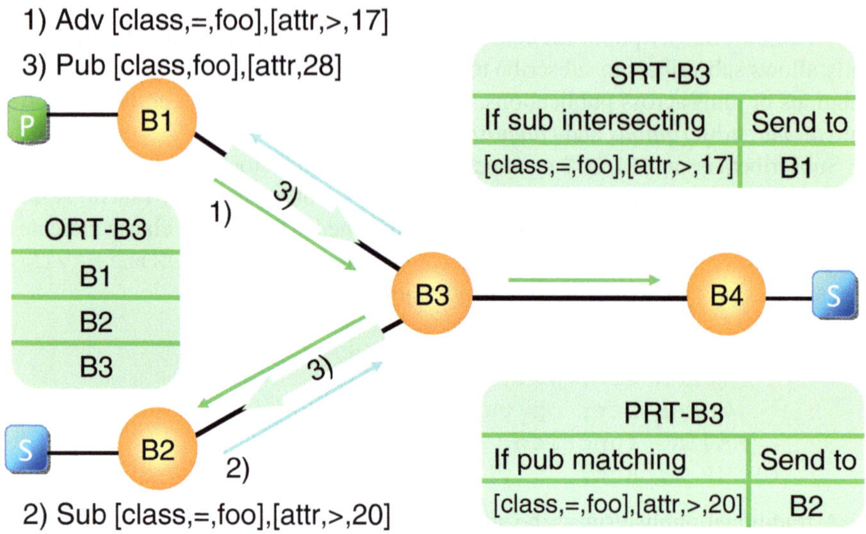

Fig. 18.2 PADRES Broker Network.

```
SELECT e2.time, e2.duty_cycle, e3.degrees
  FROM e1 AND e2 AND e3
 WHERE e1.class = thermostat,
       e1.duty_cycle > 90%,
       e1.start_time = NOW - 2 months
       e2.class = thermostat,
       e2.id = e1.id,
       e3.class = weather
       e3.location = Paris,
       e3.degrees > 30,
       e3.time = e2.time
```

The identifier in the `FROM` clause specifies that three different publications are required to satisfy this query, and each publication must match the `WHERE` constraints. The three publications may come from different publishers, and may conform to different schemas, as long as they match the specified constraints.

The subscription above queries air conditioner duty cycles over the past two months (event `e1`), correlates these with both duty cycle readings (event `e2`) and weather reports (event `e3`) in the future, and reports the performance of historically inefficient air conditioners during future warm weather situations. Manufacturers can use this information to study the behaviour of their poorly performing products.

Notice that a composite subscription can collect, correlate, and filter publications in the event processing network. Without this feature, a user must retrieve all future publications and then query databases for associated historic data. This would be expensive (both for the user and in terms of network traffic) in cases where the future publications are generated frequently.

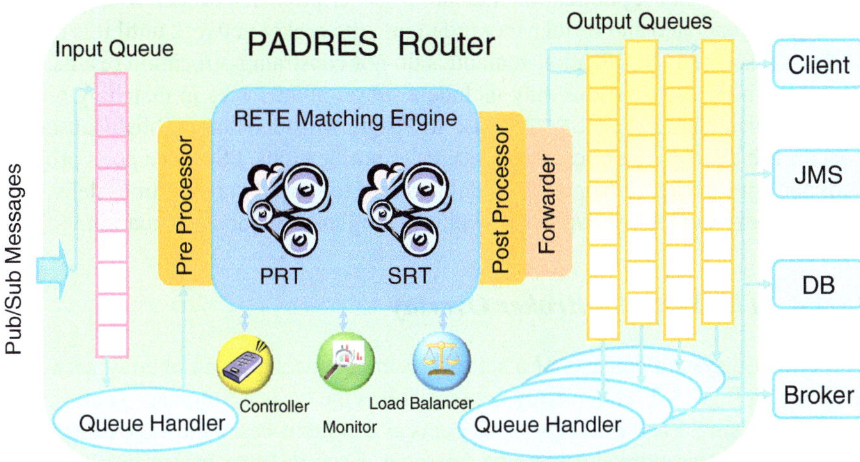

Fig. 18.3 PADRES Broker Architecture.

Event aggregation is supported in PSQL as well. The `HAVING` clause can specify constraints across a set of matching publications. The functions `AVG(`a_i`,N)`, `MAX(`a_i`,N)`, and `MIN(`a_i`,N)` compute the appropriate aggregation across attribute a_i in a window of N matching publications. The window may either slide over matching publications, or be reset when the `HAVING` constraints are satisfied. For example, the following subscription matches if the average duty cycle across 10 readings of a particular air conditioner exceeds maximum fault 80%.

```
  SELECT *
   WHERE class = thermostat
  HAVING AVG(duty_cycle, 10) > 80%
GROUP BY id
```

Any attributes specified by functions in the `HAVING` clause must appear in the publication. So, an implicit `duty_cycle = *` condition is added to the `WHERE` clause above. Also, the `GROUP BY` clause has the same semantics as in SQL and serves to constrain the set of publications over which the `HAVING` clause operates.

Although advertisements have the same format as atomic subscriptions, they have different semantics. Matching publications of the subscription must have all the attributes specified in the subscription, while publications induced from an advertisement may have only subset of the attributes defined in the advertisement. Another difference is a subscription may specify the notification semantics (e.g., in PADRES SQL). That is, what information(e.g., a subset of attributes) should be delivered to the subscriber if there is a match. A subscription *intersects* a advertisement if the sets of publications matching the advertisement and the subscription intersect.

Notifications: When a publication matches a subscription at a broker, a *notification* message is generated and further forwarded into the broker network until it is delivered to subscribers. Notification semantics do not constrain notification results, but transform them. Notifications may include a subset of attributes in matching publications indicated in the `SELECT` clause in PSQL. Most existing publish/subscribe systems use matching publication messages as notifications. PSQL supports projections and aggregations over matching publications to simplify notifications delivered to subscribers and reduce overhead by eliminating unnecessary information.

18.4.4 The PADRES *Broker Overlay*

The PADRES system consists of a set of brokers connected in an overlay network, as shown in Fig. 18.2. The overlay network forms the basis for message routing and event processing. Each PADRES broker acts as a content-based message router that routes and matches publish/subscribe messages. Each PADRES broker is essentially an event processing engine. The PADRES overlay constitutes an event processing network that can filter events published by many sources, distribute events to many subscribers, correlate and aggregate events from multiple sources to detect composite events, match composite subscriptions and match subscriptions for future events, historic events, and combinations of future and historic events [22, 23, 24].

A PADRES broker only knows its direct neighbours. The overlay information is stored in the Overlay Routing Tables (ORT) at each broker. Clients connect to brokers using various binding interfaces such as Java Remote Method Invocation (RMI) and Java Messaging Service (JMS). Publishers and subscribers are clients to the overlay. A publisher issues an advertisement before it publishes. Advertisements are effectively flooded to all brokers along the overlay network. A subscriber may subscribe at any time. The subscriptions are processed according to the Subscription Routing Table (SRT), which is built based on the advertisements. The SRT

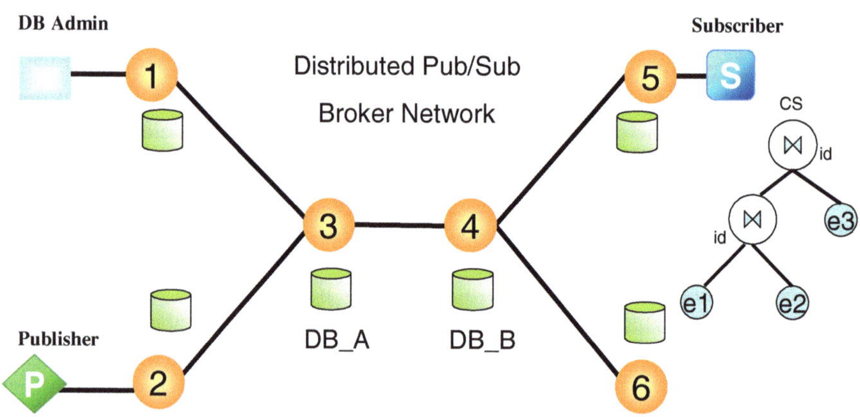

Fig. 18.4 Historic Data Access.

is essentially a list of [advertisement,last hop] tuples. If a subscription intersects an advertisement in the SRT, it will be forwarded to the last hop broker the advertisement came from. Subscriptions are routed hop by hop to the publisher, who advertises information of interest to the subscriber. Subscriptions are used to construct the Publication Routing Table (PRT). Like the SRT, the PRT is logically a list of [subscription,last hop] tuples, which is used to route publications. If a publication matches a subscription in the PRT, it is forwarded to the last hop broker of that subscription until it reaches the subscriber. A diagram showing the overlay network, SRT and PRT is provided in Fig. 18.2. In this figure in Step *1)* an advertisement is published at B_1. In Step *2)* a matching subscription enters from B_2. Since the subscription overlaps the advertisement at broker B_3, it is sent to B_1. In Step *3)* a publication is routed along the path established by the subscription to B_2.

Each broker consists of an input queue, a router, and a set of output queues, as shown in Fig. 18.3. A message arrives in the input queue. The router takes the message from the input queue, matches it against existing messages according to the message type, and puts it in the proper output queues representing different destinations. Other components are provided to support advanced features. For example, the controller component provides an interface for a system administrator to manipulate a broker (e.g., shut down a broker, inject a message into a broker etc.); the monitor component collects operational statistics (e.g., incoming message rate, average queueing time, and matching time etc.). If a broker is overloaded (e.g., the incoming message rate is above a certain threshold), a load balancer triggers offload algorithms [11] to balance the traffic among brokers. A failure detector is monitors the broker network. If a failure is detected, a recovery procedure is triggered in order to guarantee message delivery in presence of failures [20]. PADRES also proposes a novel policy model and framework for content-based publish/subscribe systems that benefits from the scalability and expressiveness of existing content-based publish/subscribe matching algorithms [33].

In PADRES, we explore optimized content-based routing protocols to provide more efficient and robust message delivery for the content-based publish/subscribe model, including covering-based routing [21] and adaptive content-based routing in cyclic overlays [23]. The goal of covering-based routing is to guarantee a compact routing table without information loss, so that the performance of a matching algorithm can be improved based on the concise routing table and no redundant information is forwarded into the network. The adaptive routing protocol takes advantage of multiple paths available in a cyclic network, balances publication traffic among alternative paths, and provide more robust message delivery.

18.4.5 Historic Data Access Architecture

Subscriptions for future publications are routed and handled as usual [14, 10, 13, 26, 23]. To support historic subscriptions, databases are attached to a subset of brokers as shown in Fig. 18.4. The databases are provisioned to sink a specified subset of publications, and to later respond to queries. The set of possible publications, as determined by the advertisements in the system, is partitioned and these partitions assigned to the databases. A partition may be assigned to multiple databases to achieve replication, and multiple partitions may be assigned to the same database if database consolidation is desired. Partition assignments can be modified at any time, and replicas will synchronize among themselves. The only constraint is that each partition be assigned to at least one database so no publications are lost. Partitioning algorithms, partition selection and partition assignment policies are described and evaluated in [24].

Each database subscribes to DB_CONTROL publications addressed to it, and the administrator assigns partitions to databases by sending publications with STORE commands to the appropriate database. For example, the following publications assigns to database *DB_A* a partition of weather reports for a particular city.

```
INSERT (class, command, db, partition_spec)
VALUES (DB_CONTROL, STORE, DB_A,
        'SELECT * WHERE class = weather, id = *,
         time = *, location = Paris,
         degrees = *, humidity = *')
```

The partition specification is itself a subscription with the selection formula expressed in the WHERE clause. A database that receives the STORE command will extract the partition specification and issue it as an ordinary future subscription. Matching publications will then be delivered to the database which will store them.

When the first broker receives a historic subscription issued by a subscriber, it assigns it a unique query identifier and then routes the subscription as usual towards publishers whose advertisements intersect the subscription. This ensures that the subscription will arrive at databases whose partitions intersect the subscription. The database(s) convert the subscription into a SQL query, retrieve matching publications from the database, and publish the results. These "historic" publications are annotated with the subscription's unique query identifier so they are only delivered to the requesting subscriber. After the result set has been published, the database will issue an END publication, which is used to *unsubscribe* the historic subscription.

The interaction with the databases fully leverages the content-based publish/subscribe model, and the databases are never addressed directly. In fact, it is impossible for publishers to discover where their publications are being stored, or for subscribers to know which databases process their queries. This simplifies management since databases can be moved, added or removed, and partitions reassigned at will.

To improve availability, fault-tolerance and query performance, a partition may be replicated. Partition assignment strategies include partitioning, partial replication and full replication.

With partitioning, a database may be assigned several partitions, but each partition is assigned to only one database. That is, there is only one replica per partition. With partial replication, a given partition may be replicated by assigning it to multiple databases. With full replication, every database maintains replicas of all partitions. That is, each database stores all publications.

The various strategies have tradeoffs and are appropriate under different circumstances. The partitioning policy is simple and avoids replica consistency issues, but is sensitive to failures. Partial replication can tolerate failures of all but one replica, but requires logic to ensure the historic subscription is answered by only one of the replicas. Full replication is even more robust, and historic subscriptions can always be answered fully by the nearest database, minimizing network traffic. However, the high degree of replication imposes greater overall traffic and storage costs, as well as larger synchronization overhead. The partition assignment policies allow an administrator to tradeoff storage space, routing complexity, query delay, network traffic, parallelism of queries, and robustness. Detailed discussion, algorithms, and evaluations of these strategies can be found in [24].

18.4.6 Subscription Routing

Subscriptions in PADRES can be *atomic* expressing constraints on single publications, or *composite* expressing correlation constraints over multiple publications.

Atomic subscription routing: When a broker receives an atomic subscription, it checks the *start_time* and *end_time* attributes. A future subscription is forwarded to potential publishers using standard publish/subscribe routing [14, 10, 13, 26, 23]. A hybrid subscription is split into future and historic parts, with the historic subscription routed to potential databases as described next.

For historic subscriptions, a broker determines the set of advertisements that *overlap* the given subscription, and for each partition, selects the database with the minimum routing delay. The subscription is forwarded to only one database per partition to avoid duplicate results. When a database receives a historic subscription, it evaluates it as a database query, and publishes the results as publications to be routed back to the subscriber. Upon receiving an END publication, after the final result is published, the subscriber's host broker unsubscribes the historic subscription. This broker also unsubscribes future subscriptions whose *end_time* has expired.

Composite subscription routing: Topology-based composite subscription routing evaluates correlation constraints in the network where the paths from the publishers to the subscribers merge [22]. If a composite subscription correlates with a historic data source and with a publisher, where the former produces more publications, correlation detection would save network traffic if moved closer to the database, thereby filtering potentially unnecessary historic publications earlier in the network. We propose the adaptive composite subscription routing protocol in [24], which determines the locations of event correlation, refereed to as the *join points*, based on a routing cost model. The cost model minimizes the network traffic and the notification routing delay.

When network conditions change, join points may no longer be optimal and should be recomputed. A join point broker periodically evaluates the cost model, and upon finding a broker able to perform detection cheaper than itself, initiates a join point movement. The state transfer from the original join point to the new one includes routing path information and partial matching states. Each part of the composite subscription should be routed to the proper destinations so routing information is consistent. Publications that partially match composite subscriptions stored at the join point broker must be delivered to the new join point.

18.5 Conclusions

To address concerns such as rapidly growing electricity demand, energy independence, and global warming, efforts are underway to develop a smart grid infrastructure to generate, transmit, deliver, and consume electricity. Among other benefits, a smart grid could use real-time knowledge of electricity providers and consumers to deliver cheap, reliable, and clean energy. Furthermore, preferences about household energy usage targets, fine-grained control over devices such as air conditioners, and weather pattern information can be used to manage aggregate energy demand, supply, and storage. Leveling the peaks and troughs of the power infrastructure in this way can potentially reduce costs for both energy producers and consumers, and result in a more stable and reliable system.

To carry out any kind of coordinated action, a smart grid requires an incredibly sophisticated information systems infrastructure, or *green middleware*, to monitor, manage, and control the huge number of energy generation, storage, transmission, and consumption devices present in the system. This green middleware architecture must not only handle the unprecedented size and volume of data in these ultra-large scale systems, but also be resilient to attacks, preserve privacy constraints, and support interoperability among existing components as well as those built with unanticipated requirements decades from now.

While we have presented the benefits of a publish/subscribe system for smart grids, the architecture of a green middleware is by no means settled. What is needed is an organization, say a Green Middleware Consortium, to agree on the architecture and standardize the core protocols and interfaces of this green middleware. This body will need to understand and accommodate the requirements and visions of all the key stakeholders in a smart grid, including but not limited to, public entities that manage power transmission and delivery infrastructure, operators of large scale power sources such as nuclear power plants, manufacturers of smaller scale power sources such as wind turbines, consumer electronics manufacturers, private energy brokers, and environmental and governmental organizations.

A green middleware designed according to some of the principles we have advocated in this chapter, such as the decoupling of components, will help manage the complexity of gradually migrating the current power grid to a more distributed smart grid that will continuously evolve over the decades to come. Moreover, standardization of this open information systems architecture would foster innovation at all levels, including the energy producers and consumers, the power generator and device manufacturers, and entirely new players in the smart grid landscape.

It is instructive to consider the example of the Internet. By agreeing on a set of core open standards governing the routing of packets among machines, the Internet has grown and evolved in ways never envisioned by the designers of these standards.

It has allowed innovation at every layer from the underlying physical networks, such as fibre optic and wireless transmission, to the applications delivered over these networks, such as telephony and social networking services.

Similarly, in order to lay the foundations for a truly smart grid, it is crucial that we develop an open, standards-based green middleware; a middleware that will serve as the core information platform to connect the energy generation, storage, transmission, and consumption devices of today and of the future; a middleware that will support new business models using intelligent applications to monitor, reason about, and manage these devices; and ultimately a middleware that can deliver a better consumer experience and improved standard of living to everyone.

Acknowledgements

We would like to thank the members of the PADRES team for their insights into designing middleware platforms for large scale, enterprise applications. In particular, we further acknowledge Balasubramaneyam Maniymaran and Guoli Li for their efforts in preparing portions of this document, especially the sections about the PADRES system.

References

1. Padres web site, `http://padres.msrg.org`
2. Distributed Generation Wikipedia Entry, `http://en.wikipedia.org/wiki/Distributed_generation`
3. E-energy, `http://www.e-energy.de/`
4. MUTE, `http://www.mute-automobile.de/`
5. Smart Grid Newsletter IEEE, `http://smartgrid.ieee.org/`
6. Smart grid Wikipedia Entry, `http://en.wikipedia.org/wiki/Smart_grid`
7. The IEA – The International Energy Agency, `http://www.iea.org/`
8. Mukherjee, B., Heberlein, L.T., Network, K.L.: Intrusion detection. IEEE Network 8(3), 26–41 (1994)
9. Carzaniga, A., Rosenblum, D.S., Wolf, A.L.: Design and evaluation of a wide-area event notification service. ACM Transactions on Computer Systems 19(3), 332–383 (2001)
10. Carzaniga, A., Wolf, A.L.: Forwarding in a content-based network. In: ACM Special Interest Group on Data Communication (SIGCOMM), pp. 163–174 (2003)
11. Cheung, A., Jacobsen, H.A.: Dynamic load balancing in distributed content-based publish/subscribe. In: ACM/IFIP/USENIX International Middleware Conference, pp. 249–269 (2006)
12. Cheung, A.K.Y., Jacobsen, H.A.: Load balancing content-based publish/subscribe systems. ACM Transactions on Computer Systems 28, 9:1–9:55 (2010)
13. Cugola, G., Di Nitto, E., Fuggetta, A.: The JEDI event-based infrastructure and its application to the development of the OPSS WFMS. IEEE Transactions on Software Engineering 27, 827–850 (2001)
14. Fabret, F., Jacobsen, H.A., Llirbat, F., Pereira, J., Ross, K.A., Shasha, D.: Filtering algorithms and implementation for very fast publish/subscribe systems. In: ACM Special Interest Group on Management of Data (SIGMOD), vol. 30, pp. 115–126 (2001)
15. Fawcett, T., Provost, F.: Activity monitoring: Noticing interesting changes in behavior. In: ACM Conference on Knowledge Discovery and Data Mining (SIGKDD), pp. 53–62 (1999)

16. Feiler, P., Gabriel, R.P., Goodenough, J., Linger, R., Longstaff, T., Kazman, R., Klein, M., Northrop, L., Schmidt, D., Sullivan, K., Wallnau, K.: Ultra-large-scale systems (2008)
17. Fiege, L., Awasthi, P., Mühl, G., Buchmann, A.: Engineering event-based systems with scopes. In: Deng, T. (ed.) ECOOP 2002. LNCS, vol. 2374, pp. 309–333. Springer, Heidelberg (2002)
18. Hu, S., Muthusamy, V., Li, G., Jacobsen, H.A.: Distributed automatic service composition in large-scale systems. In: ACM/IEEE/IFIP/USENIX International Conference on Distributed Event-Based Systems (DEBS), pp. 233–244 (2008)
19. Hu, S., Muthusamy, V., Li, G., Jacobsen, H.A.: Transactional mobility in distributed content-based publish/subscribe systems. In: IEEE International Conference on Distributed Computing Systems (ICDCS), pp. 101–110 (2009)
20. Kazemzadeh, R.S., Jacobsen, H.A.: Reliable and highly available distributed publish/subscribe service. In: IEEE Symposium on Reliable Distributed Systems (SRDS), pp. 41–50 (2009)
21. Li, G., Hou, S., Jacobsen, H.A.: A unified approach to routing, covering and merging in publish/subscribe systems based on modified binary decision diagrams. In: IEEE International Conference on Distributed Computing Systems (ICDCS), pp. 447–457 (2005)
22. Li, G., Jacobsen, H.A.: Composite subscriptions in content-based publish/subscribe systems. In: ACM/IFIP/USENIX International Middleware Conference, pp. 249–269 (2005)
23. Li, G., Muthusamy, V., Jacobsen, H.A.: Adpative content-based routing in general overlay topologies. In: ACM/IFIP/USENIX International Middleware Conference, pp. 1–21 (2008)
24. Li, G., Muthusamy, V., Jacobsen, H.A.: Subscribing to the past in content-based publish/subscribe. Tech. Rep. CSRG-585, Middleware Systems Research Group, University of Toronto (2008)
25. Li, G., Muthusamy, V., Jacobsen, H.A.: A distributed service-oriented architecture for business process execution. ACM Transactions on the Web (TWEB) 4, 2:1–2:33 (2010)
26. Opyrchal, L., Astley, M., Auerbach, J., Banavar, G., Strom, R., Sturman, D.: Exploiting IP multicast in content-based publish-subscribe systems. In: IFIP/ACM International Middleware Conference, pp. 185–207 (2000)
27. Petrovic, M., Liu, H., Jacobsen, H.A.: G-ToPSS: fast filtering of graph-based metadata. In: International World Wide Web Conference (WWW), pp. 539–547 (2005)
28. Pietzuch, P.R., Bacon, J.: Hermes: A distributed event-based middleware architecture. In: International Workshop on Distributed Event-Based Systems (DEBS) (2002)
29. Rose, I., Murty, R., Pietzuch, P., Ledlie, J., Roussopoulos, M., Welsh, M.: Cobra: Content-based Filtering and Aggregation of Blogs and RSS Feeds. In: USENIX Symposium on Networked Systems Design and Implementation, NSDI (2007)
30. Schuler, C., Schuldt, H., Schek, H.J.: Supporting reliable transactional business processes by publish/subscribe techniques. In: International Workshop on Technologies for E-Services (TES), pp. 118–131 (2001)
31. U.S. Environmental Protection Agency ENERGY STAR Program: Report to congress on server and data center energy efficiency public law 109-431 (2007)
32. Webb, M.: Smart 2020: Enabling the low carbon economy in the information age. Tech. Rep., The Climate Group on behalf of the Global eSustainability Initiative, GeSI (2008)
33. Wun, A., Jacobsen, H.A.: A policy management framework for content-based publish/subscribe middleware. In: ACM/IFIP/USENIX International Middleware Conference, pp. 368–388 (2007)
34. Yan, W., Hu, S., Muthusamy, V., Jacobsen, H.A., Zha, L.: Efficient event-based resource discovery. In: ACM International Conference on Distributed Event-Based Systems (DEBS), pp. 19:1-19:12 (2009)

Chapter 19
Semantic Information Integration for Smart Grid Applications*

Yogesh Simmhan, Qunzhi Zhou, and Viktor Prasanna

Abstract. The Los Angeles Smart Grid Project aims to use informatics techniques to bring about a quantum leap in the way demand response load optimization is performed in utilities. Semantic information integration, from sources as diverse as Internet-connected smart meters and social networks, is a linchpin to support the advanced analytics and mining algorithms required for this. In association with it, semantic complex event processing system will allow consumer and utility managers to easily specify and enact energy policies continuously. We present the information systems architecture for the project that is under development, and discuss research issues that emerge from having to design a system that supports 1.4 million customers and a rich ecosystem of Smart Grid applications from users, third party vendors, the utility and regulators.

19.1 Introduction

The Los Angeles Smart Grid Demonstration Project is a Department of Energy Sponsored project[1] to investigate Smart Grid technology and innovative use of information technology to improve power usage within the largest municipal utility in the United States as it deploys smart meters across its service area [1]. The project, a collaboration between the Los Angeles Department of Water and Power (DWP), University of Southern California (USC), University of California–Los Angeles (UCLA) and NASA's Jet Propulsion Lab (JPL), is conducting research on five themes: software architecture for demand response optimization (USC and

Yogesh Simmhan · Viktor Prasanna
Ming Hsieh Department of Electrical Engineering

Yogesh Simmhan · Qunzhi Zhou · Viktor Prasanna
Center for Energy Informatics

Qunzhi Zhou · Viktor Prasanna
Computer Science Department, Viterbi School of Engineering,
University of Southern California, Los Angeles CA

* This material is based upon work supported by the Department of Energy under Award Number DE-OE0000192.

UCLA), consumer behavior (USC), electric vehicles (USC and UCLA), and cyber-security (USC and JPL), with the eventual target of demonstrating these within the Los Angeles smart power grid in 2014 and helping identify best practices for other utilities.

Demand response optimization (DR) is one of the key goals of the project [2]. DR is the ability for utilities to manage the demand for power to within the current available power generation capacity – particularly during the middle of the day when electricity consumption is at its peak and approaches power generation limits. The difference between peak and base electricity load is often as high as 50%, causing excess capacity to be provisioned by the utility.

Traditionally, DR is performed at a coarse time granularity and statically (Figure 19.1, left). Monthly usage data that is physically collected by utility personnel from individual consumer electricity meters is combined with the total electricity usage available centrally to the utility to forecast monthly trends. These are used to statically set annual incentives like time of use (TOU) pricing for different time periods of the day and shared with the consumers in the hope that they change their daily activities in the subsequent year.

Smart meters or Advanced Metering Infrastructure (AMI) form the enabling technology for Smart Grids, and allow realtime, bi-directional communication between the utility and the consumer's home or building area network (HAN and BAN) using Internet and other protocols. Smart meters allow the utility to measure realtime electricity usage for each customer, and communicate pricing or load reduction signals to the consumer's appliances or HAN/BAN, which can be acted upon automatically (Figure 19.1, right). This dramatically reduces the time taken for the DR feedback to be set in motion and can be activated for finer granularities of consumers.

Our ongoing research into *software architecture for demand response optimization* aims to bring about a quantum leap in the way DR is performed in smart power utilities [2]. Rather than considering the fine grained information and control available through smart meters in isolation, as was done previously, we approach DR from an informatics perspective: where information from diverse sources – AMIs being just one of them – are integrated together and analyzed to

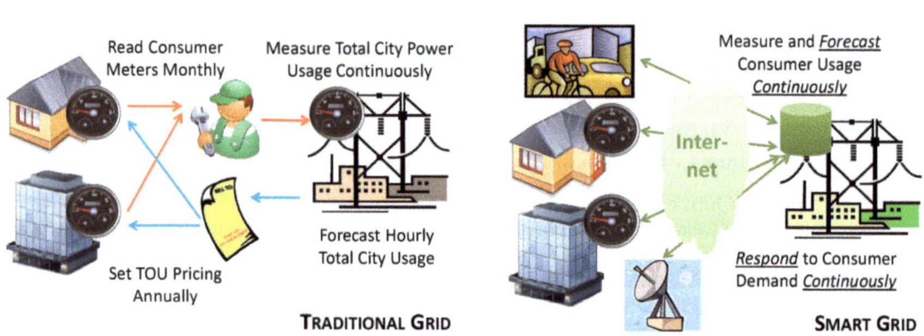

Fig. 19.1 Demand Response Optimization in Traditional and Smart Power Grids.

detect both direct predictors and indirect influencers of electricity usage. Load curtailment responses are similarly targeted both at electrical equipment and consumer behavior, and their effectiveness measured.

An effective *demand forecasting* and *load curtailment* software solution needs to meet three key characteristics for a large utility such as LA DWP, which serves 1.4 million consumers and consumes almost 1% of the total US power [1, 3, 4].

- **Intelligent, Adaptive DR:** Forecasting and response algorithms need to adapt to a dynamic city, where features that influence power usage constantly change, people move, traffic patterns change, and new electricity sources and sinks, like electric vehicles and solar panels, drastically change the energy landscape.
- **Translation of Policy to Practice:** A human in the loop of such a dynamic system with complex and multi-disciplinary attributes that impact energy use cannot keep up. Both customers and utility manager should be able to specify high level goals (e.g. "Keep load to within 90% of capacity", "Monthly electricity budget is $100") and delegate their enactment to software agents.
- **Securely Scale on Emerging Platforms:** The information analyzed by utilities is set for a dramatic increase, and will continue to grow on the order of tera- and even peta- bytes. The information architecture should deploy on scalable platforms such as Clouds. Ensuring data security and privacy will be of utmost importance as consumers get accustomed to an information-driven utility that can closely monitor their energy use.

Semantic information integration and *complex event processing* are two major components of our information architecture that address these requirements for effective DR.

Rapidly incorporate diverse information sources

As information attributes that directly and indirectly influence load forecasting and curtailment change over time, the information architecture must be able to incorporate new sources and deprecate old ones quickly. In addition, it needs to keep up with changing formats, qualities, and access policies of the sources. Semantics are needed to correlate diverse information that may refer to the same concept, and relate their impact on each other. Continuous arrival of data and online analysis by compute-intensive algorithms that mine for energy use patterns for power load forecasting means that information is constantly processed within the architecture and the warehouse contents is constantly changing.

Intuitive platform for goal-based policy specification

Information systems operate on queries. Utility managers and consumers operate on goals and policies. Means to specify higher level policies and automatically translate them into meaningful (semantic) queries needs to be present. The goals can be based on several constraints: energy conservation, social pressure, monetary benefits, ensuring quality of service, or scheduling maintenance. While the policies remain the same, their equivalent queries can change as information sources or forecast models change.

Besides the requirements of the utility for DR, other rich applications are enabled by an information-driven software architecture and will access the utility's information repository. These include consumer home automation software, third party vendors providing data analytics to consumers, mobile apps that notify users of peak demand signals and pricing incentives, data aggregators and regulators monitoring utility activities, and developers who wish to invent value added tools for customers.

In the ensuing sections, we present the design of the information systems architecture for the LA DWP Smart Grid project that is ongoing. In particular, we discuss the need for semantic information integration, and semantic complex event processing to support Smart Grid applications, and highlight the research issues they entail. This project is under active research, with initial prototyping underway to validate our architectural choices. In later sections, we present related work on these topics and present our future directions.

19.2 Semantic Information Integration

Smart Grid applications such as DR require a complete change in the energy information management paradigm. Traditional energy information systems tend to be vertically integrated, closed architectures that limit interoperability, extensibility and reuse. Customers are viewed as passive users of electricity (and information related to it), while utilities are charged with providing electricity as a commodity with fixed tariffs. As power utilities migrate to Smart Grids, information from heterogeneous sources including smart meters that report near real-time power usage and quality, intelligent thermostats that measure and control buildings' heat and humidity, and weather forecasts and traffic reports from online services are needed for accurate power *demand forecasting* and identification of effective *curtailment strategies* using data mining and analysis algorithms [2]. The information used by these DR algorithms is diverse in terms of the structures and semantics of data, software and hardware platforms used, types of interactions that they can support, and the size of the data items.

Figure 19.2 shows our information architecture that is being prototyped. Information sources at the bottom can communicate with the information integration framework through a service bus. Semi-structured data from micro-blogs and text comments are mined to extract structured data. These and other structured information arriving are semantically annotated using a Smart Grid ontology model using wrappers. The semantically meaningful instances of concepts and relationships can then be recorded in a Cloud-hosted information repository for use by DR applications. Separately, the complex event processing (CEP) system extracts events from the semantically annotated information, constructs new abstract events based on rules, and reasons over the event cloud to match patterns based on policy and trigger actions for DR. A mining process additionally looks for new energy forecast and power curtailment pattern structures for future interest. Utility and external applications, shown on the left, access this semantic repository using the semantic model through the service bus. The features of the Smart Grid ontology model are described below, and the information repository schema conforms to it.

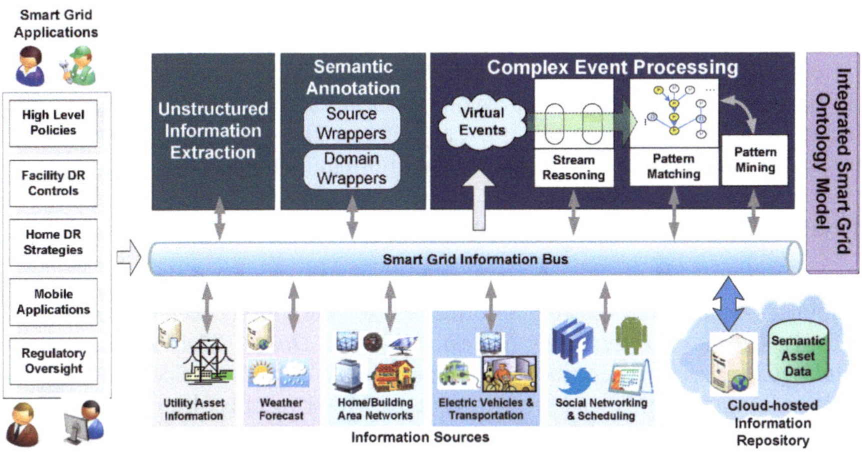

Fig. 19.2 Information Integration and CEP Achitecture for the Smart Grid.

The information integration architecture should support many-to-many points of interaction and information exchange [31]. The space of information sources and types is broad and will change as algorithms adapt. Utilities, software and hardware vendors are developing novel applications and products to meet consumer, utility and regulatory needs. While they are directly responsible for the functioning of their own software systems, they also depend on exchanging information with other systems. They need to negotiate agreements with the entities they interact with on how this information is to be exchanged. Conducting these negotiations between pairwise information sources and sinks is burdensome.

Semantic Web provides an ontology-based extensible framework for information to be shared and reused across application and domain boundaries [5]. It has been successful for information integration in domains such as eHealthCare [6], biology [7], and transportation [8]. Smart Grid data can be effectively used and interpreted when grounded in semantically meaningful terms. It is not just the utility managers who will use this information to determine policies, but also data mining and analysis applications, third party tools with which the data is shared, transmitted to other utilities and regulators, and even the consumer's application.

Applying Semantic Web for Smart Grid information integration goes beyond just a normalized form for exchanging meter, sensor or appliance data, and suggests integrating ontologies from different relevant domains – weather, traffic, and social networks, among others – and using sematic definitions for policies and software APIs using well understood semantic standards.

19.2.1 Integrated Smart Grid Ontology Model

We use the Semantic Web Ontology Language (OWL) to build the Integrated Smart Grid Ontology Model which stores the information about entities and their relationships in a structured format (Figure 19.2, right). We distinguish two types of ontologies: *domain ontologies* that are commonly used by utilities and power systems, and *external ontologies*, which describe concepts outside the electric grid such as weather and traffic, yet are required for information-driven energy use forecasting and curtailment by novel DR algorithms. These ontologies form the entities and relationships in the schema for our information repository.

19.2.1.1 Domain and External Ontologies

Existing utility warehouses contribute concepts and relationships to the domain ontologies from three existing information sources: *Meter Data Management (MDM)*, *Customer Information System (CIS)* and *Outage Management System (OMS)*. CIS is used to describe the customers in the utility's service area and their address, billing records, and service requests. MDM records electricity usage information from electric meters, including their location, correlated with the customer who is being metered and for what time period. Earlier, this usage was populated by utility personnel who physically visit the customer location and record this data once a month. In a smart utility, this information is transmitted up to once a minute by the AMI head-end over a communication network, potentially using internet protocols, to a gateway at the utility that maps the AMI-vendor specific data to a standard form. OMS is used to monitor the operation of the electricity distribution and transmission network. It helps analyze transmission reliability and narrow down sources of service outages to resolve them. Consequently, it captures the description of the entire utility's electric network itself [10].

In addition to the utility's information sources, other domain ontologies are contributed by the home and building area networks. These help capture the existence and activities of home appliance and electric vehicles (EVs), building plans such as surface area and year of construction, Heating, Ventilation and Air Conditioning (HVAC) equipment and their duty cycles, and so on [11].

Among external ontologies, we have identified four areas to initially support that affect or help forecast energy usage indirectly: *weather*, *traffic*, *scheduling* and *social networks*. Weather includes historical, current and forecast information about temperature, humidity, precipitation, cloud cover and wind speed. The first two affect the use of air-conditioning, precipitation additionally impacts traffic patterns, and cloud cover and wind speed affect power generation by solar and wind renewables. Traffic data shows the traffic flow and the rate of traffic exiting from arterial roads onto specific neighborhoods and can help predict impending energy use as customers arrive home. Additionally, movement of individual EVs is also modeled as they are a major energy sink when plugged in. Scheduling applies to both people (*when*, *where*, *with whom*) and facilities like classrooms, office rooms, and convention centers (*when*, *where*, *attendees* or *their count*). Social network data models people, their peers and friends, and their micro-blogs that is used for mining.

19.2.1.2 Ontology Normalization

Modeling the complete Smart Grid ontology is in itself an incremental process and a work in progress. It consists of identifying agreed upon standards for information modeling in that area, recognizing equivalent relationships between concepts in different domains, and defining additional relationships between concepts. Often, the ontologies do not exist in a directly usable form such as OWL. Sometimes, they do not exist as an information model as such, and need to be drawn out of communication protocols for message exchanges. This requires a keen understanding of the domain concepts by the modeler to establish their relationship.

However, this understanding does not have to extend to actually determining the exact influence that each concept has on the other with respect to energy use. That is something that will emerge from pattern specification and mining over observational data that will populate instances of the model concepts and relationships.

For example, the ontology model captures the concepts of and relationships between an office room in a building, its lighting system, its energy usage, and the infrared "people sensor" in the room, and that people in the office have a calendar schedule. These concepts and relationships are relatively easy to describe with limited knowledge of the different domains. This model captures the link between the person entering the room, the "person sensor" detecting it, triggering the lights to turn them on, and hence increasing the energy usage. This sequence is triggered at the time the person enters the room. What may be absent from the ontology model and discovered over time based on observational data for this model is: that the person's calendar has the room location and time of the meeting, that this person usually arrives early for most meetings, and that the fact that a meeting exists on her calendar means that the energy use for that room is going to increase at that point in future. This sequence is triggered at the time the calendar event is created – potentially hours or days in advance.

Information model standards for some of the domain ontologies exist and will be reused. The IEC Common Information Model (CIM) [12] provides an initial starting point to model power systems, including circuits and utility operations. It covers several aspects of MDM and OMS, but has to be extended with concepts for CIS. Also, the UML model of CIM needs to be mapped to OWL, which introduces issues that are documented [13]. ISO standards for building area network communications provide message protocols and APIs that can be used to derive concepts and relationships for an information model [14]. Weather concepts can be captured by JPL's SWEET ontology [15] and those for NextGen Network Enabled Weather (NNEW) [16]. Our prior work on ontologies for transportation networks is leveraged for the traffic models [17]. There is active work on defining standards for social network models, including a W3C incubator group [18][19]. Internet calendar standards provide the basic building blocks for scheduling [20].

19.2.2 Ingesting Information into Repository

Many information sources that contribute to our information repository do not provide semantic information; some do not even provide structured data. Two pre-processing steps of *information extraction* and *semantic annotation* need to take place before data from external sources is inserted into the repository in a semantically queryable form. These steps are exacerbated by information constantly arriving from different sources into our repository, thus distinguishing it from a traditional data-warehouse where the extract-transform-load (ETL) process is batched.

19.2.2.1 Information Extraction from Semi-structured Social Network Data

One of the primary sources of unstructured and semi-structured information useful for DR is from social networks. Information about friends on Facebook can be used for studying peer-pressure as a response mechanism for curtailing load (e.g. *"Your friends have signed up to reduce electricity usage by 10%. Would you like to?"*). Similarly, micro-blogging posts on Twitter and Facebook can be analyzed to detect user activity and indirectly forecast energy usage (e.g. *"Sick with #flu. Working from home tomorrow. #wfh"*). These are on the presumption that the utility has access and rights to this information, which is feasible through providing incentives for "friending" the utility and registering their Twitter feed with it.

Two approaches for extracting structured information from unstructured social network data are being investigated. The first approach is to just use the structured content that exists in social network data and using statistics to draw *classifications* that can be fed into the information repository rather than the raw content. User IDs, peer network, hashtags and text tags, timestamps, Twitter followers, and occasionally, location information are common structured information. It is possible to analyze the tags and Twitter followers to classify users according to, say, their environmental consciousness or technology uptake based on specific issues they follow. This classification can help tune energy conservation messaging and technology tools for target users. Similarly, frequency, time of day and location of tweets/posts can be used to map energy use patterns of individuals.

Another approach for analyzing micro-blobs is to use Natural Language Processing (NLP) to evaluate user intent on energy use and curtailment. NLP techniques have been used for data extraction and curation from literature [21] and their use for examining Twitter feeds for emergency response is being actively researched [22][23]. Challenges to applying NLP to micro-blogging are introduced by the compact space available for content and the consequent use of unique abbreviations and sentence formulation. Too, posts may be related to conversation threads that span multiple mediums (IM, text, voice). Despite this, it may be possible to determine a small class of topic terms that affect energy use, such as a user's schedule or purchase of a smart appliance, to shape DR analysis.

19.2.2.2 Online Semantic Annotation

Structured information from external sources can have static mappings defined with related semantic ontology(ies). These mappings may be specific to an information source (e.g. *NOAA* for weather data) or an information type (e.g. *electricity usage* from Smart Meters). In our architecture, wrapper libraries that can access these individual mappings are responsible for processing information as it arrives from external sources, and enhancing them with a semantic concepts and relationships. It is these semantic instances that are actually recorded in the repository.

Individual agents may be responsible for specific data sources, which makes it easier to identify the right mapping to use. These agents can either pull information from the sources, or have it pushed to them. Part of the annotation process should include performing data transformations (such as unit conversion) to provide consistent information in the repository. Software vendors such as Oracle have specialized software solutions for utilities that help perform such conversions for smart meter data from different types of meters as part of a gateway service[2].

19.2.3 Cloud-Hosted Information Repository

The need to support accretive information ingest and its sharing among thousands, potentially millions, of users means that the repository needs to be hosted on a scalable platform [24]. Initial estimates put the upper bound on the size of annual meter data for 1.4 Million Los Angeles consumers at 1 minute intervals at hundreds of Terabytes per year. In addition, a host of metadata and semantic information about the consumers is gathered in support of DR. We adopt a service oriented architecture deployed on Clouds to host the information repository, DR information processing and analysis tools, and information sharing endpoints.

19.2.3.1 Information Integration on Clouds

Clouds provide scalable resources in the form of Virtual Machines (VMs) and reliable storage services like queues, tables and blobs. These elastic resources with a pay-as-you-go model provide the flexibility needed for information acquisition, processing, storage, query and dissemination for large repositories like the ones we are considering for Smart Grids [24]. For example, it is possible to scale out clients on multiple VMs to access different online service providers for information in parallel. A number of commercial Cloud vendors and open source Cloud software fabrics exist, including Amazon AWS, Microsoft Azure, and Eucalyptus. There are, however, some issues that need to be resolved for Clouds to be used effectively for the Smart Grid information architecture.

One, current Clouds do not provide the ability to host large databases as a platform service. Microsoft SQL Azure[3] is one of the major providers of SQL server as a Cloud platform, but limits the database size to under 100GB which is

[2]www.oracle.com/us/industries/utilities

[3]www.microsoft.com/en-us/sqlazure/database.aspx

insufficient for hosting meter data. We will either have to consider a SQL database installed on a Cloud VM as an infrastructure service – with manually configured reliability features, or distribute the data into multiple databases, partitioned either by consumers or the type of information. Alternatively, a combination of public and private Clouds is possible.

In addition, the semantic features required by our architecture are not available out-of-the-box for datasets of the scale we need to process. Lastly, acquiring streaming data from millions of Smart Meters is expected to be a challenge when considering the network resource constraints. Our recent work has highlighted the absence of streaming as a Cloud storage service as problem for Smart Grids [25], and suggested algorithms to throttle bandwidth using application QoS needs [26].

19.2.3.2 Data Security and Privacy

Cyber-security has been identified as a key concern for Smart Grid adoption and is being examined in detail [27][28]. In addition to securing the communication between smart meters and the utility, it is also important to secure the actual information hosted in the information repository [26]. For particularly sensitive data, this may even amount to encrypting the data when storing it in a public Cloud to ensure the Cloud service provider themselves cannot access it. Splitting data across public and private Clouds is another approach to ensure sensitive data is kept within the confines of the utility's private Cloud, but introduces challenges when querying data that span the silos.

Another issue that is less understood but of heightened concern is one of information leakage as a result of integrating such diverse data [26]. *Data privacy* issues have been in the limelight for social networks and can severely compromise trust in the utility if measures are not put in place to clearly define information sharing and use policies that are enforced. While users are increasingly willing to post personal information in public, there is backlash when such knowledge is used by organizations to price commodities or in unconventional ways. Tracking the provenance of data used for demand forecasting and load curtailment, and distinguishing it from pricing policies is also necessary.

19.3 Semantic Complex Event Processing

New types of DR applications locate patterns among a large class of realtime information to detect power consumption behaviors and predict usage in realtime. The requirement of timely processing and response to power usage situations makes complex event processing (CEP) an attractive component of a DR information system architecture. CEP deals with computation, transformation and pattern detection over large volumes of partially ordered events and messages [29]. An event is essentially a data object that represents something that occurs or changes the current state of affairs. In Smart Grids, continuous data from realtime information sources can be abstracted as events. These may be from sensors and appliances (*ThermostatChange* event, *ToasterOn* event), smart meters (*Power Usage* event), weather phenomena (*HeatWave* event) or consumer activity

(*ClassScheduleChange* event, *HomeArrive* event). CEP correlates distributed events in order to detect meaningful event patterns or situations, thus supplying Smart Grid applications with an accumulated view of the incoming data, and allowing them to react to detected patterns.

Some limitations of current CEP systems impede their effective use for Smart Grids. Existing systems process events as plain data tuples. As such, complex event patterns can only be defined as a combination of attributes presented in event data [30]. Users have to know the details of event structures and sources, and define patterns over low level specifications. Also, most CEP systems only support well-defined and precise pattern specification and matching, without any leeway to relax pattern constraints. However, uncertainty is an intrinsic feature of real world cyber-physical applications, where potentially incomplete, imprecise and even incorrect data exist, but still need to be matched within certain bounds.

An effective CEP solution for DR needs to meet several requirements. First, it must be extensible to meet the organic growth of the Smart Grid information diversity with the provision to easily model, specify and identify new events, relationships, and event patterns. Second, it should be easy to use both by domain experts and by non-domain users like developers and consumers with limited power systems background. Third, it should support uncertainties in events and patterns. Naturally, the CEP system should also scale to large sizes and event rates.

Here, we describe our initial design of Semantic Complex Event Processing for DR that supports the above requirements [31].

19.3.1 Semantic Event Modeling

The state-of-the-art CEP systems process events as plain data tuples with structure. For example, power demand events may be modeled as *demand(appliance id, value, timestamp)*. As such, complex event patterns can only be defined as a combination of attributes presented in the event model. Data level specifications of event patterns are too complex for domain experts in an information rich space like Smart Grids.

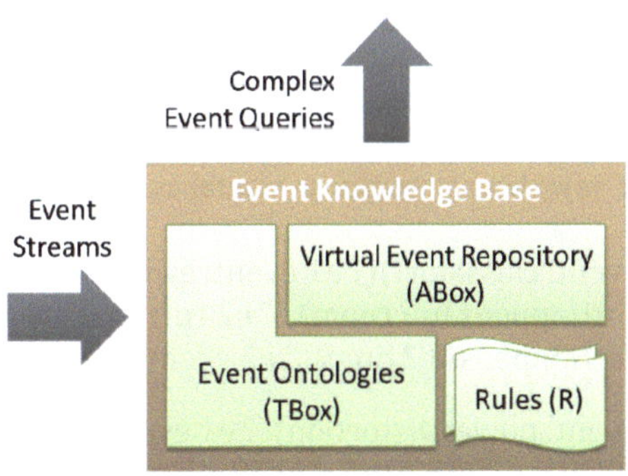

Fig. 19.3 Conceptual event model for CEP applications [1].

We propose an event model that captures the semantics of events, forming the foundation for expressive, high level complex event patterns [31]. By incorporating semantics, event data becomes meaningful information that enables automated mediation between domains and abstraction levels, and mapping goals to patterns.

We adopt a description logic based knowledge representation to model events (Figure 19.3). The event model captures the event types, type hierarchies (i.e., specialization and generalization), relations between events and other domain entities. Using description logic, a semantic event knowledge base K can be represented for the event model as:

$$K = (T, R, A)$$

where T is the *TBox* which introduces the vocabulary (or *T*erminology) of events and domain concepts; A is the *ABox*, a virtual event repository that "stores" *A*ssertions about named individual events using the vocabulary; R is a *R*ule set that represents the correlations between events, and is used to derive virtual events into the event repository based on observed events [32].

The event model maps to an OWL schema, and instances about events and domain entities can be stored in a structured OWL representation. Event ontologies need to be organized in a modular and layered manner for easy extension. The top level event ontology captures concepts and relationships between events, such as the time the event occurs and which domain specific classes it corresponds to (e.g. *ThermostatChange*, *AirConditionerTurnedOn*). The second layer of the model contains the subjective domain ontologies introduced earlier in the information integration section. The lowest level of the event model relates to external ontologies, also introduced before. Connections between the event ontology and the domain ontologies are made using properties like "eventHappensTo" and "eventHappensAt", whose domain is an event and value is a domain object. For example, the "thermostat" has a "thermostatChange" happen to is a concept defined in the *appliance* domain ontology.

Semantic Web Rule Language (SWRL) can be applied to encode correlation rules between events (R in the event knowledge base). SWRL is a proposed rule language for Semantic Web grounded in horn logic. A SWRL rule consists of a condition (rule body) and a consequence (rule head), each of which consists of a set of atoms. Event correlation rules can be defined based on domain knowledge or based on analysis of historic data. Several simple correlation rules are showed below, which derive new demand response virtual events based on known events.

```
r1: (? e1 rdf: type event: classStart)(? e1 event: eventHappensAt ? time)
    (? e1 event: eventHappensTo ? room) (? e2 rdf: type event: loadIncrease)
    (? e2 event: eventHappensAt ? time)(? e2 event: eventHappensTo ? room)

r2: (? e1 rdf: type event: powerDistortion)(? e1 event: eventHappensAt ? time)
    (? e1 event: eventHappensTo ? building)(? e2 rdf: type event: heavyDemand)
    (? e2 event: eventHappensAt ? time)(? e2 event: eventHappensTo ? building)
```

19.3.2 Reasoning over Realtime Events

In our proposed CEP system, time critical situations are detected not only based on raw event data, but also upon additional knowledge such as semantics and correlations of events [31]. The performance of the reasoning technique is time sensitive. Current research into reasoning systems have improved performance over a static or slowly growing knowledge base, but reasoning over realtime events backed by ontologies and rules is a novel challenge.

A naive implementation of event reasoning would store the event history in a repository and perform inference over the event base as new events are observed. However, events may arrive at high frequency and are unbounded. Thus, optimizations are needed to improve its performance in terms of storage cost, latency and throughput.

Reasoning over realtime events can be parallelized to improve performance. Our prior work on parallel inferencing for OWL knowledge base using graph partitioning is applicable here [33]. In event reasoning, partitioning can be applied to both the historic event data and correlation rule set. Based on the relations to each other (edges in the semantic graph), events and rules can be grouped into partitions which are dispatched to different VMs for parallel processing. VMs communicate intermediate results for necessary synchronization. The goal of the partitioning algorithms is to balance load between different partitions by minimizing the edge-cut communications.

Another potential optimization includes using filters to drop duplicate or irrelevant events from high frequency inputs before they are processed. For realtime applications, we can restrict reasoning to a certain window which consists of a subset of recently observed events while previous information is ignored. This is meaningful for two reasons. First, ignoring older information allows us to save computing resources in terms of memory, storage and processing time, and response to important events in real time. Further, most event processing applications assume that older events eventually become irrelevant. Applying appropriate time windows will help significantly reduce the size of rules and data to consider in the inference process.

19.3.3 Semantic Mining for Energy Use Patterns

Meaningful energy use patterns for complex event processing can be defined either based on domain knowledge or from an analysis of historic data [31]. For example, event patterns that represent power consumption trends can be simply expressed by using a sequence of monotonically decreasing/increasing AMI meter readings. However, identifying event patterns from a large corpus of energy data is non-trivial. Moreover, energy use patterns may also constantly evolve over time. For example, the event patterns that predict the demand of a building can vary due to changes of owners, renovation, and so on. There is a need to enhance CEP with pattern mining to assist with automatic specification of novel patterns.

Existing data mining techniques can be applied to extract static energy use patterns. The most relevant approaches are association rule learning and sequential data pattern mining. Algorithms for mining association rules from relational data have been well studied [34, 35]. The central idea is to find all rule patterns whose confidence and support are above corresponding thresholds. Some studies [36, 37] contribute to the mining of sequential patterns in temporal datasets. Most of these methods are based on searching algorithms using a heuristic which states the fact that any super-pattern of an infrequent pattern cannot be frequent.

Typically, data mining is done off-line. However there are examples in which event processing and mining run concurrently. In these cases, we can use a target event to constantly monitor the change of event patterns, for example, using meter readings to monitor patterns that predict peak power demand. When the event pattern deviates significantly from the target, it triggers the pattern mining process. For Smart Grid applications, mining dynamic energy use patterns from a complete set of historic data is not necessary as older events become less relevant when the energy use behaviors evolve. Data mining algorithms can be applied to the event data in a fixed time window to identify events frequently occurring with the target events.

19.4 Related Work

19.4.1 Semantic Information Integration

Semantic information integration is an active research area that has been studied in various research and application domains, including databases [38] [39], web services [40] [41], eHealthCare [6] [42], oil industry [43] and transportation [8]. The general objective is to facilitate interoperability of information systems and share information sources that are often heterogeneous and distributed [31].

Contemporary approaches for semantic interoperability can be classified into two categories. The first is the so-called Brute-force Data Conversions (BF) [44], which directly implements all necessary data transformations manually. This approach may require a large number of transformation agreements to be hardcoded that are difficult to maintain [31]. Another approach is Global Standardization, in which different information systems agree on a uniform standard. This causes the semantic differences to disappear and there is no need for data transforms between components. Unfortunately, such standardization is usually infeasible for many domains, for example, because of organizational and operational reasons.

Semantic Web provides an extensible framework that allows information to be shared and reused across application and domain boundaries using ontologies. The shortcomings of the traditional approaches can be overcome by declaratively describing data semantics using ontologies and separating the knowledge representation from the data transform implementation [44]. A widely adopted approach is to allow information sources to describe their vocabularies of information independently using ontologies. Inter-ontology mappings and reasoning services are then applied to identify semantically corresponding terms of component ontologies, e.g., which terms are semantically equal or similar.

Numerous research projects utilize ontologies to represent data semantics and facilitate information integration. For example, [42] describes the use of Semantic Web technologies for sharing knowledge in healthcare. It combines relational databases and ontologies to extract knowledge that was not explicitly declared within the database. An ontology representation of the UMLS (Unified Medical Language System) represents the basic medical concepts, and mappings and inference over the semantic knowledge are done to query and update heterogeneous databases. [8] developed a unifying traffic modeling and simulation framework into which more focused simulators can be integrated. A domain ontology was used as the common modeling language and data exchange model for integrated simulation.

Information processing has started to get more sophisticated at utilities in recent times. There have been efforts to display grid-level information integration through a consumer portal or a home energy monitoring system. These commonly take the form of a simple chart or histogram of energy usage over various time periods, such as Google's PowerMeter [45], the Pulse energy management software [46], and AgileWaves [47]. While this approach is promising, they do not (yet) support the ability to incorporate external applications.

Efforts have been taken to leverage standards [48, 49] for Smart Grid participants to integrate components, and span energy information from the "micro" (i.e. the power domain) to the "macro" (i.e. multiple relevant domains and users). The existing Smart Grid standards span multiple domains from electric power generation, electrical appliances to information technology, and are designed by a number of organizations: IEC, EPRI, NIST, IEEE and others. In the last decade, the power industry has made efforts in creating a common information model (CIM) to resolve semantic inconsistency between the different standards. IEC 61970 and IEC 61968 series standards [50] define data exchange specifications of CIM so that interoperability between various applications can be achieved.

There has been recent work on developing scalable, semantic-level Smart Grid information integration framework. In [51], the authors propose a shared ontology model to provide common semantics for Smart Grid applications. The ontology captures concepts governed by business semantics, engineering and scientific principles by transforming existing standards, such as CIM, to a uniform conceptual model. To make semantic modeling accessible to domain experts, they also developed the Semantic Application Design Language (SADL), a controlled-English language with an associated environment for building semantic models.

19.4.2 Complex Event Processing

Complex event processing deals with detecting real-time situations, represented as event patterns, from an event loud. CEP originated from the RAPIDE simulation research project in 1993 [52][31]. Recently, CEP has received attention in the research community due to its applicability to domains such as financial services

[53, 54], health care [55] and RFID data management [56]. Research prototypes and commercial systems, such as ruleCore [57], Oracle CEP Server [58] and Esper [59], are available.

Snoop [60] is a proposed event specification language for active databases. In an active database system, updates to the database are treated as events. The authors observe that the detection of database event patterns leads to monotonically increasing storage overhead as previous occurrences of events cannot be deleted. To overcome this problem, they introduce ~~the~~ event selection and consumption policies, i.e. so-called parameter contexts for precisely restricting event occurrences. The notion of event selection and consumption is also important for Smart Grid event processing systems. For example, redundancy will often be present in the high frequency event data from meters and appliances. The effect of duplicate events from the same meter or sensor can be subsumed by the most recent event.

Cayuga [61] is a high performance CEP system designed to efficiently support a large number of concurrent complex event subscriptions. In Cayuga, event streams are infinite sequences of relational tuples with interval-based timestamps. It defines an expressive event algebra that contains six operators: selection, projection, renaming, union, conditional sequence and iteration. Event algebra expressions are detected by non-deterministic finite automata (NFA). We believe the event algebra generalized in Cayuga is necessary but not sufficient to support Smart Grid applications. It can be used to specify rigid accurate patterns but is unable to express event patterns that incorporate semantics and uncertainties of data. Recently, parallelized and distributed complex event processing has received attention in literature [64, 65] and is of relevance to our architecture.

19.5 Conclusion

The information system architecture we have described for demand response optimization in the Los Angeles Smart Grid project is intended to transform the way utilities and consumers treat energy management. By proposing an information rich environment with semantically meaningful data integrated from diverse sources, our architecture will enable advanced analytical tools and algorithms to effectively and efficiently forecast energy load and identify load curtailment response. It also opens the door for other rich Smart Grid applications to be developed for desktop and mobile platforms that access the Cloud-hosted information repository.

The project is currently in the first year of a 3 year research and development cycle. We are in the process of prototyping several components of this architecture and evaluating them within the USC campus micro-grid testbed, and subsequently scaling the system to the LA DWP service area [2]. This software architecture will evolve during this period, informed by further research and evaluation. However, it establishes the advent of an informatics approach to Smart Grids that will lead to automated, intelligent and sustainable energy management.

Acknowledgement

This material is based upon work supported by the Department of Energy under Award Number DE-OE0000192and the Los Angeles Department of Water and Power. The authors would like to thank Carol Fern from the USC Facilities and Management Services, Julie Albright from the USC Annenberg School of Communication, Don Paul from the USC Energy Institute, and Aditya Sharma and Khaled Salem from the LA DWP for discussions and contributions.

Disclaimer: This report was prepared as an account of work sponsored by an agency of the United States Government. The views and opinions of authors expressed herein do not necessarily state or reflect those of the United States Government or any agency thereof.

References

[1] Electric power industry overview (2007), `http://www.eia.doe.gov/electricity/page/prim2/toc2.html`
[2] Simmhan, Y., Aman, S., Cao, B., Giakkoupis, M., Kumbhare, A., Zhou, Q., Paul, D., Fern, C., Sharma, A., Prasanna, V.: An Informatics Approach to Demand Response Optimization in Smart Grids, Technical Report, University of Southern California (2011)
[3] LD DWP 2010 integrated resources plan, Sec N.5.Resource Adequacy Gap Analysis (2010)
[4] Existing capacity by energy source (2009), `http://www.eia.doe.gov/cneaf/electricity/epa/epat1p2.html`
[5] W3C Semantic Web Activity, `http://www.w3.org/2001/sw`
[6] Niekerk, J.C., Griffiths, K.: Advancing Health Care Management with the Semantic Web. In: Proc. of IEEE BroadCom 2008 (2008)
[7] Taswell, C.: Doors to the semantic web and grid with a portal for biomedical computing. IEEE Transactions on Information Technology in Biomedi-cine 12(2) (2008)
[8] Zhou, Q., Bakshi, A., Soma, R., Prasanna, V.K.: Towards an Integrated Model-ing and Simulation Framework for Freight Transportation in Metropolitan Areas. In: Proc. of IEEE International Conference on Information Reuse and Integration (2008)
[9] Semantic Web Ontology Language (OWL), `http://www.w3.org/TR/owl-ref`
[10] Finamore, E.P.: Integrated Outage Management: Leveraging Utility System Assets Including GIS and AMR for Optimum Outage Response. Electric Energy Magazine (2004)
[11] Aman, S., Simmhan, Y., Prasanna, V.K.: Towards Modeling and Prediction of Energy Consumption for a Campus Micro-grid. Under Submission (2011)
[12] McMorran, A. W.: An Introduction to IEC 61970-301 & 61968-11:The Common Information Model, Technical Report, University of Strathclyde (2007)
[13] Crapo, A., et al.: Overcoming Challenges Using the CIM as a Semantic Model for Energy Applications. Grid-Interop Forum, GridWise Architecture Council (2010)
[14] ANSI/ASHRAE 135-2008/ISO 16484-5 BACnet – A Data Communication Protocol for Building Automation and Control Networks

[15] Semantic Web for Earth and EnvironmentalTechnology (SWEET 2.1), NASA Jet Propulsion Lab (2010), http://sweet.jpl.nasa.gov
[16] NextGen Network Enabled Weather (NNEW): Data Models and Formats, University Corporation for Atmospheric Research, UCAR (2011), http://wiki.ucar.edu/display/NNEWD/The+NNEW+Wiki
[17] Zhou, Q., Bakshi, A., Prasanna, V.K., Soma, R.: Towards an Integrated Modeling and Simulation Framework for Freight Transportation in metropolitan Areas. In: Proc. of IEEE International Conference on Information Reuse and Integration, Las Vegas (2008)
[18] Halpin, H., Tuffield, M.: A Standards-based, Open and Privacy-aware Social Web. W3C Incubator Group Report (December 2010)
[19] Mika, P.: Ontologies are us: A unified model of social networks and semantics. Journal of Web Semantics 5(1) (2007)
[20] Internet Calendaring and Scheduling Core Object Specification (iCalendar)
[21] Alex, B., Grover, C., Haddow, B., Kabadjov, M., Klein, E., Matthews, M., Tobin, R., Wang, X.: Automating Curation Using a Natural Language Process-ing Pipeline. Genome Biology 9(Suppl 2):S1 (2008)
[22] William, J.C., Vieweg, S.: Travis Rood and Martha Palmer, Twitter in mass emergency: what NLP techniques can contribute. In: Workshop on Computational Linguistics in a World of Social Media, WSA (2010)
[23] Kireyev, K., Palen, L., Anderson, K.: Applications of Topics Models to Analysis of Disaster-Related Twitter Data. In: NIPS Workshop on Applications for Topic Models: Text and Beyond (2009)
[24] Simmhan, Y., Giakkoupis, M., Cao, B., Prasanna, V.K.: On Using Cloud Plat-forms in a Software Architecture for Smart Energy Grids. In: IEEE InternationalConference on Cloud Computing, CloudCom (2010)
[25] Zinn, D., Hart, Q., McPhillips, T., Ludascher, B., Simmhan, Y., Giakkoupis, M., Prasanna, V.K.: Towards Reliable, Performant Workflows for Streaming-Applications on Cloud Platforms. In: IEEE/ACM International Symposium on Cluster Computing and the Grid, CCGrid (May 2011)
[26] Simmhan, Y., Kumbhare, A., Cao, B., Prasanna, V.K.: An Analysis of Security and Privacy Issues in Smart Grid Software Architectures on Clouds. In: IEEE International Conference on Cloud Computing, CLOUD (2011)
[27] Burr, M.T.: Smart-grid security: Intelligent power grids present vexing cyber security problems. Public Utilities Fortnightly, 43 (January 2008)
[28] Echols, M., Sorebo, G.: Protecting Your Smart Grid. Transmission & Distribu-tion World, Penton Media (July 2010)
[29] Luckham, D., Schulte, R. (eds.): Event processing glossary- version 1.1.,Technical Report, Event Processing Technical Society (July 2008)
[30] Dong, L., Carlos, P., John, D.: Semantic enabled complex event language for business process monitoring. In: International Workshop on Semantic Business Process Management (2009)
[31] Zhou, Q., Simmhan, Y., Prasanna, V.K.: Towards an Inexact Semantic Com-plex Event Processing Framework for Demand Response. In: ACM International Conference on Distributed Event-Based Systems, DEBS (2011)
[32] Baader, F., Calvanese, D., McGuinness, D., Nardi, D., Patel-Schneider, P.: The Description LogicHandbook: Theory, Implementation and Applications. Cambridge University Press, Cambridge (2002)

[33] Soma, R., Prasanna, V.K.: Parallel Inferencing for OWLKnowledge Bases. In: International Conference on Parallel Processing (2008)

[34] Agrawal, R., Srikant, R.: Fast algorithms for mining association rules in large databases. In: International Conference on Very Large Data Bases, VLDB (1994)

[35] Zaki, M.: Scalable Algorithms for Association Mining. IEEETransactions on Knowledge and Data Engineering 12(3) (2000)

[36] Agrawal, R., Srikant, R.: Mining Sequential Patterns. In: International Confer-ence on DataEngineering, ICDE (1995)

[37] Yang, J., Wang, W., Yu, P.S.: Mining Asynchronous Periodic Patterns in Time Series Data. IEEE Transactions on Knowledge and Data Engineering 15(3) (2003)

[38] Kim, W., Seo, J.: Classifying schematic and data heterogeneity in multidatabasesystems. IEEEComputer 24(12), 12–18 (1991)

[39] Pottingerand, R., Bernstein, P.: Merging models based on given correspon-dences. In: International Conference on Very Large Databases, VLDB (2003)

[40] Lastra, J.L.M., Delamer, I.M.: Semantic web services in factory automation: fundamental insights and research roadmap. IEEE Transactions on Industrial Informatics 2(1) (2006)

[41] Acuna, C.J., Marcos, E., Gomez, J.M., Bussler, C.: Toward Web portals inte-gration through semantic Web services. In: International Conference on Next Generation Web Services Practices (2005)

[42] Nardon, F.B., Moura, L.A.: Knowledge sharing and information integration in healthcare using ontologies and deductive databases. Medinfo, 62–66 (2004)

[43] Soma, R., Bakshi, A., Prasanna, V.K.: A Semantic Framework for Integrated Asset Management in Smart Oilfields. In: IEEE InternationalConference on Cluster Computing and the Grid (2007)

[44] Gannon, T., Madnick, S., Moulton, A., Siegel, M., Sabbouh, M., Zhu, H.: Se-mantic Information Integration in the Large: Adaptability, Extensibility, and Scalability of the Context Mediation Approach, Technical Report, CISL# 2005-04. MIT, Cambridge, MA (2005)

[45] Google PowerMeter, `http://www.google.com/powermeter`

[46] Pulse energy management software, `http://www.pulseenergy.com`

[47] AgileWaves, `http://www.agilewaves.com`

[48] NIST Framework and Roadmap for Smart Grid Interoperability Standards, `http://www.nist.gov/smartgrid`

[49] Becker, D., Falk, H., Gillerman, J., Mauser, S., Podmore, R., Schneberger, L.: Standards-based approach integrates utility applications. IEEE Computer Ap-plications in Power 13 (2000)

[50] IECStandards, `http://www.iec.ch`

[51] Crapo, A., Wang, X., Lizzi, J., Larson, R.: The semantically enabled Smart Grid. In: The semantically enabled Smart Grid. GridWise Forum (2009)

[52] Luckham, D.C., Frasca, B.: Complex Event Processing in Distributed Systems, Technical Report (1998)

[53] Adi, A., Botzer, D., Nechushtai, G., Sharon, G.: Complex Event Processing for Financial Services. In: IEEE Services Computing Workshops (2006)

[54] Magid, Y., Adi, A., Barnea, M., Botzer, D., Rabinovich, E.: Application gen-eration framework for real-time complex event processing. In: IEEE International Computer Software and Applications Conference (2008)

[55] Churcher, G.E., Foley, J.: Applying and extending sensor web enablement to a telecare sensor network architecture. In: International ICST Conference on COMmunication System softWAre and middlewaRE (2009)
[56] Wu, E., Diao, Y., Rizvi, S.: High-performance complex event processing over streams. In: ACM SIGMOD international conference on Management of data (2006)
[57] RuleCore, http://www.rulecore.com
[58] Oracle, CEP Engine, http://www.oracle.com/technetwork/middleware/complex-event-processing/overview/index.html
[59] Esper, http://esper.codehaus.org
[60] Chakravarthy, S., Mishra, D.: Snoop: an expressive event specification lan-guage for active databases. Data & Knowledge Engineering 14, 1–26 (1994)
[61] Demers, A.J., Gehrke, J., Panda, B., Riedewald, M., Sharma, V., White, W.M.: Cayuga: A general purpose event monitoring system. In: Conference on Inno-vative Data Systems Research (CIDR), pp. 412–422 (2007)
[62] Gyllstrom, D., Wu, E., Chae, H.-J., Diao, Y., Stahlberg, P., Anderson, G.: SASE: Complex event processing over streams. In: Conference on Innovative Data Systems Research, CIDR (2007)
[63] Cherniack, M., Balakrishnan, H., Balazinska, M., Carney, D., Cetintemel, U., Xing, Y., Zdonik, S.B.: Scalable distributed stream processing. In: Conference on Innovative Data Systems Research, CIDR (2003)
[64] Brenna, L., Gehrke0, J., Hong, M., Johansen, D.: Distributed event stream processing with non-deterministic finite automata. In: ACM International Conference on Distributed Event-Based Systems, DEBS (2009)
[65] Akdere, M., Cetintemel, U., Tatbul, N.: Plan-based complex event detection across distributed sources. In: VLDB Endowment (2008)

Chapter 20
Markov Chain Based Emissions Models: A Precursor for Green Control

E. Crisostomi, S. Kirkland, A. Schlote, and R. Shorten

Abstract. In this chapter we propose a new method of modeling urban pollutants arising from transportation networks. The efficacy of the proposed approach is demonstrated by means of a number of examples. Our models give rise to a number of surprising observations that are relevant for the regulation of pollution in urban networks: different actions are required for the control of different pollutants and low speed limits do not necessarily lead to low pollution.

20.1 Introduction

Contemporary discussion relating to *Green IT* has focussed to a large extent on energy awareness and the minimisation of energy consumption of devices and systems. Notwithstanding the importance of these energy-related issues, the Green agenda is, in fact, much more comprehensive. From our perspective, Green describes practices and strategies whose aggregate behaviour: (i) promotes sustainability of resources and the planet; (ii) is not harmful to human health; and (iii) recognises that overprovisioning of resources is unsustainable, and so utilises resources in a fair and cooperative manner. Thus, being *Green* refers to the sharing of resources between activities in such a way that promotes these goals. In this context, distributed optimisation and control techniques are likely to become ever more important in achieving these goals, and we refer to feedback based activities that are designed to support green solutions and products as *Green Control*.

The issue of Green Control is likely to become a major research topic over the next decade. Roughly speaking, this term refers to the development of control strategies for deployment in large scale networks that can be used to regulate, share, and optimize the use of quantities related to "green systems". Generally

---

E. Crisostomi
Department of Energy and Systems Engineering, University of Pisa, Italy
e-mail: emanuele.crisostomi@gmail.com

S. Kirkland · A. Schlote · R. Shorten
Hamilton Institute, National University of Ireland – Maynooth, Co. Kildare, Ireland
e-mail: {stephen.kirkland,arieh.schlote,robert.shorten}@nuim.ie

speaking, these "green variables" might include atmospheric pollutants, carbon emissions, energy usage and reuse. One area, in which activity in "green control" is already underway, is in the regulation of pollution and emissions in urban transportation networks. In the automotive industry, public awareness of the link between greenhouse gasses and road transportation is great. While CO_2 emissions are ultimately harmful to humans through ozone depletion, road transportation is also very harmful to humans in a direct manner. According to a UK study [13], road transportation's percentage contribution to air pollution in 1999 was: 80% in the case of CO; 75% in the case of benzene; 50% in the case of NOx; 40% for hydro carbons (producing ozone); and 25% of particulates, all of which are extremely harmful to humans. See [26] for details of established side-effects of these pollutants. It is also important to note recent advances in health science where the link between heart attack triggers and urban pollution is hypothesized [29].

Traditionally, researchers have advocated two approaches to reducing air pollution: the production of greener cars; and an overall reduction in road transportation. These involve not only introducing laws that produce ever more stringent regulations on engine manufacturers, but also some cities are trying to regulate pollution in certain areas through control policies. The city of Berlin, Germany, is one such example, where strict open-loop control policies based on clean engines are implemented.

An alternative approach is to investigate advances in ICT for the automobile sector to control and regulate emissions and pollution in our cities. This latter approach builds on three policy directions coming from regulatory bodies (such as the EU and the US government) aimed at (1) reducing greenhouse gasses, (2) reducing pollution peaks in our cities, and (3) developing instrumentation, cooperative control strategies, and modeling techniques to enable the development of proactive traffic management systems, i.e., systems that predict traffic flow and take pre-emptive measures to avoid incidents (traffic build up, pollution peaks, etc). It is also consistent with strategic developments in the automotive and networking industries where the control and regulation of large scale systems are seen as priority objective in company roadmaps. Initiatives in this direction include the IBM smarter city initiative [11] and the CISCO smart and connected communities programme[32], as well as the many initiatives underpinning vehicle to vehicle communications and vehicle to infrastructure communications.

Our objective in this work is more modest. While our ultimate objective is control and optimization, we are motivated by the fact that underpinning every good control strategy is an appropriate model. Our starting point is a recently proposed Markovian approach to model road network dynamics [8]. As discussed in [8], a Markovian framework is particularly appealing since it makes important information regarding the road network available in a convenient form to the road network designer (congestion, average travel times, sensitive links in the road network). It is also useful for control and optimization applications. Our objective in this work is to demonstrate how this model can be extended to build emissions models, and to use this model to inform what might and might not be possible in a control setting. As we shall see, inference from the models will give rise to counter intuitive facts that must be taken into account when developing policy to reduce air pollution and emissions.

20.2 A Markov Chain Approach to Describe Road Network Dynamics and Emissions

The most popular mobility model is undoubtedly the so-called Constant Speed Motion model where vehicles follow random paths over a graph corresponding to the road topology, at a constant speed (see for instance [12] for a description). Such a model can be made more accurate by adjusting speeds taking into account interactions with other vehicles as well. Despite such corrections, these models do not usually reach a sufficient level of realism for many applications of interest.

For this reason, flow models were introduced to provide a more realistic modeling of urban networks, at a different level of detail; for instance, it is possible to distinguish between microscopic, mesoscopic and macroscopic models [16, Chapter 5]. Mesoscopic levels are at an intermediate level of detail, as traffic flows are described at an aggregated level (e.g., through probability density functions), but interactions are described at an individual level. Flow models can be described employing very different mathematical backgrounds, for instance Partial Differential Equations (PDEs), discrete time equations or Cellular Automata (CA) models. A recent application of fluid-dynamic traffic model on road networks is given in [5].

Although flow models have reached a high level of realism and are able to capture many of the typical traffic phenomena (e.g., queueing or bottlenecks at junctions regulated with traffic lights or priority rules), they still suffer from two main drawbacks: (a) it is not simple to describe road network dynamics both at a microscopic and at a macroscopic level (congestion at a single junction and in the whole network), (b) they miss the ability to predict how traffic would change in reaction to the road network modifications (e.g., removal of a road, change of speed limits or priority rules).

Due to such motivations, many road network engineers prefer to use mobility simulators (e.g. SUMO [23]) as a support to urban network management decisions, rather than mathematical models. Anyway, simulators are good at describing traffic flows (both at a macroscopic and at a microscopic level), but also suffer from the inability of making accurate predictions should the original urban network change.

A new paradigm to describe road network dynamics was proposed in [8], where a Markov chain based model was used as an alternative to flow models; such an approach was proved to give the same results of a mobility simulator, at least every time the mobility simulator was able to give an answer. Moreover, the mathematical background of Markov chains could be used to make predictions in case of changes to the nominal road network, thus providing, at least in principle, a useful tool to road engineers to understand in advance what changes to the original network can provide benefits to the drivers (e.g., in terms of travel times).

The main motivation of this chapter is to show that the same background of [8] can be used to predict useful modifications of the road network that might help to decrease harmful emissions, thus paving the way to what has been called "Green Control". In fact, emissions are related to congestion, and in particular to vehicle speeds, as emphasised from many emissions models.

There are many several models that evaluate pollution at a microscopic level (e.g., single vehicle), at very different levels of accuracy and complexity [19]. A first basic model is the so-called *aggregated emission factor model*, where a single emission factor is used to represent a particular type of vehicle and a general type of driving, with a usual distinction between urban roads, rural roads and motorways. This model is rather rudimentary and is not realistic for small scale networks, as it omits several phenomena, such as congestion, which are known to significantly affect emissions. A more refined model is the so-called *average-speed model*, which will be adopted in this chapter, where emission factors are calculated as a function of average speed [2-4]. The average-speed approach is described in the UK Design Manual for Roads and Bridges (DMRB) [17] and the European Environment Agency's COPERT model [15].

Although the average-speed model has been extensively used for many applications, it suffers from a drawback that very different vehicle operational behaviours (in terms of accelerations, decelerations, maximum speed, gear-change pattern), and therefore different emission levels, can be characterized by the same average speed. A more realistic model is for instance the *comprehensive modal emissions model* (CMEM) described in reference [1]. According to this model, second-by-second exhaust emissions and fuel consumption are predicted, for a wide range of vehicle categories and ages. For the sake of simplicity, in this work the average-speed model is employed. However, the same proposed methodology can be applied in combination to any other (more accurate) vehicle emissions model.

We present no theoretical justification for our Markov chain model, other than to remind here that Markov chains have a long history of providing compact representations of large scale systems described by very complicated sets of dynamical equations. As an example, the planar restricted circular three-body problem [21-22], involving a planet, a moon orbiting on a circle around the planet, and an asteroid lying on the same plane of the moon orbit, can be solved by using advanced methods from dynamic systems theory. Yet, is was proved in [10] that similar results can be obtained by using Markov chains. Similarly, to obtain a good macroscopic description of a complicated dynamic system, the Frobenius-Perron operator can be used. The Frobenius-Perron operator describes the evolution of probability measures in time, and can be conveniently approximated using Markov chain techniques [9]. Successful applications of this idea have been obtained in the modeling of complex bio molecules [18-31]. Finally, a comprehensive introduction to modeling complex dynamic systems as Markov chains can be found in [14].

Inspired by the previous examples, we expect that as a network planner is usually interested in aggregation effects (pollution, congestion, travel times, etc.), then a Markov chain model should be able to capture such properties of interest with accuracy and high speed, even for large networks. This was first envisaged in [8] for the network dynamics, and is further developed here for the case of pollution dynamics.

20.3 A Primer on Markov Chains

The objective of this section is to present the mathematical tools that will be used in the rest of the chapter. The preliminary definitions are standard and can be found in classic references like [28] or, in a short summary, given in [24, Chapter 15]. Here we give only the most basic concepts that are needed for our discussion.

Recall that a *Markov chain* is a stochastic process characterised by the equation

$$p\left(x_{k+1} = S_{i_{k+1}} \mid x_k = S_{i_k}, \ldots, x_0 = S_{i_0}\right) = p\left(x_{k+1} = S_{i_{k+1}} \mid x_k = S_{i_k}\right) \tag{20.1}$$

where $p(E/F)$ denotes the conditional probability that event E occurs given that event F occurs. Equation (1) states that the probability that the random variable x is in state $S_{i_{k+1}}$ at time step $k+1$ only depends on the state of x at time step k and not on preceding values. Throughout the chapter only discrete-time, finite-state, homogeneous Markov chains will be considered. We present no theoretical justification for our model, other than to state that Markov chains have a long history of providing compact representations of large scale systems described by very complicated sets of dynamical equations, as already illustrated in the previous section.

The Markov chain is completely described by the $n \times n$ transition probability matrix **P** whose entries $\mathbf{P}_{ij}$ denote the probability of passing from state S_i to state S_j, and n is the number of states. The matrix **P** is a row-stochastic non-negative matrix, as the elements of each row are probabilities and they sum up to 1. Within Markov chain theory, there is a close relationship between the transition matrix **P** and its corresponding *graph*. A graph is represented by a set of *nodes* that are connected through *edges*. Therefore, the graph associated with the matrix **P** is a *directed graph*, whose nodes are represented by the states $S_i, i = 1,\ldots,n$ and there is a directed edge leading from S_i to S_j if and only if $\mathbf{P}_{ij} \neq 0$.

A graph is *strongly connected* if and only if for each pair of nodes there is a sequence of directed edges leading from the first node to the second one. The matrix **P** is *irreducible* if and only if its directed graph is strongly connected. Let us now state the well-known Perron-Frobenius theorem [24] which summarises important properties of irreducible transition matrices:

- The *spectral radius* of **P** is 1
- 1 also belongs to the spectrum of **P**, and it is called the *Perron root*
- The Perron root has an algebraic multiplicity of 1

- The left-hand Perron eigenvector $\boldsymbol{\pi}$ is the unique vector defined by $\boldsymbol{\pi}^{\mathrm{T}}\mathbf{P} = \boldsymbol{\pi}^{\mathrm{T}}$, such that $\boldsymbol{\pi} > \mathbf{0}, \|\boldsymbol{\pi}\|_1 = 1$. Except for positive multiples of $\boldsymbol{\pi}$ there are no other non-negative left eigenvectors for **P**.

In the last statement, by saying that $\boldsymbol{\pi} > \mathbf{0}$, it is meant that all entries of vector $\boldsymbol{\pi}$ are strictly positive. One of the main properties of irreducible Markov chains is that the ith component $\boldsymbol{\pi}_i$ of the vector $\boldsymbol{\pi}$ represents the long-run fraction of time that the chain will be in state S_i. The row vector $\boldsymbol{\pi}^{\mathrm{T}}$ is also called the *stationary distribution* vector of the Markov chain.

20.3.1 Mean First Passage Times and the Kemeny Constant

A transition matrix **P** with 1 as a simple eigenvalue gives rise to a singular matrix **I-P** (where the identity matrix **I** has appropriate dimensions) which is known to have a group inverse $(\mathbf{I}-\mathbf{P})^{\#}$. The group inverse is the unique matrix such that $(\mathbf{I}-\mathbf{P})(\mathbf{I}-\mathbf{P})^{\#} = (\mathbf{I}-\mathbf{P})^{\#}(\mathbf{I}-\mathbf{P})$, $(\mathbf{I}-\mathbf{P})(\mathbf{I}-\mathbf{P})^{\#}(\mathbf{I}-\mathbf{P}) = (\mathbf{I}-\mathbf{P})$ and $(\mathbf{I}-\mathbf{P})^{\#}(\mathbf{I}-\mathbf{P})(\mathbf{I}-\mathbf{P})^{\#} = (\mathbf{I}-\mathbf{P})^{\#}$. More properties of group inverses and their applications to Markov chains can be found in [27]. The group inverse $(\mathbf{I}-\mathbf{P})^{\#}$ contains important information on the Markov chain and it will be often used in this chapter. For this reason, it is convenient to denote this matrix as $Q^{\#}$.

The mean first passage time m_{ij} from the state S_i to S_j is the expected number of steps to arrive at destination S_j when the origin is S_i. If we denote $q_{ij}^{\#}$ as the ij entry of the matrix $Q^{\#}$, then the mean first passage times can be computed easily according to the equation below (see [7])

$$m_{ij} = \frac{q_{jj}^{\#} - q_{ij}^{\#}}{\pi_j}, \qquad i \neq j \tag{20.2}$$

where it is intended that $m_{ii} = 0, i = 1,\ldots,n$. The *Kemeny constant* is defined as

$$K = \sum_{j=1}^{n} m_{ij}\pi_j, \tag{20.3}$$

where the right hand side is (surprisingly) independent of the choice of i [20]. Therefore the Kemeny constant K is an intrinsic measure of a Markov chain, and if the transition matrix **P** has eigenvalues $\lambda_1 = 1, \lambda_2, \ldots, \lambda_n$ then another way of computing K is (see [25])

$$K = \sum_{j=2}^{n} \frac{1}{1-\lambda_j}. \tag{20.4}$$

Equation (4) emphasizes the fact that K is only related to the particular matrix **P** and that it increases if one or more eigenvalues of **P** is real and close to 1.

20.4 From a Road Network Model to a Pollution Model

The use of Markov chains to model road network dynamics has been described in detail in [8]. The resulting networks are fully characterized by a transition matrix **P**, which has the following form:

$$\mathbf{P} = \begin{bmatrix} P_{S_1 \to S_1} & P_{S_1 \to S_2} & \cdots & P_{S_1 \to S_n} \\ P_{S_2 \to S_1} & P_{S_2 \to S_2} & \cdots & P_{S_2 \to S_n} \\ \vdots & \vdots & \ddots & \vdots \\ P_{S_n \to S_1} & P_{S_n \to S_2} & \cdots & P_{S_n \to S_n} \end{bmatrix}. \tag{20.5}$$

The matrix **P** is a square matrix whose size is given by the number of road segments. The off-diagonal elements $P_{S_i \to S_j}$ are related to the probability that one passes directly from the road segment S_i to the road segment S_j. They are zero if the two road segments are not directly connected (i.e. at the end of road S_i it is impossible to take directly road S_j). The diagonal terms are proportional to travel times; namely, they are close to 1 if it takes a long time to cover the corresponding road while they are closer to 0 if travel times are short. The diagonal terms take into account several factors that affect travel times; such as speed limits, road surface conditions, presence of priority rules or traffic lights, weather conditions, heavy traffic, etc. The diagonal elements can be computed from average travel times through the following equation

$$P_{S_j \to S_j} = \frac{tt_j - 1}{tt_j}, \qquad j = 1, \ldots, n, \tag{20.6}$$

where tt_j indicates the average travel time along the jth road.

The main idea of this work is to use the same framework to model pollutants, This can be achieved by replacing time in the original chain described above, by a unit of pollutant (e.g. benzene, NOx, etc…). In this framework, a car is moving in the same road network, and changes (or remains in the same) state anytime a unit of pollutant is released, according to the entries of a second transition matrix $\mathbf{P_P}$, where the subscript P stands for pollution. The quantity of emissions released along a particular road S_j does not only depend on travel times, but also on other quantities like the length of the road, speed profile along the road (i.e. number of times the car accelerates or decelerates) and average types of car in the particular road (fuel type, engine capacity, etc). As a consequence, the main difference between the chain described in [8] and the chain here are the diagonal entries of matrix $\mathbf{P_P}$.

20.4.1 Construction of the Markov Chain Transition Matrix

The emissions model is completely determined by a transition matrix $\mathbf{P_P}$, whose diagonal terms indicate how many units of pollution a fleet of vehicles releases along a road segment, and by off-diagonal terms that indicate the probabilities of turning left, right, etc, when arriving at a junction. These entries are computed as follows.

Diagonal terms: According to average-speed models, the emission factor $f(t,p)$ is computed as

$$f(t,p) = \frac{k \cdot \left(a + bv + cv^2 + dv^3 + ev^4 + fv^5 + gv^6\right)}{v}, \tag{20.7}$$

where t denotes the type of vehicle (and depends on fuel, emission standard, category of vehicle, engine power), p denotes the particular type of pollution of interest (e.g. CO, CO_2, NOx, Benzene), v denotes the average speed of the vehicle, and the parameters a, b, c, d, e, f, g and k depend on both the type of vehicle and the pollutant p under consideration. For the purpose of this work, the values of the parameters are taken from Appendix D, in reference [4]. In Equation (7) it is assumed that speeds are measured in km/h and emission factors in g/km. Therefore, by assuming that the average speed of a fleet of vehicles along a road segment is v, and the length of the road segment is l, then the diagonal terms are given by

$$P_{S_j \to S_j} = \frac{f(t,p)_j \cdot l_j - 1}{f(t,p)_j \cdot l_j}, \quad j = 1, \ldots, n, \tag{20.8}$$

by analogy with Equation (6). For simplicity we first normalize all terms $f(t,p)_j l_j$ so that the minimum one has unit value. In Equation (8) note that the emission factor depends on the road segment as its primary dependence is on the average speed along the particular road segment.

Off-diagonal terms: The Markov chain models an *average car* that travels in the urban network while releasing units of pollution. Therefore, it is necessary to measure the average junction turning probabilities to build the off-diagonal terms.

20.4.2 The Role of Information Technology in the Model

The proposed emission model requires the following items to be useful:

- The **categories and types of vehicles** present in each road of the network, as they are required to use the appropriate parameters in Equation (7).
- The **average travel times** of the vehicles along each road of the network; once travel times are known, by using the knowledge of the length of the road segments, it is possible to compute the average speeds to be used in Equation (7).
- The **junction turning probabilities**, which are required to build the off-diagonal terms of matrix $\mathbf{P_P}$.
- Centralized (or decentralized) number crunching ability to calculate properties of the large scale matrix.

There are two possible ways of obtaining the information required for the Markov chain:

1. All cars are instrumented to behave as **mobile sensors** and store the data relative to their travelling history. At regular times, they communicate their data to a central database that collects all the important information.
2. Urban networks are equipped with **loop detectors** that are positioned at each junction, and they measure the required information to build the Markov transition matrix.

We emphasise here that most of the required information can be easily collected (when it is not already available) by the cars themselves, and all that is required is to collect and integrate such information to construct the model. Current vehicles are memory-less. However, given the widespread penetration of vehicle positioning systems (GPS), it is easy to imagine geospatial tagging of vehicle route information, and storing this information locally in vehicle memory. It is also possible to obtain information of the type required from special classes of vehicles (buses, taxis etc.) – though this probably contains less useful information. Note such instrumented fleets already exist in cities such as Stockholm [3]. Furthermore, recent advances in the development of Vehicular Adhoc NETworks (VANET) [16] are expected to further facilitate the proposed emissions model, as it is very easy to collect the information required to build the transition matrix.

Remark: CMEM (see Section 20.4) is one of the most accurate microscopic emission models, as it takes into account the instantaneous speed and the engine operational mode. The proposed Markovian approach can easily integrate such an emissions model. However, a high price must be paid in terms of communication between vehicles and infrastructure, as the whole speed profile of individual vehicles is required to be transmitted.

20.5 Green Interpretation of the Markov Chain Quantities

Once the transition matrix has been constructed, it is very easy to infer several quantities of potential interest to the designer of a road network. These include the Perron eigenvector, the mean first passage time matrix and the Kemeny constant.

The proposed model has the property that it is driven by a unit of pollution rather than by a unit of time as in conventional Markov chains. For ease of exposition in the present discussion we use a generic unit of pollutant, while in the simulation examples we will compute CO, CO_2, NOx and Benzene. We remind the reader that the difference between the models is in the choice of the parameters in Equation (7). We also note that CO_2 can be used as an indicator of energy efficiency, as it is the principal product of complete fuel combustion, and it is directly proportional to the fuel consumption rate [6]. Table 20.1 summarises the interpretation of the Markov chain quantities in the emission framework.

Table 20.1 Interpretation of Markov chain quantities in the emission framework

Markov chain quantity	Green interpretation
Left-hand Perron Eigenvector	This *vector* has as many entries as the number of road segments. Each entry represents the ***long run fraction of emissions*** that a fleet of vehicles will emit along the corresponding road segment. It can be used as an indicator of pollution peaks.
Mean first passage emissions	This is a *square matrix* with as many rows as the number of the road segments. ***The entry ij represents the expected quantity of emissions that a vehicle releases to go from i to j.*** The average is with respect to all possible paths from *i* to *j*.
Kemeny constant	This *number* is the average number of emissions released in a random route. ***It is an indicator of pollution in the entire network.***

There are other quantities that can be computed within the proposed framework:

Density of emissions along each road (g/km): They can be easily computed from Equation (7): it is only required to know the average fleet of vehicles and the average speed along the road.

Emissions along each road (g): They can be easily computed by multiplying the density of emissions along a road by the length of the road.

Total Emissions (g): It is sufficient to sum the emissions along each road for all the roads inside the area of interest (e.g. the urban network).

20.6 Examples

In this section the proposed approach is described in detail through several simulations. The information required to build the Markov transition matrix is recovered from simulating traffic within an urban network using the well-known mobility simulator SUMO (Simulation of Urban MObility) [23]. SUMO is an open source, highly portable microscopic traffic simulation package that was developed at the Institute of Transportation Systems at the German Aerospace Center, and is licensed under the GPL. Simulations are required to extract junction turning probabilities and average travel times along the road segments composing the urban network. Three steps must be performed to achieve this goal:

1. Creation of the urban network: this includes the topology of the network, the use of traffic lights or priority rules, fixing speed limits, choosing the number of lanes, etc.
2. A random pair of origin/destination roads is assigned to each car. The car will travel toward its destination according to the minimum length path.
3. Traffic statistics, namely junction turning probabilities and average travel times, are collected from the urban network and used as the data from which we build the Markovian emissions model.

We now give four examples of road networks to demonstrate the usefulness of our approach.

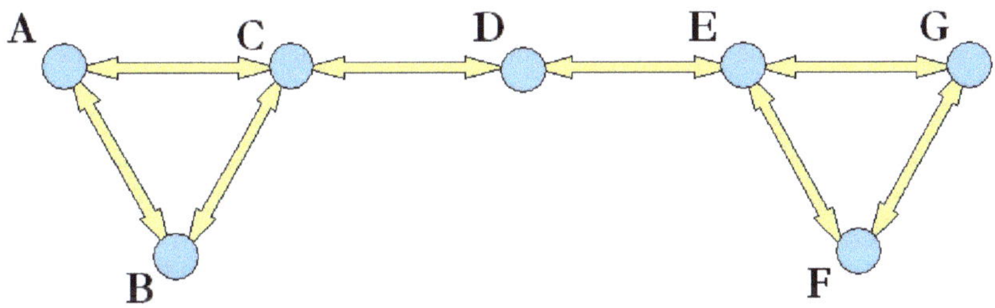

Fig. 20.1 Graphical representation of a simple urban network. Nodes and edges correspond to junctions and road segments respectively. Bidirectional arrows show that both travelling directions are allowed.

Consider the road network depicted in Figure 20.1. Here the road network is represented as a graph with nodes corresponding to road intersections and edges corresponding to streets between the intersections, we will call this representation the *primal graph* [30]. To use the analytical tools as described in Section 20.3, we need a different representation of the network called the *dual graph*. In the dual graph the nodes correspond to streets and there is an edge between two nodes if it is possible to continue from the first road to the second road. The dual graph thus contains information that is not accessible from the primal graph. As an example, a dual representation of the primal graph shown in Figure 20.1 is illustrated in Figure 20.2. Unless stated otherwise all road segments have a length of 500 meters.

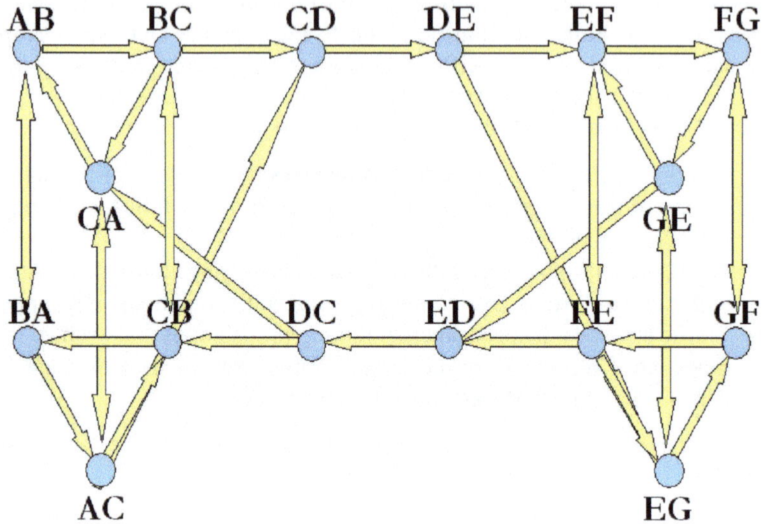

Fig. 20.2 Dual representation of the primal network shown in Figure 20.1. For instance, the road segment AB is the (directed) road that goes from junction A to B in the primal network.

First simulation – Importance of speed dependent emission factors: SUMO is now used to create traffic inside the urban network described in Figures 20.1 and 2, in two different operating conditions. In the first case the number of cars in the road network is small, and cars are allowed to travel freely at maximum allowed speed; in the second case heavy congestion is artificially simulated by increasing the number of cars, so that actual speeds are slower than the maximum allowed by speed limits. The computation of emissions is performed according to Equation (7), using the data corresponding to Euro 4, Engine Capacity < 1400 cc petrol cars and minibuses with weight below 2.5 tonnes (Code R005/U005 from Boulter et al. 2009).

Figure 20.3 shows that if the emissions are computed using speed-independent emission factors, then they are not affected by different levels of congestion. This is clearly absurd. In fact, the stationary distribution of emissions is the same, both in the figure on the left (no congestion), and in the one on the right (congestion). On the other hand, the stationary distribution of cars along the roads (represented with the solid line) changes completely in the case of congestion.

In Figure 20.4 the pollution factors are assumed to depend on velocity, and therefore the stationary distribution of emissions changes from the non-congested (left) to the congested scenario (right), and it remains consistent with the distribution of cars. In both Figures 20.3 and 20.4, and in both scenarios, the junction turning probabilities are the same. Clearly, it is not sensible that the distribution of pollutants does not change with traffic load. Therefore, this simple simulation suggests that speed-dependent emission factors should be used to obtain realistic results. In Figure 20.4, we can also see that the manner in which pollutants are influenced by the traffic volume is not homogeneous with respect to congestion (i.e., different pollutants have different density).

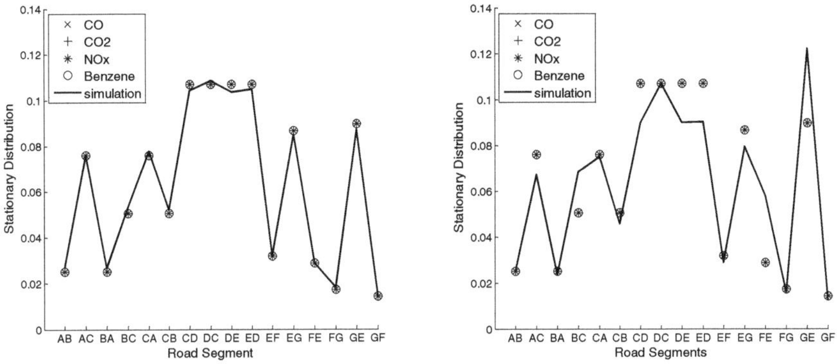

Fig. 20.3 Stationary distribution of cars (solid line) and all pollutants. Pollutant factors are constant with speed. No congestion on the left, congestion on the right. All emission factors coincide because of density normalisation (their sum is 1).

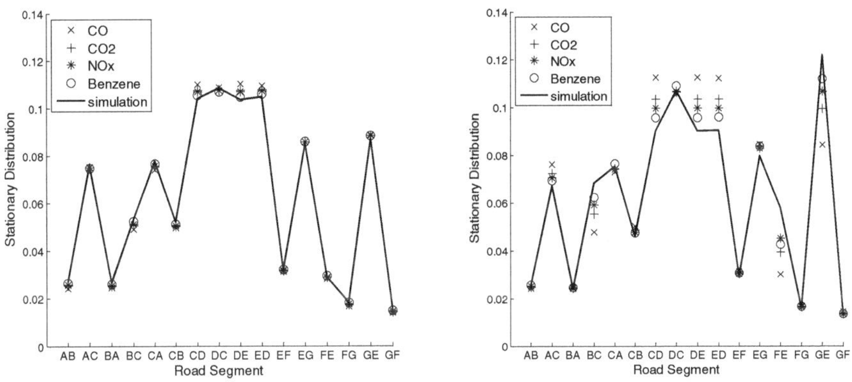

Fig. 20.4 Stationary distribution from simulation and model for time and all pollutants. Pollutant factors are dependent on speed. In this example, the pollutants density follows the car density both in the non-congested scenario (left) and in the congested one (right).

Second simulation – A large-scale road network: We now demonstrate the scalability of our approach through a significantly larger network. We generated the road network in Figure 20.5 using SUMO random network generation facilities.

Figure 20.6 shows the density of pollutants in the road network, together with the distribution of cars resulting from the SUMO simulation. While the road network of Figure 20.5 is composed of 618 road segments, only the subset of roads with ID between 250 and 300 was randomly chosen to be displayed in Figure 20.6, to improve readability of the result; qualitatively the distribution is the same for any selection of streets. In Figure 20.6 we see again that the distribution of pollutants depends on the particular pollutant, but it is always concordant with the distribution of cars. This is in accordance with common knowledge that congestion should affect emissions.

Fig. 20.5 A more complicated realistic road network.

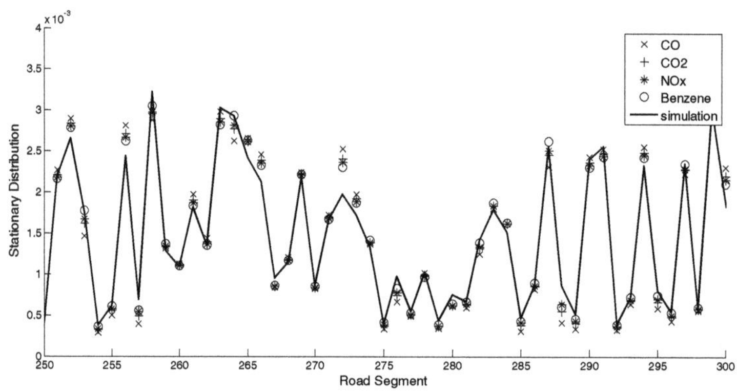

Fig. 20.6 Stationary distribution of time and pollutants in the more complicated example with heavy congestion. Only a selection of streets is shown to improve readability of the figure.

Third simulation – How to fix optimal speed limits: An additional simulation is performed to establish *optimal* speed limits. Optimality is with respect to minimum emissions. The results are shown in Table 20.2. Within each simulation we varied the speed limit uniformly over all streets from 20 to 120 km/h and calculated the corresponding Kemeny constant. As previously described, the Kemeny constant can be interpreted as an efficiency indicator, in terms of emissions, of the overall road network.

Table 20.2 Kemeny Constants for Different Global Speed Limits.

Speed [km/h]	Time [sec]	CO [g]	CO_2 [kg]	NOx [g]	Benzene [g]
20	1304	2674	1241	395.9	2.89
40	631	**2350**	875	243.4	2.05
60	437	2830	**812**	207.8	1.44
80	352	3758	815	197.2	1.03
100	311	4888	833	**194.6**	**0.91**
120	**291**	6065	853	194.9	1.07

Two lessons can be observed from Table 20.2.

(i) Too low, and too high a speed limit, both lead to high levels of pollution (although travel times are reduced with high speed limits, as expected)

(ii) Different pollutants have different corresponding optimal speed limits.

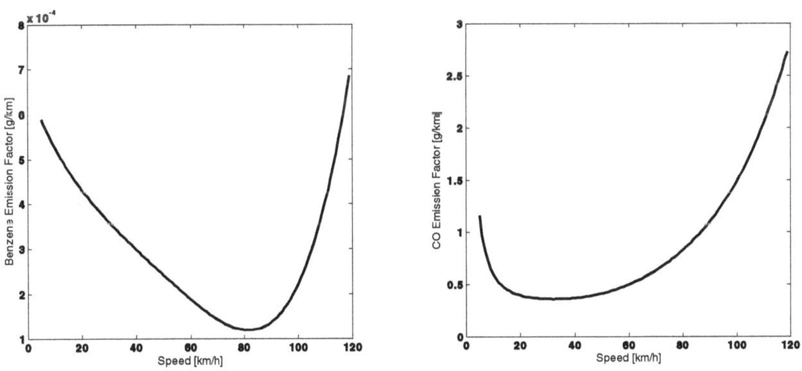

Fig. 20.7 Dependence of emission factor on average speed for Benzene and CO (based on data from Boulter et al. 2009).

To better interpret the results shown in Table 20.2, Figure 20.6 shows how the Benzene (left) and CO (right) emission factors depend on the average speed. The speeds corresponding to minimum Benzene emissions in Figure 20.6 and Table 20.2 do not coincide. This is caused by the fact that the average speed of a car is not equal to the speed limit (e.g. due to acceleration and breaking entering and leaving the road). In this example, high speed limits helped the car to accomplish an average speed closer to the optimal value for minimum emissions.

Table 20.3 Kemeny Constants for Different Speed Limits on the Bridge.

Speed [km/h]	Time [sec]	CO [g]	CO2 [kg]	NOx [g]	Benzene [g]
20	1929	5104	2109	637.4	4.76
40	1126	**4746**	1673	455.7	3.73
60	876	5606	**1605**	412.0	2.87
80	756	7831	1635	401.2	**2.29**
100	686	12119	1710	**403.0**	2.77
120	**640**	20045	1811	411.1	5.51

The above experiment was repeated with one minor change. Now we vary the speed limit at certain important points in the network, and observe the corresponding global change in pollution. Specifically, we vary the speed limit of the roads CD, DC, DE and ED from Figures 20.1 and 20.2. In spirit, this corresponds in considering the urban network of Figures 20.1 and 20.2 as composed of two clusters of roads, one on the left and one on the right, connected through some "bridging" roads. Here we changed the length of the bridging roads from 500m to 5 km - to reduce the gap between the actual average speed from the actual speed limit. Then, we vary the speed limit on the bridge from 20 to 120 km/h while keeping the other speed limits constant, thus obtaining the results summarised in Table 20.3.

Again we can see that both high and low speed limits on the bridge correspond to high overall pollution levels and that optimal speed limits differ for different pollutants.

Remark: This simulation corresponds to a realistic road engineering problem, where the optimal speed limits for a subset of roads must be established to minimise an utility function of interest. This application is further investigated in the next example.

Fourth simulation – An overall utility function: This last simulation aims at proposing an overall utility function that blends all (or some) of the previous efficiency indicators at the same time. Indeed, as shown in the last simulation, it is possible to compute a Kemeny constant that accounts for travel times, one for CO_2 (which as previously reminded is related to energy efficiency), and one for each other pollutant. In principle, each one of them is minimised by a different choice of speed limits; in this section, we answer to the question, which speed limit is optimal for all (or many) of the previous parameters?

For this purpose we normalise all Kemeny constants between 0 and 1, and we consider a utility function

$$F = \alpha \cdot K_T + \beta \cdot K_{CO_2} + \gamma, \tag{20.9}$$

where K_T is the Kemeny constant for time (in seconds), K_{CO_2} is the Kemeny constant for energy efficiency (in Kilograms), and α , β and γ are appropriate scaling coefficients that convert the measurement unit into a price unit (e.g. €). This can be done, for instance, by taking into account the hourly cost of driving (e.g., as a missed opportunity for working), the cost of energy inefficiency (e.g., fuel costs and pollution costs) plus a possible constant factor as a consequence of the normalisation procedure. For instance, Figure 20.8 (on the left) shows an example of the utility function F when speed limits are varied, according to the results given in Table 20.2, where the coefficients are chosen in order to given 10 times more importance to the energy efficiency component.

Then we assume that a road engineer solves the following optimization problem to find optimal speed limits:

$$\min K_T \quad \text{such that} \begin{cases} F \le F^* \\ K_{CO} \le K_{CO}^* \end{cases}, \tag{20.10}$$

where F^* is the maximum allowed price for urban network management (e.g., due to budget constraints) and K_{CO}^* is the maximum allowed quantity of CO emissions allowed for the urban network (e.g., due to environmental issues). The

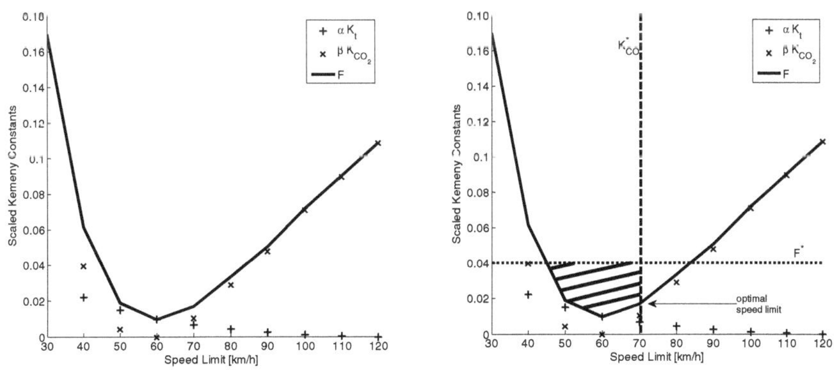

Fig. 20.8 Aggregate Kemeny constants as a function of global speed limits and optimal speed limits under given constraints.

idea of equation (10) is that travel times should be minimised, provided that the utility function does not exceed a threshold price, and the *CO* emissions are kept under a safety threshold. An example of such optimal solution is shown in Figure 20.8 (right), where the same utility function of Figure 20.8 (left) is considered.

20.7 Suggested Further Applications

The primary focus of this work has been modeling. However, our principal objective remains the control of large scale urban networks. From the perspective of the road engineer, information in the Markov chain can be used for this purpose. For example, identification and avoidance of pollution peaks is a major concern in our cities. The existence of peaks can in "some sense" be identified from the Perron eigenvector of the chain, and one may reengineer the network through adjusting speed limits, traffic light sequencing, etc., so as to keep peaks away from certain sensitive spots (hospitals etc.). One may also use this information to route vehicles along low (expected) emissions paths.

Other areas to be investigated include balancing of emissions, identification of critical roads (if they close they have a bad effect on the emissions profile of the city), and the effect of fleet mixing (electric vehicles) and priority zones on the emissions profile of the urban area. Interesting observations from our model are that different pollutants require different control strategies, that low speeds do not necessarily imply low emissions, and that high speeds do not always correspond to high emissions. Future work will investigate how the conflicting objectives of low emissions for all pollutants and low congestion can be addressed in a unified framework so as to facilitate control and optimisation strategies.

20.8 Conclusions

In this chapter we have proposed a new method of modeling urban pollutants arising from transportation networks. The main benefit of the proposed approach is that not only the distribution of different types of pollutants is given as a result, but that the same framework can be used to proactively affect and modify emissions, thus paving the way to a new application that was called *Green Control*.

The efficacy of the proposed approach is demonstrated by means of a number of examples. Some of them show how optimal road network decisions, namely speed limits, can be taken in order to minimise an utility function of interest.

Future work will investigate control and optimisation over Markov chains (with application to road network engineering), experimental evaluation of the proposed methods, and extension of the ideas to modeling city energy profiles (with respect to routing of electric and hybrid vehicles).

Acknowledgments. SK is supported in part by the Science Foundation Ireland under Grant No. SFI/07/SK/I1216b. RS, AS and EC are supported in part by Science Foundation Ireland under grant number PI Award 07/IN.1/1901.

References

1. Barth, M., An, F., Younglove, T., Scora, G., Levine, C., Ross, M., Wenzel, T.: Development of a comprehensive modal emissions model, NCHRP Project 25-11. Final report (2000)
2. Barlow, T.J., Hickman, A.J., Boulter, P.G.: Exhaust emission factors 2000: Database and emission factors, Project report PR/SE/230/00. TRL Limited, Crowthorne (2001)
3. Biem, A., Bouillet, E., Feng, H., Ranganathan, A., Riabov, A., Verscheure, O., Koutsopoulos, H., Rahmani, M., Güç, B.: Real-Time Traffic Information Management using Stream Computing. IEEE Data Engineering Bulletin 33(2) (2010)
4. Boulter, P.G., Barlow, T.J., McCrae, I.S.: Emission factors 2009: Report 3 – exhaust emission factors for road vehicles in the United Kingdom, Published project report PPR356, TRL Limited (2009)
5. Bretti, G., Natalini, R., Piccoli, B.: A fluid-dynamic traffic model on road networks. Archives of Computational Methods in Engineering 14(2), 139–172 (2007)
6. Cappiello, A., Chabini, I., Nam, E.K., Luè, A., Abou Zeid, M.: A statistical model of vehicle emissions and fuel consumption. In: Proc. of the IEEE 5 International Conference on Intelligent Transportation Systems (2002)
7. Cho, G.E., Meyer, C.D.: Comparison of perturbation bounds for the stationary distribution of a Markov chain. Linear Algebra and its Applications 335, 137–150 (2001)
8. Crisostomi, E., Kirkland, S., Shorten, R.: A Google-like model of road network dynamics and its application to regulation and control. International Journal of Control 84(3), 633-651 (2011)
9. Dellnitz, M., Junge, O.: On the approximation of complicated dynamical behaviour. SIAM J. Numer. Anal. 36(2), 491–515 (1999)
10. Dellnitz, M., Junge, O., Koon, W.S., Lekien, F., Lo, M.W., Marsden, J.E.: Transport in dynamical astronomy and multibody problems. International Journal of Bifurcation and Chaos in Applied Sciences and Engineering 3, 699–727 (2005)
11. Dirks, S., Gurdgiev, C., Keeling, M.: Smarter cities for smarter growth, How cities can optimize their systems for the talent-based economy. IBM Global Business Services Executive Report (2010), `http://www.zurich.ibm.com/pdf/isl/infoportal/IBV_SC3_report_GBE03348USEN.pdf`
12. Fiore, M., Härri, J.: The networking shape of vehicular mobility. In: Proceeding of ACM MobiHoc, Hong Kong, China (2008)
13. FoE (Friends of the Earth), Road Transport, Air Pollution and Health, Briefing Paper (1999)
14. Froyland, G.: Extracting dynamical behavior via Markov models. In: Nonlinear dynamics and statistics (Cambridge, 1998), pp. 281–321. Birkhäuser Boston, Boston (2001)
15. Gkatzoflias, D., Kouridis, C., Ntziachristos, L., Samaras, Z.: COPERT IV. Computer program to calculate emissions from road transport, User Manual (version 5.0), Published by ETC-ACC (European Topic Centre on Air and Climate Change) (2007)

16. Hartenstein, H., Laberteaux, K.: VANET – Vehicular applications and internetworking technologies. Wiley, Chichester (2010)
17. Highways Agency, Scottish Executive Development Department, Welsh Assembly Government, Department for Regional Development Northern Ireland. Design manual for Roads and Bridges, Volume 11, Section 20.3, Part 1, Air Quality, Highways Agency, Scottish Industry Department, The Welsh office and the Department of Environment for Northern Ireland, Stationary Office, London (2007)
18. Huisinga, W.: Metastability of Markovian Systems: A Transfer Operator Based Approach in Application to Molecular Dynamics, PhD thesis, Freie Universität Berlin (2001)
19. Jaaskelainen, J., Boethius, E.: Methodologies for assessing the impact of ITS applications on CO_2 Emissions, Technical Report v1.0 (2009)
20. Kemeny, J.G., Snell, J.L.: Finite Markov Chains. Van Nostrand, Princeton (1960)
21. Koon, W.S., Lo, M.W., Marsden, J.E., Ross, S.D.: Heteroclinic connections between periodic orbits and resonance transitions in celestial mechanics. Chaos 10(2), 427–469 (2000)
22. Koon, W.S., Marsden, J.E., Ross, S.D., Lo, M.W.: Constructing a low energy transfer between Jovian moons. In: Celestial Mechanics, Evanston, IL. Contemp. Math., vol. 292, pp. 129–145. Amer. Math. Soc., Providence (2002)
23. Krajzewic, D., Bonert, M., Wagner, P.: The open source traffic simulation package SUMO. In: RoboCup 2006 Infrastructure Simulation Competition, Bremen, Germany (2006)
24. Langville, A.N., Meyer, C.D.: Google's PageRank and beyond – The science of search engine rankings. Princeton University Press, Princeton (2006)
25. Levene, M., Loizou, G.: Kemeny's constant and the random surfer. American Mathematical Monthly 109, 741–745
26. Levy, J.I., Buonocore, J.J., von Stackelberg, K.: Evaluation of the public health impacts of traffic congestion: a health risk assessment. Environmental Health 9(65) (2010)
27. Meyer, C.D.: The role of group generalized inverse in the theory of finite Markov Chains. SIAM Review 17, 443–464 (1975)
28. Meyn, S.P., Tweedie, R.L.: Cambridge University Press. Cambridge University Press, Cambridge (2009)
29. Nawrot, T.S., Perez, L., Künzli, N., Munters, E., Nemery, B.: Public health importance of riggers of myocardial infarction: a comparative risk assessment. Lancet 377, 732–740 (2011)
30. Porta, S., Crucitti, P., Latora, V.: The network analysis of urban streets: a dual approach. Physica A 369, 853–866 (2006)
31. Schütte, C., Huisinga, W.: Biomolecular conformations can be identified as metastable sets of molecular dynamics. In: Handbook of Numerical Analysis, vol. x, pp. 699–744. North-Holland, Amsterdam (2003)
32. Villa, N., Mitchell, S.: Connecting cities: achieving sustainability through innovation. In: Fifth Urban Research Symposium 2009: Cities and Climate Change: Responding to an Urgent Agenda, Marseille, France (2009)

Chapter 21
Long-Endurance Scalable Wireless Sensor Networks (LES-WSN) with Long-Range Radios for Green Communications

Bo Ryu and Hua Zhu

Abstract. Two paradigm-shifting innovations are presented that have potential to enable an affordable Green Communications (GC) class: large-scale, low-cost environmental and endangered species monitoring using a generation of new unattended, long-duration wireless sensors. We first describe why existing solutions based on short-range radios have failed to address this key GC regime, and how the proposed LES-WSN innovations, namely long-range radios and two-tier (one-radio) architecture, makes low-cost wide-area environmental monitoring a reality. By utilizing high transmission power, long-range radios drastically reduce overall system cost when covering large areas. By combining the latest Long-Term Evolution waveform with this high transmission power radio hardware, we achieve both long communication range and high data rate (2+Mbps) simultaneously and at extremely high battery efficiency. The two-tier, one-radio architecture allows for only a small number of nodes to relay the sensor data, relieving vast number of sensor nodes from the burden of relaying. In addition, it eliminates the need for sensor nodes to wake up frequently and periodically, enabling mobile sensors for endangered species monitoring. Several new GC applications that become possible by the proposed LES-WSN paradigm are described, along with field experiment results that validate the claimed enabling benefits.

21.1 Introduction

Green Communications (GC) is a nebulous terminology that often refers to a subset of Information Technologies (IT) designed to enable people or devices to exchange information (voice, data, video) consuming as minimum energy as possible (subject to maintaining "reasonable" communication quality). The ever-pervasive penetration of wireless portable devices such as cell phones, smartphones, tablets, MP3 music players, and laptops into our daily lives has

Bo Ryu · Hua Zhu
Argon ST
A Wholly-Owned Subsidiary of the Boeing Company
e-mail: {bo.ryu,hua.zhu}@argonst.com

accelerated the advances in GC technologies since each and every device is expected to maximize its battery life. Naturally, maximizing the efficiency of energy usage is the single most important metric for GC devices and systems. On the other hand, GC is also frequently used to represent an initiative aimed at preserving the nature via environmental monitoring with unattended sensors. For example, "Planetary Skin" is a multi-year joint effort by NASA and Cisco to capture, analyze, and interpret *global* environmental data using satellites, airborne-based, and land-based sensors[1,2].

Wireless Sensor Networks (WSN) is a unique intersect of these two GC aspects (both as GC technology and GC application), since its primary objective is to deliver vital information about the environment (collected by sensors) using as little energy as possible. As a GC technology, a WSN system seeks long operational lifetime (months or years) so that ultimately, no battery change is needed during the entire monitoring period. As a GC application, a well-designed, strategically deployed WSN system in a remote area can collect near-real-time information about endangered species and the overall health of the environment in a large scale[3]. Consequently, it is no surprise that the growing interests in GC from policy makers, engineers, entrepreneurs and the public have produced tremendous activities in the design, development, and commercialization of emerging WSNs for various coverage sizes and applications, ranging from periodic monitoring with home-networked sensors to wide-area remote environment sensing.

A fundamental requirement for WSNs to remain a successful and enabling force in the GC marketplace is to adopt the *highest energy efficiency* wherever possible (ultra-low-power sensors, clocks, communication components, high-capacity batteries, etc.) because unattended environmental sensors are typically battery powered and an energy-inefficient sensor system requires frequent battery change or recharging. But, battery replacement can be extremely costly or impossible because of the labor involved in accessing the deployment sites that are either unsafe or hostile to humans. Therefore, energy efficiency has served a fundamental consideration in the WSN system design metrics, and consequently, led to miniaturization of sensor systems, including its communications components and protocols (e.g., smart dust [4]).

However, maximizing the energy efficiency of individual sensor nodes alone is not sufficient to address one major GC sector: a growing list of emerging WSN applications such as *wide-area long-term monitoring and tracking applications.* For example, U.S. federal and state governments have various mandates to monitor the health of the environment such as air quality, water quality, vegetation growth, and tracking of endangered species. Reliable, self-organizing, and scalable WSNs capable of remote operations have multiple Department of Defense (DoD), Department of Homeland Security (DHS), national and state parks, and

---

[1] http://www.nytimes.com/gwire/2009/03/03/03greenwire-nasacisco-project-to-flash-planetary-skin-9959.html

[2] http://www.planetaryskin.org

[3] Terrestrial Ecology Observing Systems, Center for Embedded Networked Sensing, UCLA, http://research.cens.ucla.edu/areas/2007/Terrestrial

[4] http://robotics.eecs.berkeley.edu/~pister/SmartDust

commercial applications. Specifically, DoD requirements exist for distributed sensor arrays for intelligence and situational awareness applications while commercial requirements include environmental monitoring for regulatory compliance (e.g., Army Regulation 200-1, 200-2, and 200-3), land management (e.g., various federal and state-specific land regulations) and multiple security applications. There are approximately 762M acres of Federal land that could potentially benefit from market-proven WSN solutions[5,6]. This represents a projected ~$1B market opportunity for low cost, energy efficient, and expandable network technologies.

Despite the potentially huge market demands, almost every WSN system design solution available today ends up utilizing low-power (short-range) radios by focusing on the communication energy efficiency of individual nodes as the primary design principle. This has two important ramifications that arguably have prevented the wide adoption of WSNs in our daily lives (other than home-area networking applications such as ZigBee). First, their applications for covering a wide area (e.g., large forest in remote locations) are limited because of the lack of robust protocol stack that can scale to thousands or tens of thousands of nodes. Only limited deployment of WSNs are found that depend on satellite and cellphone connections, electricity infrastructure, or high-cost renewable energy sources for monitoring large, remote areas. New communication radios and protocols aimed at addressing the scalability covering several square kilometers have not yielded satisfactory performance gains regarding battery lifetime, end-to-end throughput, and end-to-end latency. What's more, the majority of protocol algorithms and system solutions have focused on fixed sensors and periodic events [1, 2, 3, 4], and not readily applicable to mobile sensors (e.g., wild-life tracking) and random event-driven applications (e.g., intrusion detection and border security). Second, and more importantly, it has helped propagate a myth that short-range radios are fundamental to the high energy efficiency. This cannot be further from the truth: in most WSN applications, transmission energy comprises only a very small fraction of the total energy consumption due to the nature of low duty cycle. Sleep energy (primarily caused by leakage current) and overhead associated with time or clock synchronization are two dominant sources of energy consumption.

When each and every sensor node utilizes short-range communication radios, is required to participate in relaying information generated by other sensor nodes, and still wishes to maintain extremely long battery life when deployed in a wide-area field, its aggregated impact on overall performance is devastating for the following reasons. For one, the fact that a monitoring is tens of miles away from the data collection center (e.g., a commercial Point of Presence with a Internet connection and access to electricity) requires potentially thousands of nodes since the communication range is short (50~100ft), implying that the communication protocols must be able to scale to handle such a large network size. This forces the network to use most of the precious communication resources (e.g., bandwidths) for synchronization of the network control information (e.g., clock and routing entry

[5] Geoindicators Scoping Report for White Sands National Monument," available at http://www.nature.nps.gov/geology/monitoring/ib4/reports/whsa_geoindicators_report.pdf

[6] Fort Benning Environmental Program Management Branch, available at https://www.benning.army.mil/garrison/dpw/emd/content/emd_mission.htm

information), leaving very little for actual sensor data transmission. In addition, because the long battery life requirement forces individual nodes to sleep most of the time (e.g., low duty cycled operations), the time delay it takes from the source where the information is collected to the destination may add up to tens of hours even because they are likely to travel over hundreds of intermediate sensor nodes who also go through extended sleep periods.

Recognizing that simultaneously achieving wide-area coverage, mobile sensor support, low overhead, long endurance, and the desired timeliness of response for event-driven applications still remains a significant challenge, this chapter presents a new paradigm-shifting WSN approach that employs long-range radio with power-efficient high-data-rate waveform (physical-layer protocol or PHY), and two-tier architecture (medium access control layer, or MAC). By using high transmission power (e.g., 1W as approved by FCC rules [5]) as opposed to low-power methods favored by the majority of existing systems (e.g., 10mW in ZigBee), it reduces the number of nodes needed to cover a wide area, thus leading to substantial saving in both deployment and overall system costs. On the other hand, the two-tier architecture enables a natural scalability by minimizing the number of nodes which participate in the peer-to-peer ad hoc routing while utilizing the same radio for both types of the nodes in order to maintain the low system cost.

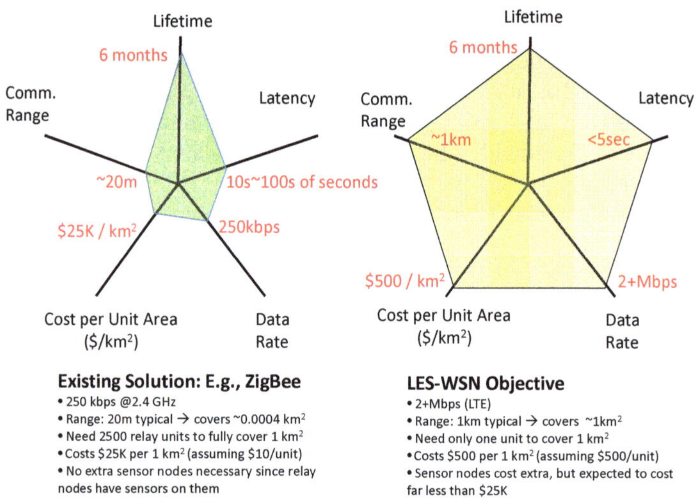

Fig. 21.1 Illustration of the total cost saving realized by long-range radios while drastically improving other performance metrics

Figure 21.1 compares the two WSN systems, one based on ZigBee as short-range radios and the other on the proposed LES-WSN with long-range radios, under a hypothetical scenario (sensor coverage of one1 square-kilometer area) in terms of five factors: battery life, end-to-end latency, data rate, cost of coverage, and communication range. The main purpose of this comparison is to illustrate the total system cost saving enabled by employing long-range radios for communication relay despite much higher unit cost. It is assumed that a short-range radio unit costs only $10 whereas long-range radio unit costs $500. While the price difference in reality may not be as big as in this example, it is there to emphasize that even the 50x saving in unit price is not sufficient to make up for the deficiency in communication range.

We claim that the biggest impact of the proposed paradigm-shifting innovations (use of long-range radio and the two-tier communications architecture with the same radio hardware) is that *it enables a generation of completely new sensing regimes* (especially for the *remote monitoring of the environment*) in both massively large scale and cost-effective manner, with little or no dependence on the high-cost assets (e.g., satellites and cellphones with recurring costs). First, by drastically reducing the number of nodes participating in the information relay by means of long-range radios, one will be able to reach areas of the nature that were once deemed too costly to install environment sensors. This reduces "total" energy costs per information delivery per area as illustrated in the example above (Figure 21.1). Second, by completely eliminating the need for sensor nodes to wake up periodically and frequently for maintaining an order in a duty-cycled WSN system (clock synchronization, transmit/receive rendezvous for relaying, etc.), sensor nodes can now dedicate its energy management task primarily on its "sensing" needs, and not burdened with communication overhead. This also makes it possible for sensors to be "mobile" for endangered species tracking, a feature that is not available in existing short-range radio based WSN systems. Last, but not least, the new power-efficient PHY protocol based on 4G Long-Term Evolution (LTE) wireless standard (entitled Wideband Sensor Waveform) will offer plenty of bandwidth to allow event-driven high-data-rate video streaming application. Imagine the excitement of an environmentalist who watches an endangered animal roam in a remote forest area sent by a nearby webcam triggered by either vibration and motion detection sensors or the mobile tag radio installed a prior on the animal without ever setting a foot on the challenging terrain! Table 21.1 summarizes the two innovative features of the proposed LES-WSN system and how it makes all these dream GC scenarios possible.

This chapter is organized as follows. Section 21.2 briefly summarizes related work in the areas of time synchronization and duty-cycled MAC protocol that have received heightened attention recently with focus on their energy consumption and end-to-end performance impact. Section 21.3 describes the two-tier (one-radio) duty-cycle architecture for the MAC protocol, one of the two core features of the LES-WSN. Section 21.4 presents the overview of WSW, a 4G-deribed high-data-rate waveform with extremely high power efficiency, further expanding the long-range communication capability. Section 21.5 covers prototype development efforts and field test results of the LES-WSN technologies in two separate demonstrations. Section 21.6 lists new WSN applications in three market segments – commercial, federal, and tactical – that become feasible with the introduction of the LES-WSN technologies. Section 21.7 concludes the chapter with a brief summary of the new capabilities and applications being made possible by the LES-WSN paradigm.

Table 21.1 Innovation and Benefit Summary of the New WSN System with Long-Range Radio and Two-Tier Architecture

Paradigm-Shifting Innovation	Energy (Green) Benefit
One-Radio, Two-Tier System Architecture – Adaptive switching between Relay and Leaf modes – Low, bounded end-to-end latency even at low duty cycle ratio	Greatly extends the overall battery life by adaptively deciding its mode (either Relay or Leaf) based on QoS, topology, and battery life metrics. Makes near-real-time monitoring possible by keeping end-to-end latency for a nominal 5-hop route to be less than 2 sec on average at 10% Relay-mode duty cycle. Reduces manufacturing energy cost by maintaining extremely low implementation complexity
High-Data-Rate, Extended Range, Power-Efficient Waveform: – Based on commercial technology (Long-Term Evolution standard) – Small FPGA footprint – Greatly simplified transmit RF front end	Allows remote users and monitoring personnel to watch streaming video by achieving high data rate (2+Mps) and robust communication (near OFDM quality against multipath) while substantially reducing the cost of RF front end. Uses considerably less energy by allows flash-based FPGA implementation even with less computational resources than typical communication-purpose FPGAs

21.2 Related Work

Over the last five years, we conducted four projects that are highly relevant to the proposed technology development: Connectionless Networks (CN) sponsored by DARPA, SensorBone sponsored by the U.S. Army Small Business Innovative Research (SBIR) program, Robust and Extremely Energy Efficient Sensor Network (REEESN) sponsored by Office of Naval Research (ONR), and Mesh-networked Tagging, Tracking, and Locating (Mesh-TTL) sponsored by a Department of Defense agency. Table 21.2 summarizes the key innovations accomplished under each project, and how they have been leveraged for the LES-WSN.

Table 21.2 List of prior projects and their key accomplishments leveraged for the LES-WSN capability development

Project	Key Accomplishment	Contribution to LES-WSN Design
DARPA CN (2005-2006)	Designed, implemented, and prototyped two Hail radios to assist BBNs CN effort	Extremely rapid duty cycling (sub msec) of RF front end
Army SensorBone (2006-2009)	Developed, implemented, and prototyped Single-Carrier Frequency Domain Equalizer (SC-FDE) receiver on Xilinx low-power FPGA platform (SPARTAN 3), and extremely rapid on/off RF front end circuitry with minimum leakage current	Fixed-point MATLAB models and baseline VHDL code of SC-FDE (Phases 2 and 3) Prototype system (915 MHz band) for preliminary link testing Extremely rapid on/off RF front end circuitry design with high-efficiency PA
ONR REEESN (2006-2009)	Developed and demonstrated energy-adaptive mesh networking system consisting of Low-Latency, Low-Energy MAC (LL-MAC), Distributed Time Synchronization, Energy-aware Routing, and Adaptive Topology Control (CARE) [6, 7]	QualNet simulation models of LL-MAC Energy-aware ad hoc routing protocol with adaptive topology control for Relay/Leaf decision logic
Mesh-TTL (2008-2010)	Developed and successfully demonstrated a low-cost COTS-based Mesh-TTL System for Tagging, Tracking, and Locating (TTL) and Unattended Ground Sensor (UGS) [8]	Low-latency, long-endurance duty cycling algorithm and software (with hard-coded Relay and Leaf modes) implemented in C

21.2.1 Time Synchronization

In a generic sensor network, one node's clock is different from its neighbors' due to time offset, frequency offset, and randomness. Time synchronization among neighboring nodes is necessary for two reasons. For environmental monitoring or intrusion detection, it is important to know when the event exactly occurs. Sensor should record/report each event with an accurate time stamp. When duty cycling is used, for synchronous MAC protocols, time synchronization is necessary for the nodes to rendezvous at the right time instants for successful transmission and receiving. To rendezvous, each node must exchange information with its neighbors about when it will wake up, and each node's schedule is according to its own clock.

There are two types of time synchronization protocols: instant synchronization and predictive synchronization. In instant time synchronization protocols, a node corrects its clock reading by adopting its neighbor's. If the two clocks run at the same rate, a single synchronization point will suffice. However, clocks run at different rates, and the resulting difference in clock readings accumulates over time [9]. Thus to maintain high precision, many synchronization points are needed. Instant time synchronization protocols include RBS [10], TPSN [11], and time diffusion [12]. In predictive time synchronization protocols, each node tries to establish a relationship between its local clock and a target clock. A simple model for this relationship is a linear model, whose optimal solution is given by the linear regression theory [7][13]. Predictive synchronization greatly reduces the number of synchronization points, even for clocks exhibiting dramatic differences. Predictive synchronization protocols include flooding time synchronization protocol

(FTSP) [14] and rate adaptive time synchronization (RATS) [15]. In our implementation, we modify the RATS algorithm, which is suitable only for pair-wise local time synchronization, and add a global time synchronization mechanism with negligible communication overhead. We implement a modified version of the global synchronization mechanism, Flooding Time Synchronization Protocol (FTSP) [14], due to its multi-hop synchronization capability. In [13], we analyzed the impact of two parameters on the precision of linear regression based time synchronization protocols: (1) the frequency at which the time stamp data are collected; (2) the window size, i.e., the number of time stamps used for linear regression. Based on rigorous theoretical analysis and extensive experimental and simulation results, we show that counter to intuition, given a fixed prediction interval, a more frequent synchronization may result in worse time synchronization performance. This discovery suggests that a linear regression based time synchronization protocol can achieve both high precision and good energy efficiency by operating at a low synchronization frequency.

21.2.2 Low-Power Adaptive MAC

To save energy, the radios stay in the sleep mode most of the time and wake up periodically to check if they should stay awake and prepare to receive data. That is, the radios are duty cycled. There has been a fairly large variety of research work on duty-cycling MACs published recently.

One class of popular early works includes SMAC [16], TMAC [17], DMAC [18] and PMAC [19]. These MAC protocols basically follow a straightforward concept of *macro synchronous duty-cycling*. In macro duty-cycling, nodes cycle through large sleep and active intervals in synchronized pace. The sleep and active interval is in fairly a large time scale comparing to the actual transmission duration of a typical packet. For example, the active duration could be 200 ms while the sleep duration 800 ms (i.e. a duty-cycle of 20%). The advantage of this macro duty-cycling approach is that there is no change to underlying PHY communication mechanism. However, there is substantial MAC latency involved, which may lead to highly undesirable performance in a multi-hop communication paradigm. Since the accumulation of such large per-hop latency may be rendered as failure for many existing upper layer protocols and applications, it will significantly limit the applicability of these MAC schemes to only a small portion of sensor applications. Another weakness of these schemes is the synchronous duty cycles. SMAC, TMAC and PMAC require synchronous starting time of the active period among nodes while DMAC proposed staggered, yet synchronous, duty-cycles. Note that the DMAC approach is less generic and only works in highly "customized" topology and traffic scenarios. Although all works claim that only loose synchronization is needed given the large time-scale of active duration, one should be aware of the challenge of network clock synchronization in large scale ad hoc networks. It's also worth noting that there are plenty of works that investigate the macro duty-cycling from high-level theoretical perspective, such as the scheduling schemes in [20, 21].

Another class of schemes adopts the concept of *micro duty-cycling*, also referred to as *low power listening* in some literature. Examples include BMAC [22], BMAC+ [23], XMAC [24], CMAC [25], UBMAC [26], and SCP-MAC [27]. Different from macro duty-cycling schemes, micro duty cycling schemes attempt to minimize the time-scale of the idling active duration. Therefore, compared to macro schemes with same duty cycle level, the active-sleep cycle is minimized, which in turn mitigates the latency as much as possible. Furthermore, by keeping the minimum idling active interval and increasing the sleep interval, micro schemes can reach ultra-low duty cycle that is impractical for macro schemes.

The idling active interval is usually set to the minimum duration feasible for hardware and system implementation. If a radio detects/senses a signal during the idling active interval, it will stay active until reception is completed. Otherwise it will go back to sleep immediately. For example, if the active interval is set 50 us and the duty cycle interval is 5 ms, the *idling duty-cycle* will be 1%. Idling duty-cycle means the percentage of active time if no data communication event occurs. The actual duty-cycle of the radio may be higher than the idling duty-cycle due to extended active time for data communication needs. The short duty-cycle interval enables prompt response to traffic. As a result, unlike macro duty-cycling, micro duty-cycling can be seen as near-transparent to upper layers.

The benefit of micro duty-cycling does not come free of cost, however. To enable duty cycling, it is necessary for the radios to know when to wake up and when to transmit since communication is possible only if the transmitter and the receiver can rendezvous at the same time. There are three main approaches based on whether synchronization information is available or not: synchronized, asynchronous and hybrid approaches. On synchronized protocols, such as SCP-MAC, nodes exchange a synchronized schedule that specifies when nodes are awake and asleep. Knowing when nodes will be awake in order to communicate reduces the time and energy wasted in prolonged preamble transmission and potential idle listening. However, it introduces extra traffic and energy overhead due to synchronization. Asynchronous protocols such as B-MAC [22], WISEMAC [28], and X-MAC [24] use preamble sampling to link together a sender with data to a receiver. Idle listening is reduced in asynchronous protocols at the expense of the sender. For example, when a sender has data, in B-MAC, the sender transmits a preamble that is at least as long as the sleep period of the receiver. The receiver will wake up, detect the preamble, and stay awake to receive the data. This allows low power communication without the need of explicit synchronization between the nodes. However, there is an overhearing problem in BMAC where receivers who are not the target of the sender also stay awake during the long preamble. X-MAC addresses the overhearing overhead associated with long preambles by using a strobe sequence of short packets including the target ID allowing for fast shutdown and response. In a hybrid approach, such as UBMAC [26], nodes maintain their own asynchronous schedules yet exchange the schedule with other nodes for communication.

Which duty cycle and sync scheme should be used so that the energy consumed is near minimum for a given traffic load? We conducted a thorough investigation on the relationship among duty cycle, synchronization scheme, traffic load and energy saving, through extensive network simulation and the results are reported in [6]. Simulation results show that it is not necessarily true that the less the duty cycle, the less the energy consumed.

21.3 One-Radio, Two-Tier Architecture for Duty-Cycled MAC Protocol

While it is tempting to leverage prior work in duty cycling for WSNs such as those surveyed in Section 21.2, the majority of them impose a flat (single-tier) duty-cycling architecture (i.e., every node follows the same wakeup/sleep rule, and serves as a potential router). Such architecture is very inflexible in dealing with the often conflicting yet essential objectives of minimizing energy consumption of a set of nodes and limiting latency of another set of nodes. In addition, the cost of duty cycling (either synchronization or prolonged preamble) is not scalable for some of the existing schemes. For schemes based on prolonged preamble, a single transmission with the prolonged preamble places a significant jamming effect on all other nodes within its interference range. Such jamming effect aggravates the already limited network capacity as the network grows larger. The negative impact on multi-hop latency is even worse. On the other hand, those schemes based on time synchronization, exchanging synchronization messages across large scale ad hoc networks incurs substantial overhead and often results in low overall energy efficiency.

However, what is viable even with the radios not designed for long-endurance WSNs is to duty cycle the entire node for as many nodes as possible, provided that: (i) the remaining nodes can form and maintain robust, low-latency network in an energy-efficient manner; (ii) the tight time synchronization is not required between the up nodes and sleeping nodes in order to bring the nodes back into the network; and (iii) the number of the remaining nodes is kept small (say 30% or less) on average.

To do so, we have adopted a pseudo-TDMA-based, two-tier duty-cycling channel sharing scheme that provides a flexible control over how fast the delivery can be at what energy cost. Unlike commercial applications such as environmental monitoring dealing with periodic, scheduled data transmissions, we have added the near-real-time responsiveness requirement in order to broaden the applicability of the solution, and thus, certain nodes must be up more frequently than others. Yet, we want to minimize the number of nodes that need to be up more frequently, and leave others off as long as possible. That is precisely what our two-tier, self-organizing duty cycled MAC protocol enables by allowing each node to be either Relay or Leaf, but with the same hardware (RF).

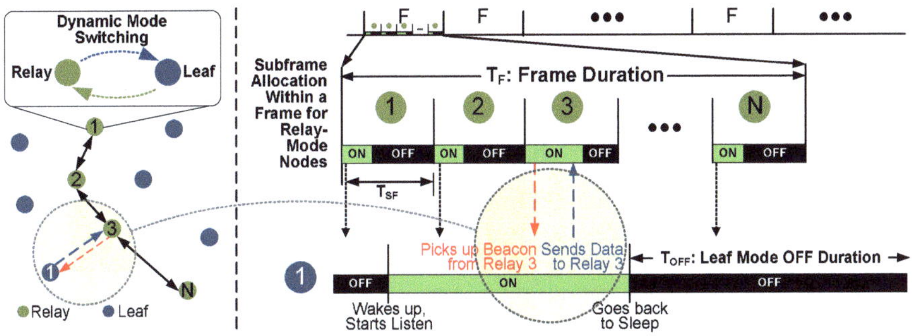

Fig. 21.2 Illustration of adaptive two-tier duty cycling scheme (Medium Access Control Protocol). Each node adaptively selects between Relay and Leaf modes depending on remaining battery reserves and route constraints

Figure 21.2 illustrates a high-level operation of this two-tier mesh-networking protocol and how it interacts among Relays as well as between Relays and Leafs. First, the TDMA structure for Relays consists of two time scales: frame and subframe. When a Relay owns a subframe, it is called a Master of that subframe, and the rest of Relays within the communication range with this Master Relay are called Slave Relays. A Master Relay receives data generated or relayed by nearby Relays or Leafs. The allocation of Master/Slave subframes is based on a self-organizing Medium Access Protocol algorithm (Link State Protocol), and the resulting allocation is collision-free (for static Relay nodes). The principle of the algorithm has been discussed in depth in [29].

When a Leaf wakes up, it listens for a beacon signal generated by nearby Master Relay(s) over the duration of frame. In doing so, the Leaf is guaranteed to pick up a beacon signal from a nearby Relay (if within communication range, of course) without its time synchronized to the network. Once it correctly decodes the received beacon signal, it immediately sends its data to that Master Relay (which always waits for short period of time right after transmitting a beacon signal). The Master Relay goes back to sleep once it finishes listening or receiving packets from Leaf or nearby Relay. Each slave Relay, on the other hand, listens for a beacon from the corresponding Master Relay of that subframe. If there is no packet to relay to that Master Relay, the slave Relay go back to sleep.

The above design principle yields the following salient features:

- Two configurable system parameters dominate energy-performance tradeoff: T_F and T_{OFF}
- No time synchronization between Relays and Leafs is required, greatly simplifying duty cycling software design and reducing energy usage through saved overhead messages.
- Relay transmits beacon at the beginning of its dynamically allocated master subframe with spatial reuse.

- Relay listens per every slave subframe and does: (i) dynamic subframe allocation; (ii) route computation; (iii) time synch adjustment with its peers; and (iii) forward packets to next-hop Relay if necessary.
- Average latency over a stable N-hop route = $N * T_F / 2$.
- Maximum latency over a stable N-hop route = $N * T_F$.

A candidate WSN system solution based on the above two-tier duty cycling protocol has been developed and demonstrated with a low-cost commercial radio (CC11xx series by Texas Instrument) for the tracking and monitoring applications identified in Section 21.6. While this particular implementation was successfully demonstrated, it lacks one important performance metric: bandwidth. The data rate of this radio is only 250kbps A future, high-data-rate physical layer solution based on Long-Term Evolution (LTE) Single-Carrier FDMA (SC-FDMA) is presented in the next section, along with its advantages and system challenges.

21.4 Wideband Sensor Waveform (WSW): High-Data-Rate, Power-Efficient Radio for LES-WSN

At the core of the proposed waveform for the LES-WSN, the Wideband Sensor Waveform (WSW), is the Single-Carrier FDMA (SC-FDMA) scheme used by the Long-Term Evolution (LTE) standard for its uplink (handset to base station). WSW based on SC-FDMA utilizes single carrier modulation at the transmitter and frequency domain equalization (thus dubbed often as SC-FDE) at the receiver, and is a technique that has similar performance and essentially the same overall structure as those of an OFDM system. One prominent advantage over OFDM is that the SC-FDMA signal has lower peak-to-average power ratio (PAPR). SC-FDMA has drawn great attention as an attractive alternative to OFDM, especially in the uplink communications where lower PAPR greatly benefits the mobile terminal in terms of transmit power efficiency. Consequently, SC-FDMA was adopted for the uplink multiple access scheme in 3GPP Long Term Evolution (LTE).

One of the primary drivers of energy consumption in mobile radio is the Power Amplifier (PA) used for transmitting RF data. Further, the OFDM waveform commonly in use in high data rate communication systems such as 802.11a/g/n requires highly linear class A or AB amplifiers used in a manner which wastes much of the power being drawn by the amplifier. Indeed, this is one of the advantages of FM (a constant envelope waveform) over AM. The FM transmitter allows for efficient (non-linear) amplifiers to be used, similar to the WSW built upon SC-FDMA. The use of WSW then provides the ability to significantly extend the communication range while marginally improving battery life of Relay nodes due to a more efficient use of the PA. Because of the high PAPR, existing OFDM waveforms must combat the following problems:

- The PA at the transmitter has to back off more than in a single carrier transmitter, which results in reduced output power and lower energy efficiency. Because there are four RF chains, any power efficiency lost at a single RF slice is multiplied by four.

- The A/D and D/A converters have to provide larger dynamic ranges for achieving the same performance as in a single-carrier system, which increases hardware complexity and cost.

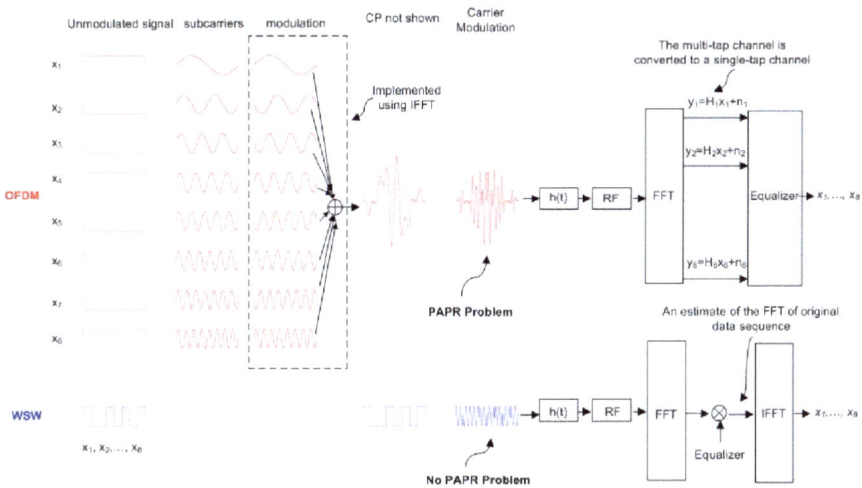

Fig. 21.3 WSW does not suffer from high PAPR problem as in OFDM, ideal for LES-WSN requiring power-efficient transmissions for unattended sensor radios.

In contrast, the WSW waveform, derived from SC-FDE/SC-FDMA, does not suffer from these problems, and yet exhibits similar performance to the OFDM waveform in terms of spectrum vs. data rate. This increased efficiency can be utilized to extend the range of the radio or with a different PA supply markedly extend the battery life for Relay nodes which operate at a relatively high duty cycle (e.g., 10% or so). Consequently, we envision the overwhelming advantage of the WSW waveform for the LES-WSN thanks to the plethora of high-efficiency PAs available in the commercial marketplace. Note that the power used by a PA is highly correlated to the maximum output which is close to the compression region. The power used by the PA is mostly independent of the signal going through it. Therefore, the OFDM PA uses much more power than the WSW PA on the right **independent of the signal being transmitted.** The WSW waveform can then be used with either PA, exhibiting extended range with the existing OFDM PA or extended battery life with the WSW amplifier whose lower power draw (P_{supply}) is implicit based on the lower non-linearity region. More interested readers are encouraged to read additional materials on SC-FDMA[7]

We conducted extensive simulation study of the WSW waveform in MATLAB in order to validate the claimed advantages of the WSW. Figure 21.4 describes the MATLAB model structure of the WSW specification.

[7] http://en.wikipedia.org/wiki/SC-FDMA may be a good starting point as it contains links to seminar publications and tutorials on this topic.

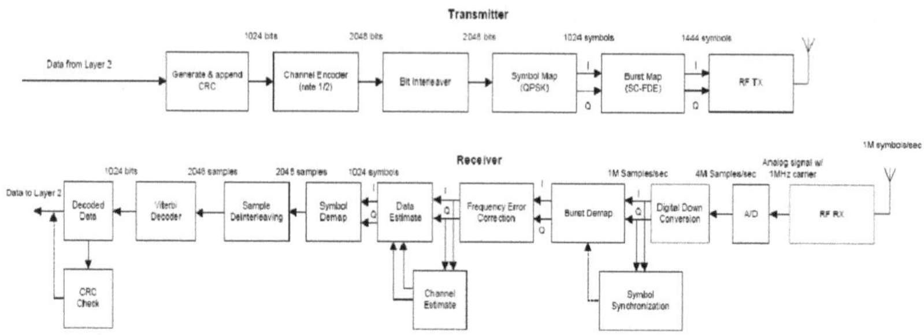

Fig. 21.4 Baseline WSW physical layer MATLAB model

Figure 21.5 shows the BER performance of OFDM and WSW for the following scenario: Rayleigh fading, IEEE 802.16a SUI-5 channel power profile, QPSK, 5M symbols/sec, FFT block size 2048, CP length 128, convolutional code of rate 1/2 and constraint length 7, interleaving, and two types of channel estimates: perfect channel knowledge and Least-Squared channel estimate. It is easy to see that the performance of WSW is slightly worse than that of OFDM, but the difference is within 1 dB. In short, WSW offers similar BER performance to that of OFDM.

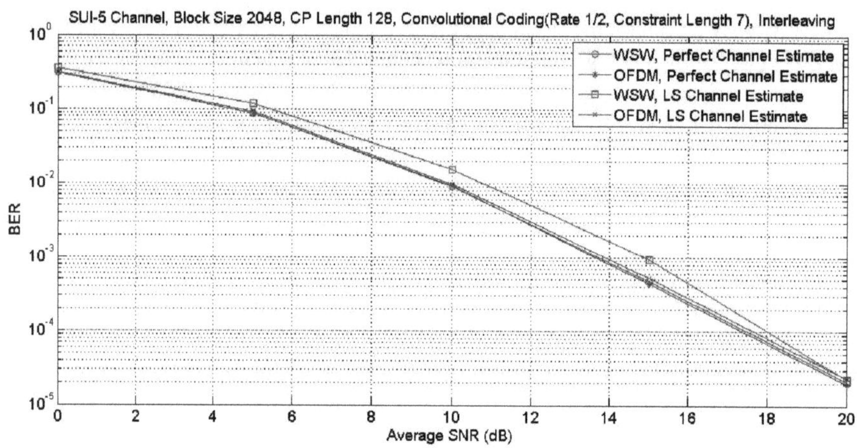

Fig. 21.5 WSW exhibits similar BER performance to OFDM. WSW can use much less transmit power than its OFDM counterpart

21.5 Experimental Results

This section presents the results of the two field experiments conducted with the initial prototypes of the LES-WSN system. The first experiment, described in Section 21.5.1, validates the core concepts and benefits of the LES-WSN with low-data-rate, commercial radios. The second experiment, described in

Section 21.5.2, demonstrates the feasibility of the WSW implementation on resource-constrained FPGAs for ultra-low-power operations.

21.5.1 Two-Tier Architecture with High-Power, Low-Data-Rate Radio

A three-week-long field experiment was conducted near Norfolk, VA, area during April 15 – May 7 as part of a large-scale demonstration hosted by a DoD agency. The demonstration network topology is shown in Figure 21.6.

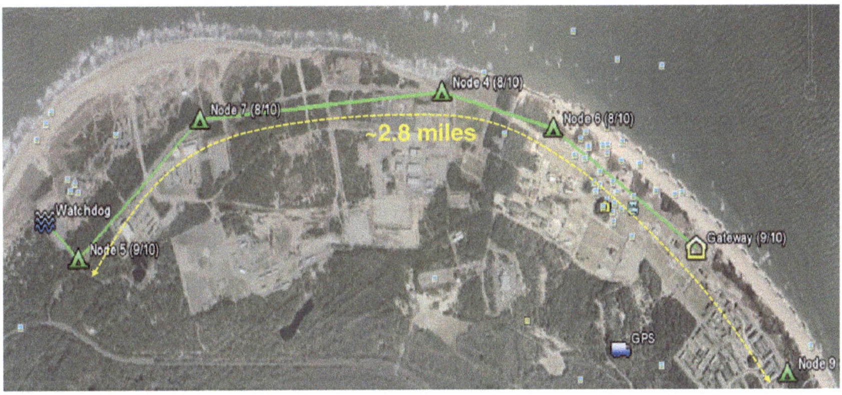

Fig. 21.6 Long-endurance field experiment network topology with power-enhanced CC1110 radio at Norfolk, VA. About 3 miles of shoreline was covered with six (static) Relay-mode nodes to track target objects (either vehicle or a boat). The numbers next to the Node ID (9/10 or 8/10) indicate remaining battery reserves after 3-week experiment, confirming that their power consumption behaviors offer at least 6 months of lifetime with 250kJ batteries.

The testbed wireless network consisted of five Relay-mode nodes, one Gateway, one tag node (with GPS tag), and one sensor Leaf-mode node (weather sensor). Specifically, we have developed two form factors (one for tag Leaf node and one for Relay node) that are derived from the same RF specification as shown in Figure 21.7. Note that CC1110's CPU speed is at 26MHz with 32kB flash memory but only 4kB RAM. The system allowed us to queue about 10 short packets maximum with a packet size of 30 bytes in RAM before leveraging the onboard EEPROM. Nonetheless, they provided sufficient cushion for retransmission upon detecting the collisions (as part of the two-tier duty-cycled MAC algorithm). CC1110 provides AES encryption capability via a 128-bit AES security coprocessor, which is sufficient for the authentication and encryption needs of the sensor networks as long as symmetric, pairwise keys can be established between communicating nodes. There are many research works recently on key pre-distribution and distribution in sensor networks [30, 31, 32, 33, 34, 35, 36], which can be easily applied to our framework and prototypes.

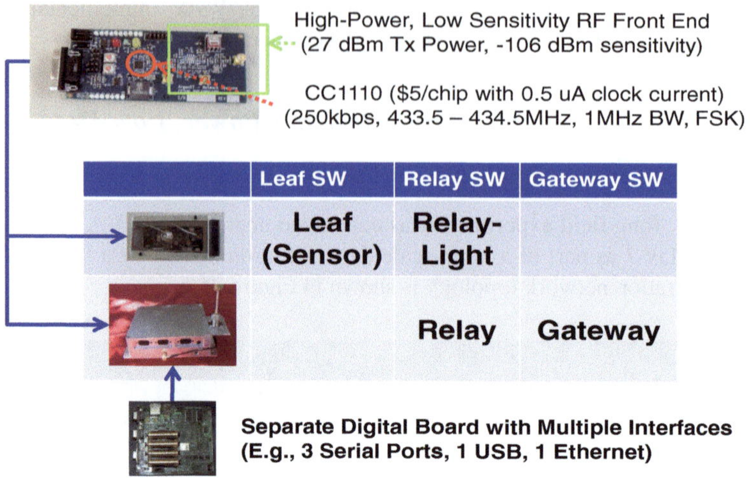

Fig. 21.7 One Radio, Two Form Factors, Three SW Modules (Leaf or Sensor, Relay, and Gateway), and Four Node Types used during the experiment. The same radio is used for both Relay and Leaf (Sensor) modes. Only software differs among the modes.

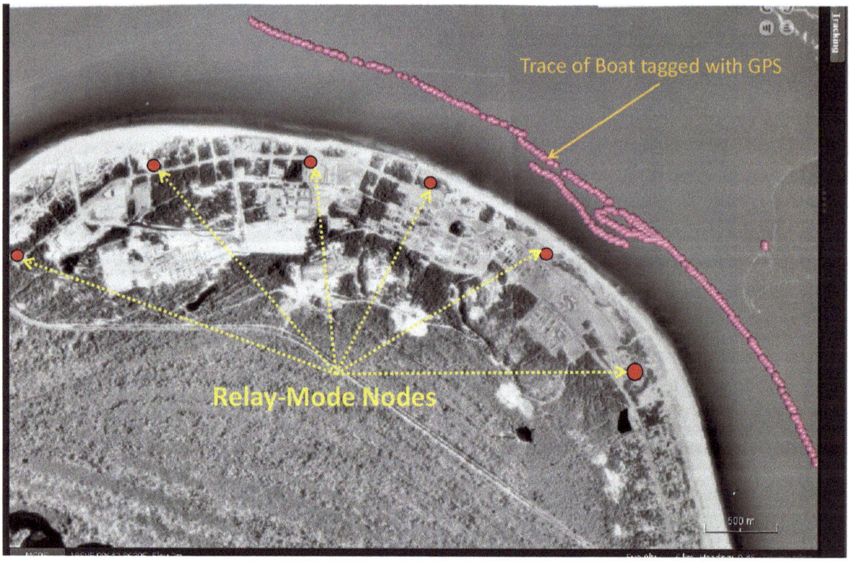

Fig. 21.8 Trajectory of tagged object with GPS tag (boat) detected and stored by the two-tier duty cycled wireless network. The final demonstration as well as the 3-week experiment was performed with no incidents or interruptions of the network.

Figure 21.8 shows a real-time snapshot of the trajectory of the tagged object (boat) detected by our long-endurance, low-latency wireless network. This particular demonstration resembles endangered species tracking, because we did not know when the boat would be coming within the coverage of the network. The final demonstration at the end of the 3-week experiment showed near flawless performance. No long-range node failed while delivering data in real-time from both GPS tag Leaf node and Leaf node with very high reliability and low energy usage.

21.5.2 Experimental Results of WSW Prototype System

This section presents the summary results of the field experiments conducted with the WSW prototype implemented under prior projects (sponsored by the U.S. Army and ONR), which confirmed that WSW works properly with high-efficiency PA, and that resource-constrained FPGAs such as flash-based FPGA can host the WSW.

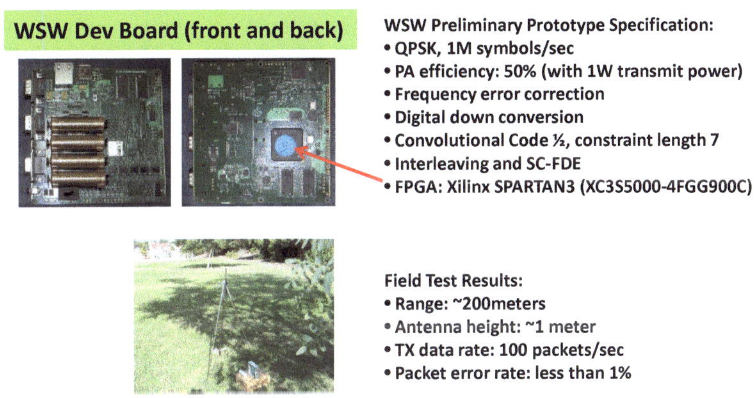

Fig. 21.9 WSW Preliminary Prototype Boards, Specification, and Link Test Results

The RF of the WSW prototype transceiver operates in the 915MHz band, and transmits 1M symbols/sec with a maximum transmission power of 1Watt. The block diagram of the digital section is shown in Figure 21.9. The prototype has a few advanced features. It utilizes digital down conversion to separate I and Q dimensions in the digital domain, which reduces the number of RF chains from 2 to 1. It uses the carrier frequency offset correction algorithm developed by Moose for wireless fading channels, and can handle a maximum Doppler shift of 2kHz. The digital section is implemented on a low-cost Xilinx Spartan3 FPGA chip. There are a number of additional optimizations on the prototype FPGA firmware that were not implemented due to budget and resource constraints, with which could further improve the performance of the WSW prototype transceiver. The main accomplishment made from this implementation exercise is that the implementation complexity of WSW is sufficiently low for resource-constrained FPGAs designed for extremely low-power operations. The next step of WSW is to utilize these

FPGAs (e.g., Actel's Egloo series), and be integrated with the two-tier duty-cycling protocol (Section 21.3). This will produce the first fully developed LES-WSN system as originally envisioned.

21.6 Applications

The market opportunity for the LES-WSNs targeting wide-area GC application is extremely bright, including its potential to open up the following new and untapped market categories, each with stated requirements.

Federal - The Federal (non-tactical) market is defined as real time remote environmental monitoring of military and federal lands administered by the DoD and other government agencies. These lands require the capabilities provided by a low cost, low power communications architecture using multiple unattended sensors to monitor and collect data on surface, subsurface, and littoral areas. This includes border operations and critical federal infrastructures where environmental and surveillance data gathering is both a compliance and security necessity. For example, the Army's Integrated Training Area Management (ITAM) program includes monitoring activities for ground water level at remote wells in White Sands, Arizona, or weather data and water samples from in accessible river banks in Fort Benning. As a whole, military bases, as a result of various Army regulations and training needs, require land and environmental management activities to maintain facilities and satisfy environmental regulations. Examples of such regulations that have created a need for the LES-WSN systems include Army regulation 350-19 (Sustainable Range Program - SRP) from Army Environmental Command, which calls for "information excellence" regarding data and science used to monitor and manage ranges and training land assets. In addition, an Army requirement exists for the "Real-time monitoring of land use conditions and usage" from the Environmental Quality Technology Research and Development Program. A second major target within this segment is The Department of the Interior. This Federal department manages large tracts of land through the Bureau of Land Management, Fish and Wildlife Service, National Park Service, and several other agencies.

Commercial - The Commercial opportunity is defined as the market for real time remote monitoring for environmental purposes of privately held land, facilities, or infrastructure monitoring such as remote oil fields and oil and gas pipelines. In addition to popular GC applications such as Wildlife Tracking and Wide-Area Environmental Monitoring, the same solution can be used to detect environmentally harmful accidents such as oil pipe leakages, forest fires, and sudden vegetation changes caused by environmental changes.

Tactical –The Tactical opportunity is defined as the market for real time remote monitoring for intelligence gathering in the battlefield, most notably Unattended Ground Sensors and Unattended Littoral Sensors (UGS and ULS). Though not part of the class of GC applications, these sensor systems have very similar requirements as they require long-term operations and must be expandable and dispensable (low-cost). Examples of this category include, but not limited to: Tagging, Tracking, and Locating (TTL), Intrusion Detection, and persistent ground

Intelligence, Surveillance, and Reconnaissance (ISR) such as detection and monitoring of Improvised Explosive Devices (IEDs). Figure 21.10 illustrates these examples.

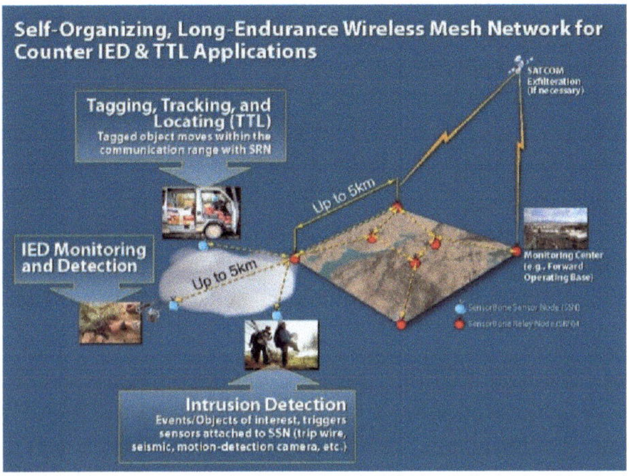

Fig. 21.10 Illustration of non GC applications enabled by the LES-WSN paradigm

21.7 Summary

Experiences with the fielded system and interaction with the current and prospective GC customers clearly indicate tremendous market opportunities for a high-data-rate version of the low-latency, long-endurance system such as the LES-WSN solution proposed in this effort. This is largely attributed to the capability to support near-real-time event detection and response while supporting occasional video streaming in an affordable and scalable manner. Though not part of GC applications, this new paradigm opens a new market for infrastructure security and border protection applications. Collectively, the new applications enabled by the proposed paradigm include, but not limited to, Bike Tracking, Children Tracking in urban areas, Wildlife Tracking, Wide-Area Environmental Monitoring, and Infrastructure Protection, to name a few. These applications share common research challenges that are different from existing solutions. Sensor nodes with short-range radios that are popular with existing commercial wireless products will be more costly for these wide-area applications due to the sheer number of nodes required. In addition, few are able to scale with respect to the network size. Consequently, radically different approaches are needed to produce commercially viable solutions and products to address these emerging applications. The LES-WSN presented in this chapter meets these daunting challenges through two key innovations.

In this chapter, we presented a novel WSN system design with several innovative features tackling fundamental challenges for long-endurance, wide-area sensing with unattended sensors such as environmental monitoring, tactical ground ISR operations, and security/intrusion detection. Unlike existing sensor networking systems which are heavily focused on local-area coverage only, persistent wide-area sensing, a fundamental requisite for GC applications, have stringent and yet conflicting requirements:

- Long Lifetime: Since environmental sensors are likely to be located in hostile/remote areas, making physical access on a regular basis becomes impractical. This forces sensor nodes to be in sleep mode as much as possible.
- Fast Detection Time: Time-stringent event-response applications (e.g., forest fire detection) force the nodes to be in wakeup mode as much as possible.
- Wide Adaptation Range: In remote or restricted areas, it is not practical or possible to deploy a large number of nodes due to safety or cost concerns. Consequently, it is more desirable for sensor nodes to have adaptive data rates and large communication for wide-area sensing, forcing the use of high-power components.

We presented a novel two-tier duty-cycling architecture that overcomes the above requirements: the Relay-mode nodes form a pseudo TDMA backbone without explicit time synchronization overhead, and Leaf-mode nodes are allowed to remain off for an arbitrary duration of time since they are not part of the routing topology. In addition, our LES-WSN design adopts extremely power-efficient Wideband Sensor Waveform (WSW) derived from the 3GPP LTE commercial wireless standard.

On the commercial side, the low-data-rate solution using COTS radio is already under a medium-volume production for various customers, justifying the benefits of the presented innovations. When WSW, the high-data-rate physical layer solution based on Long-Term Evolution (LTE) Single-Carrier FDMA (SC-FDMA), is completed, the resulting LES-WSN solution will enable even more exciting capability such as event-driven video streaming while maintaining long endurance and scalable, and opens a new market for both GC and non-GC applications.

References

[1] Boukerche, A., Pazzi, R.W.N., Araujo, R.B.: HPEQ A Hierarchical Periodic, Event-driven and Query-based Wireless Sensor Network Protocol. In: Proc. of The IEEE Conference on Local Computer Networks, LCN 2005 (2005)

[2] Kumar, A.N., Thyagarajah, K., Anbu, J.I.: Fault Tolerance and energy efficient protocol for cluster communication in sensor networks. In: Proc. of IEEE 2nd International Advance Computing Conference (IACC), pp. 91–96 (2010)

[3] Zahmati, A.S., Abolhassani, B., Shirazi, A.A.B., Bakhtiari, A.S.: An Energy-Efficient Protocol with Static Clustering for Wireless Sensor Networks. World Academy of Science, Engineering and Technology (2007)

[4] Kar, S., Moura, J.M.F.: Consensus Based Detection in Sensor Networks: Topology Optimization under Practical Constraints. In: First International Workshop on Information Theory for Sensor Networks (WITS), Santa Fe, NM, vol. 20, June 18-20 (2007)

[5] FCC 47 C.F.R Subpart C, Part 15.247: Frequency Hopping and Digitally Modulated Intentional Radiators
[6] Zhang, Z., Ryu, B., Zhu, H., Huang, Z., Ma, L.: Robust Extreme Energy Efficient Sensor Networks. In: IEEE MILCOM (October 2008)
[7] Ma, L., Zhu, H., Nallamothu, G., Ryu, B., Zhang, Z.: Impact of Linear Regression on Time Synchronization Accuracy and Energy Consumption for Wireless Sensor Networks. IEEE MILCOM (October 2008)
[8] Ryu, B.: Robust, Persistent and Mesh-Networked Tagging, Tracking and Locating Systems (Mesh-TTL). Final Technical Report, Contract No.H92222-08-C-0040 (December 2009)
[9] Allan, D.W.: Time and frequency (time-domain) characterization, estimation, and prediction of precision clocks and oscillators. IEEE Trans. Ultrasonics, Ferroelectrics, and Frequency Control, 647–654 (1987)
[10] Elson, J., Girod, L., Estrin, D.: Fine-grained network time synchronization using reference broadcasts. In: Proceedings of the 5th symposium on Operating systems Design and Implementation, OSDI (2002)
[11] Ganeriwal, S., Kumar, R., Srivastava, M.B.: Timing-sync protocol for sensor networks. In: The 1st ACM Conference on Embedded Networked Sensor Systems, Los Angeles, California (November 2003)
[12] Li, Q., Rus, D.: Global clock synchronization in sensor networks. In: INFOCOM (2004)
[13] Ma, L., Zhu, H., Nallamothu, G., Ryu, B., Howard, H.R.: Understanding linear regression for wireless sensor network time synchronization. In: International Conference on Wireless Networks, ICWN, Las Vegas, Nevada (2007)
[14] Marti, G.S.M., Kusy, B., Ldeczi: The flooding time synchronization protocol. In: The 2nd ACM Conference on Embedded Networked Sensor Systems, Baltimore, Maryland (November 2004)
[15] Ganeriwal, S., et al.: Estimating Clock Uncertainty for Efficient Duty-Cycling in Sensor Networks. In: ACM Conf. Embedded Networked Sensor Sys., pp. 130–141 (2005)
[16] Ye, W., Heidemann, J., Estrin, D.: An energy efficient MAC protocol for wireless sensor networks. In: 21st International Annual Joint Conference of the IEEE Computer and Communications Societies (INFOCOM 2002), New York, NY, USA (2002)
[17] van Dam, T., Langendoen, K.: An adaptive energy efficient MAC protocol for wireless sensor networks. In: 1st ACM Conference on Embedded Networked Sensor Systems (SenSys), pp. 171–180 (2003)
[18] Lu, G., Krishnamachari, B., Raghavendra, C.S.: An Adaptive Energy-Efficient and Low-Latency MAC for Data Gathering in Sensor Networks. In: International Workshop on Algorithms for Wireless, Mobile, Ad Hoc and Sensor Networks, WMAN (2004)
[19] Zheng, T., Radhakrishnan, S., Sarangan, V.: PMAC: An adaptive energy-efficient MAC protocol for Wireless Sensor Networks. In: Proceedings of the 19th IEEE International Parallel and Distributed Processing Symposium (April 2005)
[20] Keshavarzian, A., Lee, H., Venkatraman, L.: Wakeup scheduling in wireless sensor networks. In: Proc. of MobiHoc 2006, Florence, Italy, (May 2006)
[21] Lu, G., Sadagopan, N., Krishnamachari, B., Goel, A.: Delay Efficient Sleep Scheduling in Wireless Sensor Networks. In: Proc. Of INFOCOM 2005, vol. 4, pp. 2470–2481 (March 2005)

[22] Polastre, J., Hill, J., Culler, D.: Versatile low power media access for wireless sensor networks. In: The Second ACM Conference on Embedded Networked Sensor Systems (SenSys), pp. 95–107 (November 2004)
[23] Avvenuti, M., Corsini, P., Masci, P., Vecchio, A.: Increasing the efficiency of preamble sampling protocols for wireless sensor networks. In: Proceedings of the First Mobile Computing and Wireless Communication International Conference, (MCWC 2006), Amman, pp. 117–122 (September 2006)
[24] Buettner, M., Yee, G., Anderson, E., Han, R.: X-MAC: A Short Preamble MAC Protocol For Duty-Cycled Wireless Networks. University of Colorado at Boulder, CU-CS-1008 06 (2006)
[25] Liu, S., Fan, K.-W., Sinha, P.: CMAC: An Energy Efficient MAC Layer Protocol Using Convergent Packet Forwarding for Wireless Sensor Networks. In: 4th Annual IEEE Communications Society Conference on Sensor, Mesh and Ad Hoc Communications and Networks (SECON 2007), San Diego CA, pp. 11–20 (June 2007)
[26] Ganeriwal, S., et al.: Estimating Clock Uncertainty for Efficient Du-ty-Cycling in Sensor Networks. In: ACM Conf. Embedded Networked Sensor Sys., pp. 130–141 (2005)
[27] Ye, W., Silva, F., Heidemann, J.: Ultra-Low Duty Cycle MAC with Scheduled Channel Polling. In: Proceedings of the 4th International Conference on Embedded Networked Sensor Systems (Sensys 2006), Boulder, Colorado, USA, pp. 321–334 (2006)
[28] El-Hoiydi, A., Decotignie, J.: Low power downlink MAC proto-cols for infrastructure wireless sensor networks. ACM Mobile Networks and Applications 10(5), 675–690 (2005)
[29] Zhu, H., Tang, Y., Chlamtac, I.: Unified Collision-Free Coordinated Distributed Scheduling (CF-CDS) in IEEE 802.16 Mesh Networks. IEEE Transactions on Wireless Communications 7(10) (October 2008)
[30] Qian, Y., Lu, K., Rong, B., Zhu, H.: Optimal Key Management for Secure and Survivable Heterogeneous Wireless Sensor Network. In: IEEE Globecom 2007, Washington DC (November 2007)
[31] Liu, D., Ning, P.: Location-based pairwise key establishments for static sensor networks. In: Proc. of the 1st ACM workshop on Security of ad hoc and sensor networks (2003)
[32] Durresi, A., Bulusu, V., Paruchuri, V.: SCON: Secure management of continuity in sensor networks. Computer Communications, 2458–2468 (2006)
[33] Eschenauer, L., Gligor, V.: A key-management scheme for distributed sensor networks. In: ACM CCS, Washington, DC (November 2002)
[34] Chan, H., Perrig, A., Song, D.: Random key predistribution schemes for sensor networks. In: IEEE Symposium on Security & Privacy, Oakland, CA (May 2003)
[35] Du, W., Deng, J., Han, Y., Varshney, P.: A pairwise key predistribution scheme for wireless sensor networks. In: ACM CCS, Washington, DC (October 2003)
[36] Zhang, Y., Liu, W., Lou, W., Fang, Y.: Securing sensor networks with location-based keys. In: Proc. of IEEE Wireless Communications and Networking Conference (WCNC 2005), pp. 1909–1914 (2005)

Chapter 22
Standardization Activities for Green IT

Gahng-Seop Ahn, Jikdong Kim, and Myung Lee

Abstract. The problem of global climate change has evolved to a political and economic issue beyond a mere environmental issue and becomes critical for the survival of mankind as well as the stabilization of world economy.

Information Technology (IT) is recognized as an effective means to cope with the climate change issue expediting low carbon world. IT can be utilized as a key technique for the reduction of the greenhouse gas (GHG) emissions for buildings, transportation, logistics, and power grid. These industries are the major sources of GHG. IT can reduce GHG emissions from these sectors five times as much as IT sector produces [1].

IT sector itself is becoming a major source of GHG. Although the proportion of GHG emission by IT sector is only 2% as of 2007 [2], it will be increased to 10 to 15% by 2025 [3] due to the increased use of IT. Making IT green is becoming important.

For these reasons, various IT standardization activities have been initiated since 2008 for improving the environmental friendliness of the IT sector itself as well as other sectors.

Standardization is essential in putting green IT solutions into practical use because it enables the interoperability of various IT products and services. The overview of the standardization activities for green IT is presented in this chapter. In section 22.1, the standardization activities to reduce the impact of telecommunication on climate change are presented. In section 22.2, the IT standards for smart grid are presented as a typical example of using IT to improve the environmental friendliness of other industries.

22.1 Standards for Green Communications

The meteoric growth of voice and data communications usage and its impact on the climate change have motivated many standardization organizations to develop eco-friendly green communications. The list of standardization bodies related to

---

Gahng-Seop Ahn · Myung Lee · Jikdong Kim
CUNY
e-mail: gahn@ccny.cuny.edu, lee@ccny.cuny.edu

green communications is presented in Table 22.1. Among the list three major standardization bodies in telecommunications and their activities are introduced in this section since these organizations are dealing with technical issues while other organizations such as UNEP, EC, and OECD are dealing with policy issues. International Telecommunication Union (ITU) [4] is a specialized agent of the United Nations for information and communication technology issues, and the global focal point for governments and the private sector in developing networks and services. The Alliance for Telecommunication Industry Solutions (ATIS) [6] is a major standard body in America and the European Telecommunications Standards Institute (ETSI) [7] is a major standard body in Europe. These organizations endeavor to mitigate the impact of communications on climate change by improving the energy efficiency of equipment/networks and recycling of equipment/facilities.

Table 22.1 Standardization bodies related to green communications [8]

Area	International Organizations	Others
Policies	UNEP, Int'l Energy Agency	EC, OECD
Indicators	WMO	OECD
Data collection	ISO TC 211	IEEE SCC 40, EC JRC
Environmental Management	ISO TC 207	-
Corporate reporting	ISO JTC1/SC7	Greenpeace, GHG Protocol Initiative
Energy efficiency of equipment	IEC, ISO, ITU-T	ATIS, ETSI, Energy star, CELELEC, EC JRC
Energy efficiency of networks	ITU-T	Ethernet Alliance, Energy Efficiency Inter-Operator Collaboration Group, EC JRC

22.1.1 International Telecommunication Union

International Telecommunication Union (ITU) [4] began to study the issue of climate change with its technical watch report in 2007. The ITU Telecommunication sector (ITU-T) held two symposiums at Kyoto and London in 2008 on information and communication technologies (ICTs) and climate change (CC). As a result of the discussion at the symposiums, ITU-T Focus Group on ICTs and Climate Change (ITU-T FG ICT&CC) [5] was formed in July 2008 by the ITU-T Telecommunication Standardization Advisory Group (TSAG) to perform an analysis of the impact of ICT on climate change. It addresses the issues in three aspects: the energy consumption of ICT equipment, the efficiencies to be gained through the use of ICTs in other sectors, and the need for behavioral change by both business and consumers.

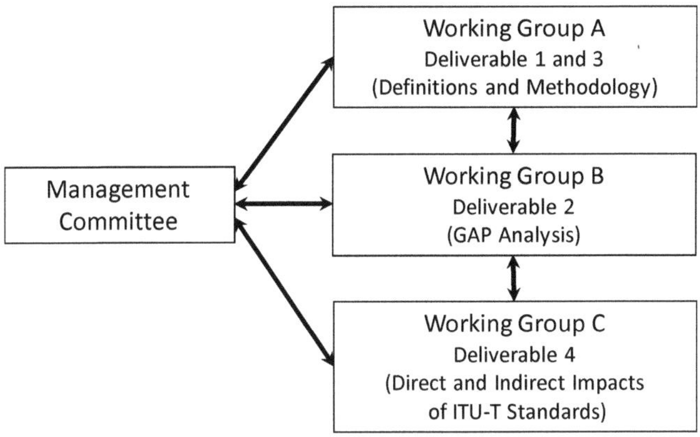

Fig. 22.1 Organization of ITU-T Focus Group on ICT & CC

The focus group has defined four deliverables and three working groups are responsible for the deliverables as illustrated in Figure 22.1. The management committee is in charge or coordination among the working groups and organizes meetings and events. The aim of work under deliverable 1 was to reach consensus on those key definitions that would be needed for work on methodologies under deliverable 3. The deliverable 3 aims to provide a method of calculating the energy usage and carbon impact arising from the ICT sector over the entire life cycle of ICT devices. It also aims to mitigate the energy usage and carbon impact by substituting ICT services and devices for intensive fossil-fuelled activities such as travel and transport through dematerialization. The aim of deliverables 2 is to identify existing standards that are relevant to ICTs and climate change so as to avoid unnecessary work. The aim of deliverables 4 is to develop guidelines to allow ITU-T Study Groups to evaluate the possible future reduction of CO_2 emission using ICT.

The Focus Group finished its tasks and was closed in April 2009. The TSAG meeting appointed ITU-T Study Group 5 (SG 5) as a lead Study Group for the issue of Environment and Climate Change.

Five Question Groups under the Study Group are established to discuss the following issues. The aim of the first group (Q.17/5) is to develop and maintain an overview of ICT&CC related recommendations, to coordinate with other SGs and other bodies on a regular basis to improve the planning of the work, and to provide and maintain an overview of key mitigation technologies and their impact on GHG emissions. The second group (Q.18/5) provides the methodology of environmental impact assessment of ICT and coordinates with other SGs and other bodies on a regular basis to collaborate effectively. The third group (Q.19/5) develops standards on the characterizations and specifications of the power feeding system, especially higher voltage DC system and on safety for humans and equipment of the power feeding system. The focus of fourth group (Q.20/5) is on establishing metrics for collecting data on energy efficiency for ICTs over the lifecycle, and collecting that data. The fifth group (Q.21/5) studies the methods to avoid using hazardous materials in ICT equipment and facilities and to mitigate waste and GHG emissions by recycling.

A Joint Coordination Activity on ICTs and Climate Change (JCA-ICT&CC) has been established under the Study Group for coordination of ICT&CC issue. Other ITU-T Study Groups are encouraged to develop energy efficiency technologies within their scopes.

22.1.2 Alliance for Telecommunications Industry Solutions

The Alliance for Telecommunication Industry Solutions (ATIS) [6] is a standardization organization in the United States of America. ATIS has more than 250 member companies including various telecommunications service providers, equipment manufacturers, and vendors.

ATIS and its members are committed to providing global leadership for the development of environmentally sustainable solutions for the information, entertainment, and communications industry. They aim to promote energy efficiencies of telecommunication systems, to reduce greenhouse gas emissions, to promote "reduce, reuse, recycle", to promote eco-aware business sustainability, and to support the potential for societal benefits. ATIS has released three documents on the assessment of telecommunication equipment energy requirements.

ATIS has created a new committee called "the Sustainability in Telecom: Energy Efficiency Committee (STEP)" that develops and recommends standards and technical reports related to power systems, electrical and physical protection for the exchange and interexchange carrier networks, and interfaces associated with user access to telecommunications networks. The committee has two subcommittees, namely TEE and NPP PWG.

The Telecommunications Energy Efficiency Subcommittee (STEP-TEE) develops standards and technical reports which define energy efficiency metrics, measurement techniques and new technologies, as well as operational practices for telecommunications components, systems and facilities. This subcommittee has published three standards used to determine telecommunication equipment's energy efficiency on March 2009. The standards introduce the Telecommunications Energy Efficiency Ratio or TEER as a measure of network-element efficiency.

The Network Physical Protection Subcommittee Pb-free Working Group (STEP - NPP PWG) proposes, develops, and recommends standards and technical reports relating to the use of lead or the restriction of lead in solder used in the manufacturing of telecommunications network equipment.

ATIS has also created an Exploratory Group on Green. The exploratory group investigates how ATIS and its members can advance environmental sustainability efforts, starting with an investigation and prioritization of items identified as important to ATIS members companies. The group released its initial report on environmental sustainability on May 2009 [13]. The group released two additional reports on ICT life cycle assessment [11] and wireless energy efficiency [12], respectively.

22.1.3 European Telecommunications Standards Institute

The European Telecommunications Standards Institute (ETSI) [6] is a standard organization in the telecommunications industry including equipment manufacturers and network operators in Europe. ETSI has developed standards for the Low Power Radio, Short Range Device, GSM cell phone system, and the TETRA professional mobile radio system.

ETSI has issued green agenda with adaptation of the ISO 14001 and 14004 standards. It operates a committee on Environmental Engineering (ETSI EE) which concerns reduction of energy consumption in telecommunication equipment and related infrastructure. The committee has published three standards for reduction of energy consumption in telecommunications equipment and related infrastructure, for better determination of equipment power and energy consumption for improved sizing of power plant, and for reduction of energy consumption in broad band telecommunication network equipment.

ETSI EE continues on the study of the use of alternative energy sources in telecommunication installations, reverse powering of small access network node by end-user equipment, energy efficiency of wireless access network equipment, and ICT energy consumption and global energy impact assessment methods.

22.2 Standards for Smart Grid

Smart grid is a form of electricity network utilizing digital technology. Smart grid delivers electricity from suppliers to consumers using two-way digital communications to control appliances at consumers' homes; this saves energy, reduces costs and increases reliability and transparency. It overlays the ordinary electrical grid with an information and net metering system, that includes smart meters. Smart grids are being promoted by many governments as a way of addressing energy independence, global warming and emergency resilience issues. Smart grid is made possible by applying sensing, measurement and control devices with two-way communications to electricity production, transmission, distribution and consumption parts of the power grid that communicate information about grid condition to system users, operators and automated devices, making it possible to dynamically respond to changes in grid condition. Smart grid includes an intelligent monitoring system that keeps track of all electricity flowing in the system. It also has the capability of integrating renewable electricity such as solar and wind. When power is least expensive the user can allow the smart grid to turn on selected home appliances such as washing machines or factory processes that can run at arbitrary hours. At peak times it could turn off selected appliances to reduce demand.

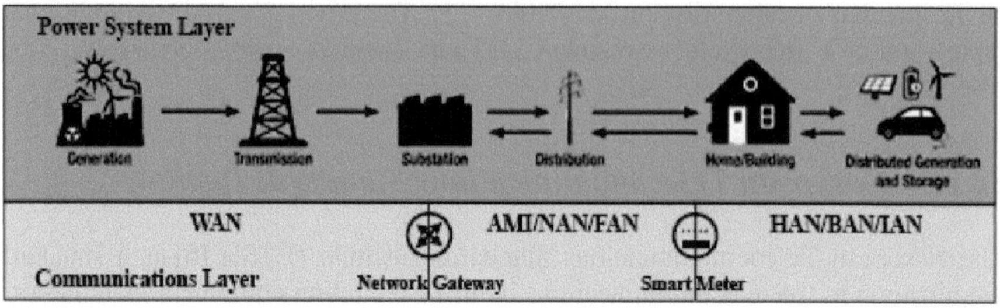

Fig. 22.2 Smart Grid Architecture

The smart grid system can be configured with power system and communications infrastructure. Communications layer interoperates with control system of smart grid through a wide range of wired and wireless communications network (see Figure 22.2). In this section, the standards for smart grid communications layer and standards bodies for Smart Grid will be introduced.

22.2.1 Standards for Smart Grid Communications Layer

A smart grid is achieved by overlaying power systems infrastructures with communications infrastructures (See Figure 22.3). Communications infrastructure can be segmented into:

- Wide Area Networks (WAN)
- Advanced Metering Infrastructure Networks, Neighborhood Area Networks (NANs), Field Area Networks (FANs)
- Home Area Networks (HANs), Building Area Networks (BANs), Industrial Area Networks (IANs)

Wide Area Networks consist of metro networks and backhaul network. Ethernet-based wired network can be used for interworking between relay systems. WiMAX (Worldwide inter-operability for Microwave Access) and cellular networks can be used for interworking between utility system and maintenance system.

Advanced Metering Infrastructure Networks, Neighborhood Area Networks (NANs), Field Area Networks (FANs) are wired and wireless networks that connect utility systems with customer premises for supporting a wide range of communication and control applications; for example, demand response and distribution automation. These networks are potentially spread over wide geographic areas. A range of technologies are relevant to these networks such as RF Mesh, WiFi, WiMAX, Cellular, ZigBee, Ethernet, and PLC (Power Line Communication).

Home Area Networks (HANs), Building Area Networks (BANs), Industrial Area Networks (IANs) are wired and wireless networks in customer premises (home, building and industry areas respectively) that support messaging between appliances, smart meters, electronics, energy management devices, applications and consumers.

Among many standards, we will focus on several key standards and present technical issues and challenges.

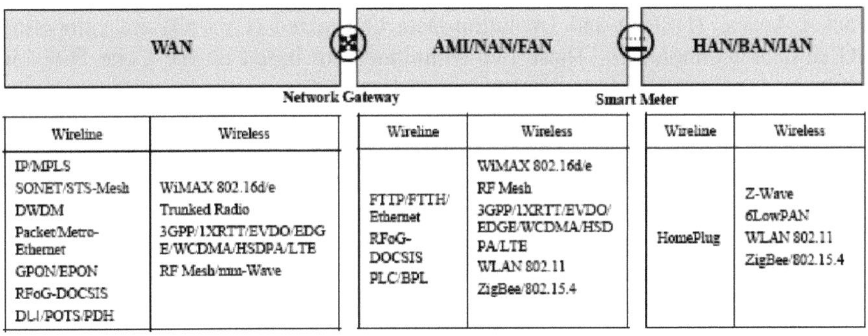

Fig. 22.3 Network Technology Standards Mapping

22.2.1.1 WiMAX

Worldwide inter-operability for Microwave Access (WiMAX) technology [30] is one of 802.16 series standards [31] designed for Wireless Metropolitan Area Network (WMAN). WiMAX's main goal is to have worldwide interoperability for microwave access. IEEE 802.16 standard first draft in 2001 defined the wide operating range of 10-66GHz for communication infrastructure. In order to achieve interoperability WiMAX forum has published a subset of the range: 3.5 and 5.8GHz bands have been dedicated for fixed communication, and 2.3, 2.5 and 3.5 GHz have been assigned for mobile communication frequency bands. 5.8GHz is unlicensed spectrum; whereas the spectrums 2.3, 2.5, 3.5GHz are licensed. In general, licensed spectrums are more suitable for long distance communication, because they can allocate higher power, compared to the unlicensed spectrums. Though distance and network speed are inversely proportional to each other, WiMAX can achieve data rate up to 70Mbps and distance up to 48km with lowering the maximum achievable date rate.

WiMAX has adopted several advanced wireless communications techniques; orthogonal frequency-division multiplexing (OFDM) to combat delay spread, multiple-input multiple output (MIMO) systems for higher data rate, and adaptive modulation and iterative coding for robustness. WiMAX provides low latency, and supports different levels of quality of service (QoS), allowing the SG operator to prioritize time-sensitive traffic. Deploying the WiMAX network is comparatively expensive, and therefore, cell planning should be done optimally to reduce the WiMAX tower.

Even though the lower frequency bands are already licensed, lower frequencies seem to be more practical for AMI applications especially in urban area, since higher frequency above 10 GHz cannot penetrate through obstacles.

22.2.1.2 Cellular Networks

Third-generation (3G) systems brought much higher data rates, compared to second-generation (2G) focused on the circuit switched voice data. High-Speed Packet Access (HSPA) and Evolution-Data Optimized (EV-DO) are competing 3G cellular technologies. These two techniques are based on the Code Division Multiple Access (CDMA) technology.

According to the downlink and uplink speed, HSPA consists of two cellular packet data protocols: High-Speed Downlink Packet Access (HSDPA) and High-Speed Uplink Packet Access (HSUPA), upgrading the downlink and uplink channels independently. HSPA has been enhanced to HSPA+ (Evolved-HSPA) which is considered to be a bridge technology to Long-Term Evolution (LTE) standard [32]. HSPA+ increases the downlink speed up to 42 Mbps. The uplink speed is same as HSUPA limited to 5.8 Mbps.

LTE is a fourth-generation cellular technology primarily designed to cope with the exponentially increasing date usage in cellular networks. To simplify interface between the cellular network and the internet, an all-Internet Protocol (all-IP) technology has been adopted. The theoretical maximum downlink speed with a 20 MHz bandwidth is 172.8 Mbps (2 x 2 MIMO) and 326.4 Mbps (4 x 4 MIMO). The theoretical maximum uplink speed is 86.4 Mbps. The enhanced version of LTE (LTE Evolution) standard, which may be considered as the real 4G cellular technology, is now being developed.

Cellular service could be expensive for individual connection and data transfer generally compared to the other service such as Wi-Fi, cellular technology may not be economical for regular data transfer or larger group of remote sites.

22.2.1.3 Satellite Communication

Point-to-point and point-to multipoint satellite communication service is operating at 2.3 GHz. Through a satellite dish and a router, it can provide coverage anywhere in the world. This mature technology uses geostationary communications satellites at 22,233 miles altitude and has been available for decades. The offered speed ranges from 64Kbps up to 8Mbps.

Satellite communication can provide better coverage but suffers 550 to 650 milliseconds of latency due to long distance between satellite and transponder. The latency may not allow the SG operator to handle time-sensitive traffic through satellite communication. Satellite communication link can be susceptible to severe winter conditions and extremely high winds.

22.2.1.4 Wireless LAN

IEEE 802.11 [28] wireless local area network (WLAN) is also known as Wi-Fi [29]. It is one of widely deployed and successful wireless data networks in the world like cellular networks because it is less expensive, easy to install and provides mobility of devices, compared to wired local area networks (LAN).

Wi-Fi has adopted the spread spectrum technology to allow multiple users to use the same frequency band, maintaining minimum interference to the other users. IEEE 802.11 standards have adopted three non-interoperable technologies: Frequency Hopping Spread Spectrum (FHSS), Direct Sequence Spread Spectrum (DSSS), and Infrared (IR). It can provide high speed point-to-point and point-to-multipoint communication. Unlike the network using licensed bands such as cellular and WiMAX, Wi-Fi operate solely on unlicensed spectrum at 2.4 GHz and 5.8Hz.

IEEE 802.11b offers the maximum data speed of 11Mbps at 2.4GHz frequency band with DSSS technique. Further, IEEE 802.11a and IEEE 802.11g can offer maximum data speed up to 54 Mbps. IEEE 802.11a, operating at 5.8GHz frequency band, has introduced Orthogonal Frequency Division Multiplexing (OFDM) modulation to enhance data rates; whereas 802.11g, operating at 2.4 GHz frequency band, still uses DSSS modulation technique. In order to increase data rates of WLAN further, Multiple Input Multiple Output (MIMO) technology has been incorporated into IEEE 802.11n standard, which can offer the maximum date rate up to 600 Mbps.

In terms of pre-established consumer comfort with the product, the prevalence of Wi-Fi can give some advantages against competitors. However, in terms of interference with other devices at home, its prevalence can be a disadvantage. The reliability and availability of wireless LAN can be less comparatively since many devices share the same frequency bands and relatively small transmission power is allowed in the unlicensed Wi-Fi band.

22.2.1.5 ZigBee and IEEE 802.15.4

ZigBee Alliance [25] developed a reliable, cost effective, and low power home area wireless network standard, ZigBee which defines the upper layers of IEEE 802.15.4 standard [26]. Zigbee's main advantage is battery life which lasts for years. ZigBee consumes battery power very little because the devices go to sleep mode when they are inactive and wake up again quickly when needed. However, even though ZigBee devices generally consume very little battery power comparatively, ZigBee devices have limited battery energy supply, internal memory and processing power due to limited physical size.

Based on the IEEE 802.15.4 standard, ZigBee can support up to 64,000 nodes and support star, tree and mesh topologies.

ZigBee can operate at the unlicensed frequency band of 868MHz, 915MHz and 2.4GHz with DSSS modulation technique. The coverage is the range of 10-100m and the supported data rate is 20-250 Kbps. ZigBee devices mostly operate in the unlicensed band of 2.4 GHz, which is very popular band many other wireless standards such as Wi-Fi also work in. The fast varying interference particularly from 802.11 WLANs can seriously affect the performance of ZigBee-based networks. To avoid harmful interference, the interoperability and coexistence between ZigBee-based networks and other wireless systems will be key issues. There are several mechanisms adopted in 802.15.4 standard; carrier sense multiple access with collision avoidance (CSMA/CA), acknowledged transmission and retries (ATR) to enhance coexistence, dynamic channel selection (DCS). The CSMA technique listens to check if a channel is occupied before transmitting and

ATR provides successful reception through an acknowledged frame delivery. The DCS technique measures the interference on a particular channel, enabling users to select the best available channel.

IEEE 802.15 working group is working on PHY (the physical layer) and MAC (the Medium Access Control layer) amendment to IEEE 802.15.4 standard to facilitate large, geographically diverse networks for smart grid with minimal infrastructure, with potentially millions of fixed endpoints. IEEE 802.15 task group 4g [27] is focusing on the PHY amendment. Current IEEE 802.15.4g draft has three complimentary PHYs; FSK/GFSK, OQPSK and OFDM. IEEE 802.15 task group 4e [34] is working on the MAC amendment to support the amended PHY. In addition, the amended MAC provides adaptive asymmetric multi-channel adaptation (AMCA) and low energy (LE) functionalities as special considerations for smart utility networks.

22.2.1.6 Power Line Communications

The power line communications (PLCs) are getting attention again as a suitable networking technology for smart grid [18]. But, there are still debates on whether PLCs are the good solutions in the smart grid. Some insist that PLCs are very good candidates for some applications, while others consider wireless technology

Table 22.2 Power Line Communications Technologies

Type	Broadband	Narrowband	
		High data rate	Low data rate
Operating bands	HF/VHF (1.8-250 MHz)	VLF/LF/MF (3 kHz - 500 kHz),	
Data rate	several Mbps ~ several hundred Mbps	tens of kbps ~ 500 kbps	several kbps
Examples	TIA-1113 (HomePlug 1.0), IEEE 1901, ITU-T G.hn (G.9960/G.9961), HomePlug AV, HomePlug Green PHY, HD-PLC, UPA Powermax.	IEEE 1901.2, ITU-T G.hnem, PRIME, G3-PLC.	ISO/IEC 14908-3 (LonWorks), ISO/IEC 14543-3-5 (KNX), CEA-600.31 (CEBus), IEC 61334-3-1, IEC 61334-5-1, Insteon, X10, HomePlug C&C, SITRED, Ariane Controls

as a more suitable alternative. There is no doubt that the smart grid will exploit multiple types of communications technologies such as fiber optics, wireless and wireline. Among the wireline technologies, PLCs may be the only candidate that has deployment cost comparable to wireless since the power lines already exist. One of the problems of the PLC is the significant deterioration of PLC signals going through the transformers. Hence, it may not be desirable for North America where the number of transformers in the power grid is relatively high. The average number of households per transformer is in the range of 1-5 in US power grid while it is around 100 in Europe. The PLC is actively considered in Europe.

The overview of PLC technologies is presented in Table 22.2. Single carrier narrowband (NB) PLC solutions, operating in the kHz region with data rates up to few kbps, have been used for decades by utility companies around the world for remote metering and load control applications. The current commercially available products have been upgrade to broadband (BB) systems, operating in the High Frequency (HF) band (2-30 MHz) with data rates up to 200 Mbps.

22.2.2 Key Interoperability Standards for Smart Grid

Interoperability in smart grid enables the utilities and consumers to seamlessly operate a hybrid mix of different connectivity solutions, software and hardware products from different providers and it is the key to success in evolving smart grid initiatives. Several national and international standard developing organizations are in charge of investigating this issue.

22.2.2.1 International Electrotechnical Commission

International Electrotechnical Commission (IEC) [14] has developed specialized communications standards for the power industry, with ongoing work to expand and enhance these standards.

IEC 61850 is for substation automation, distributed generation (photovoltaics, wind power, fuel cells, etc.), SCADA communications, and distribution automation. Work is commencing on Plug - in Hybrid Electric Vehicles (PHEV). IEC 61968 is for distribution management and Advanced Metering Infrastructure (AMI) back office interfaces. IEC 61970 (CIM) is for transmission and distribution abstract modeling. IEC 62351 is for security, focused on IEC protocols, Network and System management, and Role - Based Access Control. IEC TC 13 handles metering and may undertake a joint effort with TC57 to work on communications for metering, specifically for AMI.

22.2.2.2 Institute of Electrical and Electronic Engineers

Institute of Electrical and Electronic Engineers (IEEE) launched IEEE P2030 "Draft Guide for Smart Grid Interoperability of Energy Technology and Information Technology Operation with the Electric Power System (EPS), and End-Use Applications and Loads" [22] in June 2009 to provide an open standard for integration of information and communications technology in smart grid. When completed (estimated early 2011), P2030 will provide a base for the evolving smart grid applications, guidelines in defining SG interoperability, a knowledge-based architectural design addressing terminology, and characteristics. IEEE P2030 consists of three main task forces that are working in parallel on power engineering, information, and communications technologies.

Task Force 1 (Power Engineering Technology) focuses on functional requirements of interoperability in energy sources, transmission, and distribution in smart grid. Task Force 2 (Information Technology) is working on issues of privacy, security, data integrity, safety, management, and interfaces in smart grid. Task Force 3 (Communications Technology) is neutral to PHY / MAC standards used in smart grid and focuses on layers above PHY / MAC because the interoperability of electric power systems with end use applications should be maintained regardless of the PHY / MAC.

22.2.2.3 Internet Engineering Task Force

The Internet Engineering Task Force (IETF) [33] is responsible for Internet standards, many of which are now widely implemented in private Intranets as well. IETF is working on identifying the key infrastructure protocols for use in the Smart Grid [35].

A Request for Comment (RFC) document is the mechanism used by the IETF to develop, send out for comment, and finalize standards. Usually, RFC specifications must be implemented by more than one vendor before they can be fully accepted as standards. Some of the key IETF RFCs that can be used for smart grid are:

- RFC 791: Internet Protocol (IP)
- RFC 793: Transport Control Protocol (TCP)
- RFC 0768: User Datagram Protocol (UDP)
- RFC 1945: HyperText Transfer Protocol (HTTP)
- RFC 2571: Simple Network Management Protocol (SNMP)
- RFC 3820: Internet X.509 Public Key Infrastructure (PKI) for security
- RFC 4919: IPv6 over Low power Wireless Personal Area Networks (6LoWPAN)

6LoWPAN is a new protocol that enables energy efficient transmission of IPv6 datagrams over 802.15.4 links thereby suitable for smart metering devices in the smart grid.

22.2.2.4 National Institute of Standards and Technology

National Institute of Standards and Technology (NIST) is responsible to coordinate the development of a framework for protocols and standards of interoperability in SG system. The NIST Framework and Roadmap for Smart Grid Interoperability Standards [15] provides a conceptual reference model for SG that includes seven main domains: bulk generation, transmission, distribution, customer, markets, operations, and service provider. NIST has recognized 25 existing standards for SGC applications and 50 additional standards for further investigation. It also identifies 15 areas that urgently need new or revised standards.

NIST offers a prioritized action plan (PAP) to establish these critical standards. Among the list of PAP, the guidelines for use of Internet Protocol suite, wireless communications guidelines, and cyber security are related to IT.

22.3 Summary

This chapter has reviewed the standardization activities in telecommunications to improve the environmental friendliness of IT itself. Also, the standards for smart grid are presented as an example of using IT to improve the environmental friendliness and energy efficiency of other industries. The collaborative effort among IT industries as well as the cooperation with every sector of industries is critical in mitigating the impact on climate change and improving the eco-friendliness of the industries. Standardization activities play an important role in coordinating the collaborative effort and pushing the green IT solutions into practical use. Currently, many organizations and industries are actively working on standardization of green IT solutions besides the activities introduced in this chapter. Those activities are in the areas such as green computing, Energy Star and TCO certificates, smart transportation, and smart logistics.

References

[1] Boccaletti, G., Löffler, M., Oppenheim, J.M.: How IT can cut carbon emissions. The McKinsey Quarterly (2008)
[2] Malmodin, J., Moberg, Å., Lund´en, D., Finnveden, G., Lövehagen, N.: Greenhouse Gas Emissions and Operational Electricity Use in the ICT and Entertainment & Media Sectors. Wiley Journal of Industrial Ecology 14(5) (2010)
[3] The Global e-Sustainability Initiative (GeSI) and the Climate Group Report, SMART 2020: Enabling the low carbon economy in the information age (June 2008)
[4] International Telecommunication Union, http://www.itu.int
[5] ITU-T Focus Group on ICTs and Climate Change (FG ICT&CC), http://www.itu.int/ITU-T/focusgroups/climate/
[6] Alliance for Telecommunications Industry Solutions (ATIS), http://www.atis.org/
[7] European Telecommunications Standards Institute, http://www.etsi.org/
[8] Kim, E., Kim, Y.-w., Kim, H.J., Jung, H.W.: Standardization activities on ICTs & climate change. In: 31st International Telecommunications Energy Conference (INTELEC 2009), Incheon, Korea (October 2009)
[9] Kelly, T., Adolph, M.: ITU-T Initiatives on Climate Change. IEEE Communication Magazine (October 2008)
[10] Somemura, Y., Origuchi, T., Sugiyama, Y., Kobayashi, R., Nishi, S., Sawada, T.: Standardization Activities on ICTs and Climate Change in ITU-T. NTT Technical Review (2009)
[11] ATIS Exploratory Group on Green, ATIS Report on ICT Life Cycle Assessment (LCA) (January 2010)
[12] ATIS Exploratory Group on Green, ATIS Report on Wireless Energy Efficiency (January 2010)

[13] ATIS Exploratory Group on Green, ATIS Report on Environmental Sustainability (March 2009)
[14] International Electrotechnical Commission (IEC), http://www.iec.ch
[15] NIST Office of the National Coordinator for Smart Grid Interoperability, Framework and Roadmap for Smart Grid Interoperability Standards, Release 1.0 (January 2010)
[16] Utility Standards Board, Smart Grid: Interoperability and Standards, an Introductory Review (September 2008)
[17] Parikh, P.P., Kanabar, M.G., Sidhu, T.S.: Opportunities and challenges of wireless communication technologies for smart grid applications. In: IEEE Power and Energy Society General Meeting (2010)
[18] Galli, S., Scaglione, A., Wang, Z.: Power Line Communica-tions and the Smart Grid. In: First IEEE International Conference on Smart Grid Communications (SmartGrid-Comm 2010), Gaithersburg, MD (October 2010)
[19] Farhangi, H.: The path of the smart grid. IEEE Power Energy Magazine 8(1), 18–28 (2010)
[20] Goodman, F., et al.: Technical and system requirements for advanced distribution automation, Electrical Power Research Institute (EPRI) Technical Report 1010915 (June 2004)
[21] Gungor, V.C., Lambert, F.C.: A survey on communication networks for elec-tric system automation. Elsevier, Amsterdam (Febuary 2006)
[22] IEEE P2030, Draft Guide for Smart Grid Interoperability of Energy Technology and Information Technology Operation with the Electric Power System (EPS), and End-Use Applications and Loads, Draft. 3 (July 2010)
[23] Wheeler, A.: Commercial applications of wireless sensor networks using ZigBee. IEEE Comm. Mag. 45(4), 70–77 (2007)
[24] Anderson, M.: WiMax for Smart Grid, IEEE Spectrum Magazine (July 2010)
[25] ZigBee Alliance, http://www.zigbee.org
[26] IEEE Std 802.15.4-2006, Wireless Medium Access Control (MAC) and Physical Layer (PHY) Specifications for Low Rate Wireless Personal Area Networks, WPANs (2006)
[27] IEEE 802.15.4g, IEEE 802.15 WPAN Task Group 4g Smart Utility Net-works, http://www.ieee802.org/15/pub/TG4g.html
[28] IEEE 802.11 Wireless Local Area Networks, The Working Group for WLAN Standards, http://www.ieee802.org/11
[29] Wi-Fi Alliance, http://www.wi-fi.org
[30] WiMax Forum, http://www.wimaxforum.org
[31] IEEE 802.16 Working Group on Broadband Wireless Access Standards, http://www.wimaxforum.org
[32] Long Term Evolution (LTE), http://www.3gpp.org/LTE
[33] The Internet Engineering Task Force (IETF), http://www.ietf.org
[34] IEEE 802.15.4e, IEEE 802.15 WPAN Task Group 4e, MAC amendment for industrial applications, http://www.ieee802.org/15/pub/TG4e.html
[35] Baker, F.: Core Protocols in the Internet Protocol Suite, draft-baker-ietf-core-03.txt, internet draft (October 2009)

Author Index

Index

MIX
Papier aus verantwortungsvollen Quellen
Paper from responsible sources
FSC® C105338

If you have any concerns about our products,
you can contact us on
ProductSafety@springernature.com

In case Publisher is established outside the EU,
the EU authorized representative is:
Springer Nature Customer Service Center GmbH
Europaplatz 3, 69115 Heidelberg, Germany

Printed by Libri Plureos GmbH
in Hamburg, Germany